T0207355

Lecture Notes in Computer Science 13257

More information about this series at https://link.springer.com/bookseries/558

Volker Diekert · Mikhail Volkov (Eds.)

Developments in Language Theory

26th International Conference, DLT 2022
Tampa, FL, USA, May 9–13, 2022
Proceedings

 Springer

Editors
Volker Diekert 🆔
University of Stuttgart
Stuttgart, Baden-Württemberg, Germany

Mikhail Volkov 🆔
Ural Federal University
Ekaterinburg, Russia

ISSN 0302-9743 ISSN 1611-3349 (electronic)
Lecture Notes in Computer Science
ISBN 978-3-031-05577-5 ISBN 978-3-031-05578-2 (eBook)
https://doi.org/10.1007/978-3-031-05578-2

This Springer imprint is published by the registered company Springer Nature Switzerland AG
The registered company address is: Gewerbestrasse 11, 6330 Cham, Switzerland

Preface

The 26th International Conference on Developments in Language Theory (DLT 2022) was held in Tampa, Florida, USA during May 9–13, 2022. It was organised by the Department of Mathematics and Statistics of the University of South Florida. The preparations for DLT 2022 were overshadowed by the deeply shocking and frightening events in Ukraine. The situation of the ongoing heavy military actions made it hard to continue our work. However, as science and research are built on collaboration regardless of race, religion, gender or politics, the decision was taken to continue with the conference. The mission *Scientists-Sans-Frontière* is a manifesto for peace.

The DLT conference series provides a forum for presenting current developments in formal languages and automata. Its scope is very general and includes, among others, the following topics and areas: grammars, acceptors and transducers for words, trees and graphs; algebraic theories of automata; algorithmic, combinatorial, and algebraic properties of words and languages; variable length codes; symbolic dynamics; cellular automata; groups and semigroups generated by automata; polyominoes and multidimensional patterns; decidability questions; image manipulation and compression; efficient text algorithms; relationships to cryptography, concurrency, complexity theory, and logic; bio-inspired computing; and quantum computing.

Since its establishment by Grzegorz Rozenberg and Arto Salomaa in Turku (1993), a DLT conference had been held every other year in Magdeburg (1995), Thessaloniki (1997), Aachen (1999), and Vienna (2001). Since 2001, a DLT conference takes place in Europe every odd year and outside Europe every even year. The locations of DLT conferences since 2002 were: Kyoto (2002), Szeged (2003), Auckland (2004), Palermo (2005), Santa Barbara (2006), Turku (2007), Kyoto (2008), Stuttgart (2009), London (2010), Milano (2011), Taipei (2012), Marne-la-Vallée (2013), Ekaterinburg (2014), Liverpool (2015), Montréal (2016), Liège (2017), Tokyo (2018), Warsaw (2019), and Porto (2021). In 2020, the DLT conference was planned to be held in Tampa, Florida, but due to COVID-19 pandemics, it was canceled. However, accepted papers of DLT 2020 were published in volume 12086 of Lecture Notes in Computer Science, and the authors of these papers were invited to present their work at DLT 2021. The decision to hold DLT 2022 in Tampa was made in the anticipation of restoring the functioning of the DLT conference series in its established rhythm.

In 2018, the DLT conference series instituted the Salomaa Prize to honour the work of Arto Salomaa, as well as the success of automata and formal languages theory. The prize is funded by the University of Turku. The ceremony for Salomaa Prize 2021 took place during DLT 2022, and we congratulate the winner Juhani Karhumäki.

This volume contains invited contributions and the accepted papers of DLT 2022. There were 32 submissions by 68 authors from 18 countries: Belgium, Canada, China, Columbia, Czechia, Finland, France, Germany, India, Italy, Japan, Poland, Russia, Slovakia, South Korea, UK, and the USA. Each submission was reviewed by at least three referees. All submissions were thoroughly discussed by the Program Committee

(PC) who decided to accept 22 papers (68.75% acceptance rate) to be presented at the conference. We would like to thank the members of the Program Committee, and all external referees, for their work in evaluating the papers and for valuable comments that led to the selection of the contributed papers.

The Organizing Committee of DLT 2022 established a Best Paper Award, which the PC has awarded to the paper "Visit-bounded Stack Automata" by Jozef Jirasek and Ian McQuillan.

There were six invited talks, which were presented by:

– Paola Bonizzoni (University of Milano-Bicocca, Italy)
– Joel Day (Loughborough University, UK)
– Delaram Kahrobaei (City University of New York, USA)
– Jarkko Kari (University of Turku, Finland)
– Volodymyr Nekrashevych (Texas A&M University, USA)
– Helmut Seidl (Technical University of Munich, Germany)

We warmly thank the invited speakers, as well as all authors of submitted papers. Their effort was the basis for the success of the conference.

The EasyChair conference system provided excellent support in the selection of the papers, the preparation of these proceedings, as well as in the making of the conference schedule. We would like to thank Springer's editorial staff, and in particular Ronan Nugent, Anna Kramer, Maree Shirota, and Guido Zosimo-Landolfo for their help during the process of publishing this volume.

We are grateful to the Organizing Committee members: Lina Fajardo Gomez, Margherita Ferrari, Nataša Jonoska, Abdulmelik Mohammed, Masahico Saito, and Dmytro Savchuk.

DLT 2022 was financially supported by National Science Foundation, National Security Agency, University of South Florida (USF), and USF Research One.

We are looking forward to DLT 2023 at the University of Umeå, Sweden.

May 2022   Volker Diekert
                                                                     Mikhail Volkov

Organization

Steering Committee

Marie-Pierre Béal	Université Gustave Eiffel, France
Cristian S. Calude	University of Auckland, New Zealand
Volker Diekert	Universität Stuttgart, Germany
Yo-Sub Han	Yonsei University, South Korea
Juraj Hromkovic	ETH Zürich, Switzerland
Oscar H. Ibarra	University of California, Santa Barbara, USA
Nataša Jonoska	University of South Florida, USA
Juhani Karhumäki	University of Turku, Finland
Martin Kutrib	Universität Giessen, Germany
Giovanni Pighizzini (Chair)	Università degli Studi di Milano, Italy
Michel Rigo	University of Liège, Belgium
Antonio Restivo	Università degli Studi di Palermo, Italy
Wojciech Rytter	University of Warsaw, Poland
Kai Salomaa	Queen's University, Canada
Shinnosuke Seki	University of Electro-Communications, Japan
Mikhail Volkov	Ural Federal University, Russia
Takashi Yokomori	Waseda University, Japan

Program Committee

Volker Diekert (Co-chair)	University of Stuttgart, Germany
Yo-Sub Han	Yonsei University, Korea
Artur Jeż	University of Wrocław, Poland
Jarkko Kari	University of Turku, Finland
Alexander Okhotin	St. Petersburg State University, Russia
Joël Ouaknine	Max Planck Institute for Software Systems, Germany
Svetlana Puzynina	St. Petersburg State University, Russia
Narad Rampersad	University of Winnipeg, Canada
Helmut Seidl	Technical University of Munich, Germany
Krishna Shankara Narayanan	Indian Institute of Technology Bombay, India
Mikhail Volkov (Co-chair)	Ural Federal University, Russia
Marc Zeitoun	University of Bordeaux, France

Organizing Committee

Lina Fajardo Gomez	University of South Florida, USA
Margherita Ferrari	University of South Florida, USA
Nataša Jonoska	University of South Florida, USA
Abdulmelik Mohammed	University of South Florida, USA
Masahico Saito	University of South Florida, USA
Dmytro Savchuk	University of South Florida, USA

Additional Reviewers

Antolin, Yago
Bienvenu, Laurent
Carton, Olivier
Charlier, Émilie
Chistikov, Dmitry
D'Alessandro, Flavio
De Luca, Alessandro
Fleischmann, Pamela
Frougny, Christiane
Göller, Stefan
Govind, R.
Guillon, Pierre
Haase, Christoph
Holub, Stepan
Holík, Lukáš
Jacquemard, Florent
Ko, Sang-Ki
Kociumaka, Tomasz
Kopra, Johan
Kufleitner, Manfred
Kutrib, Martin
Lopez, Aliaume
Makarov, Vladislav

Marin, Mircea
Martyugin, Pavel
Masopust, Tomáš
Mayr, Richard
McQuillan, Ian
Niehren, Joachim
Parshina, Olga
Peltomäki, Jarkko
Praveen, M.
Prigioniero, Luca
Richomme, Gwenaël
Roychowdhury, Sparsa
Saarela, Aleksi
Salomaa, Kai
Sangnier, Arnaud
Sarkar, Saptarshi
Sebastien, Labbe
Senizergues, Geraud
Shur, Arseny
Steiner, Wolfgang
Walukiewicz, Igor
Zetzsche, Georg

Abstracts of Invited Talks

Algebraic Methods for Periodicity in Multidimensional Symbolic Dynamics

Jarkko Kari ⓘ

Department of Mathematics and Statistics,
University of Turku, 20014 Turku, Finland
jkari@utu.fi

Abstract. A d-dimensional configuration is a coloring $c : \mathbb{Z}^d \longrightarrow A$ of the infinite grid by elements of a finite set $A \subseteq \mathbb{Z}$. It is natural to express such a configuration as a formal power series with d variables $X = (x_1, \ldots, x_d)$ where the coefficient of the X^u term is $c(u)$ for all $u \in \mathbb{Z}^d$. Invariance of c under the translation by $v \in \mathbb{Z}^d$ then means that the difference (Laurent) polynomial $X^v - 1$ *annihilates* the power series in the sense that its formal product with the series is the null series. More generally, we say that a polynomial p *periodizes* c if the formal product pc is strongly periodic. All periodizing polynomials of c form a polynomial ideal, and we can use algebraic geometry to study the structure of this ideal $\mathrm{Per}(c)$. We call a polynomial a *line polynomial* if it has at least two non-zero terms and the exponents of the terms lie on a single line. If $\mathrm{Per}(c)$ contains a line polynomial then clearly c is periodic in the direction of the line. If $\mathrm{Per}(c)$ contains a non-zero polynomial then it can be proved using a dilation lemma and Hilbert's Nullstellensatz that $\mathrm{Per}(c)$ contains a product of line polynomials [1, 2]. In the two-dimensional case $d = 2$ one can further show that $\mathrm{Per}(c)$ is a principal ideal generated by a product of line polynomials. It follows in the two-dimensional case that if $\mathrm{Per}(c)$ contains a polynomial without line polynomial factors then c is strongly periodic. Our methods can be applied, for example, on *low complexity configurations*, containing at most $|D|$ patterns of a finite shape $D \subseteq \mathbb{Z}^d$. We have shown that a two-dimensional uniformly recurrent configuration that has low complexity with respect to a convex shape D must be periodic [4, 5]. This implies that any 2D subshift containing a low complexity configuration with respect to a convex shape D also contains a periodic configuration. We have also shown that any low complexity configuration (with respect to any shape D) of the well-known *Ledrappier subshift* is periodic, and this result can be extended to many other algebraically defined subshifts [6].

Keywords: Multidimensional symbolic dynamics · Periodicity · Formal power series · Multivariate polynomials · Algebraic subshifts

References

1. Kari, J., Szabados, M.: An algebraic geometric approach to Nivat's conjecture. In: Hall-dórsson, M., Iwama, K., Kobayashi, N., Speckmann, B. (eds.) ICALP 2015. LNCS, vol. 9135, pp. 273–285. Springer, Heidelberg (2015). https://doi.org/10.1007/978-3-662-47666-6_22

2. Kari, J., Szabados, M.: An algebraic geometric approach to Nivat's conjecture. Inf. Comput. **271**, 104481 (2020). https://doi.org/10.1016/j.ic.2019.104481

3. Kari, J.: Low-complexity tilings of the plane. In: Hospodár, M., Jirásková, G., Konstantinidis, S. (eds.) DCFS 2019. LNCS, vol. 11612, pp. 35–45. Springer, Cham (2019). https://doi.org/10.1007/978-3-030-23247-4_2

4. Kari, J., Moutot, E.: Decidability and periodicity of low complexity tilings. In: Proceedings of STACS 2020, Leibniz International Proceedings in Informatics (LIPIcs), vol. 154, pp. 14:1–14:12 (2020). https://doi.org/10.4230/LIPIcs.STACS.2020.14

5. Kari, J., Moutot, E.: Decidability and periodicity of low complexity tilings. Theory Comput. Syst. (2021). https://doi.org/10.1007/s00224-021-10063-8

6. Kari, J., Moutot, E.: Nivat's conjecture and pattern complexity in algebraic subshifts. Theor. Comput. Sci. **777**, 379–386 (2019). https://doi.org/10.1016/j.tcs.2018.12.029

Non-deterministic Transducers

Volodymyr Nekrashevych🆔

Texas A&M University, 3368 TAMU, College Station, TX 77843-3368, USA
nekrash@math.tamu.edu
https://www.math.tamu.edu/ nekrash/

Abstract. A *deterministic* (synchronous) transducer is defined as a map $\pi : Q \times X \longrightarrow X \times Q$, where X is a finite set (alphabet), and Q is the *set of states*, together with a choice of the *initial state* $q_0 \in Q$. We say that the automaton is finite if Q is finite.

Here $\lambda(q,x) = (y,p)$ is interpreted as the condition that being in state q and reading the letter x on the input, the automaton transitions to the state p and prints y to the output. This way the automaton defines a transformation $\pi_{q_0} : X^* \longrightarrow X^*$ of the set of finite words and a transformation $\pi_{q_0} : X^\omega \longrightarrow X^\omega$ of the set of one-sided infinite words. Namely, if $\pi(q,x) = (y,p)$, then we set $\pi_q(xv) = y\pi_p(v)$ for every $v \in X^*$ and every $v \in X^\omega$.

Deterministic automata (transformation defined by them) are important tools in group theory. Many examples of groups with interesting properties are generated by transformations defined by finite automata (see [1]). Theory of automata is effectively used to study properties of such groups.

A group generated by transformations defined by finite automata acts on the corresponding sets X^n of finite words of given length. It follows that any such group is residually finite.

A more general class of groups is obtained by passing to a more general class of transformations. It appears naturally in the study of hyperbolic dynamical systems and provides first examples of simple finitely generated groups with various finiteness conditions (e.g., amenability or sub-exponential growth, see [2, 3]).

A *non-deterministic* transducer is a finite labeled graph given by the set of vertices Q (states of the automaton), a set of edges E, maps $s, r : E \longrightarrow Q$ (the *source* and the *range* of the edges), a labeling $\lambda : E \longrightarrow X^2$, and a set $Q_0 \subset Q$ of *initial states*. We say that a pair $(x_1 x_2 \ldots, y_1 y_2 \ldots) \in X^\omega \times X^\omega$ is *accepted* by the automaton if there exists a path $e_1 e_2 \ldots \in E^\omega$ such that $s(e_1) \in Q_0$, $r(e_i) = s(e_{i+1})$ for all $i \geq 1$, and $\lambda(e_i) = (x_i, y_i)$. We say that the automaton is ω-deterministic if for every $x_1 x_2 \ldots \in X^\omega$ there exists at most one sequence $y_1 y_2 \ldots \in X^\omega$ such that the pair $(x_1 x_2 \ldots, y_1 y_2 \ldots)$ is accepted. If the automaton is ω-deterministic, then the transformation $x_1 x_2 \ldots \mapsto y_1 y_2 \ldots$ is a continuous map between two closed subsets of X^ω.

We are interested in groups generated by homeomorphisms of a fixed closed subset $F \subset X^\omega$ defined by ω-deterministic finite automata. It seems that many well known techniques of groups generated by finite deterministic automata can be extended to this much larger class of transformations. However, many other tools (especially ones relying on the actions on the sets X^n of finite words) are missing, and generalizing them to non-deterministic automata is one of our goals. Another goal is understanding transformations defined by finite non-deterministic automata in the context of ω-deterministic languages.

Keywords: Non-deterministic transducer · ω-deterministic automaton · Group generated by homeomorphisms

References

1. Grigorchuk, R.I., Nekrashevich, V.V., Sushchanskii, V.I.: Automata, dynamical systems and groups. Proc. Steklov Inst. Math. **231**, 128–203 (2000)
2. Juschenko, K., Monod, N.: Cantor systems, piecewise translations and simple amenable groups. Ann. Math. **178**(2), 775–787 (2013). https://doi.org/10.4007/annals.2013.178.2.7
3. Nekrashevych, V.: Palindromic subshifts and simple periodic groups of intermediate growth. Ann. Math. **187**(3), 667–719 (2018). https://doi.org/10.4007/annals.2018.187.3.2

Origin-Equivalence for Macro Tree Transducers

Helmut Seidl ⓘ

Fakultät für Informatik, TU München, Germany
seidl@in.tum.de

Abstract. We consider a notion of origin for deterministic macro tree transducers with look-ahead which records for each output node, the corresponding input node for which a rule-application generated that output node. With respect to this natural notion, we show that *origin equivalence* is decidable—whenever the transducers are weakly self-nesting. The latter means that whenever two nested calls on the same input node occur, then there must be at least one other node (a terminal output node or a call on another input node) in between these nested calls. We also indicate that for monadic input alphabets, equivalence of the transducers can be reduced to origin equivalence – whenever unrestricted self-nesting is allowed.

These results have been obtained jointly with Sebastian Maneth.

Keywords: Macro tree transducers · Origin equivalence · Decidablility

Contents

Invited Talks

Can Formal Languages Help Pangenomics to Represent and Analyze Multiple Genomes?

Paola Bonizzoni[1]([envelope])[iD], Clelia De Felice[2][iD], Yuri Pirola[1][iD], Raffaella Rizzi[1][iD], Rocco Zaccagnino[2][iD], and Rosalba Zizza[2][iD]

[1] Dip. di Informatica, Sistemistica e Comunicazione, University of Milano-Bicocca, viale Sarca 336, 20126 Milan, Italy
{paola.bonizzoni,yuri.pirola,raffaella.rizzi}@unimib.it
[2] Dip. di Informatica, University of Salerno, via Giovanni Paolo II 132, 84084 Fisciano, Italy
{cdefelice,rzaccagnino,rzizza}@unisa.it

Abstract. Graph pangenomics is a new emerging field in computational biology that is changing the traditional view of a reference genome from a linear sequence to a new paradigm: a sequence graph (pangenome graph or simply pangenome) that represents the main similarities and differences in multiple evolutionary related genomes. The speed in producing large amounts of genome data, driven by advances in sequencing technologies, is far from the slow progress in developing new methods for constructing and analyzing a pangenome. Most recent advances in the field are still based on notions rooted in established and quite old literature on combinatorics on words, formal languages and space efficient data structures. In this paper we discuss two novel notions that may help in managing and analyzing multiple genomes by addressing a relevant question: how can we summarize sequence similarities and dissimilarities in large sequence data? The first notion is related to variants of the *Lyndon factorization* and allows to represent sequence similarities for a sample of reads, while the second one is that of *sample specific string* as a tool to detect differences in a sample of reads. New perspectives opened by these two notions are discussed.

1 Introduction

The 1000 Genomes Project [16] marks the beginning of new computational approaches to genomic studies involving the use of efficient data structures to represent the high variation rate among multiple genomes. Indeed, a main result of the project has been the characterization of a broad spectrum of genetic variations in the human genome, including the discovery of novel variations in the

This project has received funding from the European Union's Horizon 2020 research and innovation programme under the Marie Skłodowska-Curie grant agreement No 872539.

V. Diekert and M. Volkov (Eds.): DLT 2022, LNCS 13257, pp. 3–12, 2022.
https://doi.org/10.1007/978-3-031-05578-2_1

South Asian, African and European populations—thus enhancing the catalogue of variability within the human individuals. In particular, the question "what is an ideal human reference genome?" is becoming the focus of investigations that also involve theoreticians in the computer science community. While the literature in computational biology presents experimental evidence of the advantages of the idea of replacing a linear reference with a pangenome graph [23,37,39], still theoretical foundations of computational pangenomics is missing. A recent tutorial introduces the main theoretical background in graph pangenomics [1]. It is interesting to note that formal language theory has again played a crucial role in suggesting novel approaches to this new emerging field. The first main representation of a graph pangenome is based on building a prefix language from the interpretation of the graph as an automaton [33], while *Wheeler graphs* [22] establish an interesting connection between regular languages and compressed data structures which are fundamental in the indexing of pangenomes. Language theoretic notions that have been recently rediscovered in Bioinformatics are those of Lyndon words and of the Lyndon factorization of a word [15,21,32]. Indeed, these well-known notions intervene in a bijective transformation [29] alternative to the Burrows-Wheeler Transform for compressing sequences and in new measures of similarities between sequences [7]. The investigation of sequence similarity and dissimilarity measures is a crucial topic in Bioinformatics for comparing sequences. For example, sequence alignment is the oldest standard procedure performed to measure the distance between sequences. However, the search for alignment-free approaches to sequence comparison is the focus of deep investigations in the framework of pangenomics, since there is the need to cope with the high computational cost of the alignment and have fast approaches to compute genetic variations in a pangenome [1]. In this direction, a possible alignment-free approach may consist in applying mathematical transformations on sequences that lead easily to a fast sequence comparison. In particular, summarizing sequences by alternative representations is becoming a new paradigm for facing the huge amount of sequencing data. Factorizing a word is intuitively a way to give an alternative representation of it: thus, a main question is whether there exists a way to factorize sequences so that it may lead to a more compact representation to detect shared regions between sequences. This work is focused on the Lyndon factorization as a factorization preserving similarities among sequences. Since being able to detect dissimilarities is also important in sequence comparison, here we also present a novel notion aiming at discovering differences among similar sequences. This is the notion of *sample specific string* (SFS) [27]. We show applications of both notions in facing problems motivated by computational pangenomics [13,20].

This paper is structured as follows. After introducing preliminaries on sequences, Lyndon words and Lyndon factorization, we survey some main theoretical results on Lyndon-based factorizations motivated by Bioinformatics applications. Then, we discuss preliminary results on their application. In Sect. 4 we discuss the theoretical background of the notion of sample specific strings and then, we present its application in structural variant detection. We conclude with some open problems related to Lyndon-based factorizations.

2 Preliminaries

Throughout this paper we follow [31] for the notations. Let $w = a_1 \cdots a_m$ be a string (or word) over a finite alphabet Σ. The empty word is denoted by 1. The *length* of w (that is, the number m of its characters) will be denoted by $|w|$. A word $x \in \Sigma^*$ is a *factor* of $w \in \Sigma^*$ if there are $u_1, u_2 \in \Sigma^*$ such that $w = u_1 x u_2$. If $u_1 = 1$ (resp. $u_2 = 1$), then x is a *prefix* (resp. *suffix*) of w. A factor (resp. prefix, suffix) x of w is *proper* if $x \neq w$. We recall that, given a nonempty word w, a *border* of w is a word which is both a proper prefix and a suffix of w. The longest border is also called *the border* of w. A word $w \in \Sigma^+$ is bordered if it has a nonempty border. Otherwise, w is *unbordered*. A nonempty word w is *primitive* if $w = x^k$ implies $k = 1$. An unbordered word is primitive. Given $w, w' \in \Sigma^*$, we denote by $w < w'$ (resp. $w \leq w'$) if w is lexicographically smaller than w' (resp. smaller than or equal to w'). Furthermore, for two nonempty words w, w', we write $w \ll w'$ if $w < w'$ and additionally w is not a proper prefix of w' [4]. We recall that a *factorization* of a string w is a sequence $F(w) = (f_1, f_2, \ldots, f_n)$ of factors such that $w = f_1 f_2 \cdots f_n$.

In [6] a numeric representation of a factorization of a string is defined, named the *fingerprint* of w with respect to $F(w)$, i.e., the sequence $\mathcal{L}(w) = (|f_1|, |f_2|, \ldots, |f_n|)$ of the lengths of the factors of $F(w)$. In addition, a *k-finger* is a k-mer of $\mathcal{L}(w)$, that is, any substring $(l_i, l_{i+1}, \ldots, l_{i+k-1})$ composed of k consecutive elements of $\mathcal{L}(w)$.

In this framework, we consider strings over a the DNA alphabet and they will be simply called sequences, meaning to represent genomes or fragments of genome sequences. For preliminaries to computational pangenomics and some basic notions, we address the reader to a recent tutorial [1].

3 Lyndon Words and Lyndon-Based Factorization

The Lyndon Factorization CFL. In order to obtain read fingerprints, in [7] some special kinds of factorizations are proposed, named *Lyndon-based factorizations*, since they are defined starting from the well-known *Lyndon factorization* of a string w [32]. Each string w can be uniquely factorized into a non-increasing product (w.r.t. the lexicographic order) of *Lyndon words* [32]. A Lyndon word is a string which is strictly smaller than any of its nonempty proper suffixes. Lyndon words are primitive and unbordered. For example, suppose that $\Sigma = \{a, c, g, t\}$ and $a < c < g < t$ (in next examples, we always suppose this ordering on the alphabet). Thus, $accgctct$ is a Lyndon word, whereas cac is not a Lyndon word,

Formally, given a string w, its Lyndon factorization CFL(w) is a sequence CFL(w) = $(f_1, f_2, \ldots, f_n)$ of words such that $f_1 \geq f_2 \geq \cdots \geq f_n$ and each f_i is a Lyndon word. For example, given $w_1 = gcatcaccgctctacagaac$, we have that CFL($w_1$) = $(g, c, atc, accgctct, acag, aac)$. The notation CFL is due to the fact that stating the uniqueness of this factorization is usually credited to Chen, Fox and Lyndon [15]. We recall that CFL can be computed in linear time and constant space [21].

The notion of Lyndon words has been shown to be useful in theoretical applications, such as the well-known proof of the *Runs Theorem* [2], as well as in string compression analysis. Furthermore, the Lyndon factorization has recently revealed to be a useful tool also in investigating queries on suffixes of a word and sorting such suffixes with strong potentialities for string comparison that have not been completely explored and understood. Relations between Lyndon words and the Burrows-Wheeler Transform (BWT) have also been discovered first in [18,34] and, more recently, in [3,28,29]. A connection is found between the Lyndon factorization CFL and the Lempel-Ziv (LZ) factorization [26], where it is shown that in general the size of the LZ factorization is larger than the size of the Lyndon factorization, and in any case the size of the Lyndon factorization cannot be larger than a factor of 2 with respect to the size of LZ. This result has been further extended in [40] to overlapping LZ factorizations.

Conservation Property of CFL. In [10] a new property of the Lyndon factorization, named *Conservation Property* [6,7,13], has been proved, which is crucial in our framework, and here reported. More precisely, let $\mathsf{CFL}(w) = (\ell_1, \ell_2, \ldots, \ell_n)$. We firstly recall that x is a *simple* factor of w if, for each occurrence of x as a factor of w, there is j, with $1 \leq j \leq n$, such that x is a factor of ℓ_j. So, let $x = \ell_i'' \ell_{i+1} \cdots \ell_{j-1} \ell_j'$ be a non simple factor of w, for some indexes i, j with $1 \leq i < j \leq n$, and $\ell_i = \ell_i' \ell_i''$, $\ell_j = \ell_j' \ell_j''$.

The above-mentioned Conservation Property is stated below and it compares the Lyndon factorization of w and that of its non-simple factors.

Lemma 1 [9,10]. *Let* $w \in \Sigma^+$ *be a word and let* $\mathsf{CFL}(w) = (\ell_1, \ldots, \ell_n)$ *be its Lyndon factorization. For any* i, j, *with* $1 \leq i \leq j \leq n$, *one has* $\mathsf{CFL}(\ell_i \ell_{i+1} \cdots \ell_j) = (\ell_i, \ell_{i+1}, \ldots, \ell_j)$. *In addition, let* x *be a non-simple factor of* w *such that* x *is not a concatenation of consecutive factors of* $\mathsf{CFL}(w)$ *and let* $\ell_i'', \ell_{i+1}, \ldots, \ell_{j-1}, \ell_j'$ *be such that* $x = \ell_i'' \ell_{i+1} \cdots \ell_{j-1} \ell_j'$, *with* $1 \leq i < j \leq n$.
Let $\mathsf{CFL}(\ell_i'') = (m_1, \ldots, m_h)$ *and* $\mathsf{CFL}(\ell_j') = (v_1, \ldots, v_t)$. *We have*

$$\mathsf{CFL}(x) = (m_1, \ldots, m_h, \ell_{i+1}, \ldots, \ell_{j-1}, v_1, \ldots, v_t)$$

where it is understood that if $\ell_i'' = 1$ *(resp.* $\ell_j' = 1$*), then the first* h *terms (resp. last* t *terms) in* $\mathsf{CFL}(x)$ *vanish.*

According to Lemma 1, given two strings w and w' sharing a common overlap x, under some hypothesis, there exist factors that are in common between $\mathsf{CFL}(w)$ and $\mathsf{CFL}(w')$. Thus w and w' will have fingerprints sharing k-fingers for a suitable size k. For example, consider again $w_1 = gcatcaccgctctacagaac$ and let $w_2 = ccaccgctctacagaagcatc$. Then, $\mathsf{CFL}(w_1) = (g, c, atc, accgctct, acag, aac)$ and we have that $\mathsf{CFL}(w_2) = (c, c, accgctct, acag, aagcatc)$. Hence, we have $\mathcal{L}(w_1) = (1, 1, 3, 8, 4, 3)$ and $\mathcal{L}(w_2) = (1, 1, 8, 4, 7)$. The two common consecutive elements $(8, 4)$ are related to the same factors in the two words (8 is related to $accgctct$ and 4 is related to $acag$) and capture the common substring $accgctctacag$ given by their concatenation.

Even though the hypothesis that x is not simple with respect to $\mathsf{CFL}(w)$ cannot be dropped (see [6]), it is worthy of note that in real data this hypothesis is always satisfied. Such an interesting property suggests the possibility of using directly k-fingers as features. Indeed, in [6] it is presented an approach in which k-fingers are used for classifying sequencing reads (Sect. 3.1).

Canonical Inverse Lyndon Factorization ICFL. The *Canonical Inverse Lyndon factorization* $\mathsf{ICFL}(w) = (f_1, f_2, \ldots, f_n)$ has been introduced in [8] as a factorization of w such that $f_1 \ll f_2 \ll \cdots \ll f_n$ and each f_i is an *inverse Lyndon word*, that is, each nonempty proper suffix of f_i is strictly smaller than f_i [8]. For example, cac, $tcaccgc$ are inverse Lyndon words. Let us consider again $w_1 = gcatcaccgctctacagaac$. We have that $\mathsf{ICFL}(w_1) = (gca, tcaccgc, tctacagaac)$. Observe that, differently from Lyndon words, inverse Lyndon words may be bordered. Furthermore, this factorization is also unique and can be computed in linear time [8].

What is the motivation of introducing a new factorization? In [10] two main results are proved: *(i)* an upper bound on the length of the longest common prefix of two factors of w starting from different positions on w is provided, and *(ii)* a relation among sorting of global suffixes, i.e., suffixes of the word w, and sorting of local suffixes, i.e., suffixes of the products of factors in $\mathsf{ICFL}(w)$ is given. The latter result is the counterpart for $\mathsf{ICFL}(w)$ of the compatibility property, proved in [35] for the Lyndon factorization. However, *(ii)* is in some sense stronger than that one in [35], as we explain below. Indeed, as a preliminary result, in [10] it is proved that that the longest common prefix between f_i and f_{i+1} is shorter than the border of f_i, when w is not an inverse Lyndon word. This result is obtained thanks to the *grouping* property of ICFL proved in [8]: given a word w, the factors in $\mathsf{ICFL}(w)$ are obtained by grouping together consecutive factors of the anti-Lyndon factorization of w that form a non-increasing chain for the prefix order (the anti-Lyndon factorization of w is the Lyndon factorization w.r.t. the inverse lexicographic order).

In this framework, a natural question is whether and how the longest common extensions for arbitrary positions on w are related to the size of the factors in $\mathsf{ICFL}(w)$. It is proved that there are relations between the length of the longest common prefix $\mathrm{lcp}(x, y)$ of two factors x, y of a word w starting from different positions on w and the maximum length $\mathcal{M}$ of two consecutive factors of the inverse Lyndon factorization of w. More precisely, $\mathcal{M}$ is an upper bound on the length of $\mathrm{lcp}(x, y)$. Thus, this result is in some sense stronger than the compatibility property, proved in [35] for the Lyndon factorization and in [10] for the inverse Lyndon factorization. Roughly, the compatibility property allows us to extend to the suffixes of the whole word the mutual order between suffixes of the concatenation of (inverse) Lyndon factors.

3.1 Some Applications: Representing and Querying Read Sequences

Sequencing technologies produce the main input data for a vast majority of algorithms in Bioinformatics. For example, the only way to get the whole sequence of the genome of a single individual is to produce (by sequencing) fragmented multiple copies of the genome sequence (called *reads*), that are computationally assembled into the original sequence. The extraordinary improvements in the sequencing technologies has allowed to obtain long enough fragments w.r.t. to the original massive sequencing consisting of reads of an average length of around 100 base pairs. In this section we touch upon two applications of the notion of *fingerprint*, presented in the previous sections, related to two traditionally difficult Bioinformatics tasks: genome assembly and transcript read classification. Indeed, read fingerprints provide a compact representation of the reads and, thanks to the Conservation Property, they are effective in preserving sequence similarities. In fact, the k-fingers (sub-pieces of a fingerprint) are able to capture the similarity regions between two reads in a more flexible way with respect to the k-mers of a sequence: the length k of a k-mer is fixed, whereas the length of the read substring, undergoing a k-finger, is variable. Furthermore, fingerprints are numerical sequences shorter than the represented character sequences and we also expect that they are resilient to errors occurring in the reads (especially in long reads). The first application is related to genome assembly based on the use of an overlap graph which is constructed by detecting the overlaps between genomic reads [11,12]. When dealing with long reads this task is further complicated by the length of the reads and the high sequencing error rate. In [13] a novel alignment-free algorithm for discovering the overlaps in a set of noisy long reads is presented, which exploits the fingerprints of the reads. Indeed, the k-fingers provide anchors for computing the overlaps between reads. The algorithm takes as input a set S of genomic reads and, after factorizing them, builds a hash table of the k-fingers by performing a linear scanning of the fingerprints. Next, the hash table is used in order to compute in $O(LN)$ time the common regions between each read s and the reads previously processed, assuming that the read length is L and N is the maximum number of occurrences of a unique (that is, occurring once) k-finger of s in the reads considered at the previous iterations. At the end, the algorithm obtains the read overlaps from all the detected common regions. Observe that comparing reads in a reference-free framework often requires a pairwise comparison and is computationally demanding (refer for example to the problem of the identification of the relationships between metagenomic reads [25]). The second application of the read fingerprints is related to the problem of assigning transcriptomic reads (that is, reads sequenced from gene transcripts of RNA-Seq reads) to their origin gene. In [6] fingerprints are used as a machine-interpretable representation of sequencing data in order to define an effective feature embedding method for assigning RNA-Seq reads to the origin gene. Indeed, a fingerprint (and the sequence of k-fingers) is used to produce an embedded representation of the read. Moreover, the machine learning classifier proposed in [6] was also extended for detecting chimeric RNA-Seq reads, which is a subtask of gene-fusion finding methods [19,30,38]. In fact, the chimeric reads detection problem can be seen as a variant of the read-gene classification problem since it requires to assign a chimeric read to two genes (instead of a single gene), which have been fused after genomic rearrangement.

4   Sample Specific Strings and Structural Variations in Human Genome

A classical example of how combinatorics on words is helping comparative genomics to analyze sequences, is given by the notion of *minimal absent word* [5]: this is a word absent from y whose all proper factors occur in y. It has several applications in Bioinformatics [14,36]. Here we consider a slightly different variant based on the idea of considering minimal words that are absent in a sequence but present in another sequence: we call them *specific strings*. Recently, in [27] the notion of sample specific string has been proposed to detect signatures of variations between a reference genome R and a sample T of reads from a target individual. A sample of reads is the typical output of the sequencing of an individual and consists of a collection of strings or reads.

Let us formally recall the notion of specific strings introduced in [27].

Definition 1. *Let R (reference) and T (target) be two strings over a finite alphabet. Then a factor s of T is a T-specific string w.r.t. to R (in short specific string) if the following properties hold:*

1. *s is not a factor of R,*
2. *any proper factor of s occurs in R.*

Then given a collection of strings S and a string R, s is a sample specific string for the collection S, SFS in short, if s is a T-specific string for some target T in S. A linear-time algorithm for computing T-specific string that are not overlapping on the input sequence T is given in [27], while an extensive discussion of some algorithmic properties is reported in [27]. SFSs have been proved in [20] to be effective in detecting breakpoints of structural variants (SV) i.e. medium to large size insertions and deletions in a reference genome that are present in a human sample of high quality long reads, (e.g. PACBIO HIFI). Indeed, the main idea behind the notion of SFS is that they may be of variable length w.r.t. fixed length k-mers traditionally used to identify SVs as unique k-mers occurring in a sequence. More precisely, given a substring x of a sequence R, an insertion or deletion inside x it is likely to produce a new string y that does not occur in R. Moreover, the breakpoints of the insertion or deletions (a breakpoint in x is a position of x delimiting the insertion or deletion) are likely to be associated to two factors which may be absent from R. Behind the practical interest in SFSs they are an interesting notion from the theoretical point of view. In particular, we conjecture that the SFSs could provide bounds on the classical edit distance and on the edit distance with moves, a generalization of the edit distance allowing the exchange of blocks, i.e. factors inside the sequence [17].

5   Open Problems

The method given in [6] uses representation of reads obtained starting from Lyndon based factorizations. A natural question, faced in the same paper, is whether the corresponding representation produced by its fingerprint or by its k-fingers is unique, a property which is closely related to the collision phenomenon: distinct strings may have common k-fingers. An open problem is of how the lexicographic ordering of the alphabet may affect the collision phenomenon, The properties described in [6] show that the choice of a specific ordering of the initial alphabet can have a significant impact on the collision phenomenon. However, the problem of understanding if there exists an order that minimizes this phenomenon remains open (and, if exists, which is this order) and future investigations should be devoted to it. It is worth of note that in general, the questions of finding an optimal alphabet ordering for Lyndon factorization (*i.e.*, such that number of Lyndon factors is at most, or at least, n, for a given number n) is hard [24].

As already mentioned in Sect. 3, it could be interesting to investigate how the bound proved for the longest common prefix between suffixes of factors in ICFL may be used for efficiently sorting suffixes. Furthermore, one challenging question is whether ICFL could be used instead of CFL for defining a new bijective version of the Burrows Wheeler Transform, as done in [29].

References

1. Baaijens, J.A., et al.: Computational graph pangenomics: a tutorial on data structures and their applications. Nat. Comput. (2022). https://doi.org/10.1007/s11047-022-09882-6
2. Bannai, H.I.T., Inenaga, S., Nakashima, Y., Takeda, M., Tsuruta, K.: The "Runs" Theorem. SIAM J. Comput. **46**(5), 1501–1514 (2017)
3. Bannai, H., Kärkkäinen, J., Köppl, D., Piatkowski, M.: Indexing the bijective BWT. In: Pisanti, N., Pissis, S.P. (eds.) 30th Annual Symposium on Combinatorial Pattern Matching, CPM 2019, 18–20 June 2019, Pisa, Italy. LIPIcs, vol. 128, pp. 17:1–17:14 (2019)
4. Bannai, H., Tomohiro, I., Inenaga, S., Nakashima, Y., Takeda, M., Tsuruta, K.: A new characterization of maximal repetitions by Lyndon trees. In: Proceedings of the Twenty-Sixth Annual ACM-SIAM Symposium on Discrete Algorithms, SODA 2015, San Diego, CA, USA, 4–6 January 2015, pp. 562–571 (2015)
5. Béal, M.-P., Mignosi, F., Restivo, A.: Minimal forbidden words and symbolic dynamics. In: Puech, C., Reischuk, R. (eds.) STACS 1996. LNCS, vol. 1046, pp. 555–566. Springer, Heidelberg (1996). https://doi.org/10.1007/3-540-60922-9_45
6. Bonizzoni, P., et al.: Numeric Lyndon-based feature embedding of sequencing reads for machine learning approaches. CoRR abs/2202.13884 (2022), https://arxiv.org/abs/2202.13884
7. Bonizzoni, P., et al.: Can we replace reads by numeric signatures? Lyndon fingerprints as representations of sequencing reads for machine learning. In: Martín-Vide, C., Vega-Rodríguez, M.A., Wheeler, T. (eds.) AlCoB 2021. LNCS, vol. 12715, pp. 16–28. Springer, Cham (2021). https://doi.org/10.1007/978-3-030-74432-8_2
8. Bonizzoni, P., De Felice, C., Zaccagnino, R., Zizza, R.: Inverse Lyndon words and inverse Lyndon factorizations of words. Adv. Appl. Math. **101**, 281–319 (2018)

9. Bonizzoni, P., De Felice, C., Zaccagnino, R., Zizza, R.: Lyndon words versus inverse lyndon words: queries on suffixes and bordered words. In: Leporati, A., Martín-Vide, C., Shapira, D., Zandron, C. (eds.) LATA 2020. LNCS, vol. 12038, pp. 385–396. Springer, Cham (2020). https://doi.org/10.1007/978-3-030-40608-0_27
10. Bonizzoni, P., De Felice, C., Zaccagnino, R., Zizza, R.: On the longest common prefix of suffixes in an inverse Lyndon factorization and other properties. Theor. Comput. Sci. **862**, 24–41 (2021)
11. Bonizzoni, P., Della Vedova, G., Pirola, Y., Previtali, M., Rizzi, R.: An external-memory algorithm for string graph construction. Algorithmica **78**(2), 394–424 (2017)
12. Bonizzoni, P., Della Vedova, G., Pirola, Y., Previtali, M., Rizzi, R.: FSG: fast string graph construction for de novo assembly. J. Comput. Biol. **24**(10), 953–968 (2017)
13. Bonizzoni, P., Petescia, A., Pirola, Y., Rizzi, R., Zaccagnino, R., Zizza, R.: Kfinger: capturing overlaps between long reads by using Lyndon fingerprints. In: IWBBIO Conference, Gran Canaria, Spain, 27th–30th June 2022, Proceedings. to appear (2021)
14. Chairungsee, S., Crochemore, M.: Using minimal absent words to build phylogeny. Theor. Comput. Sci. **450**, 109–116 (2012)
15. Chen, K.T., Fox, R.H., Lyndon, R.C.: Free differential calculus, IV. The quotient groups of the lower central series. Ann. Math. **68**, 81–95 (1958)
16. Consortium,G.P., et al.: A global reference for human genetic variation. Nature **526**(7571), 68 (2015)
17. Cormode, G., Muthukrishnan, S.: The string edit distance matching problem with moves. ACM Trans. Algorithms (TALG) **3**(1), 1–19 (2007)
18. Crochemore, M., Désarménien, J., Perrin, D.: A note on the Burrows-Wheeler transformation. Theor. Comput. Sci. **332**(1), 567–572 (2005)
19. Davidson, N.M., Chen, Y., Ryland, G.L., Blombery, P., Göke, J., Oshlack, A.: JAFFAL: Detecting fusion genes with long read transcriptome sequencing. bioRxiv (2021). https://doi.org/10.1101/2021.04.26.441398
20. Denti, L., Khorsand, P., Bonizzoni, P., Hormozdiari, F., Chikhi, R.: Improved structural variant discovery in hard-to-call regions using sample-specific string detection from accurate long reads. bioRxiv (2022)
21. Duval, J.: Factorizing words over an ordered alphabet. J. Algorithms **4**(4), 363–381 (1983)
22. Gagie, T., Manzini, G., Sirén, J.: Wheeler graphs: a framework for BWT-based data structures. Theor. Comput. Sci. **698**, 67–78 (2017)
23. Garrison, E., Sirén, J., Novak, A.M., et al.: Variation graph toolkit improves read mapping by representing genetic variation in the reference. Nat. Biotechnol. **36**, 875–879 (2018)
24. Gibney, D., Thankachan, S.V.: Finding an optimal alphabet ordering for Lyndon factorization is hard. In: 38th International Symposium on Theoretical Aspects of Computer Science (STACS2021), pp. 1–15. Leibniz International Proceedings in Informatics (LIPIcs) (2021)
25. Girotto, S., Pizzi, C., Comin, M.: MetaProb: accurate metagenomic reads binning based on probabilistic sequence signatures. Bioinform. **32**(17), 567–575 (2016)
26. Kärkkäinen, J., Kempa, D., Nakashima, Y., Puglisi, S.J., Shur, A.M.: On the size of Lempel-Ziv and Lyndon factorizations. In: 34th Symposium on Theoretical Aspects of Computer Science, STACS 2017, 8–11 March 2017, Hannover, Germany. pp. 45:1–45:13 (2017)

27. Khorsand, P., Denti, L., Human Genome Structural Variant, C., Bonizzoni, P., Chikhi, R., Hormozdiari, F.: Comparative genome analysis using sample-specific string detection in accurate long reads. Bioinform. Adv. **1**(1), vbab005 (2021)

28. Köppl, D., Hashimoto, D., Hendrian, D., Shinohara, A.: In-place Bijective Burrows-Wheeler transforms. In: 31st Annual Symposium on Combinatorial Pattern Matching (CPM 2020). Leibniz International Proceedings in Informatics (LIPIcs), vol. 161, pp. 21:1–21:15 (2020)

29. Kufleitner, M.: On bijective variants of the Burrows-Wheeler transform. In: Proceedings of the Prague Stringology Conference 2009, Prague, Czech Republic, August 31–September 2, 2009. pp. 65–79 (2009)

30. Liu, Q., Hu, Y., Stucky, A., Fang, L., Zhong, J.F., Wang, K.: LongGF: computational algorithm and software tool for fast and accurate detection of gene fusions by long-read transcriptome sequencing. BMC Genomics **21**, 793 (2020). https://doi.org/10.1186/s12864-020-07207-4

31. Lothaire, M.: Algebraic combinatorics on words. Encycl. Math. Appl., vol. 90. Cambridge University Press, Cambridge (1997)

32. Lyndon, R.: On Burnside problem I. Trans. Amer. Math. Soc. **77**, 202–215 (1954)

33. Mäkinen, V., Välimäki, N., Sirén, J.: Indexing graphs for path queries with applications in genome research. IEEE ACM Trans. Comput. Biol. Bioinform. **11**(2), 375–388 (2014)

34. Mantaci, S., Restivo, A., Rosone, G., Sciortino, M.: An extension of the Burrows-Wheeler transform. Theor. Comput. Sci. **387**(3), 298–312 (2007)

35. Mantaci, S., Restivo, A., Rosone, G., Sciortino, M.: Suffix array and Lyndon factorization of a text. J. Discrete Algorithms **28**, 2–8 (2014)

36. Pinho, A.J., Ferreira, P.J., Garcia, S.P., Rodrigues, J.M.: On finding minimal absent words. BMC Bioinform. **10**(1), 1–11 (2009)

37. Rakocevic, G., et al.: Fast and accurate genomic analyses using genome graphs. Nat. Genet. **51**(2), 354–362 (2019)

38. Rautiainen, M., et al.: AERON: transcript quantification and gene-fusion detection using long reads. bioRxiv (2020). https://doi.org/10.1101/2020.01.27.921338

39. Sibbesen, J.A., Maretty, L., Krogh, A.: Accurate genotyping across variant classes and lengths using variant graphs. Nat. Genet. **50**(7), 1054–1059 (2018)

40. Urabe, Y., Kempa, D., Nakashima, Y., Inenaga, S., Bannai, H., Takeda, M.: On the size of overlapping Lempel-Ziv and Lyndon factorizations. In: 30th Annual Symposium on Combinatorial Pattern Matching, CPM 2019, 18–20 June 2019, Pisa, Italy. LIPIcs, vol. 128, pp. 29:1–29:11. Schloss Dagstuhl - Leibniz-Zentrum fuer Informatik (2019)

Word Equations in the Context of String Solving

Joel D. Day$^{(\boxtimes)}$

Loughborough University, Loughborough, UK
`J.Day@lboro.ac.uk`

Abstract. String solvers are tools for automatically reasoning about words over some finite alphabet. They are commonly used to perform analyses of string manipulating programs. A fundamental problem which string solvers need to address is solving word equations, usually in combination with additional constraints involving e.g. string lengths or regular languages. In this article, a survey of results on the topic of word equations is presented with an emphasis on recent results which are relevant to the theoretical foundations of string solvers.

Keywords: Word equations · String constraints · String solving

1 Introduction

Describing one object as a combination others, whether expressed concretely or abstractly, is one of the most fundamental things we do in mathematics. Naturally, we can also do this for words w over some alphabet Σ. By introducing a set X of variables, we can express w as the concatenation of smaller words, some of which are not known explicitly. For example, if $x, y \in X$ are variables and $a, b \in \Sigma$, we can express that w is a word containing an occurrence of aba (so consisting of an unknown word followed by aba followed by another unknown word) by writing w as $xabay$.

Word equations arise when we have multiple ways of expressing the same word in this way. For example, we might describe a word containing both ab and ba via the word equation $x_1 aby_1 \doteq x_2 bay_2$. Formally, a word equation is a pair $(U, V) \in (X \cup \Sigma)^*$ which we usually write as $U \doteq V$. Its solutions are substitutions of the variables for words in Σ^* which identify the two sides. Formally, solutions are modelled by morphisms $h : (X \cup \Sigma)^* \to \Sigma^*$ satisfying $h(a) = a$ for all $a \in \Sigma$ and such that $h(U) = h(V)$.

Perhaps unsurprisingly given their fundamental nature, many natural and well-studied problems related to words can be expressed in terms of word equations. We have already seen how to express the *pattern matching problem*, namely the property that a concrete word (aba) occurs as a factor of some larger word $w \in \Sigma^*$ via the equation $w \doteq xabay$. Expressing that a concrete word occurs as a (scattered) subsequence of w can be done by adding further variables $z_1, z_2, \ldots$ between each of the letters. In the case of aba, we get the equation

© Springer Nature Switzerland AG 2022
V. Diekert and M. Volkov (Eds.): DLT 2022, LNCS 13257, pp. 13–32, 2022.
https://doi.org/10.1007/978-3-031-05578-2_2

$w \doteq xaz_1bz_2ay$. More generally, word equations of the form $w \doteq V$ where $w \in \Sigma^*$ and $V \in (X \cup \Sigma)^*$ correspond exactly to the *membership problem for pattern languages* (also called *pattern matching with variables*). For a constant $k \in \mathbb{N}$, the relation "x is a length-k scattered subsequence of w" where x is a variable can also be expressed, meaning that the k-spectra (see, e.g. [67]) of words can also be expressed as solution-sets to word equations. On the other hand, it follows from [41] that the property "x is a scattered subsequence of y" where both x and y are variables and therefore not of bounded length cannot be expressed via word equations.

The problems mentioned above have been well studied independently from word equations and are substantial areas of research in their own right. Arguably the most interesting and complex cases of word equations are when the variables occur in such a way that they form cyclic dependencies. For example, the equation $xy \doteq yx$ for variables x and y, often referred to as the commutation equation, is solved by a substitution if and only if x and y are substituted for repetitions of the same word w [61]. The equations $xyz = zux$ and $xaby \doteq ybax$ are examples whose solution-sets' descriptions are much more involved, and which have strong connections to the Fibonacci, and Standard and Sturmian words respectively [20,49].

The explicit study of word equations (equivalently equations in a free monoid or semigroup) can be traced back as far as A. A. Markov in the 1940s, although connections to Diophantine equations mean that in some sense, they have been studied indirectly for much longer [32]. Originally it was hoped that the satisfiability problem - whether or not a given word equation has a solution - might provide an means of connecting Diophantine equations with the computations of Turing machines in such a way as to allow for a proof that solving the former is undecidable and thus settling Hilbert's famous 10th problem. However this turned out not to be feasible with Makanin famously providing an algorithm for the satisfiability problem for word equations in 1977 [64]. Since then, several further algorithms have been presented by Plandowski and Rytter [76], Plandowski [74] and Jeż [52]. The latter, based on the *Recompression* technique resulted in a considerably simpler proof of correctness and has since been use to improve the PSPACE complexity upper bound given by Plandowski's algorithm to non-deterministic linear space [53].

The positive result of Makanin became highly influential in combinatorial group theory: in a series of results ultimately resolving Tarski's Conjectures, it was first adapted to work for equations in free groups by Makanin himself [63,65], before being used by Razborov to provide a method for describing all solutions to equations in free groups via so-called Makanin-Razborov diagrams [78]. More recently, algorithms for solving word equations have been extended to work in a range of algebraic structures such as hyperbolic groups [19,23] and partially commutative groups [35].

There has also been much interest in the free monoid case (on which we concentrate in the current paper) from several perspectives. Constructions exist (see e.g. [55,61] for reducing the satisfiability of Boolean combinations of word equations to the satisfiability of a single word equation. Consequently, Makanin's result extends to the existential theory of a free semigroup or monoid. Further results on logics involving words/concatenation can be found e.g. in Büchi and Senger [14] where the notion of definability is considered, and Quine [77] and Durnev [36] who show undecidability when quantifier alternations are allowed. Further undecidability results considering logics involving words and additional predicates (such as "is a scattered subsequence of" and "is abelian-equivalent to") can be found in [26,41]. The expressive power of word equations in defining relations and languages via (projections of) their solution-sets is considered in detail in [55] where some powerful tools are given for showing inexpressibility based on the notion of a *synchronising factorisation*.

In the field of combinatorics on words, constant-free word equations (equations of the form $U \doteq V$ where $U, V \in X^*$) have been studied extensively. In addition to solving specific equations or families of equations [21,43,46,70,80], a major topic of ongoing research involves *independent systems* of word equations - systems such that removing any one of the equations leads to a strictly larger set of solutions [20,42,48,56,71]. A connection between constant-free word equations and the general case is shown in [81].

From an algorithms and complexity perspective, it is easily seen that the satisfiability problem for word equations is NP-hard. Indeed, the result follows directly from NP-completeness of the membership problem for pattern languages [8]. On the other hand, it remains a long-standing open problem as to what the true complexity of the problem is, and in particular whether or not it is NP-complete. Plandowski and Rytter [76] showed that long solutions are highly compressible, and consequently, even an exponential bound on the length of the shortest solution, when one exists, would imply inclusion in NP. Nevertheless, the best known bound remains double-exponential [52,73].

Open Problem 1 ([76]). *Is the satisfiability problem for word equations contained in NP?*

2   String Solving

In recent years, a whole new community has formed with the goal of developing automated reasoning tools capable of proving or disproving statements involving words called *string solvers*, motivated primarily by applications in formal methods. Simply put, string solvers take as input a *string constraint* which can be thought of as a (usually quantifier-free) formula comprising Boolean combinations of atomic constraints involving:

- constants from Σ^* for some finite alphabet Σ,
- variables whose potential values range over Σ^*,
- common relations and operations on words such as concatenation, equality, length-comparisons, regular language membership, and typical string-manipulation functions such as `Replace_All()` and `Index_Of()`, string-number conversion, etc.

Their task is then to automatically determine the satisfiability of that formula, so, whether or not values for the variables can be found such that the formula becomes true. A comprehensive list of common relations and operations occurring as string constraints can be found in [7]. A standardised language for specifying string constraints in string solvers based on the *Satisfiability Modulo Theories* framework has also been implemented as part of the SMT-LIB format [2,11]. Most currently available string-solvers do not cater for the whole SMT-LIB standard, but focus rather on specific subsets to which their approaches or target applications are best suited.

Growing interest in string solving is possibly due to a steady increase in string-manipulating programs which are vulnerable to exploitation or attack (e.g. due to being publicly accessible on the web) in combination with substantial improvements in string solvers' own performance when employed in static analysis tasks aimed at improving security and reliability of software. There are many ways string solvers can be deployed in the context of software analysis. We list a few below (see e.g. [5,12,15,60]):

Path Feasibility. A common task in static analysis is to break down the possible executions of a piece of software into finitely many cases (paths). The problem of *path feasibility* involves working out which combinations of conditions on internal values result in paths which might actually occur in a real execution, and conversely which combinations of conditions are contradictory and so don't need to be considered further. Path feasibility analyses are also particularly useful in automated test-case generation.

Sanitisation and Validation. Cross-site scripting (XSS) and SQL injection attacks have both regularly been listed on the OWASP list of *Top 10 Web Application Security Risks* in recent years [1]. Although the two categories have been merged into a single one: "injection" in the 2021 list, it remains in a prominent position at #3. Such injection attacks involve tricking a system to execute malicious code. In the case of SQL injection, it might be that textual input is given which later forms part of an SQL database query constructed as a string. Maliciously designed entries, when not properly sanitised, can then influence the structure and meaning of the query, allowing the user to access or erase data. In XSS attacks, a similar effect can be achieved by getting a user's browser to execute malicious code e.g. in a link. Errors in input sanitisation and validation are not uncommon and automated analysis of (parts of) software handling externally generated data can help to reduce such errors [15].

Dynamically Generated Code. It is becoming increasingly common in programming languages to be able to dynamically load code such as functions and classes from string variables meaning that the executed code depends on the values of those string variables at runtime. While extremely powerful, this is also dangerous if the strings are not constructed safely and correctly. Again, both static and dynamic analyses performed with the help of string solvers can mitigate to some extent these risks.

The are also many potential areas of application for string solvers beyond software analysis and formal methods more generally, for example in database theory [10,37,38] or as automated proof assistants for areas such as combinatorics on words (see e.g. [47]).

Many string solving tools are now available employing a wide range of strategies, including CVC5 [9], Z3Str4 [68], Norn [3], Z3-Trau [4], OSTRICH [17], Sloth [45], Woorpje [24,28], CertiStr [54] and HAMPI [57]. Some, such as CVC5 and Z3str4 are designed to be more general, while others are developed with more specific tasks in mind. Many benchmarks exist, and in addition to the MOSCA (meeting on String Constraints and Applications) workshop [40], there is also now a string track at the annual SMT competition. Several meta-tools have been developed for comparing string solvers [58], automatically producing or altering test cases and benchmarks (fuzzing) [13] and very recently also for analysing large and often opaque sets of benchmarks [27].

Nevertheless, many challenges remain, of which one is obtaining a better understanding of the theoretical foundations involving word equations in combination with other combinations of constraints. When atomic constraints are restricted to involve only concatenation and equality comparisons, string solving can be reduced to solving word equations. Several other types of constraint, like regular language membership and linear arithmetic involving string-lengths are well understood in isolation and can be tackled in practice using highly optimised tools. However a common feature of string solving applications is that types of constraints often have to satisfied in combination.

It has been known since the 1990s that satisfiability of word equations with regular language membership constraints on the variables is decidable [33,82] (see also [61]). However, many other combinations quickly lead to undecidability: for example in the presence of a `Replace_All()` operator (modelled formally as finite transducers) [60] or in the presence of functions which count the numbers of a letter occurring in a word [14,26]. Since concatenation, equality (and thus word equations), regular language membership and linear (in)equalities over lengths of variables are particularly prominent types of string constraints (which additionally can in combination be used to model several other common constraints such as `Index_Of()`), a particularly important open problem remains:

Open Problem 2. *Is the satisfiability problem decidable for quantifier-free formulas combining word equations with regular language membership and linear (in)equalities over the lengths of variables?*

Formally, the satisfiability problem for quantifier-free formulas is a natural extension of the satisfiability problem for word equations: can we find substitutions for the variables which make the formula true when evaluated according to the natural semantics?

We call atomic constraints consisting of regular language membership *regular constraints*. They have the form $x \in L$ where x is a variable and L is a regular language given e.g. by an NFA or regular expression[1]. Similarly, we call atomic constraints consisting of linear (in)equalities over lengths of variables *length constraints*. We use $|x|$ to denote the length of a word x, including the case that x is a variable and the length is then an unknown number. Length constraints can be formalised e.g. as having the form $c_1|x_1| + c_2|x_2| \ldots c_k|x_k| \oplus d_1|x_1| + d_2|x_2| \ldots d_k|x_k|$ where $\oplus \in \{>, =\}$, and for $1 \leq i \leq k$, $c_i, d_i \in \mathbb{Z}$ and x_i are variables. Using this terminology, and recalling that Boolean combinations of word equations can be rewritten as a single equation, Open Problem 2 asks whether the satisfiability problem for word equations with regular and length constraints is decidable. An example of a string constraint involving word equations, regular constraints and length constraints is given below (x, y, z are variables and $a, b \in \Sigma$.

$$(x \doteq yabz \land |y| > |z|) \lor (x \doteq ybaz \land z \in a^*)$$

Both regular constraints and length constraints can be subsumed by language-membership constraints involving *visibly pushdown languages* (VPLs) [6]. VPLs generalise regular languages while retaining many of the desirable closure and algorithmic properties. Nevertheless, it is shown in [25] that combining word equations with VPL membership constraints results in an undecidable satisfiability problem.

Theorem 1 ([25]). *For every recursively enumerable language L, there exists a quantifier free formula f combining word equations and VPL membership constraints and a variable x occurring in f, such that*

$$L = \{w \mid x \text{ may be substituted by } w \text{ as part of a satisfying assignment for } f\}.$$

It follows that the satisfiability problem for such formulas is undecidable.

Since VPLs share many properties with regular languages, VPL-membership constraints can be viewed as only a minor generalisation of regular (and length) constraints. This leads us to the following open problem, for which a negative would also yield a negative answer to Open Problem 2.

Open Problem 3. *Does there exist a recursively enumerable language L which is not expressible as the set*

$$\{w \mid x \text{ may be substituted by } w \text{ as part of a satisfying assignment for } f\}$$

for some quantifier free formula f combining word equations, regular constraints and length constraints?

A careful application of Greibach's theorem reveals that we cannot in general decide whether a property of words expressed by solutions to a string constraint

---

[1] Since it is easy to simulate the intersection of regular languages via conjunctions of regular constraints, the satisfiability problem for formulas containing regular constraints is automatically PSPACE-hard. Therefore, we can convert between any reasonable choices for specifying regular languages without affecting the computational complexity.

combining word equations, regular constraints, and length constraints can also be expressed using a combination of word equations and regular constraints alone.

Theorem 2 ([25]). *The following problem is undecidable: given a quantifier free formula f combining of word equations, regular constraints and length constraints, and a variable x occurring in f, does there exist a quantifier free formula f' containing only word equations and regular constraints and a variable y in f' such that $S_1 = S_2$ where:*

$S_1 = \{w \mid x \text{ may be substituted by } w \text{ as part of a satisfying assignment for } f\}$

and

$S_2 = \{w \mid x \text{ may be substituted by } w \text{ as part of a satisfying assignment for } f'\}$?

A weaker version of Open Problem 2, where regular constraints are omitted, is also a long standing open problem in the field of word equations.

Open Problem 4. *Is the satisfiability problem decidable for quantifier-free formulas combining word equations and linear (in)equalities over the lengths of variables?*

The difficulty of dealing with word equations in combination with length constraints is highlighted in [62], where they show that the set $\{(|h(x)|, |h(y)|) \mid h \text{ is a solution to } xaby \doteq ybax\}$, where $|w|$ denotes the length of the word w, is not definable in Presburger arithmetic.

3 A Closer Look at Solution-Sets

One of the most natural ways to try to improve our understanding of word equations, and in particular to improve techniques for solving them in combination with other constraints in practice, is to look closer at the structure of their solution-sets. Indeed, if when we are given a string constraint consisting, for example, of word equations, regular language memberships and length (in)equalities, one could try to solve the overall constraint by first providing a description of the set of solutions to the system of word equations without the additional constraints, and subsequently use that description to reason whether a solution exists satisfying the additional constraints. For example, given the string constraint

$$xy \doteq yx \wedge |y| > |x| \wedge x \in \{ab\}^* \wedge y \notin a\{a, b\}^+$$

one could first notice that a pair of words x, y is a solution to the word equation $xy \doteq yx$ if and only if they are both repetitions of the same word, or more formally, if there exist $w \in \Sigma^*$ and $p, q \in \mathbb{N}_0$ such that $x = w^p$ and $y = w^q$ where w^0 is the empty word and $w^{i+1} = w^i w$ [61]. With such an explicit description to hand, it is then not difficult to observe that since $x \in \{ab\}^+$, x and y must necessarily both be repetitions of $w = ab$. From $y \notin a\{a, b\}^*$ we further conclude that y is the empty word, and since this implies that $|y| > |x|$ cannot be met, that this string constraint is unsatisfiable.

3.1   Parametric Solutions

The description of solutions x, y to the equation $xy \doteq yx$ given in terms of an unknown word w and numbers p, q is called a *parametric solution*. Parametric solutions are particularly useful because they are a very explicit description of the solution-set. Following [20], they are defined formally as follows.

Definition 1. *Let Δ, Γ be alphabets. We call elements of Δ word parameters and elements of Γ numerical parameters. Parametric words are defined inductively as follows.*

- *Each element of Δ is a parametric word,*
- *if δ is a parametric word and $k \in \Gamma$ is a numerical parameter, then δ^k is a parametric word,*
- *if δ_1, δ_2 are parametric words, then their concatenation $\delta_1 \delta_2$ is also a parametric word.*

Every assignment φ of words in Σ^* to the word parameters and numbers from $\mathbb{N}_0$ to the numerical parameters maps a parametric word to a unique concrete word $\varphi(\delta) \in \Sigma^*$ called the value.

Given a word equation E over n variables, a parametric solution *for E is an n-tuple of parametric words such that every assignment φ induces a solution. Moreover, E is said to be* parametrizable *if the solution-set is exactly described by finitely many parametric solutions.*

The definition above is given in context of constant-free word equations. For this reason, it does not include provision for constants occurring in parametric solutions. However, it is very natural to extend Definition 1 to the more general case by simply adding the axiom that each element of $\Sigma \cup \{\varepsilon\}$ is also a parametric word, where ε denotes the empty word.

Unfortunately, although it was shown by Hmelevskii [44] that every constant-free equation with at most three variables is parametrizable, the same does not hold when a fourth variable is introduced. A more concise proof of the latter fact is given by Czeizler in [20].

Theorem 3 (Czeizler [20], Hmelevskii [44]). *Let x, y, z, v be variables. Then the word equation $xyz \doteq zvx$ is not parametrizable.*

Of course the negative parametrizability result also carries through to the general case in which constants are also allowed. In fact, it follows from [49] that the word equation $xaby \doteq ybax$ with only two variables x and y is not parametrizable.

3.2   Graph Representations of Solution-Sets

A key question then, is how to represent solution-sets to word equations if not by parametric words? One answer can be found in approaches for algorithmically solving word equations which have been extended to produce descriptions of the full solution-set.

Decision procedures for solving word equations usually revolve around some non-deterministic search for solutions, made necessary by the fact that the satisfiability problem is NP-hard. This search can often be formalised in terms of iteratively applying transformation rules (e.g. to a possibly extended representation of the equation, or a solution to it[2]) with the aim of eventually reaching some trivial case signifying that a solution exists. Such an approach yields a (possibly infinite) graph which, with the correct setup, provides a complete description of the set of solutions by virtue of accounting for the search across all possibilities. Guaranteeing that the graph is finite presents more of a challenge, although there are now several different approaches which achieve this.

In an early example of this approach, Makanin's algorithm for solving equations in a free group was used by Razborov in developing an algorithmic representation of all solutions to systems of equations in a free group [78]. In [75], Plandowski adapted his algorithm for solving word equations to produce a finite graph representing all solutions, and more recently, the same was achieved in a simpler manner using the Recompression technique of Jeż [52].

Diekert, Jeż and Plandowski generalised the Recompression approach to work in the presence of both regular membership constraints and involution [34]. This combination is significant because it allows for the extension of methods from the free monoid to the free group setting. Shortly after, Ciobanu, Diekert and Elder [18] provided a simpler representation in terms of EDT0L languages.

Theorem 4 ([18]). *Solution-sets to word equations with regular membership constraints and involution (and hence also equations in free groups) are EDT0L languages. In particular, they are also indexed languages.*

Indexed languages are a subset of the context-sensitive languages. EDT0L (Extended Deterministic Table 0-Interaction Lindermayer) languages are languages defined by a specific variety of so-called L-systems, which generate words by iterated applications of morphisms to some initial "seed" word. They are strictly contained in the indexed languages and are incomparable to context-free languages. Despite their verbose name, EDT0L languages are a natural class with a simple intuition, corresponding to the case when the application of the morphisms is constrained by some NFA-like control.

Definition 2. *Let A be an alphabet and $L \subseteq A^*$. Then L is an EDT0L language if there exists an alphabet C with $A \subseteq C$, a word $w \in C^*$ and a rational set R of morphisms $h : C^* \to C^*$ such that $L = \{h(w) \mid h \in R\}$.*

Given an equation $U \doteq V$ with variables $x_1, x_2, \ldots, x_n$, the construction from [18] essentially equates the set of solutions

$$\{(g(x_1), g(x_2), \ldots, g(x_n)) \mid g(U) = g(V)\}$$

with the set $\{(h(c_1), h(c_2), \ldots, h(c_n)) \mid h \in R\}$ for some rational set of morphisms R and additional letters $c_1, c_2, \ldots, c_k$. The latter set is then easily encoded as an EDT0L language using a separator symbol $\#$ as the set

---

[2] These two viewpoints are not mutually exclusive as a word equation can be thought of as a compact representation of a solution.

$\{h(c_1 \# c_2 \# \ldots \# c_n) \mid h \in R\}$. The set R, when represented by the underlying NFA, provides a graph representation of all solutions. This graph representation facilitates algorithmic solutions to problems other than just the satisfiability problem. In particular (in)finiteness can be determined by looking for cycles in the NFA defining R.

Corollary 1 ([18]). *It is decidable whether the solution-set to a (system of) word equations with regular language membership constraints is finite.*

Unfortunately, these graph representations are not as well suited to other canonical decision problems. Indeed, negative results can be inferred from [36, 39], both of which provide proofs of the fact that deciding the truth of logical sentences of the form

$$\forall x \exists y_1, y_2, \ldots, y_n. \varphi$$

where φ is a Boolean combination of word equations is undecidable. In particular, we get the following.

Theorem 5 ([36,39]). *It is undecidable whether or not, given a word equation E containing a variable x (and possibly others), the set $\{h(x) \mid h$ is a solution to $E\}$ is exactly Σ^*.*

Moreover, we note the following negative result from [25].

Theorem 6 ([25]). *It is undecidable, given a word equation E containing a variable x (and possibly others), the set $\{h(x) \mid h$ is a solution to $E\}$ is a regular language.*

Consequently, we cannot expect that any reasonable (computable) representation of solution-sets to word equations is sufficiently descriptive as to allow for inference of all interesting properties.

3.3   Nielsen Transformations

A disadvantage of the graph representations discussed in the previous section is that, even in for those that are guaranteed to be finite, the edge relations are complex (or, at least in the case of Recompression, highly non-deterministic) meaning it is difficult to study their structure in detail. Quadratic word equations, in which each variable occurs at most twice, offer a much simpler means of producing a finite graph describing all solutions via a rewriting process based on a well-known type of morphism called Nielsen transformations.

The rewriting relation, which we denote $\Rightarrow_{NT}$, is combinatorially simple at a local level. When applied iteratively to a given word equation E, it induces a graph $\mathcal{G}_E^{\Rightarrow_{NT}}$ describing all solutions which in the general case is usually infinite, but in the quadratic case is guaranteed to be finite. $\mathcal{G}_E^{\Rightarrow_{NT}}$ has as vertices word equations (including E) and its edges are labelled with morphisms $\psi : (X \cup \Sigma)^* \to (X \cup \Sigma)^*$. Solutions to E are obtained by composing the morphisms occurring as edge-labels on walks[3] in the graph starting at E and finishing at the trivial equation $\varepsilon \doteq \varepsilon$. The underlying idea comes from a basic fact concerning semigroups called Levi's lemma, stated as follows.

Lemma 1 (Levi's Lemma). *Let $u, v, x, y \in \Sigma^*$ be words such that $uv = xy$. Then there exists w such that either:*

- $u = xw$ and $wv = y$, or
- $x = uw$ and $wy = v$.

Levi's lemma applies to word equations in the following way: given a word equation $xU \doteq yV$ where $x, y \in X \cup \Sigma$ are the leftmost symbols on each side of the equation and $U, V \in (X \cup \Sigma)^*$ are the remaining parts, we have three possibilities for a non-erasing solution[4] h: either

- $h(x) = h(y)$ and $h(U) = h(V)$ (this corresponds to the case that $w = \varepsilon$ in Levi's lemma), or
- $h(x) = h(y)w$ and $wh(U) = h(V)$ for some $w \in \Sigma^+$
- $h(x)w = h(y)$ and $h(U) = wh(V)$ for some $w \in \Sigma^+$.

In the first case, if $x, y \in \Sigma$ with $x \neq y$, no solutions exist. If $x = y \in \Sigma$ then solutions to E are exactly solutions to $U \doteq V$. Otherwise, solutions h to E can be found by looking for solutions h' to the equation $U' \doteq V'$ obtained by replacing x everywhere by y in U and V respectively if x is a variable (or vice-versa if y is a variable).

In the second case, if x is not a variable, then no solutions exist for this case. Otherwise, by introducing a new variable z (intended to account for w, so that $h(z) = w$) we can find solutions h to E in terms of solutions h' to the equation $zU' \doteq V'$ obtained by replacing all occurrences of x by yz in U and V respectively. The third case is symmetrical to the second.

Thus, overall, solutions to E can be reduced to solutions to (at most) three further equations derived by cancelling some symbols from the left and performing a replacement of the form $x \to yz$ or $x \to y$. Notice that when performing a replacement $x \to yz$, we actually remove all occurrences of x from the equation, and so we might as well re-use the variable x in place of z to get $x \to yx$ (and similarly for $y \to xz$ we might as well use y in place of z to get $y \to xy$). These replacements can be performed via the application of morphisms from a set Ψ defined as follows: for $x \in X \cup \Sigma$ and $y \in X$, $\psi_{(x,y)}, \hat{\psi}_{(x,y)} : (X \cup \Sigma)^* \to (X \cup \Sigma)^*$ belong to Ψ such that:

$$\psi_{(x,y)}(y) = xy \qquad\qquad \hat{\psi}_{(x,y)}(y) = x$$
$$\psi_{(x,y)}(z) = z \text{ for } z \neq y, \qquad\qquad \hat{\psi}_{(x,y)}(z) = z \text{ for } z \neq y.$$

---

[3] Paths in which both vertices and edges may be repeated.

[4] Non-erasing solutions are solutions for which $h(x)$ is not the empty word for any variable x. The general case can be reduced to the non-erasing case by simply guessing in advance which variables should be mapped to the empty word and removing them from the word equation(s).

The morphisms $\psi_{(x,y)}$ are called *Nielsen transformations*, hence the name of this approach. We denote by $\Rightarrow_{NT}$ the relation consisting of pairs (E_1, E_2) such that E_2 may be derived from E_1 according to one of the three cases above.

Now suppose we have a quadratic equation E and consider E' such that $E \Rightarrow_{NT} E'$. Then the removal of the leftmost symbols means that $|E| \geq |E'|$. Similarly, the number of occurrences of each variable in E is at least as high as in E', and no new symbols are introduced. It follows that there are only finitely many equations E'' such that $E \Rightarrow_{NT}^* E''$ where $\Rightarrow_{NT}^*$ denotes the reflexive transitive closure of $\Rightarrow_{NT}$.

Let $\mathcal{G}_E^{\Rightarrow NT}$ be the graph whose vertices are equations E'' reachable from E by iteratively applying $\Rightarrow_{NT}$, and whose edges are given by $\Rightarrow_{NT}$, labelled with the appropriate corresponding morphisms. Given a word equation E' occurring as a vertex in this graph, for each solution h' to E' there exists an edge in the graph from E' to a (not necessarily distinct) equation E'' labelled with a morphism ψ, such that $h' = h'' \circ \psi$ for some strictly shorter solution[5] h'' to E''. For this reason, all solutions to the original equation E can be obtained by composing the morphisms occurring on a walk in the graph from the original equation to the trivial equation $\varepsilon \doteq \varepsilon$ as mentioned previously. If the equation $\varepsilon \doteq \varepsilon$ is not present in the graph, no solutions exist.

As an example, consider the equation E given by $Xabaa \doteq baaXa$ over the variable X and constants $a, b \in \Sigma$. The graph $\mathcal{G}_E^{\Rightarrow NT}$ (with labels) is shown in Fig. 1.

Treating the graph as an NFA $\mathcal{A}_E$ over the alphabet of morphisms Ψ whose accepting state is $\varepsilon \doteq \varepsilon$ and whose initial state is E, we obtain a rational set $L(\mathcal{A}_E)$ of morphisms exactly describing the set of solutions to E. For example one solution is given by $h = \hat{\psi}_{(a,X)} \circ \psi_{(b,X)} \circ \psi_{(a,X)} \circ \psi_{(a,X)} \circ \psi_{(b,X)}$ (note that the composition occurs in the opposite order from left to right to the "word" from Ψ^* accepted by $\mathcal{A}$), which is the substitution h given by $h(X) = baaba$, $h(a) = a$ and $h(b) = b$.

$$h(X) = \hat{\psi}_{(a,X)} \circ \psi_{(b,X)} \circ \psi_{(a,X)} \circ \psi_{(a,X)} \circ \psi_{(b,X)}(X)$$
$$= \hat{\psi}_{(a,X)} \circ \psi_{(b,X)} \circ \psi_{(a,X)} \circ \psi_{(a,X)}(bX)$$
$$= \hat{\psi}_{(a,X)} \circ \psi_{(b,X)} \circ \psi_{(a,X)}(baX)$$
$$= \hat{\psi}_{(a,X)} \circ \psi_{(b,X)}(baaX)$$
$$= \hat{\psi}_{(a,X)}(baabX)$$
$$= baaba$$

The simplicity of this approach based on Nielsen transformations and Levi's lemma, along with the fact that it is easily adapted for use with regular language membership constraints and length constraints means it is a good candidate for practical implementations. As such it has been used in the string solving tool Woorpje [28], and other string solvers make use of similar ideas. It also has

---

[5] Where the length of the solution is measured in terms of the word obtained by applying it to one side of the equation.

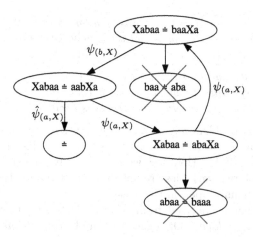

Fig. 1. The graph $\mathcal{G}_E^{\Rightarrow NT}$ in the case that E is the equation $Xabaa \doteq baaXa$. Trivially unsolvable equations in the graph are crossed out and their ingoing edge labels omitted.

several advantages from a theoretical point of view: the structural properties of the graph $\mathcal{G}_E^{\Rightarrow NT}$ provide information about the solution-set of E. For example if it is a directed acyclic graph (DAG), then it is straightforward to show that E is parametrizable. Actually, a slightly stronger statement, namely that $\mathcal{G}_E^{\Rightarrow NT}$ does not have two distinct cycles sharing a vertex, is sufficient to guarantee that E is parametrizable [72]. Similar restrictions on the structure of $\mathcal{G}_E^{\Rightarrow NT}$ have been used to identify cases where satisfiability remains decidable even when length constraints are added [62]. Several case where the graph is guaranteed to be finite even when the underlying equation E is not quadratic are considered in [69]. Moreover the simplicity of the rewriting transformations make the graphs obtained via Nielsen transformations much more accessible for more detailed combinatorial analyses such as the one given in [29].

3.4   Restricted Word Equations

In the absence of positive answers to Open Problems 1, 2 and 4, it is natural to consider them also in the context of syntactically restricted subclasses of word equations. For simplicity, we concentrate in this section on single word equations. Some care is needed when generalising to Boolean combinations: while in general it is no restriction to do so due to the constructions e.g. in [55], these constructions are not guaranteed to respect the syntactic restrictions and so cannot be used directly to generalise the results in this section. However, in most cases equivalent results hold at least for systems (conjunctions) of equations.

Solution-sets to word equations containing only one variable have a particularly restricted form and are well understood:

Theorem 7 ([59,71]). *Let x be a variable, and let $U \doteq V$ be a word equation such that $U, V \in \{x\} \cup \Sigma^*$. Then one of the following holds:*

1. *The set of solutions for $U \doteq V$ is finite and has cardinality at most three, or*
2. *There exist words $u, v \in \Sigma^*$ such that uv is primitive[6] the set of solutions for $U \doteq V$ has the form $\{h : (\{x\} \cup \Sigma)^* \to \Sigma^* \mid h(x) \in (uv)^*u\}$.*

Corollary 2. *Word equations with exactly one variable are parametrizable. Moreover, the satisfiability problem for word equations with one variable and with length constraints and regular constraints is decidable.*

The satisfiability problem for word equations is solvable in deterministic linear time [51]. Similarly, word equations with two variables are also solvable in polynomial time [16,22,49]. However, we have already seen that solution-sets to word equations with two variables are not necessarily parametrizable, and Open Problems 2 and 4 remain open in this case.

Further cases can be derived from restricted classes of string constraints. The notions of *solved form* [39], *acyclic* [5] and *straight-line* [60] constraints are all syntactic restrictions designed such that cyclic dependencies between the variables are avoided. As such, the solution-sets to constraints adhering to these definitions are generally parametrizable, and satisfiability for constraints involving word equations, length constraints and regular constraints (and even `Replace_All()` in the case of the straight-line fragment) become decidable.

We have already mentioned in the previous section that quadratic word equations - equations which contain each variable at most twice (although the number of variables is unconstrained) - possess the desirable property that the simple Nielsen transformation algorithm for producing a graph representation of all solutions is guaranteed to terminate. Nevertheless, Open Problems 1, 2 and 4 all remain open even in the quadratic case. It was shown in [79] that the satisfiability problem remains NP-hard in the quadratic case. On the other hand, word equations in which variables occur at most once are trivially parametrizable, and it is easily seen that the satisfiability problem remains decidable in the presence of various additional constraints, including length constraints and regular constraints.

In [66], the class of regular word equations was proposed as a natural subclass of the quadratic word equations. The initial idea was to consider classes of equations $U \doteq V$ for which the satisfiability problem remains NP-hard, even when the two sides U and V constitute patterns for which the membership problem can be solved in polynomial time. Regular word equations derive their name from regular patterns [50] in which each variable occurs at most once. Consequently, each variable may occur twice overall, but not twice on the same side of the equation.

Definition 3. *A word equation $U \doteq V$ is regular if each variable occurs at most once in U and at most once in V.*

Surprisingly, even severely restricted subclasses of regular word equations have an NP-hard satisfiability problem. The class of regular-ordered word equations (ROWEs) is the class of regular word equations for which the variables

---

[6] A word is primitive if it cannot be written as the repetition of a strictly shorter word.

occur in the same order from left to right on both sides of the equation (some variables may still occur on only one side). So, for example, the equation $x_1 a x_2 b x_3 \doteq x_1 b a b a x_3$ is regular-ordered, but $x_1 a x_2 b x_3 \doteq x_3 b a b a x_1$ is not.

Theorem 8 ([30]). *The satisfiability problem for ROWEs is NP-complete.*

Moreover, it was shown in [62] that the satisfiability problem for ROWEs with length constraints is decidable, extending a weaker result from [26].

In [31], Theorem 8 was extended slightly to cover regular-reversed word equations (RRWEs): equations $U \doteq V$ in which the order of variables in V is exactly the reverse of the order of the variables in U. A much more comprehensive result is given in [29], which describes in detail the structure of the graphs $\mathcal{G}_E^{\Rightarrow NT}$ obtained as the result of the Nielsen transformation algorithm described in Sect. 3.3 in the case of all regular word equations. A consequence of this description is that the minimal path between any two vertices (when it exists) has length bounded by a polynomial in the length of the original equation.

Theorem 9 ([29]). *Let E be a regular word equation. Let v_1, v_2 be vertices in the graph $\mathcal{G}_E^{\Rightarrow NT}$ such that there exists a path from v_1 to v_2. Then the shortest such path has length at most $O(|E|^{12})$.*

Consequently, in order to (non-deterministically) check whether a solution exists to a regular word equation E, it suffices to guess the equations occurring along the path in $\mathcal{G}_E^{\Rightarrow NT}$ from E to $\varepsilon \doteq \varepsilon$. Each equation on that path will be no larger than E, and it is easily verifiable in polynomial time that there is indeed an edge between each successive pair of equations. Thus, it follows that satisfiability for regular word equations is in NP. Combined with the hardness result from [30], we get the following.

Corollary 3 ([29,30]). *The Satisfiability problem for regular word equations is NP-complete.*

Of course one of the most natural open problems remains whether the techniques of [29] can be extended to work for quadratic word equations more generally.

Open Problem 5. *Does Theorem 9 also hold for all quadratic word equations?*

Moreover, the detailed analysis in [29] would be a good basis from which to try to resolve Open Problems 4 and 2 in the sub-case of regular word equations. In particular, several structures in the graphs $\mathcal{G}_E^{\Rightarrow NT}$ are described which might provide new insights into how complex the set of lengths of solutions to (regular) word equations can be.

Open Problem 6. *What is the decidability status for the satisfiability problem for regular word equations with length constraints? What about for regular word equations with length constraints and regular constraints?*

4   Conclusions

The study of word equations has yielded many significant and influential results over the past half-century, and is of interest in a variety of areas, including logic, formal languages, combinatorics on words, and combinatorial group theory. More recently, there has been substantial interest in the topic from the within the formal methods community, specifically in relation to string solvers, which aim to solve problems involving words which often incorporate word equations alongside other constraints such as regular language membership and length-comparisons. The two fields are mutually beneficial: theoretical results on word equations and related topics can provide insights and ideas for more efficient, powerful and ultimately practical algorithms implemented in string solvers while on the other hand, a better understanding of the problems that string solvers must tackle can reveal new open problems and directions to be explored in the theory. Recent results focusing on the structure and properties of solution-sets for restricted classes of word equations provide a basis from which we can hope to make progress on long standing open problems which remain central to both the theory and practice.

References

1. OWASP top ten web application security risks. https://owasp.org/www-project-top-ten/. Accessed 15 Mar 2022
2. SMT-LIB standard for unicode strings. https://smtlib.cs.uiowa.edu/theories-UnicodeStrings.shtml. Accessed 15 Mar 2022
3. Abdulla, P.A., et al.: Norn: an SMT solver for string constraints. In: Kroening, D., Păsăreanu, C.S. (eds.) CAV 2015. LNCS, vol. 9206, pp. 462–469. Springer, Cham (2015). https://doi.org/10.1007/978-3-319-21690-4_29
4. Abdulla, P.A., et al.: Efficient handling of string-number conversion. In: Proceedings of the 41st ACM SIGPLAN Conference on Programming Language Design and Implementation, pp. 943–957 (2020)
5. Abdulla, P.A., et al.: String constraints for verification. In: Biere, A., Bloem, R. (eds.) CAV 2014. LNCS, vol. 8559, pp. 150–166. Springer, Cham (2014). https://doi.org/10.1007/978-3-319-08867-9_10
6. Alur, R., Madhusudan, P.: Visibly pushdown languages. In: Proceedings of the 36th ACM Symposium on Theory of Computing (STOC), STOC 2004, pp. 202–211 (2004)
7. Amadini, R.: A survey on string constraint solving. ACM Comput. Surv. (CSUR) 55(1), 1–38 (2021)
8. Angluin, D.: Finding patterns common to a set of strings. In: Proceedings of the Eleventh Annual ACM Symposium on Theory of Computing, pp. 130–141 (1979)
9. Barbosa, H., et al.: cvc5: a versatile and industrial-strength SMT solver. In: Fisman, D., Rosu, G. (eds) TACAS 2022. LNCS, vol. 13243, pp. 415–442. Springer, Cham (2022). https://doi.org/10.1007/978-3-030-99524-9_24
10. Barceló, P., Muñoz, P.: Graph logics with rational relations: the role of word combinatorics. ACM Trans. Comput. Logic (TOCL) 18(2), 1–41 (2017)
11. Barrett, C., Stump, A., Tinelli, C., et al.: The SMT-LIB standard: Version 2.0. In: Proceedings of the 8th International Workshop on Satisfiability Modulo Theories, Edinburgh, England, vol. 13, p. 14 (2010)

12. Bjørner, N., Tillmann, N., Voronkov, A.: Path feasibility analysis for string-manipulating programs. In: Kowalewski, S., Philippou, A. (eds.) TACAS 2009. LNCS, vol. 5505, pp. 307–321. Springer, Heidelberg (2009). https://doi.org/10.1007/978-3-642-00768-2_27

13. Blotsky, D., Mora, F., Berzish, M., Zheng, Y., Kabir, I., Ganesh, V.: StringFuzz: a fuzzer for string solvers. In: Chockler, H., Weissenbacher, G. (eds.) CAV 2018. LNCS, vol. 10982, pp. 45–51. Springer, Cham (2018). https://doi.org/10.1007/978-3-319-96142-2_6

14. Büchi, J.R., Senger, S.: Definability in the existential theory of concatenation and undecidable extensions of this theory. In: Mac Lane, S., Siefkes, D. (eds.) The Collected Works of J. Richard Büchi, pp. 671–683. Springer, New York (1990). https://doi.org/10.1007/978-1-4613-8928-6_37

15. Bultan, T., Yu, F., Alkhalaf, M., Aydin, A.: String Analysis for Software Verification and Security, vol. 10. Springer, Cham (2017). https://doi.org/10.1007/978-3-319-68670-7

16. Charatonik, W., Pacholski, L.: Word equations with two variables. In: Abdulrab, H., Pécuchet, J.-P. (eds.) IWWERT 1991. LNCS, vol. 677, pp. 43–56. Springer, Heidelberg (1993). https://doi.org/10.1007/3-540-56730-5_30

17. Chen, T., Hague, M., Lin, A.W., Rümmer, P., Wu, Z.: Decision procedures for path feasibility of string-manipulating programs with complex operations. Proc. ACM Program. Lang. **3**(POPL), 1–30 (2019)

18. Ciobanu, L., Diekert, V., Elder, M.: Solution sets for equations over free groups are EDT0L languages. Internat. J. Algebra Comput. **26**(05), 843–886 (2016)

19. Ciobanu, L., Elder, M.: Solutions sets to systems of equations in hyperbolic groups are EDT0L in PSPACE. arXiv preprint arXiv:1902.07349 (2019)

20. Czeizler, E.: The non-parametrizability of the word equation xyz = zvx: a short proof. Theoret. Comput. Sci. **345**(2–3), 296–303 (2005)

21. Czeizler, E., Holub, Š, Karhumäki, J., Laine, M.: Intricacies of simple word equations: an example. Int. J. Found. Comput. Sci. **18**(06), 1167–1175 (2007)

22. Dąbrowski, R., Plandowski, W.: Solving two-variable word equations. In: Díaz, J., Karhumäki, J., Lepistö, A., Sannella, D. (eds.) ICALP 2004. LNCS, vol. 3142, pp. 408–419. Springer, Heidelberg (2004). https://doi.org/10.1007/978-3-540-27836-8_36

23. Dahmani, F., Guirardel, V.: Foliations for solving equations in groups: free, virtually free, and hyperbolic groups. J. Topol. **3**(2), 343–404 (2010)

24. Day, J.D., et al.: On solving word equations using SAT. In: Filiot, E., Jungers, R., Potapov, I. (eds.) RP 2019. LNCS, vol. 11674, pp. 93–106. Springer, Cham (2019). https://doi.org/10.1007/978-3-030-30806-3_8

25. Day, J.D., Ganesh, V., Grewal, N., Manea, F.: Formal languages via theories over strings: What's decidable? Unpublished manuscript

26. Day, J.D., Ganesh, V., He, P., Manea, F., Nowotka, D.: The satisfiability of word equations: decidable and undecidable theories. In: Potapov, I., Reynier, P.-A. (eds.) RP 2018. LNCS, vol. 11123, pp. 15–29. Springer, Cham (2018). https://doi.org/10.1007/978-3-030-00250-3_2

27. Day, J.D., Kröger, A., Kulczynski, M., Manea, F., Nowotka, D., Poulsen, D.B.: BASC: benchmark analysis for string constraints. Unpublished manuscript

28. Day, J.D., Kulczynski, M., Manea, F., Nowotka, D., Poulsen, D.B.: Rule-based word equation solving. In: Proceedings of the 8th International Conference on Formal Methods in Software Engineering, pp. 87–97 (2020)

29. Day, J.D., Manea, F.: On the structure of solution-sets to regular word equations. In: Theory of Computing Systems, pp. 1–78 (2021)

30. Day, J.D., Manea, F., Nowotka, D.: The hardness of solving simple word equations. In: 42nd International Symposium on Mathematical Foundations of Computer Science (MFCS 2017). Schloss Dagstuhl-Leibniz-Zentrum fuer Informatik (2017)
31. Day, J.D., Manea, F., Nowotka, D.: Upper bounds on the length of minimal solutions to certain quadratic word equations. In: 44th International Symposium on Mathematical Foundations of Computer Science (MFCS 2019). Schloss Dagstuhl-Leibniz-Zentrum fuer Informatik (2019)
32. Diekert, V.: More than 1700 years of word equations. In: Maletti, A. (ed.) CAI 2015. LNCS, vol. 9270, pp. 22–28. Springer, Cham (2015). https://doi.org/10.1007/978-3-319-23021-4_2
33. Diekert, V., Gutierrez, C., Hagenah, C.: The existential theory of equations with rational constraints in free groups is PSPACE-complete. Inf. Comput. **202**(2), 105–140 (2005)
34. Diekert, V., Jeż, A., Plandowski, W.: Finding all solutions of equations in free groups and monoids with involution. Inf. Comput. **251**, 263–286 (2016)
35. Diekert, V., Muscholl, A.: Solvability of equations in free partially commutative groups is decidable. In: Orejas, F., Spirakis, P.G., van Leeuwen, J. (eds.) ICALP 2001. LNCS, vol. 2076, pp. 543–554. Springer, Heidelberg (2001). https://doi.org/10.1007/3-540-48224-5_45
36. Durnev, V.G.: Undecidability of the positive $\forall\exists^3$-theory of a free semigroup. Sib. Math. J. **36**(5), 917–929 (1995)
37. Freydenberger, D.D.: A logic for document spanners. Theory Comput. Syst. **63**(7), 1679–1754 (2019)
38. Freydenberger, D.D., Peterfreund, L.: The theory of concatenation over finite models. In: 48th International Colloquium on Automata, Languages, and Programming (ICALP 2021). Schloss Dagstuhl-Leibniz-Zentrum für Informatik (2021)
39. Ganesh, V., Minnes, M., Solar-Lezama, A., Rinard, M.: Word equations with length constraints: what's decidable? In: Biere, A., Nahir, A., Vos, T. (eds.) HVC 2012. LNCS, vol. 7857, pp. 209–226. Springer, Heidelberg (2013). https://doi.org/10.1007/978-3-642-39611-3_21
40. Hague, M.: Strings at MOSCA. ACM SIGLOG News **6**(4), 4–22 (2019)
41. Halfon, S., Schnoebelen, P., Zetzsche, G.: Decidability, complexity, and expressiveness of first-order logic over the subword ordering. In: 2017 32nd Annual ACM/IEEE Symposium on Logic in Computer Science (LICS), pp. 1–12. IEEE (2017)
42. Harju, T., Nowotka, D.: On the independence of equations in three variables. Theoret. Comput. Sci. **307**(1), 139–172 (2003)
43. Harju, T., Nowotka, D.: On the equation $x^k = z_1^{k_1} z_2^{k_2} \cdots z_n^{k_n}$ in a free semigroup. Theoret. Comput. Sci. **330**(1), 117–121 (2005)
44. Hmelevskii, J.I.: Equations in free semigroups, volume 107 of Am. Math. Soc. Transl. Proc. Steklov and Insti. Mat (1976)
45. Holik, L., Janku, P., Lin, A.W., Rümmer, P., Vojnar, T.: String constraints with concatenation and transducers solved efficiently. In: Proceedings of the ACM on Programming Languages, vol. 2, pp. 1–32. ACM Digital Library (2018)
46. Holub, Š, Kortelainen, J.: On systems of word equations with simple loop sets. Theoret. Comput. Sci. **380**(3), 363–372 (2007)
47. Holub, Š., Starosta, Š.: Formalization of basic combinatorics on words. In: 12th International Conference on Interactive Theorem Proving (ITP 2021). Schloss Dagstuhl-Leibniz-Zentrum für Informatik (2021)
48. Holub, Š, Žemlička, J.: Algebraic properties of word equations. J. Algebra **434**, 283–301 (2015)

49. Ilie, L., Plandowski, W.: Two-variable word equations. RAIRO-Theoret. Inform. Appl. **34**(6), 467–501 (2000)
50. Jain, S., Ong, Y.S., Stephan, F.: Regular patterns, regular languages and context-free languages. Inf. Process. Lett. **110**(24), 1114–1119 (2010)
51. Jeż, A.: One-variable word equations in linear time. Algorithmica **74**(1), 1–48 (2016)
52. Jeż, A.: Recompression: a simple and powerful technique for word equations. J. ACM (JACM) **63**(1), 1–51 (2016)
53. Jeż, A.: Word equations in non-deterministic linear space. J. Comput. Syst. Sci. **123**, 122–142 (2022)
54. Kan, S., Lin, A.W., Rümmer, P., Schrader, M.: CertiStr: a certified string solver. In: Proceedings of the 11th ACM SIGPLAN International Conference on Certified Programs and Proofs, pp. 210–224 (2022)
55. Karhumäki, J., Mignosi, F., Plandowski, W.: The expressibility of languages and relations by word equations. J. ACM (JACM) **47**(3), 483–505 (2000)
56. Karhumäki, J., Saarela, A.: On maximal chains of systems of word equations. Proc. Steklov Inst. Math. **274**(1), 116–123 (2011)
57. Kiezun, A., Ganesh, V., Artzi, S., Guo, P.J., Hooimeijer, P., Ernst, M.D.: HAMPI: a solver for word equations over strings, regular expressions, and context-free grammars. ACM Trans. Softw. Eng. Methodol. (TOSEM) **21**(4), 1–28 (2013)
58. Kulczynski, M., Manea, F., Nowotka, D., Poulsen, D.B.: ZaligVinder: a generic test framework for string solvers. J. Softw. Evol. Process, e2400 (2021)
59. Laine, M., Plandowski, W.: Word equations with one unknown. Int. J. Found. Comput. Sci. **22**(02), 345–375 (2011)
60. Lin, A.W., Barceló, P.: String solving with word equations and transducers: towards a logic for analysing mutation XSS. In: Proceedings of the 43rd Annual ACM SIGPLAN-SIGACT Symposium on Principles of Programming Languages, pp. 123–136 (2016)
61. Lothaire, M.: Algebraic Combinatorics on Words, vol. 90. Cambridge University Press, Cambridge (2002)
62. Majumdar, R., Lin, A.W.: Quadratic word equations with length constraints, counter systems, and Presburger arithmetic with divisibility. Log. Meth. Comput. Sci. **17** (2021)
63. Makanin, G.S.: Decidability of the universal and positive theories of a free group. Math. USSR-Izvestiya **25**(1), 75 (1985)
64. Makanin, G.S.: The problem of solvability of equations in a free semigroup. Matematicheskii Sbornik **145**(2), 147–236 (1977)
65. Makanin, G.S.: Equations in a free group. Math. USSR-Izvestiya **21**(3), 483 (1983)
66. Manea, F., Nowotka, D., Schmid, M.L.: On the complexity of solving restricted word equations. Int. J. Found. Comput. Sci. **29**(05), 893–909 (2018)
67. Maňuch, J.: Characterization of a word by its subwords. In: Developments in Language Theory: Foundations, Applications, and Perspectives, pp. 210–219. World Scientific (2000)
68. Mora, F., Berzish, M., Kulczynski, M., Nowotka, D., Ganesh, V.: Z3str4: a multi-armed string solver. In: Huisman, M., Păsăreanu, C., Zhan, N. (eds.) FM 2021. LNCS, vol. 13047, pp. 389–406. Springer, Cham (2021). https://doi.org/10.1007/978-3-030-90870-6_21
69. Nepeivoda, A.: Program specialization as a tool for solving word equations. In: Electronic Proceedings in Theoretical Computer Science, EPTCS, pp. 42–72 (2021)
70. Nowotka, D., Saarela, A.: One-variable word equations and three-variable constant-free word equations. Int. J. Found. Comput. Sci. **29**(05), 935–950 (2018)

71. Nowotka, D., Saarela, A.: An optimal bound on the solution sets of one-variable word equations and its consequences. SIAM J. Comput. **51**(1), 1–18 (2022)
72. Petre, E.: An elementary proof for the non-parametrizability of the equation $xyz=zvx$. In: Fiala, J., Koubek, V., Kratochvíl, J. (eds.) MFCS 2004. LNCS, vol. 3153, pp. 807–817. Springer, Heidelberg (2004). https://doi.org/10.1007/978-3-540-28629-5_63
73. Plandowski, W.: Satisfiability of word equations with constants is in NEXPTIME. In: Proceedings of the Thirty-First Annual ACM Symposium on Theory of Computing, pp. 721–725 (1999)
74. Plandowski, W.: Satisfiability of word equations with constants is in PSPACE. J. ACM (JACM) **51**(3), 483–496 (2004)
75. Plandowski, W.: An efficient algorithm for solving word equations. In: Proceedings of the Thirty-Eighth Annual ACM Symposium on Theory of Computing, pp. 467–476 (2006)
76. Plandowski, W., Rytter, W.: Application of Lempel-Ziv encodings to the solution of word equations. In: Larsen, K.G., Skyum, S., Winskel, G. (eds.) ICALP 1998. LNCS, vol. 1443, pp. 731–742. Springer, Heidelberg (1998). https://doi.org/10.1007/BFb0055097
77. Quine, W.V.: Concatenation as a basis for arithmetic. J. Symbolic Logic **11**(4), 105–114 (1946)
78. Razborov, A.A.: On systems of equations in free groups. In: Combinatorial and Geometric Group Theory, pp. 269–283 (1993)
79. Robson, J.M., Diekert, V.: On quadratic word equations. In: Meinel, C., Tison, S. (eds.) STACS 1999. LNCS, vol. 1563, pp. 217–226. Springer, Heidelberg (1999). https://doi.org/10.1007/3-540-49116-3_20
80. Saarela, A.: Word equations with kth powers of variables. J. Comb. Theory Ser. A. **165**, 15–31 (2019)
81. Saarela, A.: Hardness results for constant-free pattern languages and word equations. In: 47th International Colloquium on Automata, Languages, and Programming (ICALP 2020). Schloss Dagstuhl-Leibniz-Zentrum für Informatik (2020)
82. Schulz, K.U.: Makanin's algorithm for word equations-two improvements and a generalization. In: Schulz, K.U. (ed.) IWWERT 1990. LNCS, vol. 572, pp. 85–150. Springer, Heidelberg (1992). https://doi.org/10.1007/3-540-55124-7_4

A Survey on Delegated Computation

Giovanni Di Crescenzo[1], Matluba Khodjaeva[2(✉)], Delaram Kahrobaei[3], and Vladimir Shpilrain[4]

[1] Peraton Labs, Basking Ridge, NJ, USA
gdicrescenzo@peratonlabs.com
[2] CUNY John Jay College of Criminal Justice, New York, NY, USA
mkhodjaeva@jjay.cuny.edu
[3] CUNY Graduate Center, New York, NY, USA
dkahrobaei@gc.cuny.edu
[4] City University of New York, New York, NY, USA
vshpilrain@ccny.cuny.edu

Abstract. In the area of delegated computation, the main problem asks how a computationally weaker client device can obtain help from one or more computationally stronger server devices to perform some computation. Desirable guarantees for the client include correctness of the computation and privacy of the inputs, regardless of any server malicious behavior. In this survey, we review techniques in the area of single-server delegated computation, focusing on a representative subset of algebraic operations that are components of many cryptographic schemes: (a) ring multiplication, (b) cyclic group exponentiation, and (c) bilinear-map pairing. We describe and analyze examples of delegation protocols for these operations, discuss the state of the art, report numeric performance results, and point at several directions for future research.

1 Introduction

Modern computation paradigms (e.g., cloud/fog/edge computing, large-scale computations over big data, Internet of Things, etc.) may involve devices of potentially very different computational resources (i.e., power, sizes and memory). Methods to share computation among devices of significantly different computational power are now attracting both practitioners and researchers for their combined theoretical insights and relevance to applications.

The area of server-aided cryptography, or delegation/outsourcing of cryptographic primitives, studies some of these questions with the focus on the following main problem: how can a computationally weaker client delegate cryptographic computations to computationally superior servers? This problem has been first discussed in [1, 29, 44] and a first formal model has been produced in [35]. In the past few years, this problem is seeing an increased interest because of the prevalence of cryptography in today's practical applications, and the momentum in the shift of modern computation paradigms.

A solution to this problem is an interactive protocol between a computationally weaker client and one or more servers, where the client holding an input x

© Springer Nature Switzerland AG 2022
V. Diekert and M. Volkov (Eds.): DLT 2022, LNCS 13257, pp. 33–53, 2022.
https://doi.org/10.1007/978-3-031-05578-2_3

wants to delegate to the server(s) computation of a function F on input x, and the main desired requirements for this delegated computation of $F(x)$ are:

1. *result correctness*: if the client and server(s) are honest, the client obtains the correct output of the function evaluated on its input;
2. *input privacy*: only minimal or no information about x should be revealed to the server(s);
3. *result security*: the server(s) should not be able, except possibly with very small probability, to convince the client to accept a result different than $F(x)$; and
4. *client efficiency*: the client's runtime should be much smaller than computing $F(x)$ without delegation.

Following, for instance [55], protocols can be partitioned into (a) an offline phase, where input x is not yet known, but somewhat expensive computation can be performed by the client or a client deployer and stored on the client's device, and (b) an online phase, where we assume the client's resources are limited, and thus the client needs the server's help to compute $F(x)$. (see Fig. 1 for a pictorial description of this interaction model.)

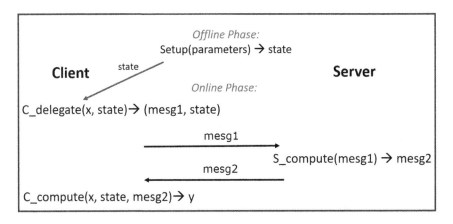

Fig. 1. Delegated computation of $y = F(x)$: interaction model

In this survey, we review techniques in the area of delegated computation to a single, possibly malicious, server, focusing on a representative subset of operations that are components of a large number of cryptographic schemes: (a) ring multiplication, (b) cyclic group exponentiation, and (c) bilinear-map pairing. We describe and analyze examples of protocols for delegated computation of these operations where the amount of work performed by the client in the online phase is less than what is needed in the best known method for non-delegated computation, as listed in Table 1. We also report numeric performance results, listed in Table 2, derived by software implementations or benchmark-based analysis, to give evidence that the achieved asymptotic improvements are also relevant for input lengths of practical interest. We conclude with some open problems on the delegated computation of these as well as other operations.

Table 1. Delegated computations and reduced client work

Delegated computation	C's dominating online computation
Ring multiplication	Reduction modulo small prime
	Multiplication modulo small prime
Group exponentiation	Small-exponent exponentiation
Bilinear-map pairing	Small-exponent exponentiation

Organization of This Survey. In Sect. 2 we review main notions and require-ments of interest in the design of delegated computation protocols. We then review the problems of delegated computation of ring multiplication in Sect. 3, group exponentiation in Sect. 4, and bilinear-map pairings in Sect. 5. In each of these sections, we describe the related algebraic setting, show an example of a delegated computation protocol, and review related literature work. Finally, we conclude in Sect. 6 by discussing directions for future research.

Table 2. Runtimes of delegated computations derived from commodity-laptop imple-mentations [19,21,25] or estimations [23] from benchmark runtimes [10], setting $\lambda = 60$, $\sigma = 2048$ for multiplication, exponentiation and product of exponentiations, $\sigma = 256$ for pairings, and $m = 10$ for product of multiplications. Parameters λ, σ, metrics t_F, t_S, t_C and input scenarios are defined in Sect. 2.

	Mod Mult $y = a * b \bmod p$		Exp $y = g^x$	Prod of Exp $y = \prod_i^m g_i^{x_i}$	Pairing $y = e(A, B)$		
	a and b pub on	a priv on b pub off	g pub off x priv on	$g_1, \ldots, g_m$ pub off $x_1, \ldots, x_m$ priv on	A pub on B pub off	A priv on B pub off	A priv on B priv off
t_F	6.64E–06	6.85E–06	4.16E–03	4.15E–02	1.15E+02	1.15E+02	1.15E+02
t_S	8.26E–06	4.82E–06	1.66E–02	9.13E–02	2.30E+02	2.30E+02	2.30E+02
t_S/t_F	1.24	0.70	3.98	2.20	2.00	2.00	2.00
t_C	1.82E–06	2.53E–06	1.82E–04	1.75E–04	6.26E+00	6.33E+01	1.15E+01
t_F/t_C	3.65	2.71	22.92	237.63	18.38	18.18	9.98

2   Model and Definitions

In this section we describe the model for delegated computation, including a discussion of the requirements of result correctness, input privacy, result security, and efficiency for delegated client-server computation protocols in the single, possibly malicious, server model.

System Scenario, Entities, and Protocol. We consider a system model with a single *client*, denoted as C, and a single *server*, denoted as S. As client's computational resources are expected to be more limited than server's, C is interested in delegating the computation of specific functions to S. We assume that the communication link between C and S is authenticated or not subject to modification attacks, and note that such attacks can be separately addressed using communication security techniques from any applied cryptography textbook (see, e.g., [45]). The interaction model admits an offline phase, where, say, precomputation can be performed and its results can be made available onto C's device. This model has been justified in several ways, all appealing to different application settings. In the presence of a trusted party, such as a deploying entity setting up C's device, the trusted party can simply perform the precomputation and store them on C's device. If no trusted party is available, in the presence of a pre-processing phase where C's device does not have significant computation constraints, C can itself perform the precomputed exponentiations and store them on its own device.

Let σ denote the computational security parameter (i.e., the length of an instance to an underlying intractable computational problem, based on the best known solution for this problem), and let λ denote the statistical security parameter (i.e., a parameter such that events with probability $2^{-\lambda}$ are extremely rare). Both parameters are expressed in unary notation (i.e., $1^\sigma, 1^\lambda$).

Let $F : Dom(F) \to CoDom(F)$ be a function, where $Dom(F)$ denotes F's domain, $CoDom(F)$ denotes F's co-domain, and $desc(F)$ denotes F's description. Assuming $desc(F)$ is known to both C and S, and input x is known only to C, we define a *client-server protocol for the delegated computation of F* in the presence of an offline phase as a 2-party, 2-phase, interactive protocol (C, S), which operates in an *offline phase* (before the function input is available) and an *online phase* (after the function input is available), and satisfies the following result correctness, result security, input privacy and efficiency requirements.

Result Correctness Requirement. Informally, the (natural) correctness requirement states that if both parties follow the protocol, then C obtains some output at the end of the protocol, and this output is, with high probability, equal to the value obtained by evaluating function F on C's input.

Result Security Requirement. Informally speaking, the most basic result security requirement would state the following: if C follows the protocol, a malicious adversary corrupting S cannot convince C to output, at the end of the protocol, some result y' different from the value $y = F(x)$, except for probability $2^{-\lambda}$, where λ is the (configurable) statistical parameter. To model an adversary's partial knowledge of the input, we define a stronger result security requirement, by allowing the adversary to even choose C's input x, before attempting to convince C of an incorrect output. We also do not restrict the adversary to run in polynomial time. The definition assumes that the communication link between C and S is not subject to modification attacks, and we note that such attacks can be independently addressed using textbook time-efficient symmetric cryptography techniques (e.g., MACs). We also augment the adversary's power so that the adversary can even take part in a number of protocol executions, where it chooses C's input before attempting to convince C to output an incorrect result.

Input Privacy Requirement. Informally speaking, the input privacy requirement guarantees the following: if C follows the protocol, a malicious adversary corrupting server S, cannot obtain any information about C's input x from a protocol execution. This is formalized by extending the indistinguishability-based approach typically used in formal definitions for encryption schemes. That is, the adversary can pick two equal-size inputs x_0 and x_1, then one of these two inputs is chosen at random and used by C in the protocol with the adversary acting as S, and then at the end of the protocol the adversary tries to guess which of the two inputs was used by C. Then the privacy requirement says that such adversary cannot correctly guess which of the two inputs was used by C better than by a random guess. We also augment the adversary's power so that the adversary can even take part in a number of protocol executions, where it chooses C's input before picking x_0 and x_1.

Efficiency Metrics and Requirements. Let (C, S) be a client-server protocol for the outsourced computation of a function F. We say that (C, S) has *efficiency parameters* $(t_F, t_P, t_C, t_S, sc, cc, mc)$ if F can be computed (without outsourcing) using $t_F(\sigma, \lambda)$ atomic operations, C can run the protocol in the offline phase using $t_P(\sigma, \lambda)$ atomic operations and in the online phase using $t_C(\sigma, \lambda)$ atomic operations, S can run the protocol using $t_S(\sigma, \lambda)$ atomic operations, C and S exchange a total of at most mc messages, of total length at most cc, C's storage complexity sc. It is obviously of interest to minimize all these protocol efficiency metrics, although the main efficiency goals in designing protocols are

1. $t_C(\sigma, \lambda) << t_F(\sigma, \lambda)$, and
2. $t_S(\sigma, \lambda)$ not significantly larger than $t_F(\sigma, \lambda)$.

Input Scenarios. The use of the result of a delegated computation within a cryptographic scheme or protocol may impose (or not) privacy and/or efficiency requirements. Accordingly, it makes sense to distinguish input scenarios when delegating computations. We say that an input x to F is

- *public online* if x is unknown in the offline phase but known by both parties in the online phase;
- *public offline* if x is known by both parties starting from the offline phase;
- *private online* if x is unknown in the offline phase but known to C in the online phase;
- *private offline* if x is known to C starting from the offline phase but unknown to S.

Generally speaking, a delegated computation protocol with a private online/offline input is also a delegated computation protocol where that input is public online/offline, but a protocol for the latter input scenario may be more efficient than a protocol for the former. Similarly, a delegated computation protocol with a public/private online input is also a delegated computation protocol where that input is public/private offline, but again a protocol for the latter input scenario may be more efficient than a protocol for the former.

3   Delegated Computation of Ring Multiplication

3.1   Algebraic Setting

Let $(R, +, \cdot)$ be a ring, where by '+', '−' and '·', we denote the addition, subtraction and multiplication operations between values in the set R. By exponentiation to an exponent t we denote the ring multiplication of an element by itself $t - 1$ times. If σ denotes a security parameter, then we let ℓ denote the length of the binary representation of elements in R and we consider the typical parameter setting of cryptographic applications, where we set $\ell = \sigma$. We say that R is *efficient* if its addition, subtraction and multiplication operation can be computed in time polynomial in ℓ.

In this section we will mainly consider the following two classes of efficient rings:

- the ring $(\mathbb{Z}, +, \cdot)$ of integers;
- the ring $(\mathbb{Z}_p, + \mod p, \cdot \mod p)$, for a prime p, where $\mathbb{Z}_p$ denotes the set of integers $\{0, \ldots, p - 1\}$.

We briefly recall the best known runtime results for relevant operations over these rings, on input of a σ_1-bit value x_1 and a σ_2-bit value x_2, where we assume, wlog, that $x_1 > x_2$:

- addition/subtraction: $T_a(\sigma_1) = O(\sigma_1)$
- modular reduction (i.e., $x_1 \mod x_2$): $T_{mr}(\sigma_1, \sigma_2) = O(\sigma_2(\sigma_1 - \sigma_2))$
- multiplication: $T_m(\sigma_1, \sigma_2)$, which is $= O(\sigma_1\sigma_2)$, using the grade school algorithm, in practical uses for elliptic-curve cryptography; or $= O(\sigma_1^{1.585})$, using Karatsuba's algorithm (see, e.g., [40,41]), in practical uses for RSA and discrete logarithm based cryptography. Algorithms with even lower asymptotic runtime expressions, include the Toom-Cook algorithm (see, e.g., Section 9.5.1 of [18]), the Schonhage-Strassen algorithm based on FFT (see, e.g., Section 9.5.6 of [18]), and Furer's algorithm [30], although according to [31], these are less recommended for applied cryptography because of practical analysis considerations. (For more details, see Section 2.2 of [31] and references therein.)

3.2   An Example Protocol

As an example of delegated computation of ring multiplication, we describe the single-server protocol $\mathcal{P}_{mul}$ from [25], for ring $(\mathbb{Z}_p, + \mod p, \cdot \mod p)$, where p is a large prime, in the input scenario 'a and b public online' (i.e., no privacy is required on values $a, b \in \mathbb{Z}_p$).

Informal Description of $\mathcal{P}_{mul}$. The protocol is based on the generalization in [25] of a classical probabilistic test for the verification of integer products, credited to Pippenger and Yao in [38]. The server computes the product w of the two inputs a, b over the ring of integers and sends to the client the quotient w_0 and remainder w_1 modulo p of this product. The client reduces a, b, w_0, w_1 modulo a small randomly chosen prime q, and verifies the correct computation of w_0, w_1 with respect to a, b modulo q. To satisfy the client efficiency and security requirements, the prime q is chosen in the offline phase so that the verification in the online phase is satisfied by an incorrect server message only with probability $\leq 2^{-\lambda}$, for a desired statistical parameter λ. To further reduce client online computation, even the value $p \mod q$ is computed in the offline phase.

Formal Description of $\mathcal{P}_{mul} = (C, S)$. By $\pi(x)$ we denote the number of prime integers less than or equal to x.

Offline Input: $1^\sigma, 1^\lambda$, prime $p \in \{0, 1\}^\sigma$

Offline phase instructions:

1. Randomly chooses a prime $q < 2^\eta$,
       where $\eta = \lceil \lambda + \log_2 \lambda + \log_2(\pi(2\sigma)) \rceil$
2. Compute $p' = p \mod q$
3. **Return:** (q, p') and store this pair on C's device

Online Input to C and S: $1^\sigma, 1^\lambda$, prime $p \in \{0, 1\}^\sigma$, $a, b \in \mathbb{Z}_p$

Online Input to C: q, p'

Online phase instructions:

1. S computes $w := a \cdot b$ (i.e., the product, over $\mathbb{Z}$, of a, b, considered as integers)
   S computes w_0, w_1 such that $w = w_0 \cdot p + w_1$ (over $\mathbb{Z}$), where $0 \leq w_1 < p$
   S sends w_0, w_1 to C
2. C computes $w'_0 := w_0 \mod q$ and $w'_1 := w_1 \mod q$
   C computes $a' := a \mod q$ and $b' := b \mod q$
   If $a' \cdot b' \neq w'_0 \cdot p' + w'_1 \mod q$ then
       C **returns:** $\perp$ and the protocol halts
   C **returns:** $y := w_1$

Protocol $\mathcal{P}_{mul}$ satisfies result correctness, result security (with probability $2^{-\lambda}$), and various efficiency properties.

The *correctness* property follows by observing that if C and S follow the protocol, then S computes w_0, w_1 as $w = a \cdot b = w_0 \cdot p + w_1$ and the equation $a \cdot b = w_0 \cdot p + w_1$ is satisfied over $\mathbb{Z}$ and is therefore satisfied also modulo the small prime q. This prevents C to return $\perp$, and allows C to return the correct output value $w_1 = w \mod p = a \cdot b \mod p$.

The *security* property follows by showing that for any malicious S, possibly deviating from the protocol instructions, the probability that S convinces C to accept a result $y \neq a \cdot b \mod p$ is $\leq 2^{-\lambda}$. The proof of this property uses the following lemma from [25], which is a 2-parameter generalization of a result underlying Pippenger's probabilistic test [38,56], on checking integer equations modulo small random primes.

Lemma 1. Let λ, σ be integers such that $\lambda \geq 28$ and $7 \leq \sigma \leq 4096$. Also, let $\pi(u)$ denote the number of prime integers $\leq u$ and let P_η be the set of prime integers $< 2^\eta$. For any integer x such that $1 \leq x \leq 2^{2\sigma}$, if $\eta = \lceil \lambda + \log_2 \lambda + \log_2(\pi(2\sigma)) \rceil$, it holds that $\mathrm{Prob}\left[q \leftarrow P_\eta : x = 0 \mod q\right] \leq 2^{-\lambda}$.

As for the *efficiency* properties, by protocol inspection we observe that

1. C's online runtime t_C is dominated by the time to compute 4 η-bit-modulus reductions of σ-bit ring values, 2 multiplications between η-bit integers, and 1 addition between η-bit values;
2. S's runtime t_S is dominated by the time to compute 1 multiplication and 1 division   mod p;
3. offline phase runtime t_P is dominated by the time to compute 1 η-bit random prime generation and 1 η-bit-modulus reduction of a σ-bit ring value;
4. communication complexity consists of $cc = 2$ σ-bit integers sent by S to C;
5. message complexity consists of a single message from S to C;
6. storage complexity on C's device consists of 2 η-bit integers.

In [25], performance results from a software implementation of $\mathcal{P}_{mul}$ on a commodity laptop were discussed. In particular, the improvement between delegated and non-delegated computation was measured as $t_F/t_C \in [5.07, 6.87]$, for values of σ, λ that are relevant to practical applications of discrete-logarithm and factoring-based cryptographic schemes (i.e., when $\sigma = 3072$ and $\lambda \in [10, 50]$).

3.3   Related Work

The ring multiplication operation is ubiquitous in cryptographic schemes.

A result in [7], building on techniques from [8], can be stated as a protocol for the delegated computation of multiplication in ring $(\mathbb{Z}_p, + \mod p, \cdot \mod p)$, in the input scenario 'a and b private online', and requiring 5 non-colluding and honest-but-curious servers. We believe this protocol can be adapted to run with 4 non-colluding and honest-but-curious servers and no offline phase, or 3 servers with an offline phase. In all these protocols the client only performs linear-time operations (e.g., additions, subtractions, random choice of a ring value).

In [25] the authors show that ring multiplication can be efficiently delegated by a client with quasilinear online runtime to a single, malicious, server, in each of these two input scenarios: (1) a and b public online; and (2) a private online and b public offline.

4   Delegated Computation of Group Exponentiation

4.1   Algebraic Setting and Preliminaries

Let $(G, \times)$ be a group, let σ be its computational security parameter (i.e., the length of an instance for the underling problem which is assumed to be computationally intractable), and let ℓ denote the length of the binary representation of elements in G. Typically, in cryptographic applications we set $\ell = \sigma$. We also assume that $(G, \times)$ has order q, for some large prime q, and is thus cyclic, and denote as g one of its generators.

By $y = g^x$ we denote the *exponentiation (in G)* of g to the x-th power; i.e., the value $y \in G$ such that $g \times \cdots \times g = y$, where the multiplication operation $\times$ is applied $x - 1$ times. Let $\mathbb{Z}_q = \{0, 1, \ldots, q - 1\}$, and let $F_{g,q} : \mathbb{Z}_q \to G$ denote the function that maps every $x \in \mathbb{Z}_q$ to the exponentiation (in G) of g to the x-th power. By $desc(F_{g,q})$ we denote a conventional description of the function $F_{g,q}$ that includes its semantic meaning as well as generator g, order q and the efficient algorithms computing multiplication and inverses in G. On input a σ_1-bit exponent x and a σ_2-bit base g, the runtime T_{exp} of exponentiation can be assumed to be $O(\sigma_1 T_m(\sigma_2))$ using the 'square-and-multiply' algorithm, where $T_m(\sigma)$ denotes the time for multiplication of two σ-bit integers. In practice, a few optimized variants of this algorithm are used (for more details, see Section 2.8 of [31] and references therein).

We say that a group is *efficient* if its description is short (i.e., has length polynomial in σ), its associated $\times$ operation and the inverse operation are efficient (i.e., they can be executed in time polynomial in σ).

We now define an *efficiently verifiable membership protocol* for G as a one-message protocol, denoted as the pair (mProve,mVerify) of algorithms, satisfying

1. *completeness*: for any $w \in G$, mVerify(w,mProve(w))=1;
2. *soundness*: for any $w \notin G$, and any mProve$'$,
        mVerify(w,mProve$'(w)$)=0; '
3. *efficient verifiability*: the runtime of mVerify is $o(T_{exp})$;
4. *efficient provability*: the runtime of mProve is $O(T_{exp})$.

While groups used in cryptography schemes are usually efficient, it is not clear whether all of them have an efficiently verifiable group membership protocol. We now show three examples of groups that are often used in cryptography and that do have efficiently verifiable membership protocols.

Example 1: $(G, \times) = (\mathbb{Z}_p^*, \cdot \mod p)$, for a large prime p. This group was one of the most recommended for early foundational cryptographic schemes like the Blum-Micali pseudo-random generator [6], etc. Note that multiplication and inverses modulo p can be computed in time polynomial in $\log p$, and an efficiently verifiable membership protocol goes as follows:

1. on input w, mProve does nothing;
2. on input w, mVerify returns 1 if $0 < w < p$ and 0 otherwise.

The completeness, soundness, efficient provability properties of this protocol are easily seen to hold. The efficient verifiability property follows by noting that mVerify runs in time linear in $\log p$, which is strictly smaller than the time for exponentiation $\mod p$ (in fact, even the time for multiplication $\mod p$).

Example 2: $(G, *) = (G_q, \cdot \mod p)$, for large primes p, q such that $p = kq + 1$, where $k \neq q$ is another prime and G_q is the q-order subgroup of $\mathbb{Z}_p^*$. This group is

one of the most recommended for cryptographic schemes like the Diffie-Hellman
key exchange protocol [26], El-Gamal encryption [28], Cramer-Shoup encryption
[17], DSA etc. It is known that by Sylow's theorem, G_q in this case is the only
subgroup of order q in the group $\mathbb{Z}_p^*$ (i.e. $g^q = 1 \mod p$ if and only if $g \in G_q$).
Also, the set of elements of G_q is precisely the set of k-th powers of elements of
$\mathbb{Z}_p^*$. Thus, an efficiently verifiable membership protocol can be built as follows:

1. on input w, mProve computes $r = w^{(q+1)/k} \mod p$ and returns r;
2. on input w, r, mVerify returns 1 if $w = r^k \mod p$ and 0 otherwise.

The completeness and soundness properties of this protocol are easily seen to
hold. The efficient provability follows by noting that mProve only performs 1
exponentiation mod p. The efficient verifiability property follows by noting
that mVerify requires one exponentiation mod p to the k-th power. We note
that mVerify is very efficient in the case when k is small (e.g., $k = 2$), which is
a typical group setting in cryptographic protocols based on discrete logarithms.

Example 3: $(G, +) = (E(\mathbb{F}_p), \text{point addition})$, for a large prime $p > 3$: an elliptic curve E over a field $\mathbb{F}_p$, is the set of pairs $(x, y) \in E(\mathbb{F}_p)$ that satisfy the
Weierstrass equation
$$y^2 = x^3 + ax + b \mod p,$$
together with the imaginary point at infinity $\mathcal{O}$, where $a, b \in \mathbb{F}_p$ and $4a^3 + 27b^2 \neq$
$0 \mod p$. The elliptic curve defined above is denoted by $E(\mathbb{F}_p)$. This group
is one of the most recommended for cryptographic schemes like Elliptic-curve
Diffie-Hellman key exchange protocol, Elliptic-curve ElGamal encryption, etc.
Moreover, many discrete logarithm based cryptographic protocols defined over
the set $\mathbb{Z}_p$ in Example 1 can be rewritten as defined over $E(\mathbb{F}_p)$. When those
protocols are rewritten using the additive operation for this group instead of
modular multiplication over $\mathbb{Z}_p$, the multiplication operation is rewritten as point
addition and the exponentiation is rewritten as scalar multiplication in the group
$E(\mathbb{F}_p)$, and the textbook "square-and-multiply" algorithm becomes a "double-
and-add" algorithm. An efficiently verifiable membership protocol for this group
simply consists of verifying the Weierstrass equation, as follows:

1. on input (x, y), mProve does nothing;
2. on input (x, y), mVerify returns 1 if $y^2 = x^3 + ax + b \mod p$ and 0 otherwise.

The completeness, soundness, efficient provability properties of this protocol are
easily seen to hold. The efficient verifiability property follows by noting that
mVerify performs only 4 multiplications mod p.

4.2   An Example Protocol

As an example of delegated computation of group exponentiation, in this section
we describe a single-server protocol $\mathcal{P}_{exp}$ for the delegated computation of exponentiation, i.e. the delegation of function $F_{g,q}(x) = g^x$, where (g, q) is public
offline and x is private online. Here, g is a generator of the q-order group G,

which is assumed to be efficient and to have an efficiently verifiable membership protocol.

Informal Description. Protocol $\mathcal{P}_{exp}$ is a slight simplification of the protocol in [19], and the main idea consists of C using a probabilistic verification equation to check S's computations, this equation being verifiable using only a much smaller number of modular multiplications than in a non-delegated computation of F. More specifically, C injects an additional random element in the inputs on which S is asked to compute the value of function F, so to satisfy the following properties: (a) if S returns correct computations of F, then C can use these random values to correctly compute y; (b) if S returns incorrect computations of F, then S either does not meet some deterministic verification equation or can only meet C's probabilistic verification equation for one possible value of the random elements; (c) C's message hides the values of the random element as well as C's input to the function. By choosing a large enough domain (i.e., $\{1, \ldots, 2^{\lambda}\}$) from which this random value is chosen, the protocol achieves a very small security probability (i.e., $2^{-\lambda}$). As this domain is much smaller than G_q, this results in a considerable efficiency gain on C's online running time.

Formal Description of $\mathcal{P}_{exp}$. Let (mProve,mVerify) denote an efficiently verifiable membership protocol for efficient prime-order group $(G, \times)$, such as any of the groups in Sect. 4.1, examples 1, 2, and 3.

Offline Input: $1^{\sigma}, 1^{\lambda}, desc(F_{g,q})$

Offline phase instructions:

1. Randomly choose $u_i \in \mathbb{Z}_q$, for $i = 0, 1$
2. Set $v_i = g^{u_i}$ and store (u_i, v_i) on C, for $i = 0, 1$

Online Input to C: $1^{\sigma}, 1^{\lambda}, desc(F_{g,q}), x \in \mathbb{Z}_q, (u_0, v_0), (u_1, v_1)$
Online Input to S: $1^{\sigma}, 1^{\lambda}, desc(F_{g,q})$

Online phase instructions:

1. C randomly chooses $c \in \{1, \ldots, 2^{\lambda}\}$
   C sets $z_0 := (x - u_0) \mod q$, $z_1 := (c \cdot x + u_1) \mod q$
   C sends z_0, z_1 to S
2. S computes $w_i := g^{z_i}$ and $\pi_0 := \text{mProve}(w_0)$
   S sends w_0, w_1, π_0 to C
3. If $\text{mVerify}(w_0, \pi_0) = 0$ then
       C **returns:** $\bot$ and the protocol halts
   C computes $y := w_0 * v_0$
   If $w_1 = y^c * v_1$ then
       C **returns:** $\bot$ and the protocol halts
   C **returns:** y

Protocol $\mathcal{P}_{exp}$ satisfies result correctness, input privacy, result security (with probability $2^{-\lambda}$), and various efficiency properties.

The *correctness* property follows by showing that if C and S follow the protocol, C always outputs $y = F_{g,q}(x) = g^x$. First, we show that the 2 tests performed by C are always passed and then the latter equality for C's output y. To show that the membership test is always passed, we note that w_i is computed by S as g^{z_i}, for $i = 0, 1$, and thus $w_0 \in G$ since g is a generator of group G. Moreover, the probabilistic test is always passed since $w_1 = g^{z_1} = g^{cx+u_1} = (g^x)^c g^{u_1} = y^c v_1$. This implies that C never returns $\bot$, and thus returns y. To see that this returned value y is the correct output, note that $y = w_0 * v_0 = g^{z_0} * g^{u_0} = g^{x-u_0} * g^{u_0} = g^x$.

The *input privacy* property follows by observing that the message (z_0, z_1) sent by C does not leak any information about x, since both z_0 and z_1 are randomly and indepedently distributed in $\mathbb{Z}_q$, as so are u_0 and u_1 chosen.

The *result security* property follows by showing that for any malicious S, possibly deviating from the protocol instructions, the probability that S convinces C to accept a result $y \neq g^x$ is $\leq 2^{-\lambda}$. The proof of this property follows as a consequence of the following facts:

1. if the pair (w_0', w_1') sent by S satisfies C's membership and probabilistic tests, then both w_0' and w_1' belong to G;
2. if $w_0', w_1' \in G$ then there exists at most one value of $c \in \{1, \ldots, 2^\lambda\}$ such that any pair $(w_0', w_1') \neq (w_0, w_1)$ sent by S satisfies C's probabilistic test;
3. the pair (z_0, z_1) sent by C does not leak any information about c.

We note that Fact 1 follows from group properties of G_q, Fact 2 follows from a discrepancy analysis on the probabilistic test (see [19] for details), and Fact 3 follows using similar arguments as in the proof of the input privacy property.

As for the *efficiency* properties, by protocol inspection we observe that

1. C's online runtime t_C is dominated by the time to compute 1 λ-bit-exponent σ-bit-base exponentiation; lower-order operations include 1 group membership verifications in G, 2 multiplications in G and 1 modular multiplication in $\mathbb{Z}_q$;
2. S's runtime t_S is dominated by the time to compute 2 σ-bit-exponent σ-bit-base exponentiations + 1 group membership proof generations in G;
3. offline phase runtime t_P is dominated by the time to compute 2 σ-bit-random-exponent σ-bit-base exponentiations in G;
4. communication complexity consists of $cc = 5$ σ-bit integers sent between S and C;
5. message complexity consists of $mc = 2$ messages between C and S;
6. storage complexity consists of $sc = 4$ σ-bit values stored on C's device.

Based on a software implementation of $\mathcal{P}_{exp}$ on a commodity laptop, the improvement between delegated and non-delegated computation was measured as $t_F/t_C = 22.92$, for values of σ, λ that are relevant to practical applications of discrete-logarithm and factoring-based cryptographic schemes (i.e., when $\sigma = 2048$, $\lambda = 60$), and using example 1 in Sect. 4.1, when $k = 2$, as a group.

4.3   Related Work

The group exponentiation operation is ubiquitous in cryptographic schemes.

A number of solutions have been proposed for delegated computation of exponentiation in some groups used in cryptography, even before the introduction of a formal privacy/security model, which were either not accompanied by proofs or broken in follow-up papers. The single-server solution from [35] assumes that the server is honest on almost all inputs. Other solutions were proposed in more recent papers, including [13–15,27,42,54], but these solutions either only consider a semi-honest server [15], or two non-colluding servers [14], or do not target input privacy [27] or do not satisfy input privacy [11,42], or do not satisfy result security [58], or only achieve constant security probability (of detecting a cheating server) [13,42,54], or require the communication of more than a constant number of group values [22].

The scheme presented in [19], allows, in the public-offline-base private-online-exponent input scenario, efficient and result-secure delegation of exponentiation in any prime-order group with an efficiently verifiable group membership protocol; including all groups in Examples 1–3 from Sect. 4.1. The scheme in [20] allows, in the private-online-base public-offline-exponent input scenario, efficient and result-secure delegation of exponentiation in the RSA group $\mathbb{Z}_n^*$, where n is the product of two safe same-length primes. The schemes in [22] achieve the same properties for exponentiation in arbitrary efficient groups, but at the expense of communicating more than a constant number of group values.

5 Delegation of Pairings

5.1 Algebraic Setting

Let $\mathbb{G}_1$, $\mathbb{G}_2$ be additive cyclic groups of order q and $\mathbb{G}_T$ be a multiplicative cyclic group of the same order q, for some large prime q. A *bilinear map* (also called *pairing* and so called from now on) is an efficiently computable map $e : \mathbb{G}_1 \times \mathbb{G}_2 \to \mathbb{G}_T$ with the following properties:

1. *Bilinearity:* for all $A \in \mathbb{G}_1$ and $B \in \mathbb{G}_2$, and for any $r, s \in \mathbb{Z}_q$, it holds that $e(rA, sB) = e(A, B)^{rs}$
2. *Non-triviality:* if U is a generator for $\mathbb{G}_1$ and V is a generator for $\mathbb{G}_2$ then $e(U, V)$ is a generator for $\mathbb{G}_T$

The last property rules out the trivial scenario where e maps all of its inputs to 1. We denote a conventional description of the bilinear map e as $desc(e)$.

The currently most practical *pairing realizations* use an ordinary elliptic curve E defined over a field $\mathbb{F}_p$, for some large prime p, as follows. Group $\mathbb{G}_1$ is the q-order additive subgroup of $E(\mathbb{F}_p)$; group $\mathbb{G}_2$ is a specific q-order additive subgroup of $E(\mathbb{F}_{p^k})$ contained in $E(\mathbb{F}_{p^k}) \setminus E(\mathbb{F}_p)$; and group $\mathbb{G}_T$ is the q-order multiplicative subgroup of $\mathbb{F}_{p^k}^*$. Here, k is the embedding degree; i.e., the smallest positive integer such that $q|(p^k - 1)$; $\mathbb{F}_{p^k}$ is the extension field of $\mathbb{F}_p$ of degree k; and $\mathbb{F}_{p^k}^*$ is the field composed of non-zero elements of $\mathbb{F}_{p^k}$. After the Weil pairing was considered in [9], more efficient constructions have been proposed as variants of the Tate pairing, including the more recent ate pairing variants (see, e.g., [3,49,53] for more details on the currently most practical pairing realizations). Even after these improvements and much work on optimization of pairing algorithms, non-delegated computation of a pairing is more expensive than exponentiation in any of $\mathbb{G}_1, \mathbb{G}_2, \mathbb{G}_T$.

Moreover, motivated by reducing the chances of low-order attacks in cryptographic protocols, in [3] the authors proposed the notion of *subgroup-secure* elliptic curves underlying a pairing, in turn extending the notion of $\mathbb{G}_T$-*strong* curves from [49]. They observe that in the currently most practical pairing realizations, $\mathbb{G}_T$ is a subgroup of a group $\mathcal{G}_T$, also contained in $\mathbb{F}_{p^k}^*$, with the following two properties: (1) testing membership in $\mathcal{G}_T$ is more efficient than testing membership in $\mathbb{G}_T$;(2) all elements of $\mathcal{G}_T$ have order $\geq q$. They also suggest to set $\mathcal{G}_T = G_{\Phi_k(p)}$, where $G_{\Phi_k(p)}$ is the cyclotomic subgroups of order $\Phi_k(p)$ in $\mathbb{F}_{p^k}^*$, and where $\Phi_k(p)$ denotes the k-th cyclotomic polynomial. Satisfaction of above property (2) is directly implied by the definitions of both $\mathbb{G}_T$-strong and subgroup-secure curves. Satisfaction of above property (1) when $\mathcal{G}_T = G_{\Phi_k(p)}$ is detailed in Section 5.2 of [3] for the curve families BN-12, BLS-12, KSS-18, and BLS-24, in turn elaborating on Section 8.2 of [49]. There, testing membership in $\mathcal{G}_T$ is shown to only require one multiplication in $\mathbb{G}_T$ and a few lower-order Frobenius-based simplifications. This is a significant improvement with respect to testing membership in $\mathbb{G}_T$, for which currently best methods require an exponentiation in $\mathbb{G}_T$ with a large or somewhat large exponent (see, e.g. [50]).

For *parameterized efficiency* evaluation of pairing delegated computation protocols, we can use the following definitions:

- $desc(e)$ denotes a conventional description of the bilinear map e;
- a_1 (resp. a_2) denotes the runtime for addition in $\mathbb{G}_1$ (resp. $\mathbb{G}_2$);
- $m_1(\ell)$ (resp. $m_2(\ell)$) denotes the runtime for scalar multiplication of a group value in $\mathbb{G}_1$ (resp. $\mathbb{G}_2$) with an ℓ-bit scalar value;
- m_T denotes the runtime for multiplication of group values in $\mathbb{G}_T$;
- $e_T(\ell)$ denotes the runtime for an exponentiation in $\mathbb{G}_T$ to an ℓ-bit exponent;
- p_T denotes the runtime for the bilinear pairing e;
- t_M denotes the runtime for testing membership of a value to $\mathcal{G}_T = G_{\Phi_k(p)}$.

We summarize some known facts about these quantities, of interest when evaluating the efficiency of protocols. First, for large enough $\ell \leq q$,

$$a_1 << m_1(\ell), \ a_2 << m_2(\ell), \ m_T(\ell) << e_T(\ell), \ e_T(\ell) < p_T.$$

Also, using a double-and-add (resp., square-and-multiply) algorithm, one can realize scalar multiplication (resp., exponentiation) in additive (resp., multiplicative) groups using, for random scalars (resp., random exponents), about 1.5ℓ additions (resp., multiplications). Membership of a value w in $\mathbb{G}_T$ can be tested using one exponentiation in $\mathbb{G}_T$ to the q-th power (i.e., checking that

$w^q = 1$). For some specific elliptic curve families, including some of the most recommended in practice (i.e., BN-12, BLS-12, KSS-18, BLS-24), membership of a value w in $\mathcal{G}_T = G_{\Phi_k(p)}$ can be tested using about 1 multiplication in $\mathbb{G}_T$ and lower-order Frobenius-based simplifications (see, e.g., [3,49]).

5.2   An Example Protocol

As an example of delegated computation of a bilinear-map pairing, in this section we describe a single-server protocol $\mathcal{P}_{pair}$ from [23] for delegated computation of a pairing e, when input A is public online and B is public offline in the efficient cyclic group G. In this protocol, the client's online complexity is dominated by a single exponentiation to a short exponent. The main idea in this protocol is that since both inputs A and B are publicly known, S can compute $w_0 = e(A, B)$ and send w_0 to C, along with some efficiently verifiable 'proof' that w_0 was correctly computed. This proof is realized by the following 3 steps: first, C sends to S a randomized version Z_1 of value A, then S computes and sends to C pairing value $w_1 = e(Z_1, B)$; and finally C verifies that $w_0 \in \mathcal{G}_T$ and uses w_1 and a pairing value computed in the offline phase in an efficient probabilistic verification for the correctness of S's message (w_0, w_1). A formal description follows.

Formal Description of $\mathcal{P}_{pair}$.

Offline Input: $desc(e), B \in \mathbb{G}_2$

Offline phase instructions:

1. Randomly choose $U_1 \in \mathbb{G}_1$
2. Set $v_1 := e(U_1, B)$
3. Store U_1, v_1 on C's device

Online Input to C: $1^\sigma, 1^\lambda, desc(e), U_1, v_1, A \in \mathbb{G}_1, B \in \mathbb{G}_2$
Online Input to S: $1^\sigma, 1^\lambda, desc(e), A \in \mathbb{G}_1, B \in \mathbb{G}_2$
Online phase instructions:

1. C randomly chooses $c \in \{1, \dots, 2^\lambda\}$
   C sets $Z_1 := c \cdot A + U_1$ and sends Z_1 to S
2. S computes $w_0 := e(A, B)$, $w_1 := e(Z_1, B)$ and sends w_0, w_1 to C
3. (Group $\mathcal{G}_T$'s Membership Test:) C checks that $w_0 \in \mathcal{G}_T$
   (Probabilistic (w_0, w_1)-Correctness Test:) C checks that $w_1 = w_0^c \cdot v_1$
   If any of these tests fails then C **returns** $\perp$ and the protocol halts
   C **returns** $y = w_0$

Protocol $\mathcal{P}_{pair}$ satisfies result correctness, result security (with probability $2^{-\lambda}$), and various efficiency properties.

The *result correctness* property follows by showing that if C and S follow the protocol, C always outputs $y = e(A, B)$. We first show that the 2 tests performed by C are always passed. The membership test is always passed since the value w_0 sent by S belongs to $\mathbb{G}_T \subseteq \mathcal{G}_T$. The probabilistic test is always passed since

$$w_1 = e(Z_1, B) = e(c \cdot A + U_1, B) = e(A, B)^c \cdot e(U_1, B) = w_0^c \cdot v_1.$$

This implies that C never returns $\perp$, and thus always returns $y = w_0 = e(A, B)$.

The *result security* property follows by showing that for any malicious S, possibly deviating from the protocol instructions, the probability that S convinces C to accept a result $y \neq e(A, B)$ is $\leq 2^{-\lambda}$. The proof of this property follows as a consequence of the following facts:

1. if the pair (w_0', w_1') sent by S satisfies C's membership and probabilistic tests, then both w_0' and w_1' belong to $\mathcal{G}_T$;
2. if $w_0', w_1' \in \mathcal{G}_T$ then there exists at most one value of $c \in \{1, \ldots, 2^\lambda\}$ such that any pair $(w_0', w_1') \neq (w_0, w_1)$ sent by S satisfies C's probabilistic test;
3. the value Z_1 sent by C does not leak any information about c.

We note that Fact 1 follows from group properties of $\mathcal{G}_T$, Fact 2 follows from a discrepancy analysis on the probabilistic test (see [23] for details) and critically uses the property that all elements of $\mathcal{G}_T$ have order $> 2^\lambda$, and Fact 3 follows by observing that Z_1 is uniformly distributed in $\mathbb{G}_1$ since so is U_1.

As for the *efficiency* properties, by protocol inspection we observe that

1. C's online runtime t_C is dominated by the time to compute 1 λ-bit-exponent exponentiation in $\mathbb{G}_T$ and 1 λ-bit-scalar multiplication in $\mathbb{G}_1$; lower-order operations include 1 multiplication in $\mathbb{G}_T$ and 1 membership verification in group $\mathcal{G}_T$;
2. S's runtime t_S is dominated by the time to compute 2 pairing computations;
3. offline phase runtime t_P is dominated by the time to compute 1 pairing computation;
4. communication complexity consists of $cc = 3$ values in $\mathbb{G}_1$ or $\mathbb{G}_T$ sent between S and C;
5. message complexity consists of $mc = 2$ messages between C and S;
6. storage complexity consists of 2 values in $\mathbb{G}_1$ or $\mathbb{G}_T$ stored on C's device.

The parameterized efficiency improvement on t_C is about $p_T/e_T(\lambda)$ over non-delegated computation t_F. The numeric efficiency improvement over non-delegated computation was estimated to range between 7.180 and 24.721 depending on the elliptic curve underlying the pairing, using benchmark results from [10].

5.3 Related Work

Pairing-based cryptography, starting with [9,37,48], has attracted much research in the past 2 decades (see, e.g., [43]).

Protocols for the delegated computation of a pairing were first studied in a work by Girault et al. [32]. However, protocols in both this work and [34] did not satisfy result security in the presence of a malicious server. Schemes with this latter property were proposed in [16,39], but these protocols turned out to require more client computation than in a non-delegated computation. Later, [12] showed that in their protocols the client's runtime is strictly lower than non-delegated computation of a pairing on the KSS-18 curve [36], but, according to [34], not on a BN curve. In [23], efficient, input-private and result-secure delegated pairing computation protocols are presented for scenarios where at least one of the two pairing inputs is known in the offline phase; there, about 1 order of magnitude improvement on client online runtime is achieved over non-delegated computation, for 4 of the currently most practical elliptic curve families. In [24], efficient, input-private and result-secure delegated pairing computation protocols are presented for scenarios where both pairing inputs are known in the online phase, again with client online runtime improving over non-delegated computation, for 4 of the currently most practical elliptic curve families.

6   Conclusions and Directions for Future Research

We reviewed techniques for a computationally weaker client to perform an efficient, private and secure delegated computation of operations often used in cryptography schemes, such as ring multiplication, group exponentiation and bilinear-map pairings, where delegation is performed to a single, possibly malicious, server. Single-server delegated computation is a relatively new area and there are several directions of research and open questions, in particular about the existence (or not) of protocols for both the computations considered in Sects. 3,4, and 5 and related computations considered in cryptography schemes. For instance, other delegated computations considered in the literature include group inverses [13,47], scalar multiplication [59], isogenies [46], as well as encryption and signature schemes (e.g., [21,35,52]). For all the discussed or mentioned computations, it is of interest to prove or disprove the existence of delegation protocols along the following directions:

1. for all input scenarios (e.g., public/private, online/offline, etc.)
2. with improved performance on efficiency metrics (i.e., client computation, communication, client storage complexity);
3. with security in the presence of malicious server(s) behavior;
4. with minimal possible number of servers;
5. with no or reduced use of client storage from the offline phase.

This survey focuses on delegated computation of specific operations often used in cryptography schemes. A survey on delegated computation of specific operations beyond cryptography (yet often using cryptographic techniques to achieve privacy and security properties) can be found in [51]. Surveys on delegated computation of arbitrary functions, with clients more computationally powerful than considered here, can be found in [2,57].

Other very related areas include the theory of program result checking [5,55], where the output of a program on a given input is being probabilistically verified by a computationally more efficient checker, and the theory of multi-prover interactive proof systems [4], where a polynomial-time verifier checks the truth of a language membership statement by interacting with multiple computationally more powerful and non-colluding provers. These, in turn, are very related to the area of interactive zero-knowledge proof systems [33], where a polynomial-time verifier checks the truth of a language membership statement by interacting with a single, computationally more powerful, prover, and the zero-knowledge property guarantees that the verifier obtains no additional information.

Acknowledgements. Part of the first author's work was supported by the Defense Advanced Research Projects Agency (DARPA), contract n. HR001120C0156. The U.S. Government is authorized to reproduce and distribute reprints for Governmental purposes notwithstanding any copyright annotation hereon. Disclaimer: The views and conclusions contained herein are those of the authors and should not be interpreted as necessarily representing the official policies or endorsements, either expressed or implied, of DARPA, or the U.S. Government.

References

1. Abadi, M., Feigenbaum, J., Kilian, J.: On hiding information from an oracle. In: J. Comput. Syst. Sci. **39**(1), 21–50 (1989)
2. Ahmad, H., et al.: Primitives towards verifiable computation: a survey. Front. Comput. Sci. **12**(3), 451–478 (2018). https://doi.org/10.1007/s11704-016-6148-4
3. Barreto, P.S.L.M., Costello, C., Misoczki, R., Naehrig, M., Pereira, G.C.C.F., Zanon, G.: Subgroup security in pairing-based cryptography. In: Lauter, K., Rodríguez-Henríquez, F. (eds.) LATINCRYPT 2015. LNCS, vol. 9230, pp. 245–265. Springer, Cham (2015). https://doi.org/10.1007/978-3-319-22174-8_14
4. Ben-Or,M., Goldwasser, S., Kilian, J., Wigderson, A.: Multi-prover interactive proofs: how to remove intractability assumptions. In: STOC, pp. 113–131 (1988)
5. Blum, M., Kannan, S.: Designing programs that check their work, In J. ACM **42**(1), 269–291 (1995). Also Proc. of ACM STOC 89
6. Blum, M., Micali, S.: How to generate cryptographically strong sequences of pseudo-random bits. SIAM J. Comput. **13**(4), 850–864 (1984)
7. Blum, M., Luby, M., Rubinfeld, R.: Program result checking against adaptive programs and in cryptographic settings. In: Distributed Computing and Cryptography, pp. 107–118 (1989)
8. Blum, M., Luby, M., Rubinfeld, R.: Self-testing/correcting with applications to numerical problems. J. Comput. Syst. Sci. **47**(3), 549–595 (1993)
9. Boneh, D., Franklin, M.: Identity-based encryption from the Weil pairing. In: Kilian, J. (ed.) CRYPTO 2001. LNCS, vol. 2139, pp. 213–229. Springer, Heidelberg (2001). https://doi.org/10.1007/3-540-44647-8_13
10. Bos, J.W., Costello, C., Naehrig, M.: Exponentiating in pairing groups. In: Lange, T., Lauter, K., LisoněK P. (eds.) SAC 2013. LNCS, vol. 8282. Springer, Cham (2013)
11. Cai, J., Ren, Y., Jiang, T.: Verifiable outsourcing computation of modular exponentiations with single server. Int. J. Netw. Secur. **19**(3), 449–457 (2017)
12. Canard, S., Devigne, J., Sanders, O.: Delegating a pairing can be both secure and efficient. In: Boureanu, I., Owesarski, P., Vaudenay, S. (eds.) ACNS 2014. LNCS, vol. 8479, pp. 549–565. Springer, Cham (2014). https://doi.org/10.1007/978-3-319-07536-5_32

13. Di Crescenzo, G., Khodjaeva, M., Kahrobaei, D., Shpilrain, V.: Efficient and secure delegation of exponentiation in general groups to a single malicious server. Math. Comput. Sci. **14**(3), 641–656 (2020). https://doi.org/10.1007/s11786-020-00462-4
14. Chen, X., Li, J., Ma, J., Tang, Q., Lou, W.: New algorithms for secure outsourcing of modular exponentiations. In: Foresti, S., Yung, M., Martinelli, F. (eds.) ESORICS 2012. LNCS, vol. 7459, pp. 541–556. Springer, Heidelberg (2012). https://doi.org/10.1007/978-3-642-33167-1_31
15. Chevalier, C., Laguillaumie, F., Vergnaud, D.: Privately outsourcing exponentiation to a single server: cryptanalysis and optimal constructions. In: Proceedings of ESORICS: 261–278, Springer, Cham (2016). https://doi.org/10.1007/978-3-030-58951-6
16. Chevallier-Mames, B., Coron, J.-S., McCullagh, N., Naccache, D., Scott, M.: Secure delegation of elliptic-curve pairing. In: Proceedings of the 9th IFIP WG 8.8/11.2 International Conference on Smart Card Research and Advanced Application, pp. 24–35 (2010). https://doi.org/10.1007/978-3-642-12510-2_3
17. Cramer, R., Shoup, V.: Design and analysis of practical public-key encryption schemes secure against adaptive chosen ciphertext attack. SIAM J. Comput. **33**(1), 167–226 (2003)
18. Crandall, R., Pomerance, C.: Prime Numbers: A Computational Perspective, 2nd edn., Springer, Cham (2005). https://doi.org/10.1007/0-387-28979-8-2005
19. Di Crescenzo, G., Khodjaeva, M., Kahrobaei, D., Shpilrain, V.: Practical and secure outsourcing of discrete log group exponentiation to a single malicious server. In Proceedings of 9th ACM CCSW, pp. 17–28 (2017)
20. Di Crescenzo, G., Khodjaeva, M., Kahrobaei, D., Shpilrain, V.: Secure delegation to a single malicious server: exponentiation in RSA-type groups. In: Proceedings of the IEEE CNS, pp. 1–9 (2019)
21. Di Crescenzo, G., Khodjaeva, M., Kahrobaei, D., Shpilrain, V.: Delegating a product of group exponentiations with application to signature schemes. In J. Math. Cryptol. **14**(1), 438–459 (2020)
22. Di Crescenzo, G., Khodjaeva, M., Kahrobaei, D., Shpilrain, V.: Efficient and secure delegation of exponentiation in general groups to a single malicious server. Math. Comput. Sci. **14**(3), 641–656 (2020). https://doi.org/10.1007/s11786-020-00462-4
23. Di Crescenzo, G., Khodjaeva, M., Kahrobaei, D., Shpilrain, V.: Secure and efficient delegation of elliptic-curve pairing. In: Proceedings of ACNS 2020, LNCS, vol. 12146. Springer, Cham (2021). https://doi.org/10.1007/978-3-030-81645-2
24. Di Crescenzo, G., Khodjaeva, M., Kahrobaei, D., Shpilrain, V.: Secure and efficient delegation of pairings with online inputs. In: Liardet, P.-Y., Mentens, N. (eds.) CARDIS 2020. LNCS, vol. 12609, pp. 84–99. Springer, Cham (2021). https://doi.org/10.1007/978-3-030-68487-7_6
25. Di Crescenzo, G., Khodjaeva, M., Shpilrain, V., Kahrobaei, D., Krishnan, R.: Single-server delegation of ring multiplications from quasilinear-time clients. In 14th International Conference on Security of Information and Networks (SIN), pp. 1–8 (2021)
26. Diffie, W., Hellman, M.E.: New directions in cryptography. IEEE Trans. Inf. Theory **22**(6), 644–654 (1976)
27. Dijk, M., Clarke, D., Gassend, B., Suh, G., Devadas, S.: Speeding up exponentiation using an untrusted computational resource. Designs Codes Cryptogr. **39**(2), 253–273 (2006)
28. El Gamal, T.: A public key cryptosystem and a signature scheme based on discrete logarithms. IEEE Trans. Inf. Theory **31**(4), 469–472 (1985)

29. Feigenbaum, J.: Encrypting problem instances: or ..., can you take advantage of someone without having to trust him? In: Proceedings of CRYPTO, pp. 477–488 (1985)
30. Fürer, M.: Faster integer multiplication. SIAM J. Comput. **39**(3), 979–1005 (2009)
31. Galbraith, S.: Mathematics of Public-Key Cryptography, Cambridge Press, Cambridge (2018)
32. Girault, M., Lefranc, D.: Server-aided verification: theory and practice. In: Roy, B. (ed.) ASIACRYPT 2005. LNCS, vol. 3788, pp. 605–623. Springer, Heidelberg (2005). https://doi.org/10.1007/11593447_33
33. Goldwasser, S., Micali, S., Rackoff, C.: The knowledge complexity of interactive proof systems. SIAM J. Comput. **18**(1), 186–208 (1989)
34. Guillevic, A., Vergnaud, D.: Algorithms for outsourcing pairing computation. In Joye M., Moradi, A. (eds.) Smart Card Research and Advanced Applications. CARDIS 2014. LNCS, vol. 8968. Springer, Berlin (2011). https://doi.org/10.1007/978-3-642-27257-8
35. Hohenberger, S., Lysyanskaya, A.: How to securely outsource cryptographic computations. In: Proceedings of the TCC, pp. 264–82 (2005)
36. Kachisa, E.J., Schaefer, E.F., Scott, M.: Constructing Brezing-Weng pairing friendly elliptic curves using elements in the cyclotomic field. In Galbraith, S.D., Paterson, K.G. (eds.) Pairing-Based Cryptography - Pairing 2008. LNCS, vol. 5209. Springer, Berlin (2008). https://doi.org/10.1007/978-3-540-85538-5
37. Joux, A.: A one round protocol for tripartite Diffie–Hellman. In: Bosma, W. (ed.) ANTS 2000. LNCS, vol. 1838, pp. 385–393. Springer, Heidelberg (2000). https://doi.org/10.1007/10722028_23
38. Kaminski, M.: A note on probabilistically verifying integer and polynomial products. J. ACM **36**(1), 142–149 (1989)
39. Kang, B.G., Lee, M.S., Park, M.S.: Efficient delegation of pairing computation. In: IACR Cryptology ePrint Archive, n. 259 (2005)
40. Karatsuba, A., Ofman, Y.: Multiplication of many-digital numbers by automatic computers. In: Proceedings of the USSR Academy of Sciences, pp. 145: 293–294. Translation in Physics-Doklady, 7, pp. 595–596 (1963)
41. Karatsuba, A.A.: The complexity of computations. In: Proceedings of the Steklov Institute of Mathematics, pp. 211: 169–183. Translation from Trudy Mat. Inst. Steklova, vol. 211, pp. 186–202 (1995)
42. Ma, X., Li, J., Zhang, F.: Outsourcing computation of modular exponentiations in cloud computing. Cluster Comput. **16**(4), 787–796 (2013). https://doi.org/10.1007/s10586-013-0252-0
43. Moody, D., Peralta, R., Perlner, R., Regenscheid, A., Roginsky, A., Chen, L.: Report on pairing-based cryptography. J. Res. Natl. Inst. Stand. Technol. **120**, 11–27 (2015)
44. Matsumoto, T., Hideki Imai, K.-K.: Speeding up secret computations with insecure auxiliary devices. In: Proceedings of CRYPTO, pp. 497–506 (1988)
45. Menezes, A., van Oorschot, P.C., Vanstone, S.A.: Handbook of Applied Cryptography. CRC Press, Location Boca Raton (1996)
46. Pedersen, R., Uzunkol, O.: Secure delegation of isogeny computations and cryptographic applications. In: Proceedings of the ACM CCSW, pp. 29–42 (2019)
47. Ping, Y., Guo, X., Wang, B., Zhou, J.: Secure outsourcing of modular inverses and scalar multiplications on elliptic curves. Int. J. Secur. Netw. **15**(2), 101–110 (2020)
48. Sakai, R., Ohgishi, K., Kasahara, M.: Cryptosystems based on pairing. In: Symposium on Cryptography and Information Security (SCIS) (2000)

49. Scott, M.: Unbalancing pairing-based key exchange protocols. In: IACR Cryptology ePrint Archive, n. 688 (2013)
50. Scott, M.: A note on group membership tests for $\mathbb{G}_1$, $\mathbb{G}_2$ and $\mathbb{G}_T$ on BLS pairing-friendly curves. In: IACR Cryptology ePrint Archive, n. 1130 (2021)
51. Shan, Z., Ren, K., Blanton, M., Wang, C.: Practical secure computation outsourcing: a survey. ACM Comput. Surv. **51**(2), 31:1–31:40 (2018)
52. Uzunkol, O., Rangasamy, J., Kuppusamy, L.: Hide the modulus: a secure non-interactive fully verifiable delegation scheme for modular exponentiations via CRT. In: ISC, pp. 250–267 (2018)
53. Vercauteren, F.: Optimal pairings. IEEE Trans. Inf. Theory **56**(1), 455–461 (2010)
54. Wang, Y., Wu, Q., Wong, D.S., Qin, B., Chow, S.S.M., Liu, Z., Tan, X.: Securely outsourcing exponentiations with single untrusted program for cloud storage. In: Kutyłowski, M., Vaidya, J. (eds.) ESORICS 2014. LNCS, vol. 8712, pp. 326–343. Springer, Cham (2014). https://doi.org/10.1007/978-3-319-11203-9_19
55. Wasserman, H., Blum, M.: Software reliability via run-time result-checking. J. ACM **44**(6), 826–849 (2019)
56. Yao, A.: A lower bound to palindrome recognition by probabilistic turing machines. In Tech. Rep. STAN-CS-77-647 (1977)
57. Yu, X., Yan, Z., Vasilakos, A.V.: A survey of verifiable computation. Mob. Netw. Appl. **22**(3), 438–453 (2017). https://doi.org/10.1007/s11036-017-0872-3
58. Zhao, L., Zhang, M., Shen, H., Zhang, Y., Shen, J.: Privacy-preserving outsourcing schemes of modular exponentiations using single untrusted cloud server. KSII Trans. Internet Inf. Syst. **11** (2) (2017)
59. Zhou, K., Ren, J.: Secure outsourcing of scalar multiplication on elliptic curves. In: ICC, pp. 1–5 (2016)

Regular Papers

Checking Regular Invariance Under Tightly-Controlled String Modifications

C. Aiswarya[1,2](✉) ⓘ, Sahil Mhaskar[1] ⓘ, and M. Praveen[1,2] ⓘ

[1] Chennai Mathematical Institute, Chennai, India
{aiswarya,sahil,praveenm}@cmi.ac.in
[2] CNRS IRL ReLaX, Chennai, India

Abstract. We introduce a model for transforming strings, that provides fine control over what modifications are allowed. The model consists of actions, each of which is enabled only when the input string conforms to a predefined template. A template can break the input up into multiple fields, and constrain the contents of each of the fields to be from pre-defined regular languages. The template can also constrain two fields to be duplicates of each other. If the input string conforms to the template, the action can be performed to modify the string. The output consists of the contents of the fields, possibly in a different order, possibly with different numbers of occurrences. Optionally, the action can also apply transductions on the contents of the fields before outputting.

For example, the sentence "`DLT will be held <cap:1>online</cap:1> if<cap:2>covid-19</cap:2> cases surge.`" conforms to the template x`<cap:`y`>`z`</cap:`y`>`w. The output of the action can be defined as $xf(z)w$, where f is defined by a transducer. If f just capitalises its input, then we can perform this action twice to get the output "`DLT will be held ONLINE if COVID-19 cases surge.`" Notice that, if we did not have the identifiers specified by y, then it will capitalise parts of the input text not intended to be capitalised.

We want to check that whenever the input comes from a given regular language, the output of any action also belongs to that language. We call this problem regular invariance checking. We show that this problem is decidable and is PSPACE-complete in general. For some restricted cases where there are no variable repetitions in the source and target templates (or patterns) and the regular language is given by a DFA, we show that this problem is co-NP-complete. We show that even in this restricted case, the problem is W[1]-hard with the length of the pattern as the parameter.

1 Introduction

We consider systems maintaining information in text format in scenarios where changes to the textual information should be tightly controlled. Often the textual information that is maintained conforms to some syntactic structure, such as that provided by context-free or regular languages. One of the desirable properties of such systems is that if the modification is applied to textual information that belongs to a specified language, the modified text should still belong to

© Springer Nature Switzerland AG 2022
V. Diekert and M. Volkov (Eds.): DLT 2022, LNCS 13257, pp. 57–68, 2022.
https://doi.org/10.1007/978-3-031-05578-2_4

that language. In other words, the specified language should be invariant under the modifications allowed by the system. A typical example of such systems is maintaining configuration information in data centers, a task for which there are dedicated softwares such as Apache ZooKeeper [1]. Resources such as GPUs can be allocated to processes running on the service and we would want to move GPUs from one process to another. Suppose we want to ensure that GPUs are not moved from high priority processes to low priority ones. Such tight control over the kind of modifications that are allowed is the main purpose of the formal model we introduce here. We study the complexity of checking whether a given regular language is invariant under the modifications allowed by a given system. Currently our model is for strings, for which the invariance-checking problem is already intractable.

One way to organise such information is to use models for business artifact systems [3,6,8–10]. These use databases to store information and have predefined actions that can modify this information. There are two issues with using database-driven systems as a model for maintaining configuration information. First, databases are an overkill for such applications. Even configuration management softwares are using more economical text files or simple data structures like trees for this purpose. Second, while these database-driven systems allow unbounded data domain, the domain is often uninterpreted and the only operation permitted on data is equality checking. Some works have considered numerical domains [2,7], however string domains have not been considered yet. Hence, transformations on strings such as the ones implemented by Mealy machines would not be supported in database-driven system models.

For modelling transformations on text files, another alternative one would think of are the traditional models of transducers. However, the main limitation of these is that they cannot restrict a transduction to be enabled at some non-regular set of strings. This is indeed needed for the configuration file example, as we may need to check that the text entered in two different fields are equal in order to enable a transformation.

We propose a formal model that is inspired by the database-driven systems and has the power of simple transducers (Mealy machines) inside. Like the database-driven systems, our model has a set of actions, that transform a text to another one. In the database-driven systems, actions are guarded by queries with free variables, which also serve as handles in the database for manipulation. In our model, actions are guarded by templates, and for an action to be enabled at a text, the text should fit the template. A template is simply a pattern over variables. An action matches the input string to a source pattern over variables, storing parts of the input string in variables. Then it performs transductions on the strings stored in the variables, and outputs the (transduced) strings in a possibly different order, according to a target pattern over variables.

We study the following problem: Given an action and a regular language, whether the regular language is invariant under the action. Invariance-check is a basic safety check. Regularity is powerful enough to express many sanity conditions and non-existence of some bugs.

We study the complexity of regular invariance-checking and show that it is PSPACE-complete. For a subclass of actions, where there are no variable repetitions in the source and target patterns and where the regular language is given by a deterministic finite-state automaton, we show it is CO-NP-complete. We further show that the problem is W[1]-hard even with the length of the pattern as the parameter. Hence, the problem is unlikely to be fixed-parameter tractable with this parameter.

Comparison with Transducers: Our model of actions is incomparable to classical models of transducers. For instance an action can match the entire input text to a single variable x and after an identity transduction, it can rewrite it as xx. After performing this action once, the text files will be of the form ww, indicating that regularity is not preserved. In another example, the input can be scanned into a pattern of the form xx, thus enabling the action only at a non-regular set of strings. Our model of actions is orthogonal to some of the expressive transducer models such as a streaming string transducers (SST) [4]. Indeed, an SST can convert a text to a concatenation of projection to odd positions followed by projections to even positions. Our actions cannot do this. On the other hand we can enable actions at some non-regular inputs, which even a two-way SST cannot do.

2   Preliminaries

The set of all finite strings or words over a finite alphabet Σ is denoted Σ^*. The empty string is denoted ϵ. For a string w, $|w|$ is its length and $w[i]$ is its ith letter.

Mealy Machines. We consider functions from words to words defined by Mealy Machines as a basic ingredient for our text-transforming actions. These functions are also called pure sequential functions. We recall the definition of Mealy Machines here, slightly simplified to our setting.

A Mealy machine defining a transduction from Σ^* to Σ^* is given by a tuple $M = (Q, q_0, \delta, \text{out})$ where Q is a finite set of states, $q_0 \in Q$ is the initial state, $\delta : Q \times \Sigma \to Q$ is the state transition function, and $\text{out} : Q \times \Sigma \to \Sigma^*$ is the output function. We naturally extend the functions δ and out to words instead of letters as follows. We let $\widehat{\delta}(q, \epsilon) = q$, and $\widehat{\delta}(q, wa) = \delta(\widehat{\delta}(q, w), a)$, where $q \in Q$, $w \in \Sigma^*$ and $a \in \Sigma$. Similarly $\widehat{\text{out}}(q, \epsilon) = \epsilon$ and $\widehat{\text{out}}(q, wa) = \widehat{\text{out}}(q, w)\text{out}(\widehat{\delta}(q, w), a)$. The function defined by M is denoted $[\![M]\!]$. $[\![M]\!] : \Sigma^* \to \Sigma^*$ is given by $[\![M]\!](w) = \widehat{\text{out}}(q_0, w)$.

3   Model

Let var be a countable set of *variables*. A *pattern* is a string in $(\text{var} \cup \Sigma)^*$. For a pattern pat, var_{pat} is the set of elements from var appearing in pat, the elements of which will be called *variables of pat*. A *valuation* is a morphism $\sigma : \text{var} \to \Sigma^*$. By abuse of notation, we extend this to a function $\sigma : (\text{var} \cup \Sigma)^* \to \Sigma^*$ by setting it to identity on Σ, and then naturally extending it to the asterate. We now define *actions* that transform an input text to an output text.

An *action* α is a 4-tuple $\alpha := (\text{srcPat}, \text{guardLang}, \text{transd}, \text{tgtPat})$, where srcPat, tgtPat are (source and target) patterns, $\text{guardLang} : \text{var}_\alpha \to \mathcal{R}eg(\Sigma)$, transd : $\text{var}_\alpha \to \{T : T \text{ is a transduction over } \Sigma\}$ are functions. Here $\text{var}_\alpha := \text{var}_{\text{srcPat}} \cup$ $\text{var}_{\text{tgtPat}}$ and $\mathcal{R}eg(\Sigma)$ is the set of regular languages over Σ.

An action $\alpha = (\text{srcPat}, \text{guardLang}, \text{transd}, \text{tgtPat})$ is *enabled* at a string w if there exists a valuation σ such that $\sigma(\text{srcPat}) = w$ and $\sigma(x) \in \text{guardLang}(x)$ for all x in $\text{var}_{\text{srcPat}}$. We call σ an *enabling* valuation of α at w. We denote by σ_α the valuation defined by $\sigma_\alpha(x) = \text{transd}(x)(\sigma(x))$ for all x in var (recall that $\text{transd}(x)$ is a transduction associated with x). The action α acts on w using α, resulting in the string $\sigma_\alpha(\text{tgtPat})$; we denote it by $w \cdot (\alpha, \sigma)$.

Example 1. Suppose configuration information about GPUs allocated to processes are maintained in a text file. One part of the file stores priorities of processes, using strings of the form "process 1 : high", "process 2 : low" etc. Another part of the file tracks GPUs allocated to processes, with strings of the form "process 1 : gpu 1, gpu 2,", "process 2 : gpu 3," and so on. We describe an action that allows to move a GPU from a low priority process to a high priority one, provided there is still at least one GPU left for the low priority process. The source and target patterns are as follows:

Source Pattern: x_1 process x_2 : low x_3 process x_4 : high x_5 process x_2 : x_6
gpu x_7 x_8 process x_4 : x_9 x_{10}.
Target Pattern: x_1 process x_2 : low x_3 process x_4 : high x_5 process x_2 : x_6
x_8 process x_4 : x_9 gpu x_7 x_{10}.

The guard languages assigned and the intended purpose of the variables are given in Table 1. This action will move the gpu x_7 from the low priority process x_2 to the high priority process x_4. If changes to the configuration file are only allowed through this action, then GPUs cannot be moved from high priority to low priority processes.

While changing strings as above, we would like to ensure that the syntactical structure of the strings is not broken. Suppose the contents of the file in the above example belong to the language $(\text{process}[0 - 9]^+ : (\text{high} + \text{low}),)^*(\text{process}[0 - 9]^+ : (\text{gpu}[0 - 9]^+,)^*)^*$. We want to verify that after applying the actions, the resulting string is still in the language. We formalise this next.

Let Σ be a finite alphabet, $L \subseteq \Sigma^*$ be a language and α be an action over Σ. We denote by $\text{post}(L, \alpha)$ the set $\{w \cdot (\alpha, \sigma) \mid w \in L, \sigma \text{ enables } \alpha \text{ at } w\}$ of results of α acting on strings in L. We study the following problem:

Invariance-checking Problem
Input: Action α, regular language L.
Question: $\text{post}(L, \alpha) \subseteq L$?

Table 1. The variables and their guard languages from Example 1.

var	guardLang	Comments
x_1	Σ^*	filler to match the prefix up to the position where changes are to be made
x_2	$[0-9]^+$	Id of the low priority process
x_3	Σ^*	filler
x_4	$[0-9]^+$	Id of the high priority process
x_5	Σ^*	filler
x_6	$(\text{gpu } [0-9]^+,)^+$	IDs of GPUs currently allocated to process x_2, not to be transferred
x_7	$[0-9]^*$	ID of the GPU currently allocated to process x_2, to be transferred to x_4
x_8	Σ^*	filler
x_9	$(\text{gpu } [0-9]^+,)^*$	IDs of GPUs currently allocated to process x_4
x_{10}	Σ^*	filler

Example 2. Consider an action α with $\mathsf{srcPat} := xy, \mathsf{tgtPat} := yx$, $\mathsf{guardLang}$ assigns Σ^* and transd assigns the identity transduction to all variables. Note that $\mathsf{post}(L, \alpha) = \{uv \mid vu \in L, u, v \in \Sigma^*\}$ is the *rotational closure* of L. A given language L is invariant under α only if L is closed under rotations.

We study the complexity of the invariance-checking problem in the rest of the paper.

4   Complexity of the Invariance-Checking Problem

Theorem 1. *The invariance-checking problem is in* PSPACE *if the regular language L is given by an NFA A, and the transduction $\mathsf{transd}(x)$ is given by a Mealy machine for each variable x.*

Proof. It is sufficient to show that checking whether $\mathsf{post}(L(A), \alpha) \not\subseteq L(A)$ is in PSPACE. Let $\alpha := (\mathsf{srcPat}, \mathsf{guardLang}, \mathsf{transd}, \mathsf{tgtPat})$ be an action. For every $x \in \mathsf{var}_\alpha$, let A_x be an NFA for $\mathsf{guardLang}(x)$. For ease of use in this proof, let P denote srcPat and P' denote tgtPat.

Let $\sigma : \mathsf{var}_\alpha \to \Sigma^*$ be a valuation and $\sigma_\alpha : \mathsf{var}_\alpha \to \Sigma^*$ be the valuation defined by $\sigma_\alpha(x) = \mathsf{transd}(x)(\sigma(x))$ for every $x \in \mathsf{var}_\alpha$. Checking if $\mathsf{post}(L(A), \alpha) \not\subseteq L(A)$ is equivalent to checking the existence of a valuation σ satisfying the following conditions.

1. $\sigma(x) \in \mathsf{guardLang}(x)$ for all $x \in \mathsf{var}$,
2. $\sigma(P) \in L(A)$ and
3. $\sigma_\alpha(P') \notin L(A)$.

Let $A = (Q, \Sigma, \Delta, Q_{\text{in}}, Q_{\text{fin}})$ be the input NFA describing the potential invariant language. Let $\mathbb{M}(A) = \{0,1\}^{Q \times Q}$ be the set of transformation matrices of A. Suppose Id denotes the identity matrix. The set $\mathbb{M}(A)$ along with matrix multiplication (over the Boolean semiring) forms a monoid, with Id as the identity element. Let $h : \Sigma^* \to \mathbb{M}(A)$ be the homomorphism given by $h(a) = \mu_a$ where $\mu_a \in \mathbb{M}(A)$ denotes the transition matrix for the letter a in A. Note that $h(w)$ is the state transformation induced by the word w—the (q, q') entry is 1 in $h(w)$ if and only if there is a path from q to q' in A on the word w. Let $e \in \{0,1\}^Q$ be the row vector whose q^{th} entry is 1 if and only if $q \in Q_{\text{in}}$ and $f^T \in \{0,1\}^Q$ be the column vector whose q^{th} entry is 1 if and only if $q \in Q_{\text{fin}}$. The string w is in $L(A)$ iff $eh(w)f = 1$.

To check whether $\text{post}(L(A), \alpha) \not\subseteq (L(A))$, in place of checking for the existence of a valuation σ as above, we can equivalently check for the existence of functions $g, g' : \Sigma \cup \text{var}_\alpha \to \mathbb{M}(A)$ satisfying the following conditions.

4. $g(a) = h(a) = g'(a)$ for all $a \in \Sigma$,
5. for all $x \in var$, there exists $w_x \in h^{-1}(g(x)) \cap \text{guardLang}(x)$ such that $g'(x) = h(\text{transd}(x)(w_x))$,
6. $eg(P)f = 1$ and
7. $eg'(P')f = 0$.

Suppose there is a valuation σ satisfying conditions 1—3. Setting $w_x = \sigma(x)$, $g(x) = h(w_x)$ and $g'(x) = h(\text{transd}(x)(w_x))$ for all $x \in var$ will satisfy conditions 4—7. Conversely, suppose there exist functions g, g' satisfying conditions 4—7. Setting $\sigma(x) = w_x$ for all $x \in var$ will satisfy conditions 1—3.

Now we give a non-deterministic PSPACE procedure for the invariant checking problem. It will guess functions g, g' and check that they satisfy conditions 4—7. Checking conditions 4, 6 and 7 can be easily done in PSPACE. We will next prove that if there exists a string w_x satisfying condition 5, there exists such a string of length at most exponential in the size of the input, so that its existence can be verified in PSPACE.

Suppose there exists a string w_x satisfying condition 5. For every i in the set $\{0, \ldots, |w_x|\}$, let e_i be a vector over $\{0,1\}$ defined as follows. The vector e_i is indexed by $Q \times Q \times Q_x \times Q_x \times T_x \times T_x \times Q \times Q$, where Q is the set of states of A, Q_x is the set of states in the NFA A_x recognising $\text{guardLang}(x)$ and T_x is the set of states in the Mealy machine M_x for $\text{transd}(x)$. Let $w[1, i]$ be the restriction of w to positions $1 \ldots i$, with $w[1, 0] := \varepsilon$, the empty word. The entry of e_i at the index $(q_s, q'_s, q_x, q'_x, t_x, t'_x, q_t, q'_t)$ is 1 iff the following four conditions are satisfied: 1) there is a path from q_s to q'_s on $w[1, i]$ in A, 2) there is a path from q_x to q'_x on $w[1, i]$ in A_x, 3) there is a path from t_x to t'_x on $w[1, i]$ in M_x and 4) there is a path from q_t to q'_t on $\text{transd}(x)(w[1, i])$ in A. Let $n = |Q|^4 |Q_x|^2 |T_x|^2$ and $N = 2^n$ be the number of distinct vectors over $\{0,1\}$ indexed by $Q \times Q \times Q_x \times Q_x \times T_x \times T_x \times Q \times Q$ that can exist. If $|w_x| > N$, then there are distinct positions i, j such that $e_i = e_j$. We can drop the portion of w_x between i and j and the resulting string will still satisfy condition 5. We can continue this until we get a string of length at most N that satisfies condition 5. Hence a non-deterministic PSPACE procedure can guess and verify the existence of a string satisfying condition 5, using space linear in $\log N = |Q|^4 |Q_x|^2 |T_x|^2$. By Savitch's theorem, there is a PSPACE procedure that does the same.

Hence, the invariance-checking problem is in NPSPACE, and again by Savitch's theorem, it is in PSPACE. $\qquad\square$

We give complexity lower bounds for the invariance-checking problem under some restrictions on the kind of patterns that are allowed and the representation used to specify the invariant language.

Definition 1. *A pattern* **pat** *over* (**var** $\cup\, \Sigma$) *is called* copyless *if every variable in* **pat** *occurs at most once. A pattern is called* copyful *if it is not copyless.*

Theorem 2. *The invariance-checking problem is* PSPACE-*hard even if* **transd** *is the constant identity transduction function,* **guardLang** *is the constant* Σ^* *function and at least one of the following is true: 1) the candidate language is given as an NFA, 2)* **srcPat** *is copyful or 3)* **tgtPat** *is copyful.*

The proof of the above theorem is split into the following three lemmas. They all give reductions from the following DFA Intersection Problem, which is known to be PSPACE-complete [14].

DFA Intersection Problem
Input: DFAs $A_1, A_2, \ldots, A_n$.
Question: $L(A_1) \cap L(A_2) \cap \cdots \cap L(A_n) \neq \emptyset$?

Lemma 1. *Let A be an* **NFA** *and α be an action with a copyless* **srcPat** *and a copyless* **tgtPat**. *Then the problem of deciding whether* **post**$(L(A), \alpha) \not\subseteq L(A)$ *is* PSPACE-*hard.*

Proof. Let $A_1, A_2, \ldots, A_n$ be a given instance of the DFA intersection problem. Let A be an NFA over $\{\#\} \cup \Sigma$ recognising the language $\{\#\} \cup \overline{L(A_1)} \cup \overline{L(A_2)} \cup \cdots \cup \overline{L(A_n)}$. Since $A_1, A_2, \ldots, A_n$ are DFAs, the NFA A can be constructed in polynomial time. For the action α, let the source pattern be $\#$, and the target pattern be x. The guard languages are given by **guardLang**$(x) := \Sigma^*$, and finally, the function **transd** assigns the identity transduction to each variable x.

For any string $w \in L(A)$, the action α is enabled at w using a valuation σ iff $\sigma(x) \in \Sigma^*$ and $w = \#$. For such a valuation σ, $w \cdot (\alpha, \sigma) = \sigma(x)$. Hence **post**$(L(A), \alpha) = \Sigma^*$. Also, $\Sigma^* \cap \{\#\} = \emptyset$ which gives us that **post**$(L(A), \alpha) \subseteq L(A)$ iff $\Sigma^* \subseteq \overline{L(A_1)} \cup \overline{L(A_2)} \cup \cdots \cup \overline{L(A_n)} \iff \bigcap_{i=1}^{n} L(A_i) = \emptyset$. This completes the reduction. $\qquad\square$

Lemma 2. *Let A be a DFA. Let α be an action with a* **copyful** **srcPat** *and a copyless* **tgtPat**. *Then the problem of deciding whether* **post**$(L(A), \alpha) \not\subseteq L(A)$ *is* PSPACE-*hard.*

Proof. Let $A_1, A_2, \ldots, A_n$ be a given instance of the DFA intersection problem. Let A be a DFA over $\{\#\} \cup \Sigma$ for the language $\Sigma^* \# L(A_1) \# L(A_2) \# \ldots \# L(A_n)$. Since A_i are all DFAs, the DFA A can be constructed in polynomial time. For the action α, let the source pattern be $y\#x\#x\#\ldots\#x$, where we have n occurrences of the variable x, and the target pattern be x. Let guardLang be the constant Σ^* function, and finally, the function transd assigns the identity transduction to each variable x.

Notice that the target pattern (and hence any result of applying α on any input string) does not contain the symbol $\#$, but every string in the invariant language contains at least one occurrence of $\#$. Hence $\mathsf{post}(L(A), \alpha) \subseteq L(A)$ iff $\mathsf{post}(L(A), \alpha) = \emptyset$ iff α is never enabled on $L(A)$ iff $\bigcap_{i=1}^{n} L(A_i) = \emptyset$. This completes the reduction. $\square$

Lemma 3. *Let A be a DFA. Let α be an action with a copyless srcPat and a copyful tgtPat. Then the problem of deciding whether $\mathsf{post}(L(A), \alpha) \not\subseteq L(A)$ is PSPACE-hard.*

Proof. Let $A_1, A_2, \ldots, A_n$ be a given instance of the DFA intersection problem. Let A be a DFA over the alphabet $\{\natural\} \cup \{\#\} \cup \Sigma$ for the language $\{\natural\} \cup \overline{(L(A_1)\#L(A_2)\#\ldots\#L(A_n))}$. For the action α, let the source pattern be $\natural$, and the target pattern be $x\#x\#\ldots\#x$. The guard languages are given by guardLang$(x) := \Sigma^*$, $\forall x$, and finally, the function transd assigns the identity transduction to each variable x.

For any string in $w \in L(A)$, the action α is enabled at w using a valuation σ iff $\sigma(x) \in \Sigma^*$ and $w = \natural$. For such a valuation σ, $w \cdot (\alpha, \sigma) = \sigma(x)$. Hence $\mathsf{post}(L(A), \alpha) := \{w\#w\#\ldots\#w \mid w \in \Sigma^*\}$. Since $\natural$ is not in $\mathsf{post}(L(A), \alpha)$, we have that $\mathsf{post}(L(A), \alpha) \not\subseteq L(A) \iff$ there exists $w_1 \in \Sigma^*$ such that $w_1\#w_1\#\ldots\#w_1 \notin L(A) \iff w_1 \in L(A_i), \forall i \leq n \iff \bigcap_{i=1}^{n} L(A_i) \neq \emptyset$. This completes the reduction. $\square$

5   Invariance-Checking : Special Case

It turns out that if the candidate language is specified using a DFA and both srcPat and tgtPat are copyless, there is a better upper bound for the invariance-checking problem.

Theorem 3. *Let A be a DFA and let $\alpha := (\mathsf{srcPat}, \mathsf{guardLang}, \mathsf{transd}, \mathsf{tgtPat})$ be an action such that srcPat, tgtPat are copyless patterns. In this case, the problem of deciding whether $\mathsf{post}(L(A), \alpha) \subseteq L(A)$ is in co-NP.*

Proof. We will show that deciding whether $\mathsf{post}(L(A), \alpha) \cap \overline{L(A)} \neq \emptyset$ is in NP, by proving a short witness property. Suppose σ is a valuation such that $\sigma(\mathsf{srcPat}) \cdot (\alpha, \sigma) \notin L(A)$. Then we will prove that there exists a valuation σ' such that $\sigma'(\mathsf{srcPat}) \cdot (\alpha, \sigma') \notin L(A)$ and $|\sigma'(x)|$ is bounded by a polynomial in the size of the input for all variables x. This suffices, since the presence of such a σ' can be guessed and verified in polynomial time.

For a valuation σ such that $\sigma(\mathsf{srcPat}) \cdot (\alpha, \sigma) \notin L(A)$ and for a variable x, suppose $\sigma(x) = w$. We will show how to drop portions of w to get a shorter string w' that still has all the following desirable properties of w. Let M_x be the Mealy machine for transd(x).

1. The run of A on $\sigma(\mathsf{srcPat})$ enters w at state p and leaves at state p'.
2. The run of A_x (an automaton for $\mathsf{guardLang}(x)$) on w starts at state q and ends at state q'.
3. The run of M_x on w starts at state r and ends at state r'.
4. The run of A on $\sigma_\alpha(\mathsf{tgtPat})$ enters $\mathsf{transd}(x)(w)$ at state s and leaves at s'.

Let the first three runs above be in states p_i, q_i, r_i respectively just before reading $w[i]$, the ith letter of w. Let the fourth run be in state s_i just before reading w_i^t, the string output by M_x at position i while reading w. For some i, j with $i \neq j$, if the tuple $\langle p_i, q_i, r_i, s_i \rangle$ is equal to the tuple $\langle p_j, q_j, r_j, s_j \rangle$, then we can drop the portion of w between i and j and the shorter string will still have all the four properties of w. We can continue this till we get a string of length at most $|A_x||A|^2|M_x|$. For every variable x, the desired valuation σ' maps x to a short string as described above, thus proving the small model property. □

The above technique fails if A is an NFA. In that case, it is not enough to ensure that fourth run in the above proof is a rejecting run—we will have to prove that *all* runs of A on $\sigma_\alpha(\mathsf{tgtPat})$ are rejecting. The above technique fails if srcPat is copyful. If a variable x occurs more than once, then the first run in the proof above will pass through w more than once and we will have to take care of all those passes. Hence, in place of checking that the four tuple $\langle p_i, q_i, r_i, s_i \rangle$ repeats, we will have to consider tuples whose dimension depends on the input and we will not get a small model property. Similarly, the above technique breaks down if tgtPat is copyful.

Next we prove a matching lower bound for this case.

Lemma 4. *Let α be an action with copyless source and target patterns. Then the problem of checking whether a given DFA describes an invariant for α is* CO-NP-*hard.*

Proof. We will give a reduction from the complement of the problem of finding a clique in a graph. Given a number k and a graph over n vertices, the problem is to check if the graph contains a k-clique as a subgraph. We know that this problem is NP-complete [13].

Suppose we are given a graph G with vertex set $V := \{v_1, v_2, \ldots, v_n\}$ and a number $k \leq n$. We will design an action α and the candidate invariant language L with the following goal: The action α is enabled on an input string from L only if it is of the form $I a_{i_1} \underbrace{v_{i_1} v_{i_1} \ldots v_{i_1}}_{k} a_{i_2} \underbrace{v_{i_2} v_{i_2} \ldots v_{i_2}}_{k} \ldots a_{i_k} \underbrace{v_{i_k} v_{i_k} \ldots v_{i_k}}_{k}$ with $1 \leq i_1 < i_2 < \ldots i_k \leq n$. That is, it guesses k vertices, repeating each of them k times in the string. This action will produce the output string

$$\theta\ a_{i_1}\ v_{i_1} v_{i_2} v_{i_3} \ldots v_{i_k}\ a_{i_2}\ v_{i_1} v_{i_2} v_{i_3} \ldots v_{i_k}\ a_{i_3}\ \ldots\ a_{i_k}\ v_{i_1} v_{i_2} \ldots v_{i_k}.$$

Note that this output string has all the k guessed vertices in every block (delimited by a_i). The output string will belong to L only if at least two of these vertices do not have an edge between them, i.e., these k vertices do not form a

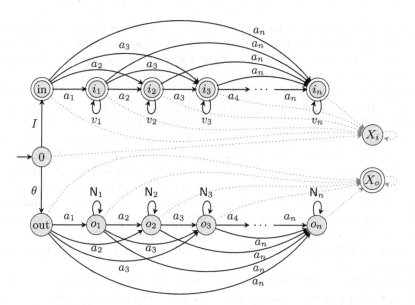

Fig. 1. The automaton A used in the polynomial reduction from the problem of finding a clique in a graph. Here N_i denotes the set of nodes at a distance at most one from v_i. That is, it includes v_i and all the neighbours of v_i. The transitions on unspecified letters go to the respective sink states, as depicted by the dotted edges.

clique. The output string will not belong to L (that is, L is not an invariant) if the k guessed vertices form a clique.

We require the source and target patterns of the action to be copyless. For this we will use k^2 variables of the form x_i^j to represent the vertices in the string. Further there are variables S_i to represent the delimiters. Let us describe the action α formally.

Let $\Sigma = V \cup \{a_1, a_2, \ldots, a_n, \theta, I\}$ and $\mathsf{var} = \{S_1, \ldots, S_k\} \cup \{x_i^j \mid 1 \le i \le k, 1 \le j \le k\}$. The action α is given by

- **Source pattern:** $I \, S_1 \, x_1^1 \, x_1^2 \ldots x_1^k \, S_2 \, x_2^1 \, x_2^2 \ldots x_2^k \, S_3 \ldots S_k \, x_k^1 \, x_k^2 \ldots x_k^k$.
- **Guard languages:** $x_i^j \mapsto \{v_1, v_2, \ldots, v_n\}, S_i \mapsto \{a_1, a_2, \ldots, a_n\}$
- **Transduction** transd maps every variable to the identity transduction.
- **Target Pattern:** $\theta \, S_1 \, x_1^1 \, x_2^1 \ldots x_k^1 \, S_2 \, x_1^2 \, x_2^2 \ldots x_k^2 \, S_3 \ldots S_k \, x_1^k \, x_2^k \ldots x_k^k$.

Now let us define the DFA for the potential invariant L. The automaton A is depicted in Fig. 1. It has a top part and a bottom part. Any string that starts with the letter I, and hence could potentially match the input pattern, runs in the top part. Any string that starts with θ, and hence all the output strings, runs in the bottom part. All the states in the top part are accepting, except for the sink state X_i. All the states in the bottom part are rejecting, except for the sink state X_o.

The top part of the DFA makes sure that in a block following a_i, it is the same vertex v_i that gets repeated. The length of each block and the number of blocks are taken care of by the source pattern. Thus the source pattern together with this automaton A ensures that the input string is of the required form. Notice that this part does not look at the edges in the graph, and in particular does not verify that these vertices form a clique.

The bottom part of the automaton reads strings that start with a θ and in particular the potential output strings of action α. This part of the automaton makes sure that in the block following a_i, all the vertices appearing in the block are neighbours of v_i or v_i itself. The set of permitted vertices is denoted N_i, the set of all vertices at a distance at most 1 from v_i. Note that if the block after a_i contains a vertex v_j which is not in N_i then it goes into the accepting sink state X_o. Thus if the guessed vertices do not form a clique, the output string will be accepted by A. On the other hand, if the guessed vertices form a clique, then in every block the vertices encountered belong to the respective N_i and the run remains in the bottom part where all states are rejecting.

Notice that the automaton A depends only on the input graph G, and not on the number k.

If the input graph G has a clique of size k, then there is an input string that matches the input pattern, and that produces the output string which is rejected by the automaton A. Thus $L(A)$ is not an invariant in this case. If the input graph does not have a clique of size k, then for every input string that enables the action α, the output string will have a block starting at some a_i containing a vertex that does not belong to the neighbour set N_i, and the run of A on the output string will end up in the accepting sink state. Thus $L(A)$ will be an invariant in this case. $\qquad\square$

The k-clique problem is unlikely to be fixed-parameter tractable with the clique-size k as the parameter, as it is W[1]-hard [11]. Thus we get:

Corollary 1. *Let α be an action with copyless source and target patterns. Then the problem of checking whether a given DFA describes an invariant for α is W[1]-hard with the length of the patterns being the parameter.*

6   Conclusions and Future Work

We have introduced a formal model for fine-controlled text modifications. We have shown that the regular invariance-checking problem is PSPACE-complete. For a restriction, the problem is shown to be CO-NP-complete.

It is interesting to see whether we can lift our results to other transducer models inside an action, instead of Mealy machines. An important next step is to see whether we can have actions that act on trees or structured text instead of simple texts. Trees are the main structures used to store configuration data by Apache ZooKeeper, one of our motivations for the work. Further it could also serve as a syntax checker for code translators etc., if we have structured text (source codes, XML). It is worth investigating whether we can lift the actions to form a sort of visibly pushdown transducer [12], and whether visibly pushdown languages [5] can be checked for invariance.

References

1. Apache ZooKeeper^{TM}. https://zookeeper.apache.org/. Accessed 21 Jan 2021
2. Abdulla, P.A., Aiswarya, C., Atig, M.F., Montali, M.: Reachability in database-driven systems with numerical attributes under Regency bounding. In: Proceedings of the 38th ACM SIGMOD-SIGACT-SIGAI Symposium on Principles of Database Systems, PODS 2019, pp. 335–352. ACM (2019)
3. Abdulla, P.A., Aiswarya, C., Atig, M.F., Montali, M., Rezine, O.: Recency-bounded verification of dynamic database-driven systems. In: Proceedings of the 35th ACM SIGMOD-SIGACT-SIGAI Symposium on Principles of Database Systems, PODS 2016, pp. 195–210. ACM (2016)
4. Alur, R.: Streaming string transducers. In: Beklemishev, L.D., de Queiroz, R. (eds.) WoLLIC 2011. LNCS (LNAI), vol. 6642, p. 1. Springer, Heidelberg (2011). https://doi.org/10.1007/978-3-642-20920-8_1
5. Alur, R., Madhusudan, P.: Visibly pushdown languages. In: Proceedings of the 36th ACM Symposium on Theory of Computing, STOC 2004, pp. 202–211. ACM (2004)
6. Hariri, B.B., Calvanese, D., De Giacomo, G., Deutsch, A., Montali, M.: Verification of relational data-centric dynamic systems with external services. In: Proceedings of the 32nd ACM SIGMOD-SIGACT-SIGAI Symposium on Principles of Database Systems, PODS 2013, pp. 163–174. ACM (2013)
7. Damaggio, E., Deutsch, A., Vianu, V.: Artifact systems with data dependencies and arithmetic. ACM Trans. Database Syst. **37**(3), 1–36 (2012)
8. Deutsch, A., Hull, R., Li, Y., Vianu, V.: Automatic verification of database-centric systems. ACM SIGLOG News **5**(2), 37–56 (2018)
9. Deutsch, A., Hull, R., Patrizi, F., Vianu, V.: Automatic verification of data-centric business processes. In: Proceedings of the 12th International Conference on Database Theory, ICDT 2009, pp. 252–267. ACM (2009)
10. Deutsch, A., Li, Y., Vianu, V.: Verification of hierarchical artifact systems. In: Proceedings of the 35th ACM SIGMOD-SIGACT-SIGAI Symposium on Principles of Database Systems, PODS 2016, pp. 179–194. ACM (2016)
11. Downey, R.G., Fellows, M.R.: Fixed-parameter tractability and completeness ii: on completeness for W[1]. Theor. Comput. Sci. **141**(1), 109–131 (1995)
12. Filiot, E., Raskin, J.-F., Reynier, P.-A., Servais, F., Talbot, J.-M.: Visibly pushdown transducers. J. Comput. Syst. Sci. **97**, 147–181 (2018)
13. Karp, R.M.: Reducibility among combinatorial problems. In: Symposium on the Complexity of Computer Computations 1972, pp. 85–103, March 1972
14. Kozen, D.: Lower bounds for natural proof systems. In: 18th Symposium on Foundations of Computer Science, SFCS 1977, pp. 254–266 (1977)

Deciding Atomicity of Subword-Closed Languages

Aistis Atminas[1([✉])] and Vadim Lozin[2]

[1] Department of Mathematical Sciences, Xi'an Jiaotong-Liverpool University,
111 Ren'ai Road, Suzhou 215123, China
`Aistis.Atminas@xjtlu.edu.cn`
[2] Mathematics Institute, University of Warwick, Coventry CV4 7AL, UK
`V.Lozin@warwick.ac.uk`

Abstract. We study languages closed under non-contiguous (scattered) subword containment order. Any subword-closed language L can be uniquely described by its anti-dictionary, i.e. the set of minimal words that do not belong to L. A language L is said to be *atomic* if it cannot be presented as the union of two subword-closed languages different from L. In this work, we provide a decision procedure which, given a language over a finite alphabet defined by its anti-dictionary, decides whether it is atomic or not.

Keywords: Subword-closed language · Joint embedding property · Decidability

1 Introduction

Throughout this paper, A is a finite alphabet and A^* is the set of all finite words over A. A word α is a *subword* of a word β if α can be obtained from β by erasing some (possibly none) letters. We say that a language L is *subword-closed* if $\beta \in L$ implies $\alpha \in L$ for every subword α of β. According to the celebrated Higman's lemma [6], the subword order is a well-quasi-order, and hence every subword-closed language L over a finite alphabet can be uniquely described by a *finite* set of minimal words not in L, called the *anti-dictionary* of L. We will denote the language defined by an anti-dictionary D by $Free(D)$ and call the words in D the *minimal forbidden words* for L.

A subword-closed language L is said to be *atomic* if L cannot be expressed as the union of two non-empty subword-closed languages different from L. It is well-known that atomicity is equivalent to the *joint embedding property* (JEP), which, in case of languages, can be defined as follows: for any two words $\alpha \in L$ and $\beta \in L$ there is a word $\gamma \in L$ containing α and β as subwords. Atomicity, or JEP, is a fundamental property, which frequently appears in the study of various combinatorial structures, for instance, growth rates of permutation classes [11] or hereditary classes of graphs, which are critical with respect to some parameters [1].

© Springer Nature Switzerland AG 2022
V. Diekert and M. Volkov (Eds.): DLT 2022, LNCS 13257, pp. 69–77, 2022.
https://doi.org/10.1007/978-3-031-05578-2_5

The main problem we study in this paper is deciding whether a subword-closed language given by its anti-dictionary is atomic or not. Decidability of atomicity, or of JEP, is a question, which was addressed in various contexts. In particular, in [3] Braunfeld has shown that this question is undecidable for hereditary classes of graphs defined by finitely many forbidden induced subgraphs. One more undecidability result appeared in [2], where Bodirsky et al. have shown that the joint embedding property is undecidable for the class of all finite models of a given universal Horn sentence. On the other hand, several positive results have been obtained by McDevitt and Ruškuc in [9], where the authors studied classes of words and permutations closed under taking consecutive subwords, also known as factors, and consecutive subpermutations. In both cases, atomicity of classes of words or permutations defined by finitely many forbidden factors or consecutive subpermutations has been shown to be decidable. We observe that every subword-closed language is also factor-closed. However, for languages defined by finitely many forbidden factors or subwords the two families are incomparable. There are languages defined by finitely many forbidden factors that are not subword-closed, and there are subword-closed languages that are not defined by *finitely many* forbidden factors. For instance, for the subword-closed language $Free(101)$ the set of minimal forbidden factors is infinite and contains all words of the form $10\ldots01$.

The main result of this paper, proved in Sect. 2, states that atomicity of subword-closed languages is decidable. We discuss possible applications of this result in Sect. 3.

2   Main Result

We start with some notational remarks. For a word $w \in A^*$, we denote by $|w|$ the number of letters in the word. Also, to simplify the notation $Free(D)$ we omit curly brackets when listing the elements of D. The main result is the following.

Theorem 1. *Let $L = Free(w_1, w_2, \ldots, w_n)$ be a language. It is algorithmically decidable whether L is atomic or not. In particular, there exists a decision procedure of complexity $O(n \times m^2)$ where $m = |w_1| + |w_2| + \ldots + |w_n|$.*

The proof of this theorem will be given by induction on $m = |w_1| + |w_2| + \ldots + |w_n|$, i.e. on the total number of letters in the forbidden words. If any of the forbidden words consists of a single letter, then we claim that we can remove this word from the anti-dictionary without changing atomicity, which is proved in the following lemma.

Lemma 1. *Let $L = Free(w_1, w_2, \ldots, w_n)$ be a language. If $|w_i| = 1$ for some i, then L is atomic if and only if $L' = Free(w_1, w_2, \ldots w_{i-1}, w_{i+1}, w_{i+2} \ldots, w_n)$ is atomic.*

Proof. Suppose w_i is the word consisting of only one letter $a \in A$. As the set of words defining the language is assumed to be minimal, we can see that letter

a does not appear in any of the words w_j with $j \neq i$. Suppose first that L is not atomic, i.e. $L = L_1 \cup L_2$ for some non-empty languages $L_1 \neq L$ and $L_2 \neq L$. Then clearly, L_1 and L_2 do not contain letter a, so they can be written as $Free(a, x_1, x_2 \dots, x_k)$ and $Free(a, y_1, y_2, \dots, y_l)$ for some words x_i and y_i not containing letter a. But then $L' = Free(x_1, x_2, \dots, x_k) \cup Free(y_1, y_2, \dots, y_l)$, and hence L' is not atomic either. On the other hand, suppose that L is atomic. Pick any two words $x', y' \in L'$. Let the words x and y be the subwords of x' and y' obtained by deleting all letters a in x and y, respectively. Then $x, y \in L$ and since L is atomic, by JEP there exists $z \in L$ such that z contains x and y. By adding $|x| + |y|$ copies of letter a between any two consecutive letters of z as well as in the prefix and suffix of z, we obtain a new word $z' \in L'$, which contains x' and y'. Hence L' is atomic as well. This finishes the proof. $\qquad \square$

Let $W = \{w_1, w_2, \dots, w_n\}$ be a set of incomparable words over A each of which has at least two letters, and let $L = Free(w_1, w_2, \dots, w_n)$ be the language defined by forbidding the words in W. For each $i \in \{1, 2, \dots, n\}$, we denote by w_{i1} the first letter of w_i and by $w'_i \in A^*$ the word obtained from w_i by removing w_{i1}, i.e. $w_i = w_{i1} w'_i$. Let

$A' = \{w_{i1} : i = 1, 2, \dots, n\}$ be the set of the first letters appearing in the words $w_1, w_2, \dots, w_n$.

We call the letters in A' the *leading letters*. Also, we will say that a word $w \in L$ is *leader-free* if it contains no leading letters, and that w is an *a-word* if $a \in A'$ is the first (when reading from left to right) leading letter in w. For each letter $a \in A'$, we denote by

$I_a = \{i \in \mathbb{N} : w_{i1} = a\}$ the set of indices of the words in W that start with letter a,

$S_a = \{w_i : i \in I_a\}$ the subset of words from W that start with letter a,

$S'_a = \{w'_i : i \in I_a\}$ the set of words obtained from the words in S_a by removing the first letter a,

$W_a = \{w'_i : i \in I_a\} \cup \{w_i : i \notin I_a\}$ the set of words obtained from the words in W by removing the first appearance of letter a from all the words that start with a,

$L_a = \{pws : p \in Free(A'), w \in \{a, \emptyset\}, s \in Free(W_a)\}$. Informally, L_a is the subword closure of the set of a-words in L. We observe that all leader-free words from L belong to L_a.

Clearly, each L_a is a subword-closed language and $L = Free(w_1, w_2, \dots, w_n) = \cup_{a \in A'} L_a$.

Lemma 2. *L is atomic only if $L = L_a$ for some $a \in A'$.*

Proof. Assume that for each $a \in A'$ the language L_a is a proper sublanguage of L. Then take the minimal set $A'' \subseteq A'$ such that $\cup_{a \in A''} L_a = L$. Such a set exists as $\cup_{a \in A'} L_a = L$ and has size $|A''| \geq 2$ as each L_a is a proper sublanguage of L. Fixing any $b \in A''$ we obtain two proper sublanguages L_b and $\cup_{a \in A'' \setminus \{b\}} L_a$ of L whose union is L. So L is not atomic. Hence, L can be atomic only if for some $a \in A'$ we have $L = L_a$. $\qquad \square$

To be able to determine whether $L_a = L$ we will determine the list of minimal forbidden subwords for the language L_a. For that purpose, let us define a simple binary relation $\circ : A \times A^* \to A^*$ as follows: for any letter $a \in A$ and any word $w \in A^*$ we define

$$a \circ w = \begin{cases} w, & \text{if } w \text{ starts with letter } a, \\ aw, & \text{otherwise.} \end{cases}$$

Given a letter $b \in A'$, we define $S_a^b = \{b \circ w_i' : i \in I_a\}$ to be the set of words obtained from the words in S_a' by adding letter b in front of all words that do not start with b.

Lemma 3. $L_a = Free(W \cup_{b \in A' \setminus \{a\}} S_a^b)$.

Proof. We denote $L' = Free(W \cup_{b \in A' \setminus \{a\}} S_a^b)$ and show first that L_a is a subset of L', i.e. we show that every word which is forbidden for L' is also forbidden for L_a. Since L_a is a subset of L, every word from W is forbidden for L_a. Now let $b \in A' \setminus \{a\}$ and assume, to the contrary, that a word $bw \in S_a^b$ belongs to L_a. Then, by definition, bw is contained in an a-word $w' \in L_a$. But then w' contains abw as a subword, which is impossible, because aw (if $w \in S_a'$) or abw (if $bw \in S_a'$) belongs to W and hence is forbidden for words in L_a. This contradiction proves that $L_a \subseteq L'$.

Conversely, consider a word $w \in L'$. Clearly, w belongs to L, since $L' \subseteq L$. If w is an a-word or leader-free, then it also belongs to L_a. Suppose w is a b-word for a letter $b \in A' \setminus \{a\}$. Then by inserting an a right before the leading b in w we obtain a word w', which still belongs to L, since otherwise a forbidden word from S_a^b can be found in w. Therefore, w' and hence w belong to L_a, proving that $L' \subseteq L_a$. □

By the lemma above, to check whether $L_a = L$ we only need to check whether each element of $\cup_{b \in A' \setminus \{a\}} S_a^b$ contains some of the words $w_1, w_2, \ldots, w_n$. If there is an element $w \in \cup_{b \in A' \setminus \{a\}} S_a^b$ which does not contain any of the words $w_1, \ldots, w_n$, then we can readily conclude that $L_a \neq L$, because in this case $w \in L$ and $w \notin L_a$. The result below describes a procedure which makes the checking efficient.

Lemma 4. *For every word $w \in S_a'$ perform the following procedure:*

1. *If the first letter of w is in $A' \setminus \{a\}$ then stop, $L_a \neq L$.*
2. *Otherwise, for every letter $b \in A' \setminus \{a\}$ do the following:*
   - *Check whether there exists a word $v \in S_b'$ contained in w. If yes, proceed to the next b, if no then stop, $L_a \neq L$.*

If the algorithm has successfully run through all the words $w \in S_a'$ and did not stop, then $L_a = L$. The algorithm has running time $O(|S_a'|nm)$ where $m = |w_1| + |w_2| + \ldots + |w_n|$.

Proof. Consider any word in $w \in S_a'$. If the first letter of w is b, for some $b \in A' \setminus \{a\}$, then $b \circ w = w$ and by definition of S_a^b it follows that $w \in S_a^b \subseteq$

$\cup_{b \in A' \backslash \{a\}} S_a^b$. As $w = w_i' \in S_a'$ is a proper subword of some word $w_i \in S_a$ and $w_1, w_2, \ldots, w_n$ are incomparable, w cannot contain any word w_j with $j \neq i$. Therefore,

$$L_a \subseteq Free(w_1, w_2, \ldots, w_{i-1}, w_i', w_{i+1}, \ldots, w_n) \neq L.$$

Next, consider the case when the first letter of w is not in $A' \backslash \{a\}$. Pick any $b \in A' \backslash \{a\}$. Then $b \circ w = bw$. Again, as $bw \in S_a^b$, $L \neq L_a$, unless bw contains some element of $\{w_1, w_2, \ldots, w_n\}$. Clearly bw cannot contain a word $w_j \in \{w_1, w_2, \ldots, w_n\} \backslash S_b$, since otherwise $w = w_i'$ contains w_j, which is a contradiction to the fact that w_i and w_j are incomparable for $i \neq j$. Therefore, $L = L_a$ only if bw contains a word $w_j \in S_b$, i.e. only if w contains a word $v = w_j' \in S_b'$. Note that this has to hold for each $b \in A' \backslash \{a\}$, since otherwise we obtain a word in S_a^b that does not contain any of $w_1, w_2, \ldots, w_n$, in which case L_a is a proper sublanguage of L.

Finally, note that if the procedure runs through all the words $w \in S_a'$ without deducing that $L \neq L_a$, then every word in S_a' starts with a letter in $A \backslash A' \cup \{a\}$, implying that for each letter $b \in A' \backslash \{a\}$, every word in the set $S_a^b = \{b \circ w : w \in S_a'\} = \{bw : w \in S_a'\}$ contains some word from the set $\{w_1, w_2, \ldots, w_n\}$. This means that none of the words in $\cup_{b \in A' \backslash \{a\}} S_a^b$ is minimal and hence

$$L_a = Free(\{w_1, w_2, \ldots, w_n\} \cup_{b \in A' \backslash \{a\}} S_a^b) = Free(\{w_1, w_2, \ldots, w_n\}) = L.$$

The main step of algorithm is checking whether a word $w \in S_a'$ contains a word from the set $\{w_1', w_2', \ldots, w_n'\}$. To check whether w contains w_i', one can go through the letters of w until the first appearance of the first letter of w_i' in w is found, then proceed to the first appearance of the second letter of w_i' in w and so on. It takes $O(|w|)$ steps to check whether w contains w_i', and it is performed for at most n different words $w_i's$. Hence for each $w \in S_a'$ it takes $O(|w|n)$ steps and hence in total it takes $O(|S_a'||w|n)$ steps. Noting that $|w| \leq m$, completes the proof of the lemma. □

By Lemma 2, L is atomic only if $L = L_a$ for some $a \in A'$. Rather than checking whether $L = L_a$ for each $a \in A'$, one can, in fact, quickly determine one specific letter $a \in A'$ for which it suffices to verify whether $L = L_a$. In the lemma below, for two vectors of integers $v = (v_1, \ldots, v_n)$ and $u = (u_1, \ldots, u_m)$ we say that v *majorizes* u if either $n \leq m$ and $v_i = u_i$ for all $i = 1, \ldots, n$ or there exists a p such that $v_p > u_p$ and $v_i = u_i$ for all $i = 1, \ldots, p-1$.

Lemma 5. *L is atomic only if $L = L_a$ for a letter $a \in A'$ which can be found using the following procedure:*

- *For each letter $b \in A'$, let $(w_{b1}, w_{b2}, \ldots, w_{bk})$ be the list of words in S_b ordered so that $|w_{b1}| \leq |w_{b2}| \leq \ldots \leq |w_{bk}|$. Define vector $v_b = (|w_{b1}|, |w_{b2}|, \ldots, |w_{bk}|)$.*
- *Find a letter b such that v_b majorizes all vectors v_c with $c \in A'$.*
- *Look at the second letter of each word in S_b, if any of these letters belong to A', say $c \in A'$, then choose $a = c$, otherwise choose $a = b$.*

Proof. Suppose that vector v_c does not majorize v_b and assume, for contradiction, $L = L_c$. We list the words of S_c as $(w_{c1}, w_{c2}, \ldots, w_{cl})$ with $|w_{c1}| \leq |w_{c2}| \leq \ldots \leq |w_{cl}|$ and the words of S_b as $(w_{b1}, w_{b2}, \ldots, w_{bk})$ with $|w_{b1}| \leq |w_{b2}| \leq \ldots \leq |w_{bk}|$. Let w'_{ci} and w'_{bi} denote the words obtained from w_{ci} and w_{bi} by removing first letters c and b, respectively.

Since $L = L_c$, by Lemma 4, we have that each word w'_{cj} for $j = 1, 2, \ldots, l$ contains a word w'_{bi} for some $i = 1, 2, \ldots, k$. Then $|w_{c1}| = |w_{b1}|$, since otherwise w'_{c1} is strictly shorter than any word in S'_b, in which case it cannot contain a word in S'_b. Let p be the largest integer such that $|w_{b1}| = |w_{b2}| = \ldots = |w_{bp}|$. Clearly, as v_c does not majorize v_b, we must also have $|w_{c1}| = |w_{c2}| = \ldots = |w_{cp}|$. For each $i \leq p$ and $j > p$, we have $|w'_{ci}| < |w'_{bj}|$. Therefore, for each $i \leq p$ the word w'_{ci} contains a word w'_{bj} with $j \leq p$, and since these words have the same length, we conclude that the set of words w'_{ci} for $i = 1, 2, \ldots p$ is just a permutation of the set of words w'_{bj} with $j = 1, 2, \ldots, p$. Now, take a word $w_{c(p+1)}$, which must exist, since v_c does not majorize v_b. If $w'_{c(p+1)}$ contains a word w'_{bj} with $j \leq p$, then $w'_{c(p+1)}$ must contain a word w'_{ch} with $h \leq p$, which is not possible, as the words in the set S_c are incomparable. This means, similarly as before, that the words in S'_c of length $|w'_{c(p+1)}|$ must form a permutation of words in S'_b of the same length. Continuing this way, we must conclude that S_c has the same number of words as S_b and $v_c = v_b$, which is a contradiction to the assumption that v_c does not majorize v_b.

Finally, consider the set $A'' = \{b \in A' : v_b \text{ majorizes all } v_c \text{ with } c \in A'\}$. Then for any $b, c \in A''$, we have $v_b = v_c$. Moreover, if for some letter $a \in A''$ we have $L = L_a$, then, by the arguments in the previous paragraph, for any letter $b \in A''$ we have $S'_b = S'_a$. Since for all letters $b \in A''$ we have the same set S'_b, the second condition of Lemma 4, is either satisfied or not, regardless of the choice of $b \in A''$. We need to check the first condition of Lemma 4 by looking at the first letter of each word in the set S'_b. If such letter c belongs to A', then the only chance for $L = L_a$ for some $a \in A''$ is when $a = c$, since otherwise the first condition of Lemma 4 is not satisfied. On the other hand, if none of the first letters of S'_b belongs to A', then the first condition of Lemma 4 is satisfied for all sets S'_b with $b \in A''$, and since all these sets are equal, we have that either $L = L_a$ holds for all $a \in A''$ or for none of them, so it is enough to pick one of them, say $a = b$ to check whether $L_a = L$ or not. This finishes the proof.   □

The final ingredient for our inductive argument is the following simple observation.

Lemma 6. L_a *is atomic if and only if* $Free(W_a)$ *is atomic.*

Proof. We recall that L_a can be presented as

$$L_a = \{pws : p \in Free(A'), w \in \{a, \emptyset\}, s \in Free(W_a)\}.$$

Suppose first that $Free(W_a)$ is atomic. Pick $x, y \in L_a$. Then $x = p_x w_x s_x$ and $y = p_y w_y s_y$ with $p_x, p_y \in Free(A')$, $w_x, w_y \in \{a, \emptyset\}$ and $s_x, s_y \in Free(W_a)$. Since $Free(W_a)$ is atomic, by JEP we have that there exists a word $s_z \in Free(W_a)$

containing s_x and s_y. Letting $p_z = p_x p_y$, we can define $z = p_z a s_z$. Clearly z contains both x and y and since $p_z \in Free(A')$, $s_z \in Free(W_a)$ we also have $z \in L_a$. So L_a satisfies JEP, and so it is atomic.

Now suppose L_a is atomic. Pick $x, y \in Free(W_a)$. Then since the words ax and ay both belong to L_a and L_a is atomic, by JEP there exists $z \in L_a$ which contains both ax and ay. Let us denote $z = pws$ with $p \in Free(A')$, $w \in \{a, \emptyset\}$ and $s \in Free(W_a)$. As ax is a subword of z, and a does not appear in p, we have that ax is a subword of ws, and since $w \in \{a, \emptyset\}$ we conclude that x is a subword of s. For the same reason, we have y is a subword of s. Since $s \in Free(W_a)$, we see that $Free(W_a)$ satisfies JEP, hence $Free(W_a)$ is atomic. Thus we conclude that L_a is atomic if and only if $Free(W_a)$ is atomic. $\qquad\square$

We are now ready to prove the main result of the paper.

Proof of Theorem 1. Let $L = Free(w_1, w_2, \ldots, w_n)$ be a given language with $w_1, w_2, \ldots, w_n \in A^*$ incomparable words. If $|w_i| = 1$ for some $i = 1, 2, \ldots, n$, then remove such a word, as by Lemma 1 this operation does not affect atomicity. So assume, without loss of generality, that $|w_i| \geq 2$ for all $i = 1, 2, \ldots, n$. Now perform the procedure of Lemma 5 to find a letter $a \in A'$ such that L is atomic only if $L = L_a$.

Then perform the procedure of Lemma 4 to check whether $L_a = L$. If not, then we know that L is not atomic. Now consider the case when $L = L_a$. In this case, by Lemma 6, L_a is atomic if and only if $Free(W_a)$ is atomic, and to determine whether $Free(W_a)$ is atomic we can proceed inductively, as the total number of letters in the set W_a is smaller than in the original set of forbidden words.

Note that the most expensive step in terms of algorithmic complexity is the application of the procedure in Lemma 4, which takes $O(|S'_a| nm)$ steps. After completing the induction step we have a set of forbidden words with $|S'_a|$ fewer letters than the original set of forbidden words. Since the removal of $|S'_a|$ letters takes $O(|S'_a| nm)$ steps to complete, to finish the procedure, i.e. to remove all m letters, we will have the computational complexity of order $O(m \times nm) = O(nm^2)$. This finishes the proof. $\qquad\square$

We finish this section with a couple of corollaries that follow from the proof of the main theorem. The first corollary gives a simple representation of all atomic subword-closed languages. Following the algorithm of the main theorem, one can efficiently move between this representation and the representation of the atomic language given by forbidden subwords.

Corollary 1. Let L be a subword-closed language over a finite alphabet A. Then L is atomic if and only if there exists a sequence of subsets $A_i \subseteq A$ for $i = 1, 2, \ldots, m + 1$ and letters $a_i \in A_i$ for $i = 1, 2, \ldots, m$, such that

$$L = \{w_1 a'_1 w_2 a'_2 \ldots w_m a'_m w_{m+1} : a'_i \in \{a_i, \emptyset\} \text{ for all } i \in \{1, 2, \ldots, m\} \text{ and}$$
$$w_i \in Free(A_i) \text{ for all } i \in \{1, 2, \ldots, m + 1\}\}.$$

The second corollary gives a simple description of all atomic languages defined by one or two forbidden subwords.

Corollary 2. Let $w, w_1, w_2 \in A^*$ be some words over a finite alphabet A with w_1 and w_2 incomparable. Then

- $Free(w)$ is atomic.
- $Free(w_1, w_2)$ is atomic if and only if $w_1 = pw's$, $w_2 = pw''s$ for some words $p, s \in A^*$ and some words $w', w'' \in A^*$ such that either $|w'| = 1$ or $|w''| = 1$.

Proof. Applying the algorithm for deciding atomicity to the language $Free(w)$ with $w = x_1 x_2 \ldots x_k$, for some $x_1, x_2, \ldots, x_k \in A$, we see that $Free(w)$ is atomic, if and only if $Free(x_2 x_3 \ldots x_k)$ is atomic, if and only if $Free(x_3 \ldots x_k)$ is atomic, ..., if and only if $Free(x_k)$ is atomic. Clearly, $Free(x_k)$ is atomic and hence $Free(w)$ is atomic. Moreover, we can represent this language as

$$\{ w_1 x_1' w_2 x_2' \ldots w_{k-1} x_{k-1}' w_k : x_i' \in \{x_i, \emptyset\} \text{ for all } i = \{1, 2, \ldots, k-1\} \text{ and}$$
$$w_i \in Free(x_i) \text{ for all } i = \{1, 2, \ldots, k\}\}.$$

Let us now write $w_1 = pw's$ and $w_2 = pw''s$, where p and s are the longest common prefix and the longest common suffix of w_1 and w_2, respectively. Note that $w' \neq \emptyset$ and $w'' \neq \emptyset$, as otherwise one of w_1 and w_2 would be a subword of the other, which is not allowed. Following the algorithm we see that $Free(w_1, w_2)$ is atomic if and only if $Free(w's, w''s)$ is atomic. Suppose that $|w''| \geq |w'|$. Let $w' = x_1 x_2 \ldots x_k$ and $w'' = y_1 y_2 \ldots y_l$ with $l \geq k$. Then, if $l > k$ the algorithm removes the letter from w'' and checks whether $y_2 y_3 \ldots y_l s$ contains $x_2 \ldots x_k s$, which happens if and only if $y_2 y_3 \ldots y_l$ contains $x_2 \ldots x_k$. If it does, then the length of $y_2 y_3 \ldots y_l$ is still bigger than of w', in which case it removes one more letter and checks whether $y_3 y_4 \ldots y_l$ contains $x_2 \ldots x_k$. The process continues until the length of the words $y_{l-k+2} y_{l-k+3} \ldots y_l$ and $x_2 \ldots x_k$ are the same, in which case to contain one another means to be equal. Now, if $k \geq 2$, this means $x_k = y_l$ and this contradicts the fact that s is the longest suffix. Thus if $k \geq 2$ the two words cannot contain each other, and we conclude that the language is not atomic. On the other hand, if $k = 1$, then clearly all containments are satisfied trivially and algorithm proceeds without stopping, thus showing that for $k = 1$ the language is atomic. This finishes the proof.   □

3   Concluding Remarks and Open Problems

In this paper we have proved that atomicity, or equivalently the joint embedding property, is algorithmically decidable for subword-closed languages. However, the question of computing a decomposition of a non-atomic language into two proper subword-closed sublanguages remains open.

The decidability procedure developed in this paper implies, in particular, that atomicity is decidable for hereditary subclasses of threshold graphs [8], since there is a bijection between threshold graphs on n vertices and binary words of length $n - 1$. Note that for general hereditary classes this question is undecidable [3].

Threshold graphs constitute a prominent example of graphs of bounded *lettericity* [10] and we conjecture that our result implies decidability of atomicity for all hereditary classes in this family.

Clique-width [4] is a notion which is more general than lettericity in the sense that bounded lettericity implies bounded clique-width but not necessarily vice versa. Graphs of bounded clique-width can be described by words (algebraic expressions) over a finite alphabet, and we believe that decidability of atomicity can be extended to graphs of bounded clique-width.

The main result of this paper also implies that atomicity is decidable for classes of linear read-once Boolean functions closed under renaming variables and erasing variables from linear read-once expressions defining the functions, because, similarly to threshold graphs, linear read-once Boolean functions can be uniquely (up to renaming variables) described by binary words. Linear read-once functions appeared in the literature under various other names such as nested canalyzing functions, unate cascade functions [7], 1-decision lists [5], and we conjecture that decidability of atomicity can be extended to classes of d-decision lists for any fixed d. To support this conjecture, we observe that the main result of this paper is valid for subword-closed languages over *infinite* alphabets, provided that the set of minimal forbidden words is finite.

References

1. Alecu, B., Lozin, V., de Werra, D.: The micro-world of Cographs, Discrete App. Math., 312, 3–14 (2022). https://doi.org/10.1016/j.dam.2021.11.004
2. Bodirsky, M., Rydval, J., Schrottenloher, A.: Universal Horn sentences and the joint embedding property. arXiv:2104.11123v3
3. Braunfeld, s.: The undecidability of joint embedding and joint homomorphism for hereditary graph classes. Discrete Math. Theor. Comput. Sci. **21**(2), 17 p (2019), Paper No. 9
4. Courcelle, B., Engelfriet, J., Rozenberg, G.: Handle-rewriting hypergraph grammars. J. Comput. Sys. Sci. **46**(2), 218–270 (1993)
5. Eiter, T., Ibaraki, T., Makino, K.: Decision lists and related Boolean functions. Theor. Comput. Sci. **270**(1), 493–524 (2002)
6. Higman, G.: Ordering by divisibility in abstract algebras. Proc. Lond. Math. Soc. **2**(3), 326–336 (1952)
7. Jarrah, A.S., Raposa, B., Laubenbacher, R.: Nested canalyzing, unate cascade, and polynomial functions. Phys. D Nonlinear Phenom. **233**(2), 167–174 (2007)
8. Mahadev, N.V.R., Peled, U.N.: Threshold graphs and related topics. Ann. Discrete Math. **56**. North-Holland Publishing Co., Amsterdam, 1995. xiv+543 pp
9. McDevitt, M., Ruškuc, N.: Atomicity and well quasi-order for consecutive orderings on words and permutations. SIAM J. Discrete Math. **35** (1), 495–520 (2021)
10. Petkovšek, M.: Letter graphs and well-quasi-order by induced subgraphs. Discrete Math. **244**, 375–388 (2002)
11. Vatter, V.: Growth rates of permutation classes: from countable to uncountable. Proc. Lond. Math. Soc. **119**(3), 960–997 (2019)

Prefix Palindromic Length
of the Sierpinski Word

Dora Bulgakova[1], Anna Frid[1(✉)], and Jérémy Scanvic[2]

[1] Aix Marseille Univ, CNRS, Centrale Marseille, I2M, Marseille, France
anna.frid@univ-amu.fr

[2] Unité de Mathématiques Pures et Appliquées (UMR 5669), École normale
supérieure de Lyon/CNRS/Inria, Lyon, France

Abstract. The prefix palindromic length $p_{\mathbf{u}}(n)$ of an infinite word $\mathbf{u}$ is
the minimal number of concatenated palindromes needed to express the
prefix of length n of $\mathbf{u}$. This function is surprisingly difficult to study;
in particular, the conjecture that $p_{\mathbf{u}}(n)$ can be bounded only if $\mathbf{u}$ is
ultimately periodic is open since 2013. A more recent conjecture concerns
the prefix palindromic length of the period doubling word: it seems that
it is not 2-regular, and if it is true, this would give a rare example of a
non-regular function of a 2-automatic word.

For some other k-automatic words, however, the prefix palindromic
length is known to be k-regular. Here we add to the list of those words
the Sierpinski word $\mathbf{s}$ and give a complete description of $p_{\mathbf{s}}(n)$.

1 Introduction

A palindrome is a word which does not change when read from left to right
and from right to left, like *rotator* or *abbaaaabba*. In this paper, we continue
to study decompositions of words over a finite alphabet to a minimal number
of palindromes: for example, for the word $w = ab%abbaabbbbaaa$, this number
is equal to 4, since we can factorize this word as $(aba)(bbaabb)(b)(aaa)$ or as
$(aba)(bb)(aabbbaa)(a)$, but cannot manage with less than four palindromes. So,
we can write that the *palindromic length* of w, denoted as $\mathrm{PL}(w)$, is equal to 4.

In 2013, Puzynina, Zamboni and the second author [7] conjectured that if the
palindromic length of factors of an infinite word $\mathbf{u}$ is bounded, then the word $\mathbf{u}$ is
ultimately periodic. This conjecture remains open despite a partial solution in the
initial paper [7] and later particular results [2,5,10]. Saarela [11] proved that the
conjecture is equivalent to the same statement about prefixes, not all factors, of $\mathbf{u}$.
His result makes reasonable to consider the *prefix palindromic length* $p_{\mathbf{u}}(n)$, which
is also denoted as $\mathrm{PPL}_{\mathbf{u}}(n)$ in previous papers. This function of an infinite word $\mathbf{u}$
and of $n \geq 0$, equal to the palindromic length of the prefix of length n of $\mathbf{u}$ is thus
conjectured to be unbounded for every word which is not ultimately periodic.

A natural exercise on every new function of an infinite word is to compute or
to estimate it for classical examples like the Thue-Morse word and the Fibonacci
word. The first of these problems appears to be not too complicated: the prefix
palindromic length of the Thue-Morse word, which is 2-automatic, appears to be

© Springer Nature Switzerland AG 2022
V. Diekert and M. Volkov (Eds.): DLT 2022, LNCS 13257, pp. 78–89, 2022.
https://doi.org/10.1007/978-3-031-05578-2_6

2-regular, and its first differences are described as a fixed point of a 4-uniform morphism [4]. At the same time, the question on the Fibonacci word has not been solved, moreover, it seems that its prefix palindromic length is even not Fibonacci-regular [6].

Since these first exercises, a progress has been made in computing the prefix palindromic length of some more known words, including the Rudin-Shapiro word, the paperfolding word [6] and the Zimin word [9]. Moreover, it has been proved that for every k-automatic word containing a finite number of distinct palindromes, the prefix palindromic length is k-regular [6]. But the most intriguing are the results of computational experiments suggesting that for example, for the period-doubling word, which is the fixed point of the morphism $a \to ab, b \to aa$, the prefix palindromic length is *not* a 2-regular sequence [6]. At the moment, this is the second challenging conjecture on the prefix palindromic length, since normally, all reasonable functions of k-automatic words are k-regular.

Unable to solve any of the big conjectures, we continue collecting examples when the prefix palindromic length is predictably regular. Here we prove it for the Sierpinski word, the 3-automatic fixed point of the morphism $\varphi : a \to aba, b \to bbb$. The fact that its prefix palindromic length is unbounded was proved already in the initial paper [7]. The first morphic description of that function was conjectured in the Master thesis of Enzo Laborde [8], but here we find a simpler one, which yet requires several pages of proofs.

A possible continuation of this research is to find a larger class of k-automatic words with k-regular prefix palindromic length. It could help to extract properties of automatic words which prevent the function to be regular.

The proofs omitted in this submission can be found in our arXiv preprint https://arxiv.org/abs/2201.09556. The result can also be generalized to all morphisms of the form $a \to ab^{n-2}a, b \to b^n$ for $n \geq 3$.

2   Definitions, Notation, Known Results

From now on, $\mathbf{s} = s[1]s[2] \cdots s[n] \cdots$ denotes the Sierpinski word, or the Cantor word

$$ababbbabababbbbbbbbbababbbabab^{27} \cdots ,$$

defined as the fixed point starting with a of the morphism

$$\varphi : \begin{cases} a \to aba, \\ b \to bbb. \end{cases}$$

Here $s[i] \in \{a, b\}$ for all $i \geq 1$. Clearly, for every k, the Sierpinski word starts with the palindrome $\varphi^k(a) = \varphi^{k-1}(a)b^{3^{k-1}}\varphi^{k-1}(a)$. A factor $s[i]s[i+1] \cdots s[j]$ can also be denoted as $\mathbf{s}[i..j]$.

In what follows, $\mathrm{PL}(u)$ denotes the palindromic length of a finite word u, that is, the minimal number of palindromes such that u is their concatenation. The prefix palindromic length of $\mathbf{s}$ is denoted by $p_{\mathbf{s}}(n)$ or $p(n)$ for short: $p(n) = \mathrm{PL}(s[1..n])$.

One of important general results on palindromic length is the following inequality, which we refer below as Saarela's inequality [11, Lemma 6]: for all words u, v we have

$$|\mathrm{PL}(u) - \mathrm{PL}(v)| \leq \mathrm{PL}(uv).$$

This result is especially useful when one of words u, v or uv is a palindrome and thus its palindromic length is equal to 1. If u is a prefix of a given infinite word $\mathbf{u}$ of length n, and v is its next letter, it also immediately implies that

$$|p_{\mathbf{u}}(n+1) - p_{\mathbf{u}}(n)| \leq 1,$$

meaning that the first differences of the prefix palindromic length of a word can be equal only to -1, 0, or 1.

As the name suggests, an infinite word $\mathbf{u}$ is called k-*automatic* if there exists a deterministic finite automaton $\mathbf{A}$ such that every symbol $u[n]$ of $\mathbf{u}$ can be obtained as the output of $\mathbf{A}$ with the base-k representation of n as the input [1]. We will also need and use an equivalent definition of the same notion: a word $\mathbf{u}$ is k-automatic if and only if there exists a k-uniform morphism $\varphi : \Sigma^* \to \Sigma^*$ and a 1-uniform morphism (or *coding*) $c : \Sigma^* \to \Delta^*$ such that $\mathbf{u} = c(\varphi^\infty(a))$ for a symbol $a \in \Sigma$. So, for example, the Sierpinski word is 3-automatic since its morphism φ is 3-uniform, and the coding c can be chosen as the trivial one, sending a to a and b to b.

A generalization of the notion of a k-automatic word to sequences on $\mathbb{Z}$ is the notion of k-regular sequence: formally speaking, a $\mathbb{Z}$-valued sequence is k-*regular* if the $\mathbb{Z}$-module generated by its k-kernel is finitely generated. Discussions and equivalent definitions of k-regular sequences can be found in Chapter 16 of Allouche and Shallit's monograph [1]; what we really need in this paper is the following lemma proven in [6] for the case when p is the prefix palindromic length of an infinite word but true for every sequence with bounded first differences due to exactly the same arguments.

Lemma 2.1. *A $\mathbb{Z}$-valued sequence $r(n)$ with bounded first differences $d_r(n) = r(n+1) - r(n)$ is k-regular if and only if the sequence d_r is k-automatic.*

Since the main object we study in this paper is a 3-automatic word, we need some addition notation concerning ternary representations.

Let $X \subset \{\mathbf{0}, \mathbf{1}, \mathbf{2}\}^*$ be the language of ternary expansions of non-negative integers without leading zeros. The fact that $x \in X$ is the ternary expansion of n will be denoted as $[x]_3 = n$ and $(n)_3 = x$. By a convention, we put $(0)_3 = \varepsilon$, so, the ternary representation of 0 is the empty string. It means that every non-empty representation starts with $\mathbf{1}$ or $\mathbf{2}$, so, $X = \{\varepsilon\} \cup \{\mathbf{1}, \mathbf{2}\}\{\mathbf{0}, \mathbf{1}, \mathbf{2}\}^*$. Note that we write symbols of ternary strings in boldface to distinguish concatenated strings from multiplied numbers.

When we consider ternary expansions with leading zeros, we mention that they are just strings over $\{\mathbf{0}, \mathbf{1}, \mathbf{2}\}^*$, not always from X.

For every function $f(n)$, we also use the notation $f(x)$, where x is a ternary expansion of n. Also, let x be a ternary expansion of n, where $2 \cdot 3^{k-1} \leq n \leq 3^k$; then we denote the ternary expansion of $3^k - n$ without leading zeros by $\overline{x}$.

For all k, we clearly have $\overline{\mathbf{10}}^k = \varepsilon$. For $x = \mathbf{1}x'$, where $x' \notin \mathbf{0}^*$, the function $\overline{x}$ is not defined, for any other $x \in X$ we have $\overline{\mathbf{2}x} = \overline{x}$.

3   Auxiliary Functions $q_j(n)$

To study the prefix palindromic length $p(n)$ of the Sierpinski word, we first define for every $j \geq 0$ an auxiliary function

$$q_j(n) = \mathrm{PL}(b^j s[1..n]).$$

Clearly, $p(n) = q_0(n)$, but for what follows, we need to study these functions for all j.

Proposition 3.1. *The functions q_j can be found as follows:*

- $q_0(0) = 0$*; for $j > 0$, we have $q_j(0) = 1$;*
- $q_0(1) = 1$*; for $j > 0$, we have $q_j(1) = 2$;*
- *for $3^k \leq n \leq 2 \cdot 3^k$, we have $q_j(n) = 1$ if $n = 3^k + j$ and $q_j(n) = 2$ otherwise;*
- *for $2 \cdot 3^k \leq n \leq 3^{k+1}$ and $j \leq 3^k$, we have*

$$q_j(n) = 1 + \min(q_{3^k - j}(n - 2 \cdot 3^k), q_j(3^{k+1} - n)),$$

- *at last, for $2 \cdot 3^k \leq n \leq 3^{k+1}$ and $j > 3^k$, we have*

$$q_j(n) = \min_{m \leq 3^k} q_m(n) + 1.$$

The proof of this proposition is omitted because of the length constraint.

Proposition 3.2. *For every $k \geq 0$, $j \geq 0$ and every $n \leq 3^k$ we have*

$$|q_j(n) - q_j(3^k - n)| \leq 1.$$

PROOF. It is sufficient to see that $q_j(3^k - n) = \mathrm{PL}(b^j s[1..3^k - n]) = \mathrm{PL}(s[n + 1..3^k]b^j)$, since the last two words are mirror images one of the other. Since $q_j(n) = \mathrm{PL}(b^j s[1..n])$ and $b^j s[1..n]s[n + 1..3^k]b^j = b^j s[1..3^k]b^j$ is a palindrome, the inequality is a particular case of Saarela's one.   □.

4   Function q and Its First Differences

In this section, we study another auxiliary function $q(n) = \min_j q_j(n)$.

Proposition 4.1. *For every $n \in \mathbb{N}$ the following equalities hold:*

$$q(n) = \begin{cases} 0, & \text{if } n = 0; \\ 1, & \text{if } n = 1 \text{ or } 3^k \leq n \leq 2 \cdot 3^k; \\ \min(1 + q(n - 2 \cdot 3^k), 1 + q(3^{k+1} - n)), & \text{if } 2 \cdot 3^k < n \leq 3^{k+1}, \end{cases} \tag{1}$$

meaning also for q as the function of X that

$$\begin{cases} q(\varepsilon) = 0; \\ q(\mathbf{1}y) = 1 \text{ for all } y \in \{\mathbf{0},\mathbf{1},\mathbf{2}\}^*; \\ q(\mathbf{2}y) = 1 + \min(q(y), q(\overline{\mathbf{2}y})) \text{ for all } y \in \{\mathbf{0},\mathbf{1},\mathbf{2}\}^*. \end{cases} \tag{2}$$

PROOF. First of all, note that $q(2 \cdot 3^{k-1}) = q(20^{k-1}) = 1 = 1 + q(0)$ for all $k > 0$, so, both (1) and (2) are true for such values. In all other cases, the two statements are equivalent, so, it is sufficient to prove (1). In fact, it immediately follows from Proposition 3.1 when we take the minimum for all j and notice that the minimal value of $q_j(n)$ for $n \leq 3^{k+1}$ is always attained for some $j \leq 3^k$. $\square$

Here is a list of basic properties of the function q.

Proposition 4.2. *For every $k \geq 0$ and every $n \leq 3^k$, we have $|q(n) - q(3^k - n)| \leq 1$.*

PROOF. Follows directly from the definition of $q(n) = \min_j q_j(n)$ and Proposition 3.2. Indeed, suppose that $q(n) \geq q(3^k - n)$ and j is such that $q(3^k - n) = q_j(3^k - n)$. Clearly, $q(n) \leq q_j(n)$. So, $q(n) - q(3^k - n) \leq q_j(n) - q_j(3^k - n) \leq 1$. The case of $q(n) \leq q(3^k - n)$ is symmetric. $\square$

Corollary 4.3. *For every n such that $2 \cdot 3^k < n \leq 3^{k+1}$, we have $|q(n - 2 \cdot 3^k) - q(3^{k+1} - n)| \leq 1$.*

PROOF. Follows immediately from the previous proposition and the fact that if $n' = n - 2 \cdot 3^k$, then $3^{k+1} - n = 3^k - n'$. $\square$

The next several properties of $q(x)$, $x \in X$, follow from (2) and are proved by the same type of induction.

Lemma 4.4. *For every $x \in X \cap \{0, 2\}^*$, we have $q(x1) = q(x2)$.*

PROOF. We proceed by induction on the length of x. For $x = \varepsilon$, we have $q(1) = q(2) = 1$, so the base of induction holds. Now consider $x = 2y$ where $y \in \{0, 2\}^*$ (so that y may contain leading zeros). We have $q(x1) = q(2y1) = 1 + \min(q(y1), q(\overline{2y1}))$ and $q(x2) = q(2y2) = 1 + \min(q(y2), q(\overline{2y2}))$. But $q(y1) = q(y2)$ by the induction hypothesis; moreover, by the same hypothesis, $q(\overline{2y1}) = q(\overline{2y2})$ since $\overline{2y1} = z2$ and $\overline{2y2} = z1$ for the same $z \in X$, where z is shorter than x. $\square$

Lemma 4.5. *For all $x \in X$, we have $q(x0) = q(x)$.*

PROOF. If $x = \varepsilon$, there is nothing to prove. If $x = 1y$, then $q(x0) = q(x) = 1$. Now for $x = 2y$, we proceed by induction on the length of x. The base is given by previous cases and $x = 2$ giving $q(20) = q(6) = q(2) = q(2) = 1$. For the induction step, consider $x = 2y$, where the statement is proven for y (which may start with leading zeros). It is sufficient to combine the last case of (2) with the induction hypothesis and the fact that $\overline{x0} = \overline{x}0$, so that $q(y) = q(y0)$, $q(\overline{x}) = q(\overline{x}0) = q(\overline{x0})$. $\square$

Lemma 4.6. *For every $x \in X \cap \{0, 2\}^*$ and for every $w \in \{0, 1, 2\}^*$, we have $q(x1w) = q(x1)$.*

PROOF. As above, we start from $x = \varepsilon$ giving $q(1w) = q(1) = 1$ and proceed by induction on the length of x: take $x = 2y$ and suppose that the lemma is true for all strings shorter than x. As above, it is sufficient to compare $q(y1)$ with

$q(y\mathbf{1}w)$, which are equal by the induction hypothesis, and $q(\overline{x\mathbf{1}})$ with $q(\overline{x\mathbf{1}w})$. For the latter comparison, we have to consider two cases: if $w \in \{\mathbf{0}\}^*$, then the equality holds due to the previous lemma. If w contains a non-zero symbol, then denote $\overline{x\mathbf{1}}$ as $t\mathbf{2}$ (indeed, its last symbol is equal to $\mathbf{2}$). Then $\overline{x\mathbf{1}w} = t\mathbf{1}w'$ for some w'; but we know by from Lemma 4.4 that $q(t\mathbf{2}) = q(t\mathbf{1})$ and from the induction hypothesis that $q(t\mathbf{1}) = q(t\mathbf{1}w')$. So, $q(\overline{x\mathbf{1}}) = q(\overline{x\mathbf{1}w})$ and thus $q(x\mathbf{1}w) = q(x\mathbf{1})$. □

Summarizing Lemmas 4.4 and 4.6, we observe the following

Corollary 4.7. *For every* $x \in X$ *such that* $x = y\mathbf{1}z$, *where* $y \in \{\mathbf{0}, \mathbf{2}\}^*$, *we have* $q(x) = q(y\mathbf{2})$.

So, we can concentrate on ternary representations from $\{\mathbf{0}, \mathbf{2}\}^*$. and, due to Lemma 4.5 even on those of them that end with $\mathbf{2}$.

For such a representation, that is, for a finite word on the alphabet $\{\mathbf{0}, \mathbf{2}\}$, let us call a *small group* a sequence of $\mathbf{2}$s separated from other such sequences by one or several $\mathbf{0}$s. In its turn, a *large group* is a word beginning and ending with $\mathbf{2}$ that does not contain two consecutive $\mathbf{0}$s and is separated from other such groups by at least two consecutive $\mathbf{0}$s. A large group is *dense* if it contains two consecutive $\mathbf{2}$s and *sparse* otherwise.

Example 4.8. The word $\mathbf{22202000022000202002}$ contains six small groups and four large groups $(\mathbf{22202}, \mathbf{22}, \mathbf{202}, \mathbf{2})$. The first two of these large groups are dense and the last two are sparse.

Theorem 4.9. *For every* $x \in X \cap \{\mathbf{0}, \mathbf{2}\}^*\mathbf{2}$,

1. $q(x) = q(\overline{x})$ *if and only if the first large group of* x *is sparse, that is, if and only if* $20^{k-1} \le [x]_3 \le (20)^{k/2}$; *otherwise* $q(x) = q(\overline{x}) + 1$;
2. *the value of* $q(x)$ *is equal to the number of small groups plus the number of dense large groups in* x.

Continuing the example above, we see that $q(\mathbf{22202000022000202002}) = 6 + 2 = 8$. Moreover, $\overline{\mathbf{22202000022000202002}} = \mathbf{20222200222020221}$, due to Lemma 4.4, $q(\mathbf{20222200222020221}) = q(\mathbf{20222200222020222})$, and the latter representation contains 5 small groups and two large groups, both of them dense, so that $q(\mathbf{20222200222020222}) = q(\mathbf{20222200222020221}) = 7$. It is predicted by the first part of the theorem since the first large group of the initial representation is dense.

The proof of the theorem is based on induction on the length of x and is omitted because of the length constraint. □

The first part of the theorem above will be used later for the results on the prefix palindromic length. As for the second part, it gives a formula for the function q and in particular allows to find its first differences $d_q(n) = q(n+1) - q(n)$. The following corollary of the theorem is straightforward.

Corollary 4.10. *For every $n \geq 0$ with $(n)_3 = x$ we have*

$$d_q(n) = \begin{cases} 0, & \text{if } x \text{ contains } \mathbf{1}; \text{ otherwise} \\ 1, & \text{if } x \text{ ends by } \mathbf{0} \text{ directly preceeded by } \mathbf{0} \text{ or a sparse large group;} \\ -1, & \text{if } x \text{ ends by } \mathbf{2} \text{ which is a part of a dense large group;} \\ 0, & \text{in all other cases.} \end{cases}$$

As it follows from this formula, the sequence $d_q(n)$ is automatic and here is the corresponding automaton.

Here and below, when considering first differences, we sometimes prefer to write – instead of -1, + instead of 1, and 0 in typewriter font.

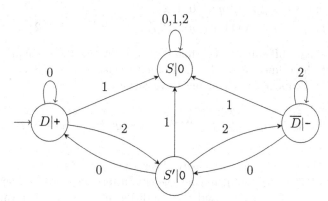

The choice of state names of this automaton will be clear from further constructions.

In its turn, this automaton is equivalent to the following morphic construction for the sequence d_q.

Theorem 4.11. *The sequence d_q is the 3-automatic word over the alphabet $\{-, 0, +\}$ given as follows:*

$$d_q = \gamma(\delta^\infty(D)),$$

where the morphism $\delta : \{D, S, S', \overline{D}\}^* \to \{D, S, S', \overline{D}\}^*$ is defined by

$$\begin{cases} \delta(D) = DSS', \\ \delta(S) = SSS, \\ \delta(S') = DS\overline{D}, \\ \delta(\overline{D}) = S'S\overline{D}, \end{cases}$$

and the coding $\gamma : \{D, S, S', \overline{D}\}^* \to \{-, 0, +\}^*$ is given by $\gamma(D) =$+, $\gamma(S) = \gamma(S') =$0, $\gamma(\overline{D}) =$-.

5 Difference Between $p(n)$ and $q(n)$

Now, after a study of the auxiliary function q, we return to the initial goal: the prefix palindromic length $p(n)$ of the Sierpinski word.

Proposition 5.1. *For every $n \geq 0$, the following holds.*

$$p(n) = \begin{cases} 0, & \text{if } n = 0; \\ 1, & \text{if } n = 1; \\ 2, & \text{if } 3^k < n \leq 2 \cdot 3^k; \\ \min(2 + q(n - 2 \cdot 3^k), 1 + p(3^{k+1} - n)), & \text{if } 2 \cdot 3^k < n \leq 3^{k+1}. \end{cases}$$

Equivalently, this formula can be written as

$$\begin{cases} p(\varepsilon) = 0; \\ p(10^k) = 1 \ \text{for all } k; \\ p(1y) = 2 \ \text{for all } y \in \{0, 1, 2\}^* \backslash 0^*; \\ p(2y) = 1 + \min(1 + q(y), p(\overline{2y})) \ \text{for all } y \in \{0, 1, 2\}^*. \end{cases} \tag{3}$$

PROOF. It is not difficult to see that the two statements are equivalent and that the first three lines of (3) hold. As for the last equality, it can be proven analogously to Proposition 4.1, using Proposition 3.1. □

Proposition 5.2. *For every $n \geq 0$ such that $2 \cdot 3^k \leq n \leq 3^{k+1}$, the equality $p(n) = q(n)$ holds if and only if $p(3^{k+1} - n) = q(3^{k+1} - n) < q(n)$. Otherwise $p(n) = q(n) + 1$.*

PROOF. For the edge values, we easily check that $q(2 \cdot 3^k) = 1 < 2 = p(2 \cdot 3^k)$, and $q(3^{k+1} - 2 \cdot 3^k) = q(2 \cdot 3^k)$, so that the condition does not hold; on the other hand, $q(3^{k+1}) = p(3^{k+1}) = 1$, and the condition holds. For other values, from the previous results, we have

$$q(n) = \min(1 + q(n - 2 \cdot 3^k), 1 + q(3^{k+1} - n)),$$

$$p(n) = \min(2 + q(n - 2 \cdot 3^k), 1 + p(3^{k+1} - n)).$$

So, if $p(3^{k+1} - n) > q(3^{k+1} - n)$, then the values compared for $p(n)$ are just greater than the respective values compared for $q(n)$, and thus $p(n) > q(n)$. Moreover, suppose that $p(3^{k+1} - n) = q(3^{k+1} - n)$. If $q(3^{k+1} - n) = q(n)$, it immediately means that $q(n) = 1 + q(n - 2 \cdot 3^k)$ and $p(n) = 1 + q(3^{k+1} - n) = 2 + q(n - 2 \cdot 3^k) > q(n)$. On the other hand, if $q(3^{k+1} - n) < q(n)$, then $q(n) = 1 + q(3^{k+1} - n) \leq 1 + q(n - 2 \cdot 3^k)$, so, $1 + q(3^{k+1} - n) = 1 + p(3^{k+1} - n) < 2 + q(n - 2 \cdot 3^k)$ and thus $p(n) = 1 + q(3^{k+1} - n) = q(n)$. The equivalence is established. □

The following statement is a direct corollary of the previous proposition and the first part of Theorem 4.9.

Proposition 5.3. *For every $x \in X$, we have $p(x) = q(x)$ if and only if $x \in 10^*$ or x starts with 2, $p(\overline{x}) = q(\overline{x})$ and $[x]_3 > (20)^{|x|/2}$.*

Now the following statement can be proven by a straightforward induction.

Proposition 5.4. *Let $S \subset X$ be the set of ternary decompositions x such that $p(x) = q(x)$. Then*

$$S = \{\varepsilon\} \cup \{10^*\} \cup \{(22^+00^+)^* . 22^+ . \{0^* \cup 0^+10^*\}.$$

In other words, $p(n) = q(n)$ if and only if $n = 0$, $n = 3^k$ for some k, or the ternary decomposition of n consists of blocks of at least two **2**s and at least two **0**s, possibly followed by one **0** or at least one **0** before 10^l for some l.

PROOF OF THE PROPOSITION 5.4. Denote by S_k the set of decompositions from S corresponding to numbers not exceeding 3^k and by D_k the difference $S_k \backslash S_{k-1}$. Clearly, Then $S_0 = \{\varepsilon, 1\}$, $D_1 = \{10\}$, $D_2 = \{22, 100\}$. Now, let us proceed by induction on k starting with this base. Due to Proposition 5.2 for every k we should look for elements of D_{k+1} among numbers of the form $3^{k+1} - m$, $(m)_3 \in S_k$. By the induction hypothesis, the elements of D_k are 10^k and some decompositions of length k starting with **22**. They correspond to the numbers m from $2 \cdot 3^{k-1} + 2 \cdot 3^{k-2}$ to 3^k. So, if $(m)_3 \in D_k$, then $3^{k+1} - m \leq 3^{k+1} - 2 \cdot 3^k - 2 \cdot 3^{k-1} < 2 \cdot 3^k$, and due to Proposition 5.2, $3^{k+1} - m \notin S$. So,

$$D_{k+1} = \{3^{k+1} - m | (m)_3 \in S_{k-1}\}.$$

It remains to check by a simple case study (whether $(m)_3$ contains **1** or not) that subtracting from 3^{k+1} numbers whose ternary decompositions are in S_{k-1} gives exactly numbers with decompositions from S, as described in the assertion, of length $k + 1$, plus 3^{k+1}. □

Note also that the above expression for D_{k+1} implies that

$$|D_{k+1}| = |S_{k+1}| - |S_k| = |S_{k-1}|,$$

and thus we can easily prove that every $|S_k|$ is a Fibonacci number: $|S_k| = F_{k+3}$ (if we start with $F_0 = 0, F_1 = F_2 = 1$).

The above proposition characterizes the function $t(n) = p(n) - q(n)$ which is equal to 0 if $(n)_3 \in S$ and to 1 otherwise. It also allows to find precisely its first differences $d_t(n) = t(n + 1) - t(n)$:

Corollary 5.5. *The first differences of the function $t(n)$ are*

$$d_t(n) = \begin{cases} 0, & if \ (n)_3 \in S \ does \ not \ contain \ \mathbf{1} \ and \ ends \ with \ \mathbf{00} \ or \ \mathbf{22}; \\ 1, & if \ (n)_3 \in S \ contains \ \mathbf{1} \ or \ ends \ with \ \mathbf{220}; \\ -1, & if \ (n)_3 \in (\mathbf{22^+00^+})^*.\{\mathbf{2} \cup \mathbf{2^+12^*}\}; \\ 0, & in \ all \ other \ cases. \end{cases}$$

Here the first case corresponds to $t(n) = t(n + 1) = 0$ and the last case to $t(n) = t(n + 1) = 1$.

The corresponding automaton for $d_t(n)$ is depicted below.

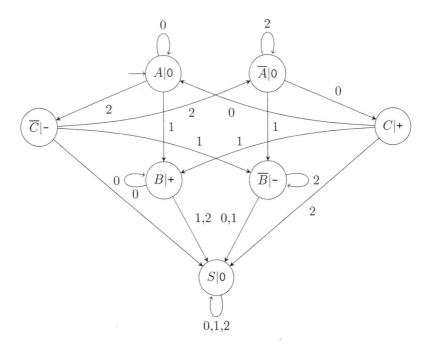

This automaton is equivalent to the following morphic construction for the sequence d_t.

Theorem 5.6. *The sequence d_t is the 3-automatic word over the alphabet $\{-,0,+\}$ given as follows:*

$$d_t = \xi(\nu^\infty(A)),$$

where the morphism $\nu : \{A,B,C,\overline{A},\overline{B},\overline{C},S\}^* \rightarrow \{A,B,C,\overline{A},\overline{B},\overline{C},S\}^*$ is defined by

$$\begin{cases} \nu(A) = AB\overline{C}, \\ \nu(B) = BSS, \\ \nu(C) = ABS, \\ \nu(\overline{A}) = C\overline{B}\overline{A}, \\ \nu(\overline{B}) = SS\overline{B}, \\ \nu(\overline{C}) = S\overline{B}\overline{A}, \\ \nu(S) = SSS, \end{cases}$$

and the coding $\xi : \{A,B,C,\overline{A},\overline{B},\overline{C},S\}^* \rightarrow \{-,0,+\}^*$ is given by $\xi(A) = \xi(\overline{A}) = \xi(S) = 0$, $\xi(B) = \xi(C) = +$, $\xi(\overline{B}) = \xi(\overline{C}) = -$.

6   First Differences of $p(n)$

By the definition of $t(n)$, the first differences of the function $p(n)$ are

$$d_p(n) = d_q(n) + d_t(n).$$

The functions $d_q(n)$ and $d_t(n)$ are completely described in Theorems 4.11 and 5.6 and by respective automata. It remains just to combine them, and one of the natural ways to do it is to define a new morphism $\psi = \binom{\delta}{\nu}$ just as a direct product of δ and ν on the direct product of alphabets. We start with both starting symbols and get $\psi\binom{A}{D} = \binom{A}{D}\binom{B}{S}\binom{\overline{C}}{S'}$; here the upper line is δ and the lower is ν. Then we define ψ on all the pairs of symbols that appeared, and continue this process while they continue to appear. We observe that only ten pairs appear in the fixed point of ψ starting with $\binom{A}{D}$: the alpha-

bet is $\mathcal{A} = \left\{ \binom{A}{D}, \binom{\overline{A}}{\overline{D}}, \binom{B}{S}, \binom{\overline{B}}{S}, \binom{C}{S'}, \binom{\overline{C}}{S'}, \binom{S}{D}, \binom{S}{\overline{D}}, \binom{S}{S}, \binom{S}{S'} \right\}$. Since we investigate the sum of the two first difference functions, each of these double letters is coded by $c\binom{X}{Y} = \gamma(X) + \xi(Y)$, where we recall that the symbols $-, 0, +$ are in fact numbers $-1, 0, 1$. So, for example, we have $c\binom{A}{D} = 0 + 1 = 1$.

It remains to simplify the notation: the first six symbols of $\mathcal{A}$ can be denoted by just their upper letters, and the last four, starting with S, are defined by their lower letters. All this gives the following

Theorem 6.1. *The sequence d_p of first differences of the prefix palindromic length of the Sierpinski word is the 3-automatic word over the alphabet $\{-, 0, +\}$ defined as*

$$d_p = c(\psi^\infty(A)),$$

where the morphism $\psi : \mathcal{B}^* \to \mathcal{B}^*$, where $\mathcal{B} = \{A, B, C, D, \overline{A}, \overline{B}, \overline{C}, \overline{D}, S, S'\}$, is defined by

$$
\begin{cases}
\psi(A) = AB\overline{C}, \\
\psi(B) = BSS, \\
\psi(C) = AB\overline{D}, \\
\psi(D) = DSS', \\
\psi(\overline{A}) = C\overline{B}A, \\
\psi(\overline{B}) = SS\overline{B}, \\
\psi(\overline{C}) = DB\overline{A}, \\
\psi(\overline{D}) = S'S\overline{D}, \\
\psi(S) = SSS, \\
\psi(S') = DS\overline{D},
\end{cases}
$$

and the coding $c : \mathcal{B}^* \to \{-, 0, +\}^*$ is given by $c(A) = c(B) = c(C) = c(D) = +$, $c(\overline{A}) = c(\overline{B}) = c(\overline{C}) = c(\overline{D}) = -$, $c(S) = c(S') = 0$.

The corresponding DFAO is depicted below.

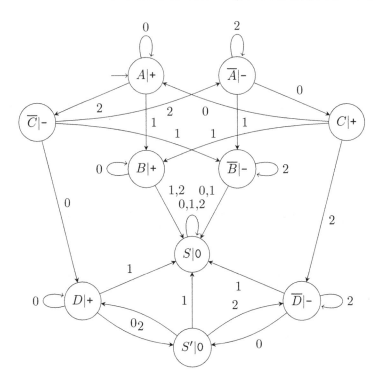

References

1. Allouche, J.-P., Shallit, J.: Automatic Sequences: Theory, Applications Generalizations. Cambridge University Press, Cambridge (2003)
2. Bucci, M., Richomme, G.: Greedy palindromic lengths. Int. J. Found. Comput. Sci. **29**, 331–356 (2018)
3. Cobham, A.: Uniform tag sequences. Math. Syst. Theory **6**, 164–192 (1972)
4. Frid. A.E.: Prefix palindromic length of the Thue-Morse word. J. Integer. Seq. **22**, Article 19.7.8 (2019)
5. Frid, A.E.: Sturmian numeration systems and decompositions to palindromes. Eur. J. Combin. **71**, 202–212 (2018)
6. Frid, A.E., Laborde, E., Peltomäki, J.: On prefix palindromic length of automatic words. Theoret. Comput. Sci. **891**, 13–23 (2021)
7. Frid, A.E., Puzynina, S., Zamboni, L.: On palindromic factorization of words. Adv. Appl. Math. **50**, 737–748 (2013)
8. Laborde, E.: Sur la longueur palindromique du préfixe de suites k-automatiques, Master thesis, Aix-Marseille Université (2020)
9. Li, S.: Palindromic length sequence of the ruler sequence and of the period-doubling sequence (2020). https://arxiv.org/abs/2007.08317
10. Rukavicka, J.: Palindromic length and reduction of powers (2021). https://arxiv.org/abs/2103.14609
11. Saarela, A.: Palindromic length in free monoids and free groups. In: Brlek, S., Dolce, F., Reutenauer, C., Vandomme, É. (eds.) WORDS 2017. LNCS, vol. 10432, pp. 203–213. Springer, Cham (2017). https://doi.org/10.1007/978-3-319-66396-8_19

Preservation of Normality
by Unambiguous Transducers

Olivier Carton[✉]

IRIF, Université de Paris, Paris, France
Olivier.Carton@irif.fr

Abstract. We consider finite state non-deterministic but unambiguous transducers with infinite inputs and infinite outputs, and we consider the property of Borel normality of sequences of symbols. When these transducers are strongly connected, and when the input is a Borel normal sequence, the output is a sequence in which every block has a frequency given by a weighted automaton over the rationals. We provide an algorithm that decides in cubic time whether an unambiguous transducer preserves normality.

Keywords: Functional transducers · Weighted automata · Normal sequences

1 Introduction

More than one hundred years ago Émile Borel [3] gave the definition of *normality*. A real number is normal to an integer base if, in its infinite expansion expressed in that base, all blocks of digits of the same length have the same limiting frequency. Borel proved that almost all (in a measure theoretic sense) real numbers are normal to all integer bases. However, very little is known on how to prove that a given number like $\sqrt{2}$ or π has this property.

The definition of normality was the first step towards a definition of randomness. Normality formalizes the least requirements about a random sequence. It is indeed expected that in a random sequence, all blocks of symbols with the same length occur with the same limiting frequency. Normality, however, is a much weaker notion than the one of purely random sequences defined by Kolmogorov, Martin-Löf and others [11].

The motivation of this work is the study of transformations preserving randomness, hence preserving normality. The paper is focused on very simple transformations, namely those that can be realized by finite-state machines. We consider automata with outputs, also known as transducers, mapping infinite sequences of symbols to infinite sequences of symbols. Input deterministic transducers were considered in [6] where it was shown that preservation of normality can be checked in polynomial time for these transducers. This paper extends the results to unambiguous transducers, that is, transducers where each sequence is the input label of at most one accepting run. These machines are of great importance because they coincide with functional transducers in the following sense. Each unambiguous transducer is indeed functional as there is at most one output for each input but it was shown conversely that each functional transducer is equivalent to some unambiguous one [8].

An auxiliary result involving weighted automata is introduced to obtain the main result. It states that if an unambiguous and strongly connected transducer is fed with a normal sequence then the frequency of each block in the output is given by a weighted automaton on rational numbers. It implies, in particular, that the frequency of each block in the output sequence does not depend on the input nor the run labeled by it as long as this input sequence is normal. As the output of the run can be the used transitions, the result shows that each finite run has a limiting frequency in the run.

Our result is connected to another strong link between normality and automata. Agafonov's theorem [1] states that if symbols are selected in a normal sequence using an oblivious finite state machine, the resulting sequence is still normal. Oblivious means here that the choice of selecting a symbol is based on the state of the machine after reading the prefix of the sequence before the symbol but not including the symbol itself. Our result allows us to recover Agafonov's theorem about preservation of normality by selection but this application is not detailed in this short version of the paper.

The paper is organized as follows. Notions of normal sequences and transducers are introduced in Sect. 2. Results are stated in Sect. 3. The main ingredients of the construction are given in Sect. 4 while the algorithm is given in Sect. 5.

2   Basic Definitions

2.1   Normality

Before giving the formal definition of normality, let us introduce some simple definitions and notation. Let A be a finite set of *symbols* that we refer to as the *alphabet*. We write $A^{\mathbb{N}}$ for the set of all sequences on the alphabet A and A^* for the set of all (finite) words. The length of a finite word w is denoted by $|w|$. The positions of sequences and words are numbered starting from 1. To denote the symbol at position i of a sequence (respectively, word) w we write $w[i]$, and to denote the substring of w from position i to j inclusive we write $w[i{:}j]$. The empty word is denoted by λ. The cardinality of a finite set E is denoted by $\#E$.

Given two words w and v in A^*, the number $|w|_v$ of occurrences of v in w is defined by $|w|_v = \#\{i : w[i{:}i + |v| - 1] = v\}$. Given a word $w \in A^+$ and a sequence $x \in A^{\mathbb{N}}$, we refer to the *frequency of w in x* as

$$\mathrm{freq}(x, w) = \lim_{n \to \infty} \frac{|x[1{:}n]|_w}{n}$$

when this limit is well-defined. A sequence $x \in A^{\mathbb{N}}$ is *normal* on the alphabet A if for every word $w \in A^+$, $\mathrm{freq}(x, w) = (\#A)^{-|w|}$

An occurrence of v is called *aligned* if its starting position i (as above) is such that $i - 1$ is a multiple of the length of v. An alternative definition of normality can be given by counting aligned occurrences, and it is well-known that they are equivalent (see for example [2]). We refer the reader to [5, Chap. 4] for a complete introduction to normality.

The most famous example of a normal sequence is due to Champernowne [7], who showed in 1933 that the sequence obtained from concatenating all the natural numbers in their usual order (spaces are added for the reader's convenience):

$$0\ 1\ 2\ 3\ 4\ 5\ 6\ 7\ 8\ 9\ 10\ 11\ 12\ 13\ 14\ 15\ 16\ 17\ 18\ 19\ 20\ 21\ 22\ 23\ 24\ 25\ 26\ 27\ 28\ 29\ 30\ldots$$

is normal on the alphabet $\{0, 1, \ldots, 9\}$.

2.2 Automata and Transducers

In this paper we consider automata with outputs, also known as transducers. We refer the reader to [12] for a complete introduction to automata accepting sequences. Such finite-state machines are used to realize functions mapping words to words and especially sequences to sequences. We mainly consider transducers in which each transition consumes exactly one symbol of their input and outputs a word which might be empty. As many reasonings ignore the outputs of the transitions, we first introduce automata.

A *(Büchi) automaton* $\mathcal{A}$ is a tuple $\langle Q, A, \Delta, I, F \rangle$ where Q is the finite state set, A the alphabet, $\Delta \subseteq Q \times A \times Q$ the transition relation, $I \subseteq Q$ the set of initial states and F is the set of final states. A transition is a tuple $\langle p, a, q \rangle$ in $Q \times A \times Q$ and it is written $p \xrightarrow{a} q$. A *finite run* in $\mathcal{A}$ is a finite sequence of consecutive transitions,

$$q_0 \xrightarrow{a_1} q_1 \xrightarrow{a_2} q_2 \cdots q_{n-1} \xrightarrow{a_n} q_n$$

Its *input* is the word $a_1 a_2 \cdots a_n$. An *infinite run* in $\mathcal{A}$ is a sequence of consecutive transitions,

$$q_0 \xrightarrow{a_1} q_1 \xrightarrow{a_2} q_2 \xrightarrow{a_3} q_3 \cdots$$

A run is *initial* if its first state q_0 is initial, that is, belongs to I. An infinite run is called *final* if it visits infinitely often a final state. Let us denote by $q \xrightarrow{x} \infty$ the existence of a final run labeled by x and starting from state q. An infinite run is *accepting* if it is both initial and final. As usual, an automaton is *deterministic* if it has only one initial state, that is $\#I = 1$ and if $p \xrightarrow{a} q$ and $p \xrightarrow{a} q'$ are two of its transitions with the same starting state and the same label, then $q = q'$. An automaton is called *unambiguous* if each sequence is the label of at most one accepting run. By definition, deterministic automata are unambiguous but they are not the only ones as is shown by the one pictured in Fig. 1 (right).

Each automaton $\mathcal{A}$ can be seen as a directed graph $\mathcal{G}$ by ignoring the labels of its transitions. We define the *strongly connected components* (SCC) of $\mathcal{A}$ as the strongly connected components of $\mathcal{G}$. An automaton $\mathcal{A}$ is called *strongly connected* if it has a single strongly connected component.

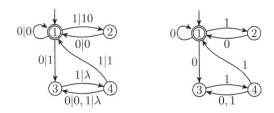

Fig. 1. An unambiguous transducer and its input automaton

A transducer with input alphabet A and output alphabet B is informally an automaton whose labels of transitions are pairs (u, v) in $A^* \times B^*$. The pair (u, v) is usually written $u|v$ and a transition is thus written $p \xrightarrow{u|v} q$. The words u and v are respectively called the *input* and *output label* of the transition. More formally a *transducer* $\mathcal{T}$ is a tuple $\langle Q, A, B, \Delta, I, F \rangle$, where Q is a finite set of states, A and B are the input and output alphabets respectively, $\Delta \subseteq Q \times A^* \times B^* \times Q$ is a finite transition relation and $I \subseteq Q$ is the set of initial states and F is the set of final states of the Büchi acceptance condition. The transducer is called *real-time* if the transition relation is contained in $Q \times A \times B^* \times Q$, that is, the input label of each transition is a symbol. The *input automaton* of a real-time transducer is the automaton obtained by ignoring the output label of each transition. A real-time transducer is called *input deterministic* (respectively, *unambiguous*) if its input automaton is deterministic (respectively, unambiguous).

A *finite run* in $\mathcal{T}$ is a finite sequence of consecutive transitions,

$$q_0 \xrightarrow{u_1|v_1} q_1 \xrightarrow{u_2|v_2} q_2 \cdots q_{n-1} \xrightarrow{u_n|v_n} q_n$$

Its *input* and *output labels* are the words $u_1 u_2 \cdots u_n$ and $v_1 v_2 \cdots v_n$ respectively. An *infinite run* in $\mathcal{T}$ is an infinite sequence of consecutive transitions,

$$q_0 \xrightarrow{u_1|v_1} q_1 \xrightarrow{u_2|v_2} q_2 \xrightarrow{u_3|v_3} q_3 \cdots$$

Its *input* and *output labels* are the sequences of symbols $u_1 u_2 u_3 \cdots$ and $v_1 v_2 v_3 \cdots$ respectively. Accepting runs are defined as for automata. The relation realized by the transducer is the set of pairs (x, y) where x and y are the input and output labels of an accepting run. It is a classical result that if the relation realized by some transducer $\mathcal{T}$ is a function, it is also realized by a real-time transducer which is easily obtained from $\mathcal{T}$ [13, Prop. 1.1, p. 646].

If $\mathcal{T}$ is an unambiguous transducer, each sequence x is the input label of at most one accepting run in $\mathcal{T}$. When this run does exist, its output is denoted by $\mathcal{T}(x)$. This output might be finite but it is always possible to modify the transducer in such a way that this output is infinite for each accepting run. Therefore it is always assumed from now on that this output is infinite. We say that an unambiguous transducer $\mathcal{T}$ *preserves normality* if for each normal sequence x, $\mathcal{T}(x)$ is also normal.

An automaton (respectively, transducer) is said to be *trim* if each state occurs in an accepting run. Automata and transducers are always assumed to be trim since useless states can easily be removed.

We end this section by stating very easy but useful facts about unambiguous automata. If $\langle Q, A, \Delta, I, F \rangle$ is an unambiguous automaton then each automaton $\langle Q, A, \Delta, \{q\}, F \rangle$ obtained by taking state q as initial state is also unambiguous. Similarly, removing states or transitions from an unambiguous automaton yields an unambiguous automaton. Combining these two facts gives that each strongly connected component, seen as an automaton, of an unambiguous automaton is still an unambiguous automaton.

2.3   Weighted Automata

We now introduce weighted automata. In this paper we only consider weighted automata whose weights are rational numbers with the usual addition and multiplication (see [13, Chap. III] for a complete introduction).

A *weighted automaton* $\mathcal{A}$ is a tuple $\langle Q, B, \Delta, I, F \rangle$, where Q is the finite state set, B is the alphabet, $I : Q \to \mathbb{Q}$ and $F : Q \to \mathbb{Q}$ are the functions that assign to each state an initial and a final weight and $\Delta : Q \times B \times Q \to \mathbb{Q}$ is a function that assigns to each transition a weight.

As usual, the weight of a run is the product of the weights of its transitions times the initial weight of its first state and times the final weight of its last state. Furthermore, the weight of a word $w \in B^*$ is the sum of the weights of all runs with label w and it is denoted $\mathrm{weight}_{\mathcal{A}}(w)$.

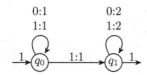

Fig. 2. A weighted automaton

A transition $p \xrightarrow{a} q$ with weight x is pictured $p \xrightarrow{a:x} q$. Non-zero initial and final weights are given over small incoming and outgoing arrows. A weighted automaton is pictured in Fig. 2. The weight of the run $q_0 \xrightarrow{1} q_1 \xrightarrow{0} q_1 \xrightarrow{1} q_1 \xrightarrow{0} q_1$ is $1 \cdot 1 \cdot 2 \cdot 2 \cdot 2 \cdot 1 = 8$. The weight of the word $w = 1010$ is $8 + 2 = 10$. More generally the weight of a word $w = a_1 \cdots a_k$ is the integer $n = \sum_{i=1}^{k} a_i 2^{k-i}$ (w is a binary expansion of n with possibly some leading zeros).

A weighted automaton can also be represented by a triple $\langle \pi, \mu, \nu \rangle$ where π is a row vector over $\mathbb{Q}$ of dimension $1 \times n$, μ is a morphism from B^* into the set of $n \times n$-matrices over $\mathbb{Q}$ with the usual matrix multiplication and ν is a column vector of dimension $n \times 1$ over $\mathbb{Q}$. The weight of a word $w \in B^*$ is then equal to $\pi \mu(w) \nu$. The vector π is the vector of initial weights, the vector ν is the vector of final weights and, for each symbol b, $\mu(b)$ is the matrix whose (p, q)-entry is the weight x of the transition $p \xrightarrow{b:x} q$. The weighted automaton pictured in Fig. 2 is, for instance, represented by $\langle \pi, \mu, \nu \rangle$ where $\pi = (1,0)$, $\nu = \binom{0}{1}$ and the morphism μ is given by

$$\mu(0) = \begin{pmatrix} 1 & 0 \\ 0 & 2 \end{pmatrix} \quad \text{and} \quad \mu(1) = \begin{pmatrix} 1 & 1 \\ 0 & 2 \end{pmatrix}.$$

3   Results

We now state the main results of the paper. The first one states that when a transducer is strongly connected, unambiguous and complete, the frequency of each finite word w in the output of a run with a normal input label is given by a weighted automaton over $\mathbb{Q}$. The second one states that it can be checked whether an unambiguous transducer preserves normality.

Theorem 1. *Given an unambiguous and strongly connected transducer, there exists a weighted automaton $\mathcal{A}$ such that for each normal sequence x in the domain of $\mathcal{T}$ and for any finite word w, $\mathrm{freq}(\mathcal{T}(x), w)$ is equal to $\mathrm{weight}_{\mathcal{A}}(w)$.*

Furthermore, the weighted automaton $\mathcal{A}$ can be computed in cubic time with respect to the size of the transducer $\mathcal{T}$.

Theorem 1 only deals with strongly connected transducers, but Proposition 5 deals with the general case by showing that it suffices to apply Theorem 1 to some strongly connected components to check preservation of normality.

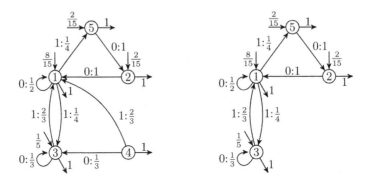

Fig. 3. Two weighted automata

To illustrate Theorem 1 we give in Fig. 3 two weighted automata which compute the frequency of each finite word w in $\mathcal{T}(x)$ for a normal input x and the transducer $\mathcal{T}$ pictured in Fig. 1. The leftmost one is obtained by the procedure described in the next section. The rightmost one is obtained by removing useless states from the leftmost one.

The decidability in cubic time of the equivalence of transducers over a (computable) field yields the following theorem. The *size* of a transducer is the sum of the sizes of its transitions where the *size* of a transition $p \xrightarrow{a|w} q$ is $|aw|$.

Theorem 2. *It can be decided in cubic time whether an unambiguous transducer preserves normality or not.*

From the weighted automaton pictured in Fig. 3, it is easily computed that the limiting frequencies of the digits 0 and 1 in the output $T(x)$ of a normal input x are respectively $9/15$ and $6/15$. This shows that the transducer T pictured in Fig. 1 does not preserve normality. To illustrate the previous theorem, we show that the transducer pictured in Fig. 4 is unambiguous and does preserve normality. The output of each transition is either the input symbol or the empty word. Therefore, the output is always a subsequence of the input sequence. It can be checked that a symbol is selected, that is copied to the output, if the number of 0 until the next 1 is finite and even, including zero.

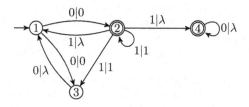

Fig. 4. Another unambiguous transducer

By Proposition 5 below, it suffices to check that the strongly connected component made of the states $\{1, 2, 3\}$ does preserve normality. The weighted automaton given by the algorithm is represented by the triple $\langle \pi, \mu, \mathbf{1} \rangle$ where π is the row vector $\pi = (3/4, 1/4)$, $\mathbf{1}$ is the column vector $\binom{1}{1}$ and the morphism μ is defined by

$$\mu(0) = \begin{pmatrix} 1/4 & 1/12 \\ 3/4 & 1/4 \end{pmatrix} \quad \text{and} \quad \mu(1) = \begin{pmatrix} 1/2 & 1/6 \\ 0 & 0 \end{pmatrix}.$$

The vector π satisfies $\pi\mu(0) = \pi\mu(1) = \frac{1}{2}\pi$ and therefore $\pi\mu(w)\mathbf{1}$ is equal to $2^{-|w|}$ for each word w. This shows that the transducer pictured in Fig. 4 does preserve normality.

4   Markov Chain of an Unambiguous Automaton

In this section, we introduce the main tools used to prove Theorems 1 and 2. Let $\mathcal{A}$ be an automaton with state set Q. The *adjacency matrix* of $\mathcal{A}$ is the $Q \times Q$-matrix M defined by $M_{p,q} = \#\{a \in A : p \xrightarrow{a} q\}/\#A$. Its entry $M_{p,q}$ is thus the number of transitions from p to q divided by the cardinality of the alphabet A. The factor $1/\#A$ is just a normalization factor to compare the spectral radius of this matrix to 1 rather than to the cardinality of the alphabet. By a slight abuse of notation, the spectral radius of the adjacency matrix is called the spectral radius of the automaton.

The adjacency matrix of the input automaton of the transducer pictured in Fig. 1 is the matrix M given by

$$M = \frac{1}{2} \begin{pmatrix} 1 & 1 & 1 & 0 \\ 1 & 0 & 0 & 0 \\ 0 & 0 & 0 & 1 \\ 1 & 0 & 2 & 0 \end{pmatrix}$$

It can be checked that the spectral radius of this matrix is 1.

Let us recall that the *uniform measure* μ is the measure on $A^{\mathbb{N}}$ such that the measure $\mu(wA^{\mathbb{N}})$ of each cylinder $wA^{\mathbb{N}}$ is $(\#A)^{-|w|}$. For each state q, let F_q be the *future* set, that is the set $F_q = \{x : q \xrightarrow{x} \infty\}$ of accepted sequences if q is taken as the only initial state. Let α_q be the measure of the set F_q. The spectral radius of the adjacency matrix of an unambiguous automaton is at most 1. The following proposition characterizes when it is equal to 1 or strictly less than 1.

Proposition 1. *Let $\mathcal{A}$ be a strongly connected and unambiguous automaton and let ζ be the spectral radius of its adjacency matrix. If $\zeta = 1$ then $\mathcal{A}$ accepts at least one normal sequence and each number α_q is positive. If $\zeta<1,<$ then $\mathcal{A}$ accepts no normal sequence and each number α_q is equal to zero.*

Now we sketch the proof of Proposition 1. It is well-known that the spectral radius of a strongly connected automaton is related to the entropy of the corresponding sofic shift [10, Thm 4.3.1]. If this spectral radius is less than 1, at least one finite word w is the label of no run in $\mathcal{A}$. Therefore, no normal sequence x is the label of a run in $\mathcal{A}$ because w occurs in x with limiting frequency $(\#A)^{-|w|}$. If the spectral radius is 1, then each finite word w is the label of at least one run in $\mathcal{A}$. By compacity of $A^{\mathbb{N}}$, it follows that each sequence, including the normal ones, is the label of a run in $\mathcal{A}$. This run might be neither initial nor final. Modifying the run at the beginning and at sparse positions (say positions 2^n for instance) yields an accepting run whose label is still normal if its label was already normal before.

Now, we introduce a Markov chain associated with an unambiguous automaton. The use of the ergodic theorem applied to this Markov chain is the main ingredient in the proof of Theorem 1. Let $\mathcal{A}$ be a strongly connected and unambiguous automaton and let p be one of its states. We also suppose that the spectral radius of its adjacency matrix M is 1. By Proposition 1, the measure α_q of each set $F_q = \{x : q \xrightarrow{x} \infty\}$ is non-zero.

We define a stochastic process $(X_n)_{n \geqslant 0}$ as follows. Its sample set is the set $F_p \subseteq A^{\mathbb{N}}$ equipped with the uniform measure μ. For each sequence $x = x_1 x_2 x_3 \cdots$ in F_p, there exists a unique accepting run

$$(p = q_0) \xrightarrow{x_1} q_1 \xrightarrow{x_2} q_2 \xrightarrow{x_3} q_3 \cdots$$

The process is defined by setting $X_n(x) = q_n$ for each $x \in F_p$. The following proposition states the main property of this process.

Proposition 2. *The process $(X_n)_{n \geqslant 0}$ is a Markov chain.*

The proof of this propostion is a mere verification. However it allows us to use the ergodic theorem for Markov chains [4, Thm 4.1].

The $Q \times Q$-matrix P of probabilities for the introduced Markov chain is given by $P_{p,q} = \#\{a \in A : p \xrightarrow{a} q\}\alpha_q/(\#A)\alpha_p$. Note that the matrix P and the adjacency matrix M of $\mathcal{A}$ are related by the equalities $P_{p,q} = M_{p,q}\alpha_q/\alpha_p$ for each states $p, q \in Q$. The stationary distribution of the stochastic matrix P is the vector $(\pi_q\alpha_q)_{q \in Q}$ where $\pi = (\pi_q)_{q \in Q}$ is the left eigenvector of the matrix M for the eigenvalue 1 and π has been normalized such that $\sum_{q \in Q} \pi_q\alpha_q = 1$. The following property states the main property of this stationary distribution.

Proposition 3. *Let $\mathcal{A}$ be a strongly connected and unambiguous automaton such that the spectral radius of its adjacency matrix is 1. Let ρ be an accepting run whose label is a normal sequence. Then, for any state r*

$$\lim_{n \to \infty} \frac{|\rho[1{:}n]|_r}{n} = \pi_r\alpha_r$$

where $\rho[1{:}n]$ is the finite run made of the first n transitions of ρ.

Note that the result of Proposition 3 implies that the frequencies of states do not depend on the input as long as this input is normal. Note also that the result is void if the spectral radius of the adjacency matrix is less than 1 because, by Proposition 1, no accepting run is labeled by a normal sequence. This assumption could be removed because the statement remains true but this is our choice to mention explicitly the assumption for clarity.

Now we sketch the proof of Proposition 3. For each real numbers $\varepsilon, \delta > 0$, there exists, by the ergodic theorem, an integer n such that for each integer $k \geqslant n$

$$\#\{w : |w| = k \text{ and } \exists p, q, r \in Q \; \big||p \xrightarrow{w} q|_r/k - \pi_r\alpha_r\big| > \delta\} < \varepsilon(\#A)^k.$$

This shows that, for k great enough, the cardinality of the set of *bad* words of length k is small where a word w is *bad* if there are three states p, q, r such that the number of occurrences of r in the run $p \xrightarrow{w} q$ is far from the expected value $\pi_r\alpha_r k$. It suffices then to split the run ρ into blocks of length k and use the fact that all words of length k have the same limiting frequency in the input sequence.

The following proposition extends to finite runs the statement of Proposition 3 about states. It is obtained by applying Proposition 3 to a new automaton whose states are the runs of length n of the starting one. The Markov chain associated with this new automaton is called the *snake* Markov chain. See Problems 2.2.4, 2.4.6 and 2.5.2 (p. 90) in [4] for more details.

Proposition 4. *Let $\mathcal{A}$ be a strongly connected and unambiguous automaton such that the spectral radius of its adjacency matrix is 1. Let ρ be an accepting run whose label is a normal sequence. For any finite run $\gamma = q_0 \xrightarrow{a_1} q_1 \cdots q_{k-1} \xrightarrow{a_k} q_k$ of length k, one has*

$$\lim_{n \to \infty} \frac{|\rho[1{:}n]|_\gamma}{n} = \frac{\pi_{q_0}\alpha_{q_k}}{(\#A)^k}$$

where $\rho[1{:}n]$ is the finite run made of the first n transitions of ρ.

5   Sketches of Proofs

In this section we skech the proofs for Theorems 1 and 2. The proofs are organized in three parts. First, the transducer $\mathcal{T}$ is normalized into another transducer $\mathcal{T}'$ realizing the same function. Then this latter transducer is used to define a weighted automaton $\mathcal{A}$. Second, the proof that the construction of $\mathcal{A}$ is correct is carried out. Third, the algorithms computing $\mathcal{A}$ and checking whether $\mathcal{T}$ preserves normality or not are given.

Each infinite accepting run in a transducer is eventually trapped in a strongly connected component with at least one final state. If the input sequence is normal, the spectral radius of this component must be one by Proposition 1. Conversely, each strongly connected component with spectral radius one contains a run labeled by a normal sequence. The next proposition follows.

Proposition 5. *An unambiguous transducer $\mathcal{T}$ preserves normality if and only if each strongly connected component of $\mathcal{T}$ with a final state and spectral radius 1 preserves normality.*

Consider for instance the transducer pictured in Fig. 4. It has two strongly connected components: the one made of states $1, 2, 3$ and the one made of state 4. The corresponding adjacency matrices are

$$\frac{1}{2} \begin{pmatrix} 0 & 1 & 1 \\ 1 & 1 & 1 \\ 1 & 0 & 0 \end{pmatrix} \quad \text{and} \quad \left(\tfrac{1}{2} \right)$$

whose spectral radii are respectively 1 and $1/2$. It follows that the transducer preserves normality if and only if the transducer reduced to the states $1, 2, 3$ does preserve normality.

In what follows we only consider strongly connected transducers. Propositions 3 and 4 have the following consequence. Let $\mathcal{T}$ be an unambiguous and strongly connected transducer. If each transition has an empty output label, the output of any run is empty and then $\mathcal{T}$ does not preserve normality. Therefore, we assume that transducers have at least one transition with a non empty output label. By Propositions 3 and 4, this transition is visited infinitely often if the input is normal because the stationary distribution $(\pi_q \alpha_q)_{q \in Q}$ is positive. This guarantees that if the input sequence is normal, then the output sequence is infinite and $\mathcal{T}(x)$ is well-defined.

Note that the output labels of the transitions in $\mathcal{T}$ from Theorem 1 may have arbitrary lengths. We first describe the construction of an equivalent transducer $\mathcal{T}'$ such that all output labels in $\mathcal{T}'$ have length at most 1. We call this transformation *normalization* and it consists in *replacing* each transition $p \xrightarrow{a|v} q$ in $\mathcal{T}$ such that $a \in A$ and $|v| \geqslant 2$ by n transitions:

$$p \xrightarrow{a|b_1} q_1 \xrightarrow{\lambda|b_2} q_2 \cdots q_{n-1} \xrightarrow{\lambda|b_n} q$$

where $q_1, q_2, \ldots, q_{n-1}$ are new states and $v = b_1 \cdots b_n$. We refer to p as the parent of $q_1, \cdots, q_{n-1}$.

The last step is the construction of a weighted automaton from the normalized transducer. This is performed by replacing the input of each transition by a weight. This weight is either $1/\#A$ if the input is a symbol or 1 if the input is the empty word λ. The transitions of the latter case have been added by the normalization: the transition is then the only transition leaving that state. The output of the transition in the transducer becomes then the input in the weighted automaton. The last problem is that this new input is either a symbol or the empty word λ. The last step consists in removing these λ-transitions. This is done as for usual automata by replacing a path of λ-transitions followed by a symbol by only one transition. This transformation must be carried out by preserving weights of runs. The transitions and their associated weights are put into two $Q \times Q$-matrices M and E as follows. Transitions labeled by a symbol are put in M while λ-transitions are put in E. The entries of M are sum of weighted symbols while entries of E are just weights. The transitions are thus represented by $M + E\lambda$. They are replaced by the transitions represented by E^*M where E^* is of course the matrix $\sum_{n \geqslant 0} E^n$. Note that this matrix E^* can be effectively computed since it is the solution of the equation $X = EX + I$ where I is the identity matrix.

6   Conclusion

The first result of the paper provides a weighted automaton which gives the limiting frequency of each block in the output of a normal input. This automaton can be used to check another property of this invariant. It can be decided, for instance, whether this measure is a Bernoulli measure. This boils down to checking whether the minimal automaton has only a single state.

In this work, it is assumed that the input of the transducer is normal, that is generic for the uniform measure. It seems that the results can be extended to the more general setting of Markovian measures. The case of hidden Markovian measure, that is, measures computed by weighted automata, seems however more involved [9].

Acknowledgments. The author would like to thank Verónica Becher for many fruitful discussions and suggestions. The author is a member of the IRP SINFIN, CONICET/Universidad de Buenos Aires–CNRS/Université de Paris and he is supported by the ECOS project PA17C04. The author is also partially funded by the DeLTA project (ANR-16-CE40-0007).

References

1. Agafonov, V.N.: Normal sequences and finite automata. Soviet Math. Doklady **9**, 324–325 (1968)
2. Becher, V., Carton, O.: Normal numbers and computer science. In: Berthé, V., Rigo, M. (eds.) Sequences, Groups, and Number Theory. TM, pp. 233–269. Springer, Cham (2018). https://doi.org/10.1007/978-3-319-69152-7_7

3. Émile Borel, M.: Les probabilités dénombrables et leurs applications arithmétiques. Rendiconti del Circolo Matematico di Palermo (1884-1940) **27**(1), 247–271 (1909). https://doi.org/10.1007/BF03019651
4. Brémaud, P.: Markov Chains: Gibbs Fields, Monte Carlo Simulation, and Queues. Springer, Cham (2008). https://doi.org/10.1007/978-3-030-45982-6
5. Bugeaud, Y.: Distribution Modulo One and Diophantine Approximation, Cambridge Tracts in Mathematics, vol. 193. Cambridge University Press, Cambridge (2012)
6. Carton, O., Orduna, E.: Preservation of normality by transducers. Inf. Comput. **282**, 104650 (2022)
7. Champernowne, D.G.: The construction of decimals normal in the scale of ten. J. London Math. Soc. **1**(4), 254–260 (1933)
8. Choffrut, C., Grigorieff, S.: Uniformization of rational relations. In: Karhumäki, J., Maurer, H.A., Paun, G., Rozenberg, G. (eds.) Jewels are Forever, Contributions on Theoretical Computer Science in Honor of Arto Salomaa, pp. 59–71. Springer, Heidelberg (1999). https://doi.org/10.1007/978-3-642-60207-8_6
9. Hansel, G., Perrin, D.: Mesures de probabilité rationnelles. In: Lothaire, M. (ed.) Mots, pp. 335–357. Hermes, Paris (1990)
10. Lind, D., Marcus, B.: An Introduction to Symbolic Dynamics and Coding. Cambridge University Press, Cambridge (1995)
11. Nies, A.: Computability and Randomness. Oxford University Press, Oxford (2009)
12. Perrin, D., Pin, J.É.: Infinite Words. Elsevier, Amsterdam (2004)
13. Sakarovitch, J.: Elements of Automata Theory. Cambridge University Press, Cambridge (2009)

A Full Characterization of Bertrand Numeration Systems

Émilie Charlier(ID), Célia Cisternino(ID), and Manon Stipulanti$^{(\boxtimes)}$(ID)

Department of Mathematics, University of Liège, Liège, Belgium
{echarlier,ccisternino,m.stipulanti}@uliege.be

Abstract. Among all positional numeration systems, the widely studied Bertrand numeration systems are defined by a simple criterion in terms of their numeration languages. In 1989, Bertrand-Mathis characterized them via representations in a real base β. However, the given condition turns out to be not necessary. Hence, the goal of this paper is to provide a correction of Bertrand-Mathis' result. The main difference arises when β is a Parry number, in which case two associated Bertrand numeration systems are derived. Along the way, we define a non-canonical β-shift and study its properties analogously to those of the usual canonical one.

Keywords: Numeration systems · Bertrand condition · Real bases expansion · Dominant root · Parry numbers · Subshifts

1 Introduction

In 1957, Rényi [14] introduced representations of real numbers in a real base $\beta > 1$. A *β-representation* of a nonnegative real number x is an infinite sequence $a_1 a_2 \cdots$ over $\mathbb{N}$ such that $x = \sum_{i=1}^{\infty} \frac{a_i}{\beta^i}$. The most commonly used algorithm in order to obtain such digits a_i is the greedy algorithm. The corresponding distinguished β-representation of a given $x \in [0,1]$ is called the *β-expansion* of x and is obtained as follows: set $r_0 = x$ and for all $i \geq 1$, let $t_i = \lfloor \beta r_{i-1} \rfloor$ and $r_i = \beta r_{i-1} - t_i$. The β-expansion of x is the infinite word $d_\beta(x) = t_1 t_2 \cdots$ written over the alphabet $\{0, \ldots, \lfloor \beta \rfloor\}$. In this theory, the β-expansion of 1 and the *quasi-greedy β-expansion* of 1 given by $d_\beta^*(1) = \lim_{x \to 1^-} d_\beta(x)$ play crucial roles, as well as the *β-shift*

$$S_\beta = \{w \in \{0, \ldots, \lceil \beta \rceil - 1\}^{\mathbb{N}} : \forall i \geq 0, \ \sigma^i(w) \leq_{\text{lex}} d_\beta^*(1)\}$$

where $\sigma(w_1 w_2 \cdots)$ denotes the shifted word $w_2 w_3 \cdots$. Parry [12] showed that the β-shift S_β is the topological closure (w.r.t. the prefix distance) of the set of infinite words that are the β-expansions of some real number in $[0,1)$ and Bertrand-Mathis [1] characterized the real bases β for which S_β is sofic, i.e., its

Émilie Charlier, Célia Cisternino and Manon Stipulanti are supported by the FNRS grants J.0034.22, 1.A.564.19F and 1.B.397.20F respectively.

V. Diekert and M. Volkov (Eds.): DLT 2022, LNCS 13257, pp. 102–114, 2022.
https://doi.org/10.1007/978-3-031-05578-2_8

factors form a language that is accepted by a finite automaton. Expansions in a real base are extensively studied under various points of view and we can only cite a few of the many possible references [1, 5, 10, 12, 15].

In parallel, other numeration systems are also widely studied, this time to represent nonnegative integers. A *positional numeration system* is given by an increasing integer sequence $U = (U(i))_{i \geq 0}$ such that $U(0) = 1$ and the quotients $\frac{U(i+1)}{U(i)}$ are bounded. The *greedy U-representation* of $n \in \mathbb{N}$, denoted $\mathrm{rep}_U(n)$, is the unique word $a_1 \cdots a_\ell$ over $\mathbb{N}$ such that $n = \sum_{i=1}^{\ell} a_i U(\ell - i)$, $a_1 \neq 0$ and for all $j \in \{1, \ldots, \ell\}$, $\sum_{i=j}^{\ell} a_i U(\ell - i) < U(\ell - j + 1)$. These representations are written over the finite alphabet $A_U = \{0, \ldots, \sup_{i \geq 0} \lceil \frac{U(i+1)}{U(i)} \rceil - 1\}$. The *numeration language* is the set $\mathcal{N}_U = 0^* \mathrm{rep}_U(\mathbb{N})$. Similarly, the literature about positional numeration systems is vast; see [2–4, 9, 11, 13, 16] for the most topic-related ones.

There exists an intimate link between β-expansions and greedy U-representations. Its study goes back to the work [2] of Bertrand-Mathis. A positional numeration system U is called *Bertrand* if the corresponding numeration language $\mathcal{N}_U$ is both prefix-closed and *prolongable*, i.e., if for all words w in $\mathcal{N}_U$, the word $w0$ also belongs to $\mathcal{N}_U$. These two conditions can be summarized as

$$\forall w \in A_U^*, \ w \in \mathcal{N}_U \iff w0 \in \mathcal{N}_U. \tag{1}$$

The usual integer base numeration systems are Bertrand, as well the Zeckendorf numeration system [19]. This form of the definition of Bertrand numeration systems, as well as their names after Bertrand-Mathis, was first given in [3], and then used in [4, 11, 13, 17]. Bertrand numeration systems were also reconsidered in [9]. Moreover, the normalization in base $\beta > 1$ in [3, 7] deals with these Bertrand numeration systems.

In [2], Bertrand-Mathis stated that a positional numeration system U is Bertrand if and only if there exists a real number $\beta > 1$ such that $\mathcal{N}_U = \mathrm{Fac}(S_\beta)$. In this case, $A_U = \{0, \ldots, \lceil \beta \rceil - 1\}$ and for all $i \geq 0$,

$$U(i) = d_1 U(i-1) + d_2 U(i-2) + \cdots + d_i U(0) + 1 \tag{2}$$

where $(d_i)_{i \geq 1} = d_\beta^*(1)$. This result has been widely used, see for example [3, 4, 10]. Note that the condition stated above is trivially sufficient. However, it is *not necessary* (see Sect. 3). The mistake that occurs in the proof of [2] is a confusion between $d_\beta^*(1)$ and $d_\beta(1)$ while describing the set $\mathrm{Fac}(S_\beta)$ (which corresponds to $L(\theta)$ in the notation of [2]). This mistake is then repeated in [10, Theorem 7.3.8]. Therefore, in this work, we propose a correction of this famous theorem by fully characterizing Bertrand numeration systems.

The authors of [11, 17] distinguish what they call Parry numeration systems (which will be our *canonical* Bertrand systems associated with a Parry number) among general Bertrand numeration systems. In fact, the only possible Bertrand systems with a regular numeration language that are not Parry (in their sense) are very specific and they will be clearly identified within our characterization.

The paper is organized as follows. We first fix some notation in Sect. 2. In Sect. 3, we illustrate the fact that the Bertrand-Mathis theorem stated above does not fully characterize Bertrand numeration systems and we correct it. Then, in Sect. 4, we investigate Bertand numeration systems based on a sequence that satisfies a linear recurrence relation. In Sect. 5, we obtain a second characterization of Bertrand numeration systems in terms of the lexicographically greatest words of each length in $\mathcal{N}_U$. This provides a refinement of a result of Hollander [8]. Finally, seeing the importance of the newly introduced non-canonical β-shift, we study its main properties in Sect. 6.

2   Basic Notation

We make use of common notions in formal language theory, such as alphabet, letter, word, length of a word, prefix distance, convergence of words, language, code and automaton [10]. In particular, the length of a finite word w is denoted by $|w|$. The notation w^ω means an infinite repetition of the finite word w. The set of factors of a word w is written $\mathrm{Fac}(w)$ and the set of factors of words in a set L is written $\mathrm{Fac}(L)$. Given a finite word w and $n \in \{1, \ldots, |w|\}$, the *prefix* and *suffix* of length n of w are respectively written $\mathrm{Pref}_n(w)$ and $\mathrm{Suff}_n(w)$. Similarly, for an infinite word w and $n \geq 0$, we let $\mathrm{Pref}_n(w)$ denote the prefix of length n of w. If $(A, <)$ is a totally ordered alphabet, then $\leq_{\mathrm{lex}}$ denotes the usual induced lexicographic order on both A^* and $A^\mathbb{N}$.

3   Characterization of Bertrand Numeration Systems

The goal of this section is to give a full characterization of Bertrand numeration systems defined by (1). In doing so, we correct the result of Bertand-Mathis stated in the introduction.

First, we note that both implications in (1) are relevant. This observation is illustrated in the following example.

Example 1. Consider the numeration system U defined by $(U(0), U(1)) = (1, 3)$ and $U(i) = U(i-1) + U(i-2)$ for all $i \geq 2$. It is not Bertrand as its numeration language is not prolongable: for instance, $2 \in \mathcal{N}_U$ but $20 \notin \mathcal{N}_U$.

Now, consider U defined by $(U(0), U(1)) = (1, 2)$ and $U(i) = 5U(i-1) + U(i-2)$ for all $i \geq 2$. It is not Bertrand since the corresponding language $\mathcal{N}_U$ is not prefix-closed. Indeed, $50 \in \mathrm{rep}_U(\mathbb{N})$ but $5 \notin \mathrm{rep}_U(\mathbb{N})$.

Then, let us show that the condition given in the original Bertrand-Mathis result characterizing the Bertrand numeration systems is *not necessary*.

Example 2. Let U be the positional numeration system defined by $U(0) = 1$ and $U(i) = 3U(i-1) + 1$ for all $i \geq 1$. This example was already considered in [11]. It is easy to see that $\mathcal{N}_U = \{0, 1, 2\}^* \cup \{0, 1, 2\}^* 30^*$. The minimal automaton of this language is depicted in Fig. 2b. Therefore, U is Bertrand. However, for all $\beta > 1$, we have $\mathcal{N}_U \neq \mathrm{Fac}(S_\beta)$, in contradiction to the result from [2] (which has been transcribed in the introduction). This can be seen by observing that for all

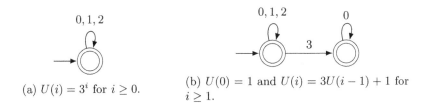

(a) $U(i) = 3^i$ for $i \geq 0$.

(b) $U(0) = 1$ and $U(i) = 3U(i-1) + 1$ for $i \geq 1$.

Fig. 1. The minimal automata of the languages $\mathcal{N}_U$ where U are respectively the canonical and non-canonical Bertrand numeration systems associated with 3.

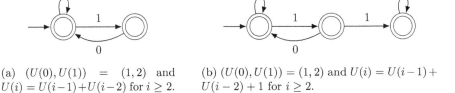

(a) $(U(0), U(1)) = (1, 2)$ and $U(i) = U(i-1) + U(i-2)$ for $i \geq 2$.

(b) $(U(0), U(1)) = (1, 2)$ and $U(i) = U(i-1) + U(i-2) + 1$ for $i \geq 2$.

Fig. 2. The minimal automata of the languages $\mathcal{N}_U$ where U are respectively the canonical and non-canonical Bertrand numeration systems associated with $\frac{1+\sqrt{5}}{2}$.

$i \geq 1$, the lexicographically maximal word of length i in $\mathcal{N}_U$ is 30^{i-1} while in $\mathrm{Fac}(S_\beta)$, this word is $\mathrm{Pref}_i(d_\beta^*(1))$. But we know that $d_\beta^*(1)$ never ends with a tail of zeroes. Also see [11, Lemma 2.5]. The sequence U satisfies (2) with $(d_i)_{i \geq 1}$ not equal to $d_3^*(1) = 2^\omega$ as prescribed in [2] but equal to $d_3(1) = 30^\omega$ instead.

Another example is the following one. We consider the positional numeration system U defined by $(U(0), U(1)) = (1, 2)$ and $U(i) = U(i-1) + U(i-2) + 1$ for all $i \geq 2$. This system is Bertrand since the corresponding numeration language is $\mathcal{N}_U = \{0, 10\}^* \cup \{0, 10\}^*1 \cup \{0, 10\}^*110^*$. The minimal automaton of this language is depicted in Fig. 2b. Similarly as in the previous paragraph, we see that $\mathcal{N}_U \neq \mathrm{Fac}(S_\beta)$ for all $\beta > 1$. The sequence U satisfies (2) with $(d_i)_{i \geq 1}$ equal to $d_\varphi(1) = 110^\omega$ where φ is the golden ratio $\frac{1+\sqrt{5}}{2}$.

We will show that, up to a single exception, the only possible Bertrand numeration systems are given by the recurrence relation (2) where the sequence of coefficients $(d_i)_{i \geq 1}$ is either equal to $d_\beta^*(1)$ or to $d_\beta(1)$, as is the case of the previous two systems. Before proving our characterization of Bertrand numeration systems, we need some technical results.

Lemma 1 ([10, Proposition 7.3.6]). *The language $\mathcal{N}_U$ of a positional numeration system U is equal to $\{a \in A_U^* : \forall i \leq |a|,\ \mathrm{Suff}_i(a) \leq_{\mathrm{lex}} \mathrm{rep}_U(U(i) - 1)\}$.*

Lemma 2. *The numeration language $\mathcal{N}_U$ of a Bertrand numeration system U is factorial, that is, $\mathrm{Fac}(\mathcal{N}_U) = \mathcal{N}_U$.*

Proof. The fact that $\mathcal{N}_U$ is prefix-closed comes from the definition of a Bertrand numeration system. Since any positional numeration system has a suffix-closed numeration language, the conclusion follows.

Lemma 3. *A positional numeration system U is Bertrand if and only if there exists an infinite word a over A_U such that $\mathrm{rep}_U(U(i) - 1) = \mathrm{Pref}_i(a)$ for all $i \geq 0$. In this case, we have $\sigma^i(a) \leq_{\mathrm{lex}} a$ for all $i \geq 0$.*

Proof. In order to get the necessary condition, it suffices to show that if U is a Bertrand numeration system then for all $i \geq 1$, $\mathrm{rep}_U(U(i) - 1)$ is a prefix of $\mathrm{rep}_U(U(i + 1) - 1)$. Let thus $i \geq 1$, and write $\mathrm{rep}_U(U(i) - 1) = a_1 \cdots a_i$ and $\mathrm{rep}_U(U(i + 1) - 1) = b_1 \cdots b_{i+1}$. On the one hand, since $b_1 \cdots b_i \in \mathcal{N}_U$, we get $b_1 \cdots b_i \leq_{\mathrm{lex}} a_1 \cdots a_i$. On the other hand, since $a_1 \cdots a_i 0 \in \mathcal{N}_U$, we get $a_1 \cdots a_i 0 \leq_{\mathrm{lex}} b_1 \cdots b_{i+1}$, hence $a_1 \cdots a_i \leq_{\mathrm{lex}} b_1 \cdots b_i$.

Conversely, suppose that there exists an infinite word a over A_U such that $\mathrm{rep}_U(U(i) - 1) = \mathrm{Pref}_i(a)$ for all $i \geq 0$. It is easily seen that for all $w \in A_U^*$ and all $i \in \{0, \ldots, |w|\}$, we have $\mathrm{Suff}_i(w) \leq_{\mathrm{lex}} \mathrm{Pref}_i(a)$ if and only if $\mathrm{Suff}_{i+1}(w0) \leq_{\mathrm{lex}} \mathrm{Pref}_{i+1}(a)$. Then we get that U is Bertrand by Lemma 1.

We now turn to the last part of the statement and we prove that $\sigma^i(a) \leq_{\mathrm{lex}} a$ for all $i \geq 0$. Suppose to the contrary that there exists $i \geq 0$ such that $\sigma^i(a) >_{\mathrm{lex}} a$. Then there exists $\ell \geq 1$ such that $a_i \cdots a_{i+\ell-1} >_{\mathrm{lex}} a_1 \cdots a_\ell$, where $a = a_1 a_2 \cdots$. This is impossible since $a_i \cdots a_{i+\ell-1} \in \mathcal{N}_U$ by Lemma 2.

Lemma 4. *Let a be an infinite word over $\mathbb{N}$ such that $\sigma^i(a) \leq_{\mathrm{lex}} a$ for all $i \geq 0$. If a is not periodic, then we define $d = a$; otherwise we let $n \geq 1$ be the smallest integer such that $a = (a_1 \cdots a_n)^\omega$ and we define $d = a_1 \cdots a_{n-1}(a_n + 1)0^\omega$. Then in both cases, we have $\sigma^i(d) <_{\mathrm{lex}} d$ for all $i \geq 1$.*

Proof. The case where a is not periodic is straightforward. Suppose that a is periodic. If $i \geq n$, then $\sigma^i(d) = 0^\omega <_{\mathrm{lex}} d$. For i with $1 \leq i \leq n - 1$, proceed by contradiction and suppose that $\sigma^i(d) \geq_{\mathrm{lex}} d$, that is, $a_{i+1} \cdots a_{n-1}(a_n + 1)0^\omega \geq_{\mathrm{lex}} a_1 a_2 \cdots a_{n-1}(a_n + 1)0^\omega$. Then $a_{i+1} \cdots a_{n-1}(a_n + 1) >_{\mathrm{lex}} a_1 \cdots a_{n-i}$. By hypothesis on a, we also have $a_{i+1} \cdots a_{n-1} a_n \leq_{\mathrm{lex}} a_1 \cdots a_{n-i}$. Thus, we get $a_{i+1} \cdots a_{n-1} a_n = a_1 \cdots a_{n-i}$. Moreover, by assumption on a, we have $\sigma^n(a) = a \geq_{\mathrm{lex}} \sigma^{n-i}(a)$. We then obtain that

$$\sigma^i(a) = a_{i+1} \cdots a_n \sigma^n(a) \geq_{\mathrm{lex}} a_1 \cdots a_{n-i} \sigma^{n-i}(a) = a.$$

Since $\sigma^i(a) \leq_{\mathrm{lex}} a$ by hypothesis, we get $\sigma^i(a) = a$, which is impossible since $i < n$ and n was chosen to be minimal for this property.

Lemma 5. *Let a be an infinite word over $\mathbb{N}$. We have $\sigma^i(a) \leq a$ for all $i \geq 0$ if and only if either $a = 0^\omega$, $a = 10^\omega$, $a = d_\beta^*(1)$ for some $\beta > 1$ or $a = d_\beta(1)$ for some $\beta > 1$.*

Proof. The sufficient condition follows from [12] (also see [10, Theorem 7.2.9 and Corollary 7.2.10]). Now, suppose that $\sigma^i(a) \leq a$ for all $i \geq 0$ and that $a \neq 0^\omega$ and $a \neq 10^\omega$. Let d be the sequence defined from a as in Lemma 4. Then $\sigma^i(d) <_{\mathrm{lex}} d$

for all $i \geq 1$. In particular, we get $d_i \leq d_1$ for all $i \geq 1$. Moreover, we have $d_1 \geq 1$ and $d \neq 10^\omega$ (for otherwise a would be equal to either 0^ω or 10^ω). Then there exists a unique $\beta > 1$ such that $d = d_\beta(1)$; see [12] or [10, Corollary 7.2.10]. Also, we know that $d_\beta^*(1) = (t_1 \cdots t_{n-1}(t_n - 1))^\omega$ whenever $d_\beta(1) = t_1 \cdots t_n 0^\omega$ with $n \geq 1$ and $t_n \neq 0$, and that $d_\beta^*(1) = d_\beta(1)$ otherwise; again, see [10,12]. We get that either $a = d_\beta^*(1)$ or $a = d_\beta(1)$ depending on the periodicity of a.

Finally, we recall the so-called Renewal theorem as stated in [6, Theorem 1 on p. 330]; also see [18, Theorem 0.18].

Theorem 1 (Renewal theorem). *Let $(c_n)_{n\geq 1}$ and $(d_n)_{n\geq 0}$ be sequences of nonnegative real numbers with $c_n \leq 1$ for all $n \geq 1$. Suppose the greatest common divisor of all integers n with $c_n > 0$ is 1. Let $(u_n)_{n\geq 0}$ be the sequence defined by the recurrence relation $u_n = d_n + c_1 u_{n-1} + \cdots + c_n u_0$ for all $n \geq 0$. If $\sum_{n=1}^\infty c_n = 1$ and $\sum_{n=0}^\infty d_n < \infty$ then $\lim_{n\to\infty} u_n = (\sum_{n=0}^\infty d_n)(\sum_{n=1}^\infty n c_n)^{-1}$ where this is interpreted as zero if $\sum_{n=1}^\infty n c_n = \infty$.*

For a real number $\beta > 1$, we define

$$S_\beta' = \{w \in \{0, \ldots, \lfloor \beta \rfloor\}^\mathbb{N} : \forall i \geq 0, \ \sigma^i(w) \leq_{\text{lex}} d_\beta(1)\}.$$

We are now ready to show the claimed correction of Bertrand-Mathis' result.

Theorem 2. *A positional numeration system U is Bertrand if and only if one of the following occurs.*

1. *For all $i \geq 0$, $U(i) = i + 1$.*
2. *There exists a real number $\beta > 1$ such that $\mathcal{N}_U = \text{Fac}(S_\beta)$.*
3. *There exists a real number $\beta > 1$ such that $\mathcal{N}_U = \text{Fac}(S_\beta')$.*

Moreover, in Case 2 (resp. Case 3), the following hold:

a. *There is a unique such β.*
b. *The alphabet A_U equals $\{0, \ldots, \lceil \beta \rceil - 1\}$ (resp. $\{0, \ldots, \lfloor \beta \rfloor\}$).*
c. *We have*

$$U(i) = a_1 U(i-1) + a_2 U(i-2) + \cdots + a_i U(0) + 1 \tag{3}$$

for all $i \geq 0$ and

$$\lim_{i\to\infty} \frac{U(i)}{\beta^i} = \frac{\beta}{(\beta - 1)\sum_{i=1}^\infty i a_i \beta^{-i}} \tag{4}$$

where $(a_i)_{i\geq 1}$ is $d_\beta^*(1)$ (resp. $d_\beta(1)$).
d. *The system U has the dominant root β, i.e., $\lim_{i\to\infty} \frac{U(i+1)}{U(i)} = \beta$.*

Proof. Let U be a positional numeration system. We start with the backward direction. If $U(i) = i+1$ for all $i \geq 0$, then $\mathcal{N}_U = 0^* \cup 0^* 10^*$, hence U is Bertrand. Otherwise, for the sake of clarity, write $S = \{w \in \mathbb{N}^\mathbb{N} : \forall i \geq 0, \ \sigma^i(w) \leq_{\text{lex}} a\}$ with $a = d_\beta^*(1)$ or $a = d_\beta(1)$ as in the statement. Suppose that $\mathcal{N}_U = \text{Fac}(S)$.

We show that U is Bertrand. Consider $y \in \mathcal{N}_U$. There exist words $x \in \mathbb{N}^*$ and $z \in \mathbb{N}^{\mathbb{N}}$ such that $xyz \in S$. Since $\sigma^i(xy0^\omega) \leq_{\text{lex}} \sigma^i(xyz)$ for all $i \geq 0$, we get that $xy0^\omega \in S$. Therefore $y0 \in \mathcal{N}_U$. The converse is immediate since if $y0 \in \text{Fac}(S)$ then $y \in \text{Fac}(S)$ as well.

Conversely, suppose that U is Bertrand. By Lemma 3, there exists $a = a_1 a_2 \cdots$ such that $\text{rep}_U(U(i) - 1) = \text{Pref}_i(a)$ and $\sigma^i(a) \leq_{\text{lex}} a$ for all $i \geq 0$. In particular, we have $a_1 \geq 1$. If $a = 10^\omega$ then $U(i) = i + 1$ for all $i \geq 0$. Otherwise, by Lemma 5, either $a = d_\beta^*(1)$ for some $\beta > 1$ or $a = d_\beta(1)$ for some $\beta > 1$. Let us show that $\mathcal{N}_U = \text{Fac}(\{w \in \mathbb{N}^{\mathbb{N}} : \forall i \geq 0, \ \sigma^i(w) \leq_{\text{lex}} a\})$. Consider $y \in \mathcal{N}_U$. By Lemma 1, we have $\text{Suff}_i(y) \leq_{\text{lex}} \text{Pref}_i(a)$ for all $i \leq |y|$. Therefore, $\sigma^i(y0^\omega) \leq_{\text{lex}} a$ for all $i \geq 0$. Conversely, suppose that y is a factor of an infinite word w over $\mathbb{N}$ such that $\sigma^i(w) \leq_{\text{lex}} a$ for all $i \geq 0$. Then $\text{Suff}_i(y) \leq_{\text{lex}} \text{Pref}_i(a)$ for all $i \geq 0$. By Lemma 1, we get $y \in \mathcal{N}_U$.

To end the proof, we note that $A_U = \{0, \ldots, \lceil \beta \rceil - 1\}$ if $a = d_\beta^*(1)$ and $A_U = \{0, \ldots, \lfloor \beta \rfloor\}$ if $a = d_\beta(1)$. Moreover, since $\text{rep}_U(U(i) - 1) = a_1 \cdots a_i$ for all $i \geq 0$, we get that the recurrence relation (3) holds for all $i \geq 0$. The computation of the limit from (4) then follows from Theorem 1 with $c_i = a_i \beta^{-i}$, $d_i = \beta^{-i}$ and $u_i = U(i)\beta^{-i}$. This in turn implies that $\lim_{i \to \infty} \frac{U(i+1)}{U(i)} = \beta$.

Note that in the previous statement, the second item coincides with the condition given in the original theorem of Bertrand-Mathis [2]. The main difference between these two results is that there exist two Bertrand numeration systems associated with a *simple Parry number* $\beta > 1$, i.e., such that $d_\beta(1)$ ends with infinitely many zeroes. To distinguish them, we call *canonical* the Bertrand numeration system defined by (3) when $a = d_\beta^*(1)$, and *non-canonical* that for which $a = d_\beta(1)$. For instance, the canonical Bertrand numeration system associated with the golden ratio $\frac{1+\sqrt{5}}{2}$ is the well-known Zeckendorf numeration system $U = (1, 2, 3, 5, 8, \ldots)$ defined by $(U(0), U(1)) = (1, 2)$ and $U(i) = U(i - 1) + U(i - 2)$ for all $i \geq 2$ [19]. The associated non-canonical Bertrand numeration system is the numeration system $U = (1, 2, 4, 7, 12, \ldots)$ from Example 2 defined by $(U(0), U(1)) = (1, 2)$ and $U(i) = U(i-1) + U(i-2) + 1$ for all $i \geq 2$. See Fig. 2 for automata recognizing the corresponding numeration languages. In Figs. 1a and 1b, we see the canonical and non-canonical Bertrand numeration systems associated with the integer base 3.

4   Linear Bertrand Numeration Systems

In the following proposition, we study the linear recurrence relations satisfied by Bertrand numeration systems associated with a *Parry number* β, i.e., a real number $\beta > 1$ such that $d_\beta(1)$ is ultimately periodic. As is usual, if an expansion ends with a tail of zeroes, we often omit to write it down.

Proposition 1. *Let U be a Bertrand numeration system.*

1. *If $\mathcal{N}_U = \text{Fac}(S_\beta)$ where $\beta > 1$ is such that $d_\beta^*(1) = d_1 \cdots d_m (d_{m+1} \cdots d_{m+n})^\omega$ with $m \geq 0$ and $n \geq 1$, then U satisfies the linear recurrence relation of characteristic polynomial $(X^{m+n} - \sum_{j=1}^{m+n} d_j X^{m+n-j}) - (X^m - \sum_{j=1}^{m} d_j X^{m-j})$.*

2. If $\mathcal{N}_U = \text{Fac}(S'_\beta)$ where $\beta > 1$ is such that $d_\beta(1) = t_1 \cdots t_n$ with $n \geq 1$ and $t_n \geq 1$, then U satisfies the linear recurrence relation of characteristic polynomial $(X^{n+1} - \sum_{j=1}^n t_j X^{n+1-j}) - (X^n - \sum_{j=1}^n t_j X^{n-j})$, that is, the polynomial $(X - 1)(X^n - \sum_{j=1}^n t_j X^{n-j})$.

Proof. We only prove the first item as the second is similar. Thus, we suppose that $\mathcal{N}_U = \text{Fac}(S_\beta)$ where $\beta > 1$ is such that $d_\beta^*(1) = d_1 \cdots d_m (d_{m+1} \cdots d_{m+n})^\omega$ with $m \geq 0$ and $n \geq 1$. By Theorem 2, we get that

$$U(i) - U(i - n) = \sum_{j=1}^i d_j U(i - j) + 1 - \sum_{j=1}^{i-n} d_j U(i - n - j) - 1$$

$$= \sum_{j=1}^{m+n} d_j U(i - j) - \sum_{j=1}^m d_j U(i - n - j)$$

for all $i \geq m + n$.

In the following corollary, we emphasize the simple form of the characteristic polynomial in the first item of Proposition 1 when β is simple Parry number: the coefficients can be obtained directly from the digits of $d_\beta(1)$.

Corollary 1. *Let U be a Bertrand numeration system such that $\mathcal{N}_U = \text{Fac}(S_\beta)$ where $\beta > 1$ is such that $d_\beta(1) = t_1 \cdots t_n$ with $n \geq 1$. Then U satisfies the linear recurrence relation of characteristic polynomial $X^n - \sum_{j=1}^n t_j X^{n-j}$.*

5   Lexicographically Greastest Words of Each Length

A key argument in the proof of Theorem 2 was the study of the lexicographically greatest words of each length; we see this in Lemmas 1 and 3. In this section, we investigate more properties of these words, which will allow us to obtain yet another characterization of Bertrand numeration systems.

In order to study the regularity of the numeration language of positional systems having a dominant root, Hollander proved the following result.

Proposition 2 ([8, Lemmas 4.2 and 4.3]). *Let U be a positional numeration system having a dominant root $\beta > 1$. If β is not a simple Parry number, then $\lim_{i \to \infty} \text{rep}_U(U(i) - 1) = d_\beta(1)$. Otherwise, $d_\beta(1) = t_1 \cdots t_n$ with $t_n \neq 0$ and for all $k \geq 0$, define $w_k = (t_1 \cdots t_{n-1}(t_n - 1))^k t_1 \cdots t_n$. Then for all $\ell \geq 0$, there exists $I \geq 0$ such that for all $i \geq I$, there exists $k \geq 0$ such that $\text{Pref}_\ell(\text{rep}_U(U(i) - 1)) = \text{Pref}_\ell(w_k 0^\omega)$.*

Example 3. For the integer base-b numeration system $U = (b^i)_{i \geq 0}$, we have $w_k = (b - 1)^k b$ for all $k \geq 0$ and $\text{rep}_U(b^i - 1) = (b - 1)^i$ for all $i \geq 0$.

For the Zeckendorf numeration system, it can be easily seen that $\text{rep}_U(U(i) - 1) = (10)^{\frac{i}{2}}$ if i is even, and $\text{rep}_U(U(i) - 1) = (10)^{\frac{i-1}{2}} 1$ otherwise. We have $w_k = (10)^k 11$ for all $k \geq 0$. Therefore, for all $\ell \geq 0$ and all $i \geq \ell$, the words $\text{rep}_U(U(i) - 1)$ and $w_{\lfloor \ell/2 \rfloor}$ share the same prefix of length ℓ.

Let U be the system defined by $(U(0), U(1)) = (1,3)$ and for $i \geq 2$, $U(i) = 3U(i-1) - U(i-2)$. Then U has the dominant root φ^2 and $\operatorname{rep}_U(U(i)-1) = 21^{i-1}$ for all $i \geq 1$. This agrees with Proposition 2 since $d_{\varphi^2}(1) = 21^\omega$.

As illustrated in the next example, when β is a simple Parry number, Proposition 2 does not imply the convergence of the sequence $(\operatorname{rep}_U(U(i) - 1))_{i \geq 0}$.

Example 4. Consider the numeration system $U = (U(i))_{i \geq 0}$ defined by $(U(0), U(1), U(2), U(3)) = (1,2,3,5)$ and for all $i \geq 4$, $U(i) = U(i-1) + U(i-3) + U(i-4) + 1$. The sequence U satisfies the linear recurrence relation of characteristic polynomial $X^5 - 2X^4 + X^3 - X^2 + 1$, which has the golden ratio as dominant root. Hence, as for the Zeckendorf numeration system, we have $w_k = (10)^k 11$ for all $k \geq 0$. For all $i \geq 4$, we can compute $\operatorname{rep}_U(U(i) - 1) = 110^{i-2}$ if $i \equiv 0, 1 \bmod 4$, and $\operatorname{rep}_U(U(i) - 1) = 10110^{i-4}$ otherwise. Therefore, for all $i \geq 4$, $\operatorname{rep}_U(U(i)-1) = \operatorname{Pref}_i(w_0 0^\omega)$ if $i \equiv 0, 1 \bmod 4$, and $\operatorname{rep}_U(U(i) - 1) = \operatorname{Pref}_i(w_1 0^\omega)$ otherwise. Thus, the limit $\lim_{i \to \infty} \operatorname{rep}_U(U(i) - 1)$ does not exist.

In Examples 2, 3 and 4, we illustrated that the sequence $(\operatorname{rep}_U(U(i) - 1))_{i \geq 0}$ may or may not converge. In the first two, we gave examples such that its limit is either $d_\beta(1)$ or $d_\beta^*(1)$. In the third, we illustrated that even if the recurrence relation satisfied by U gives the intuition that the sequence would converge to $w_1 0^\omega$, it is not the case. In fact, seeing Proposition 2, one might think that we can provide a positional numeration system U such that $\lim_{i \to \infty} \operatorname{rep}_U(U(i) - 1) = w_k 0^\omega$ with $k \geq 1$. We show that this cannot happen, which can be thought as a refinement of Proposition 2.

Proposition 3. *Let U be a positional numeration system with a dominant root $\beta > 1$. If the limit $\lim_{i \to \infty} \operatorname{rep}_U(U(i) - 1)$ exists, then it is either $d_\beta^*(1)$ or $d_\beta(1)$.*

Proof. If $d_\beta(1)$ is infinite, then the result follows from Proposition 2. Let us consider the case where $d_\beta(1) = t_1 \cdots t_n$ with $t_n \neq 0$. Proceed by contradiction and suppose that there exists $k \geq 1$ such that $\lim_{i \to \infty} \operatorname{rep}_U(U(i) - 1) = w_k 0^\omega$. For all i large enough, $t_1 \cdots t_{n-1}(t_n - 1)$ is a prefix of $\operatorname{rep}_U(U(i) - 1)$, hence the greedy algorithm implies that $\sum_{j=1}^n t_j U(i-j) > U(i) - 1$. On the other hand, for all i large enough, $t_1 \cdots t_n$ is a factor occurring at position $kn + 1$ in $\operatorname{rep}_U(U(i + kn) - 1)$, hence, again from the greedy algorithm, we get $U(i) > \sum_{j=1}^n t_j U(i-j)$. By putting the inequalities altogether, we obtain a contradiction.

Thanks to this result, we obtain another characterization of Bertrand numeration systems.

Theorem 3. *A positional numeration system U is Bertrand if and only if one of the following conditions is satisfied.*

1. *We have $\operatorname{rep}_U(U(i) - 1) = \operatorname{Pref}_i(10^\omega)$ for all $i \geq 0$.*
2. *There exists a real number $\beta > 1$ such that $\operatorname{rep}_U(U(i) - 1) = \operatorname{Pref}_i(d_\beta^*(1))$ for all $i \geq 0$.*

3. *There exists a real number $\beta > 1$ such that $\mathrm{rep}_U(U(i) - 1) = \mathrm{Pref}_i(d_\beta(1))$ for all $i \geq 0$.*

Proof. All three conditions are sufficient by Lemma 3. Conversely, suppose that U is a Bertrand numeration system. In Case 1 of Theorem 2, we have $\mathrm{rep}_U(U(i) - 1) = \mathrm{rep}_U(i) = 10^{i-1}$ for all $i \geq 1$. Otherwise, U has a dominant root $\beta > 1$ by Theorem 2. The result then follows from Lemma 3 combined with Proposition 3.

We note that the three cases of Theorem 3 indeed match those of Theorem 2.

6   The Non-canonical β-shift

In view of their definitions, the sets S_β and S'_β are both *subshifts* of $A^{\mathbb{N}}$, i.e., they are shift-invariant and closed w.r.t the topology induced by the prefix distance. These subshifts coincide unless β is a simple Parry number. Therefore, in the specific case where β is a simple Parry number, by analogy to the name β-*shift* commonly used for S_β, we call the set S'_β the *non-canonical β-shift*. In this section, we see whether or not the classical properties of S_β still hold for S'_β.

The following proposition is the analogue of [10, Theorem 7.2.13] that characterizes sofic (canonical) β-shifts, i.e., such that $\mathrm{Fac}(S_\beta)$ is accepted by a finite automaton.

Proposition 4. *A real number $\beta > 1$ is a Parry number if and only if the subshift S'_β is sofic.*

Proof (Sketch). If β is not a simple Parry number, then $S_\beta = S'_\beta$ and the conclusion follows by [10, Theorem 7.2.13]. Suppose that β is a simple Parry number for which $d_\beta(1) = t_1 \cdots t_n$ with $n \geq 1$ and $t_n \neq 0$. We get $d^*_\beta(1) = (t_1 \cdots t_{n-1}(t_n - 1))^\omega$. An automaton recognizing $\mathrm{Fac}(S'_\beta)$ can be constructed as a slight modification of the classical automaton recognizing $\mathrm{Fac}(S_\beta)$ given in [10, Theorem 7.2.13]: we add a new final state q', an edge from the state usually denoted q_n (that is, the state reached while reading $t_1 \cdots t_{n-1}$) to the new state q' of label t_n and a loop of label 0 on the state q'.

Example 5. The automata depicted in Figs. 1a and 1b accept $\mathrm{Fac}(S_3)$ and $\mathrm{Fac}(S'_3)$, and those of Figs. 2a and 2b accept $\mathrm{Fac}(S_\varphi)$ and $\mathrm{Fac}(S'_\varphi)$.

A subshift $S \subseteq A^{\mathbb{N}}$ is said to be of *finite type* if there exists a finite set of *forbidden factors* defining words in S, i.e., if there exists a finite set $X \subset A^*$ such that $S = \{w \in A^{\mathbb{N}} : \mathrm{Fac}(w) \cap X = \emptyset\}$. It is said to be *coded* if there exists a prefix code $Y \subset A^*$ such that $\mathrm{Fac}(S) = \mathrm{Fac}(Y^*)$. It is well known that the β-shift S_β is coded [10, Proposition 7.2.11] for any $\beta > 1$ and is of finite type whenever β is a simple Parry number [10, Theorem 7.2.15]. However, neither of these two properties is valid for the non-canonical β-shift S'_β.

Proposition 5. *For any simple Parry number β, the subshift S'_β is not of finite type.*

Proof. Suppose that β is a simple Parry number for which $d_\beta(1) = t_1 \cdots t_n$ with $n \geq 1$ and $t_n \neq 0$. We show that for all $k \geq n - 1$ and $d \in \{1, \ldots, \lfloor\beta\rfloor\}$, $t_1 \cdots t_n 0^k d$ belongs to the minimal set of forbidden factors. Let $k \geq n - 1$ and $d \in \{1, \ldots, \lfloor\beta\rfloor\}$. Clearly, $t_1 \cdots t_n 0^k d \notin \mathrm{Fac}(S'_\beta)$. Thus, in order to prove that any proper factor of $t_1 \cdots t_n 0^k d$ belongs to $\mathrm{Fac}(S'_\beta)$, it suffices to prove that both $t_1 \cdots t_n 0^k$ and $t_2 \cdots t_n 0^k d$ belong to $\mathrm{Fac}(S'_\beta)$. By [12] or [10, Corollary 7.2.10], we know that for all $j \in \{2, \ldots, n\}$, we have $t_j \cdots t_n 0^\omega <_{\mathrm{lex}} d_\beta(1)$. Thus, $t_1 \cdots t_n 0^k \in \mathrm{Fac}(S'_\beta)$ and for $j \in \{2, \ldots, n\}$, we get $t_j \cdots t_n 0^{j-1} <_{\mathrm{lex}} t_1 \cdots t_n$. We obtain that for each $j \in \{2, \ldots, n\}$, $t_j \cdots t_n 0^k d 0^\omega <_{\mathrm{lex}} d_\beta(1)$. Since moreover $0^\ell d 0^\omega \leq_{\mathrm{lex}} d_\beta(1)$ for each $\ell \in \{0, \ldots, k\}$, the conclusion follows.

In order to show that S'_β is not coded, we prove the stronger statement that it is not irreducible. A subshift S is said to be *irreducible* if for all $u, v \in \mathrm{Fac}(S)$, there exists $w \in \mathrm{Fac}(S)$ such that $uwv \in \mathrm{Fac}(S)$.

Proposition 6. *For any simple Parry number β, the subshift S'_β is not irreducible.*

Proof. Suppose that β is a simple Parry number for which $d_\beta(1) = t_1 \cdots t_n$ with $n \geq 1$ and $t_n \neq 0$. If S'_β were irreducible, then there would exist $w \in \mathrm{Fac}(S'_\beta)$ such that $t_1 \cdots t_n w t_1 \cdots t_n \in \mathrm{Fac}(S'_\beta)$, which is impossible.

The *entropy* of a subshift S of $A^{\mathbb{N}}$ can be defined as the limit of the sequence $\frac{1}{i} \log(\mathrm{Card}(\mathrm{Fac}(S) \cap A^i))$ as i tends to infinity. We refer the reader to [18, Theorem 7.13] or [10]. It is well known that the β-shift S_β has entropy $\log(\beta)$. The following proposition shows that the same property holds for S'_β.

Proposition 7. *For all real number $\beta > 1$, the subshift S'_β has entropy $\log(\beta)$.*

Proof. Let $\beta > 1$ be a real number. Let U be the Bertrand numeration system such that $\mathcal{N}_U = \mathrm{Fac}(S'_\beta)$, i.e., the numeration system defined by (3) with $(a_i)_{i \geq 1} = d_\beta(1)$. Since the number of length-i factors of S'_β is equal to $U(i)$, the entropy of S'_β is given by $\lim_{i \to \infty} \frac{1}{i} \log(U(i))$. The result now follows from (4).

We note that, mutatis mutandis, the same proof can be applied in order to show that the β-shift has entropy $\log(\beta)$.

Finally, whenever β is a Parry number, we prove a relation between the number of words of each length in the canonical and the non-canonical β-shifts.

Proposition 8. *Suppose that $\beta > 1$ is a real number such that $d_\beta(1) = t_1 \cdots t_n$ with $n \geq 1$ and $t_n \neq 0$, and let U and U' respectively be the canonical and non-canonical Bertrand numeration systems associated with β. Then $U'(i + n) = U(i + n) + U'(i)$ for all $i \geq 0$.*

Proof. Since $\mathrm{Pref}_{n-1}(d_\beta(1)) = \mathrm{Pref}_{n-1}(d^*_\beta(1))$, we have $U'(i) = U(i)$ for all $i \in \{0, \ldots, n-1\}$. Moreover, since $t_1 \cdots t_n$ is the only length-n factor of S'_β that is not present in S_β, we have $U'(n) = U(n) + 1$. Hence, the statement holds for $i = 0$ since $U(0) = U'(0) = 1$. Now we proceed by induction. Consider $i \geq 1$

and suppose that the result holds for indices less than i. By (3) of Theorem 2 and Corollary 1, we get that $U'(i+n) - U(i+n) = \sum_{j=1}^{n} t_j U'(i+n-j) + 1 - \sum_{j=1}^{n} t_j U(i+n-j) = \sum_{j=1}^{n} t_j (U'(i+n-j) - U(i+n-j)) + 1$ where $U'(i+n-j) - U(i+n-j) = 0$ if $j > i$, and by induction hypothesis, $U'(i+n-j) - U(i+n-j) = U'(i-j)$ if $j \leq i$. As a first case, assume that $i \in \{1, \ldots, n\}$. We obtain $U'(i+n) - U(i+n) = \sum_{j=1}^{i} t_j U'(i-j) + 1 = U'(i)$ where the second equality comes from Theorem 2. As a second case, assume $i \geq n$. Similarly, we get $U'(i+n) - U(i+n) = \sum_{j=1}^{n} t_j U'(i-j) + 1 = U'(i)$.

References

1. Bertrand-Mathis, A.: Développement en base θ; répartition modulo un de la suite $(x\theta^n)_{n\geq 0}$; langages codés et θ-shift. Bull. Soc. Math. France **114**(3), 271–323 (1986)
2. Bertrand-Mathis, A.: Comment écrire les nombres entiers dans une base qui n'est pas entière. Acta Math. Hungar. **54**(3–4), 237–241 (1989)
3. Bruyère, V., Hansel, G.: Bertrand numeration systems and recognizability. Theoret. Comput. Sci. **181**(1), 17–43 (1997)
4. Charlier, E., Rampersad, N., Rigo, M., Waxweiler, L.: The minimal automaton recognizing $m\mathbb{N}$ in a linear numeration system. Integers **11B**, A4, 24 (2011)
5. Dajani, K., Kraaikamp, C.: Ergodic Theory of Numbers, Carus Mathematical Monographs, vol. 29. Mathematical Association of America, Washington, DC (2002)
6. Feller, W.: An Introduction To Probability Theory and Its Applications, vol. 1, 2nd edn. John Wiley and Sons Inc.; Chapman and Hall Ltd., New York, London (1957)
7. Frougny, C., Solomyak, B.: On representation of integers in linear numeration systems. In: Ergodic theory of $\mathbb{Z}^d$ actions, London Mathematical Society Lecture Note Series, vol. 228, pp. 345–368. Cambridge University Press, Cambridge (1996)
8. Hollander, M.: Greedy numeration systems and regularity. Theory Comput. Syst. **31**(2), 111–133 (1998)
9. Loraud, N.: β-shift, systèmes de numération et automates. J. Théor. Nombres Bordeaux **7**(2), 473–498 (1995)
10. Lothaire, M.: Algebraic combinatorics on Words, Encyclopedia of Mathematics and Its Applications, vol. 90. Cambridge University Press, Cambridge (2002)
11. Massuir, A., Peltomäki, J., Rigo, M.: Automatic sequences based on Parry or Bertrand numeration systems. Adv. Appl. Math. **108**, 11–30 (2019)
12. Parry, W.: On the β-expansions of real numbers. Acta Math. Acad. Sci. Hungar. **11**, 401–416 (1960)
13. Point, F.: On decidable extensions of Presburger arithmetic: from A. Bertrand numeration systems to Pisot numbers. J. Symbolic Logic **65**(3), 1347–1374 (2000)
14. Rényi, A.: Representations for real numbers and their ergodic properties. Acta Math. Acad. Sci. Hungar. **8**, 477–493 (1957)
15. Schmidt, K.: On periodic expansions of Pisot numbers and Salem numbers. Bull. London Math. Soc. **12**(4), 269–278 (1980)
16. Shallit, J.: Numeration systems, linear recurrences, and regular sets. Inform. Comput. **113**(2), 331–347 (1994)
17. Stipulanti, M.: Convergence of Pascal-like triangles in Parry-Bertrand numeration systems. Theoret. Comput. Sci. **758**, 42–60 (2019)

18. Walters, P.: An Introduction to Ergodic Theory, Graduate Texts in Mathematics, vol. 79. Springer-Verlag, New York, Berlin (1982). https://link.springer.com/book/9780387951522
19. Zeckendorf, E.: Représentation des nombres naturels par une somme des nombres de Fibonacci ou de nombres de Lucas. Bull. Soc. Roy. Sci. Liège **41**, 179–182 (1972)

On the Decidability of Infix Inclusion Problem

Hyunjoon Cheon, Joonghyuk Hahn, and Yo-Sub Han[⊠]

Yonsei University, Seoul, Republic of Korea
{hyunjooncheon,greghahn,emmous}@yonsei.ac.kr

Abstract. We introduce the infix inclusion problem of two languages S and T that decides whether or not S is a subset of the set of all infixes of T. This problem is motivated from intrusion detection systems that identify malicious patterns according to their semantics, which are often disguised with additional information surrounding the patterns. In other words, malicious patterns are embedded as an infix of the whole patterns. We examine the infix inclusion problem, where a source S and a target T are finite, regular or context-free languages. We prove that the problem is 1) co-NP-complete when one of the languages is finite, 2) PSPACE-complete when S, T are regular, 3) EXPTIME-complete when S is context-free and T is regular and 4) undecidable for when S is either regular or context-free and T is context-free.

Keywords: Infix inclusion · Decidability · Regular languages · Context-free languages

1 Introduction

Regular expressions are an efficient tool on searching and replacing string fragments with common patterns. They are often used in several application domains such as data validation, syntax highlight and text search [9,13]. As regular expressions are getting more and more popular, there were several attempts to exploit vulnerabilities of the matching schemes that give rise to serious problems in practice. One of the most serious problems is the regular expression denial-of-service (ReDoS) problem. This problem arises because pattern matching engines run a very long time to match a malicious regular expression and inputs. In theory, the matching time should be polynomial [18,23]. Yet, most of the practical matching engines require an exponential runtime for special regular expressions (= malicious patterns) because of the backtracking feature, like backreferences [22]. An attacker can exploit a malicious pattern or an engine and cause a denial-of-service attack, making an engine to spend an unbounded amount of time. This makes a matching engine slow and unresponsive to process such regular expressions [3,12].

ReDoS occurs because of a wrongly designed, malicious pattern S. Especially, backreferences in practical patterns are the frequent causes of ReDoS behavior [4]. However, users often use the backreferences due to its flexibility and

© Springer Nature Switzerland AG 2022
V. Diekert and M. Volkov (Eds.): DLT 2022, LNCS 13257, pp. 115–126, 2022.
https://doi.org/10.1007/978-3-031-05578-2_9

expressibility. This tendency introduces a large amount of malicious patterns in use. Moreover, if S is malicious, then a pattern $T = \alpha S \beta$ is also malicious, which appears often in practice [17]. For example, $S = (([a-z])(\backslash 1)*)*$ is a known malicious pattern and the attacker may make a new malicious pattern $T = \text{abc}\ (([a-z])(\backslash 1)*)*\ \text{ETA}$ to avoid the current detection systems. A practical diagnosis for this problem is verifying the existence of known vulnerable patterns on the target as an infix [12,25]. This leads us to introduce the infix inclusion problem and investigate its decidability for different types of S and T.

Problem 1 (Infix inclusion problem). Given two languages S and T, the *infix inclusion problem* (IIP) of (S,T), denoted by IIP(S,T), is to decide whether or not S is a subset of infix(T).

The infix inclusion problem is trivial when T is given as an NFA A since infix(T) can be represented by an NFA of the same size to A—thus, the problem becomes the NFA inclusion problem, which is PSPACE-complete [10]. However, it is not the case when T is given as an acyclic DFA or a DFA because its infix language cannot be represented by an acyclic DFA or a DFA of the same size anymore [7,16,19,20]. We study the IIP with respect to regular or context-free languages, and their subfamilies together with deterministic and nondeterministic representations. Table 1 summarizes our findings on the IIP.

Table 1. Decidability of the infix inclusion problem with respect to different classes of source and target languages.

S \ T	Acyclic FA	FA	DPDA / NPDA
Acyclic FA		co-NP-complete (Theorem 6)	
FA		PSPACE-complete (Theorem 9)	Undecidable (Theorem 12)
DPDA / NPDA		EXPTIME-complete (Theorem 13)	

We define some basic notations in Sect. 2. Then, we investigate the infix inclusion problem when inputs are regular and context-free languages. We discuss IIP on finite languages in Sect. 3, and IIP on infinite languages in Sect. 4. We conclude the paper with a brief summary in Sect. 5.

2   Preliminaries

An alphabet Σ is a finite set of symbols. Given a finite string $w = w_1 w_2 \cdots w_n$ over Σ, $|w| = n$ denotes the length of w. λ is the string of length 0. A language L

is a set of strings. Given a language L, $L^0 = \{\lambda\}$, $L^1 = L$, $L^i = L^{i-1}L$ for $i \geq 2$ and $L^* = \bigcup_i L^i$. Given a set S, $|S|$ denotes the cardinality of S, $\mathcal{P}(S) = \{X \mid X \subseteq S\}$ is the power set of S. The symbol $\emptyset$ denotes an empty set such that $|\emptyset| = 0$. A string x is an infix of a string w if there exist $\alpha, \beta \in \Sigma^*$ such that $w = \alpha x \beta$. Given a language L, infix(L) denotes the set of all infixes of $w \in L$. Given two integers a and b that satisfy $a \leq b$, $[a, b]$ denotes a set $\{a, a+1, \ldots, b\}$ of integers.

A finite-state automaton (FA) A is specified by a tuple $(Q, \Sigma, \delta, I, F)$, where Q is a finite set of states, Σ is an alphabet, $\delta : Q \times \Sigma \rightarrow \mathcal{P}(Q)$ is a transition function, $I, F \subseteq Q$ are a set of initial and final states, respectively. An FA is *deterministic* if 1) $|I| = 1$ and 2) $|\delta(p, \sigma)| = 1$ for every pair of $p \in Q$ and $\sigma \in \Sigma$. We denote the language of A by $L(A)$. A language L is *regular* if there is an FA A such that $L(A) = L$.

A pushdown automaton (PDA) A is specified by a tuple $(Q, \Sigma, \Gamma, \delta, I, Z_0, F)$, where Q is a set of states, each Σ and Γ is an alphabet for input and stack, respectively, $\delta : Q \times (\Sigma \cup \{\lambda\}) \times \Gamma \rightarrow \mathcal{P}(Q \times \Gamma^*)$ is a transition function, $I, F \subseteq Q$ are a set of initial and final states, respectively and $Z_0 \in \Gamma$ is the bottom stack symbol. A PDA is *deterministic* if 1) $|I| = 1$ and 2) for every $q \in Q$, $\sigma \in \Sigma \cup \{\lambda\}$ and $\gamma \in \Gamma$, $|\delta(q, \sigma, \gamma)| = 1$. A language L is context-free if there is a PDA A such that $L = L(A)$.

A Turing machine (TM) M is specified by a tuple $(Q, \Sigma, \Gamma, \delta, q_0, q_{acc}, q_{rej})$, where Q is a set of states, Σ is an input alphabet, $\Gamma \supseteq \Sigma$ is a tape alphabet with a blank symbol $\sqcup \in \Gamma \setminus \Sigma$, $\delta : Q \times \Gamma \rightarrow \mathcal{P}(Q \times \Gamma \times \{L, R\})$ is a transition function and each of q_0, q_{acc} and $q_{rej} \in Q$ is a start, accept and reject state, respectively, where $q_{acc} \neq q_{rej}$. A *configuration* of TM M is a sequence of a state symbol and tape symbols that denotes the current configuration of M. The position of the state symbol denotes that the head points to the right symbol on the tape. For example, the start configuration is $q_0 w$, and the position of q_0 denotes that the head points to the first symbol of w. A configuration is *accepting* if its state is q_{acc}, and *rejecting* if its state is q_{rej}. A *computation* C of a TM M is a sequence of configurations and two adjacent configurations follow the transition rule of M. Given a input w, a computation C is *accepting* (*rejecting*) if 1) C starts with the start configuration $q_0 w$ and 2) C ends with an accepting (rejecting, resp.) configuration. The language $L(M)$ of a TM M is a set of strings w, where M has an accepting computation starting with $q_0 w$.

For more background in automata theory, the reader may refer to the textbooks [21, 24]

3   IIP on Finite Languages

We first consider the IIP when a source language S is finite, which is given by an acyclic DFA or an acyclic NFA. We solve this case by investigating its complement version; whether or not S is not in infix(T). We guess a certificate string w to verify that an IIP-instance (S, T) of a source S and a target T is a NO-instance.

Lemma 2. *Let S be a finite language given by an acyclic NFA A and T be a context-free language given by an NPDA B. Then, $\mathrm{IIP}(S,T)$ is in* co-NP.

Proof. Given an additional string w as certificate, we can verify that (S,T) is a NO-instance of IIP by determining the following three conditions.

1. $|w| < |Q_A|$, where Q_A is the set of states of A. (If $|w| \geq |Q_A|$, then it is immediate that $w \notin S$.)
2. $w \in S$
3. $w \notin \mathrm{infix}(T)$

It is easy to see that all three conditions can be checked in poly-time.   □

For the case when T is finite, we can prove that it is also in co-NP using a similar approach of the proof for Lemma 2.

Lemma 3. *Let S be a context-free language given by an NPDA A and T be a finite language given by an acyclic NFA B. Then, $\mathrm{IIP}(S,T)$ is in* co-NP.

We can use the same approach in Lemmas 2 and 3 when the finite language is given by an acyclic DFA. Next, we establish hardness for the IIP over finite languages. For the hardness proof, we present a poly-time reduction from a co-NP-complete problem, co-3SAT—the complement of the 3SAT problem [2]. A co-3SAT instance consists of two components: a set $X = \{x_1, x_2, \ldots, x_n\}$ of n boolean variables and a formula ϕ in conjunctive normal form of m clauses, where each clause is a disjunction of three boolean literals of X. The co-3SAT problem is to decide whether or not ϕ is unsatisfiable for all possible assignments $\psi : X \to \{\mathrm{T}, \mathrm{F}\}$ over X.

Lemma 4. *Let S be a finite language given by an acyclic DFA A and T be a finite language given by an acyclic NFA B. Then, $\mathrm{IIP}(S,T)$ is* co-NP-hard.

Proof. We show a reduction from the co-3SAT problem to the IIP. Given a co-3SAT instance consisting of a set $X = \{x_1, \ldots, x_n\}$ of n boolean variables and a formula ϕ in conjunctive normal form of m clauses, we define an alphabet Σ_X to be a set of all variables of X and their negations; $\Sigma_X = \{x \mid x \in X\} \cup \{\overline{x} \mid x \in X\}$. Then, an assignment consists of n symbols from Σ_X, where both x_i and its negation $\overline{x_i}$ cannot be chosen at the same time. We design two acyclic DFAs A and B such that $S = L(A)$ contains all assignments and $T = L(B)$ contains every assignment that falsifies ϕ. If $\mathrm{IIP}(S,T)$ holds, then (X, ϕ) is a YES-instance for co-3SAT problem.

An acyclic DFA $A = (Q_A, \Sigma_X, \delta_A, i_A, F_A)$ consists of $Q_A = [0, n]$, $\delta_A = \{(i-1, \sigma_i, i) \mid \sigma_i \in \{x_i, \overline{x_i}\}$ and $i \in [1, n]\}$, $i_A = 0$ and $F_A = \{n\}$. We assume that S contains all assignment sequences in fixed order $[x_1, \ldots, x_n]$.

An acyclic NFA $B = (Q_B, \Sigma_X, \delta_B, I_B, F_B)$ accepts an assignment that falsifies a clause of ϕ. This condition, checking whether or not three literals in a clause are all false, is sufficient since ϕ is a conjunction of disjunctive clauses. We assign $Q_B = \{q_{jk} \mid j \in [1, m], k \in [0, n]\}$, $I_B = \{q_{j0} \mid j \in [1, m]\}$ and

$F_B = \{q_{jn} \mid j \in [1, m]\}$. We use a state q_{jk} as a flag state to check whether or not the corresponding clause $C_j = \mathbf{F}$ for an input assignment. The set δ_B of transitions is

$$\delta_B = \{(q_{jk-1}, x_k, q_{jk}) \mid x_k \text{ is not in } C_j \text{ for } j \in [1, m] \text{ and } k \in [1, n]\}.$$

Figure 1 shows an example of B with $\phi = (\overline{x_1} \vee x_3 \vee \overline{x_4}) \wedge (\overline{x_1} \vee x_1 \vee x_5) \wedge \cdots \wedge (x_1 \vee x_4 \vee x_5)$.

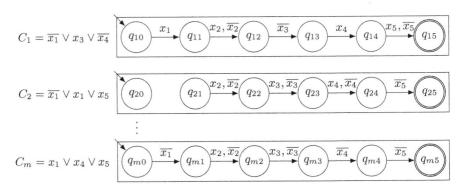

Fig. 1. A target acyclic NFA B for co-3SAT instance (X, ϕ), where $|X| = 5$ and $\phi = (\overline{x_1} \vee x_3 \vee \overline{x_4}) \wedge (\overline{x_1} \vee x_1 \vee x_5) \wedge \cdots \wedge (x_1 \vee x_4 \vee x_5)$ of m clauses. The sink state is not drawn. Note that q_{15} is reachable only if $C_1 = \mathbf{F}$ for given assignment. Also note no transitions from q_{20} to q_{21} since C_2 has both x_1 and $\overline{x_1}$.

In Fig. 1, the FA fragment for C_1 accepts every assignment sequence that does not contain $\overline{x_1}$, x_3 and $\overline{x_4}$. Thus, all accepting assignments of C_1 are unsatisfiable. Since ϕ is a conjunction of clauses, one unsatisfiable clause makes ϕ unsatisfiable. Therefore, it is easy to see that the resulting FA B, which is a union of all these FA fragments, recognizes every assignment ψ such that $\phi(\psi) = \mathbf{F}$. In other words, $T = L(B)$ is a set of all assignments that make ϕ unsatisfiable. Since S is the set of all assignments, IIP of S and T holds if ϕ is unsatisfiable. $\square$

Observe that the transition function δ_B of B is deterministic. Thus, we can make an acyclic DFA for T if we merge all initial states into a single initial state without introducing nondeterminism in B.

Lemma 5. *When two languages S and T are given by acyclic DFAs, IIP(S, T) is co-NP-hard.*

Proof. We modify the acyclic NFA B in Lemma 4 and present an acyclic DFA $B' = (Q_B \cup \{i'_B\}, \Sigma_\# \cup \Sigma_X, \delta'_B, \{i'_B\}, F_B)$ for T, where $\Sigma_\# = \{\#_i \mid i \in [1, m]\}$ is a set of m distinct symbols that are not in Σ_X. We add transitions $(i'_B, \#_j, q_{j0})$ for each clause C_j to δ'_B. Note that B' verifies an assignment with prefix $\#_j$. For instance, if $C_i = \mathbf{F}$ for a given assignment sequence $\psi \in S$, then B' accepts $\#_i \psi$.

Thus, if an assignment sequence $\psi \in S$ is unsatisfiable, which means $\phi(\psi) = F$, then B', from our constructions, accepts $\#_i\psi$, where $C_i = F$ for ψ. Therefore, IIP of S and T given by acyclic DFAs is also co-NP-hard.   □

From Lemmas 4 and 5, we establish the following result.

Theorem 6. *Given two languages S and T, if one is context-free and the other is given by an acyclic DFA or acyclic NFA, $IIP(S,T)$ is co-NP-complete.*

Proof. From Lemmas 2 and 3, we can observe that the IIP with either the source or the target being finite is in co-NP. Lemma 5 shows that IIP for two acyclic DFAs is co-NP-hard. Since an acyclic DFA is also acyclic NFA, DFA, NFA, DPDA and NPDA, we can immediately reduce the co-NP-hard problem for these cases when at least one language is given by an acyclic DFA or acyclic NFA.   □

4   IIP on Infinite Languages

Next we investigate the IIP of *infinite* languages for S and T. We first establish a connection between the NFA inclusion problem and the IIP of two NFAs.

Lemma 7. *When two languages S, T are given by NFAs, $IIP(S,T)$ is PSPACE-complete.*

Proof. (in PSPACE): The IIP of S and T is equivalent to $S \subseteq \text{infix}(T)$ by definition, and the NFA inclusion problem is PSPACE-complete [10].

(PSPACE-hardness): We can solve the NFA inclusion problem $S \subseteq T$ by solving the IIP of $\#S\#$ and $\#T\#$, where $\#$ is a symbol that does not appear in S and T.   □

We obverse that the PSPACE-completeness for the IIP of two NFAs is straightforward form the NFA inclusion result but the DFA case is not; while the inclusion problem of two DFAs is efficiently decidable, the size of a DFA for $\text{infix}(L(A))$ of an n-state DFA A can be exponential [16,19]. Lemma 8 gives a PSPACE-hard bound for IIP of infinite regular languages. We show the hardness by the reduction from the membership problem of a PSPACE TM, which is definitely PSPACE-complete.

Lemma 8. *When two languages S and T are given by DFAs A and B, $IIP(S,T)$ is PSPACE-hard.*

Proof. We give a reduction from the membership problem on the PSPACE TM [21] to the IIP. Given a PSPACE TM $M = (Q, \Sigma, \Gamma, \delta, q_0, q_{acc}, q_{rej})$ and its input string w, there is a polynomial function $l(x)$ that computes the maximum number of tape cells required for an input of length x. Thus, given an input w, each configuration of w on M has length at most $\mathcal{L} = l(|w|)$ tape symbols, which is polynomial to $|w|$. With this fact, the source DFA A recognizes a computation $C = C_0 \# C_1 \# C_2 \# \cdots \# C_n$ satisfying the following four conditions.

1. Each configuration C_i consists of a state symbol $q \in Q$ and $2\mathcal{L} + |w|$ tape symbols from Γ including the blank symbol $\sqcup$.
2. Configurations are separated by a symbol $\# \notin Q \cup \Gamma$.
3. $C_0 = \sqcup^{\mathcal{L}} q_0 w \sqcup^{\mathcal{L}}$ is a start configuration with a blank space buffer of length $\mathcal{L}$ on each end,
4. C_n is an accepting configuration and C_i $(0 \le i < n)$ is neither accepting nor rejecting.

Since the length of each configuration C_i is $N = 2\mathcal{L} + |w| + 1$, the number of states of A is also polynomial in the size of w. Note that A does not check the validity of the transitions. Due to the construction, S contains all valid, accepting computations and some invalid computations. For example, S has $\sqcup^{\mathcal{L}} q_0 w \sqcup^{\mathcal{L}} \# \sqcup^{\mathcal{L}} q_{acc} w \sqcup^{\mathcal{L}}$ even if δ has no available transitions from $\sqcup^{\mathcal{L}} q_0 w \sqcup^{\mathcal{L}}$ to $\sqcup^{\mathcal{L}} q_{acc} w \sqcup^{\mathcal{L}}$.

The target DFA B acts as a filter of S that identifies all invalid computations for w in S. A computation C is invalid for M on w if one of the following conditions is satisfied.

1. C does not start with $\sqcup^{\mathcal{L}} q_0 w \sqcup^{\mathcal{L}}$,
2. C ends with neither an accepting nor a rejecting configuration, or
3. C does not follow the transition δ.

Note that every $C \in S$ already does not satisfy the first two conditions since A rejects every computation C that satisfies either the first or the second condition. For the third condition, we construct a DFA gadget $B' = (Q_{B'}, \Sigma_{B'}, \delta_{B'}, \{i_{B'}\}, F_{B'})$, where $\Sigma_{B'} = Q \cup \Gamma \cup \{\#\}$, that reads the first three symbols from a configuration, say u, and compares them with the three symbols, say v of the same position on the next configuration. If these two parts u and v are not a possible outcome from the transition function δ, C is invalid and B' accepts the entire computation. For example, $u = 0p0$ and $v = 01q$ is a valid outcome if $(p, 0, q, 1, R) \in \delta$, while $u = 000$ and $v = 00q$ is an invalid outcome if $(p, \sigma, q, \gamma, L) \notin \delta$ with any combination of p, σ and γ, where $p, q \in Q$, $\sigma, \gamma \in \Gamma$. Note that B' validates the transition after reading v. Otherwise, B' replaces v as a new u and repeatedly reads the next three symbols on the corresponding position until the input ends. Figure 2 depicts the gadget B' identifying invalid outcomes.

In Fig. 2, state $(000, \sqcup\sqcup\sqcup)$ is final because any valid transition cannot change a symbol that is not under the head; on the other hand, $(000, 000)$ is a valid outcome and thus it is not final. Note that, if there is a transition leading to q with a left-move in the TM M, the state $(000, 00q)$ in B' denotes a valid outcome and thus it is also not final.

Since B' has a limitation that B' only compares the first three symbols of configurations over the computation, we utilize the construction of Lemma 5 to adjust where the comparison starts without introducing nondeterminism. The target DFA $B = (Q_B, \Sigma_B, \delta_B, \{i_B\}, F_{B'})$ has the following components.

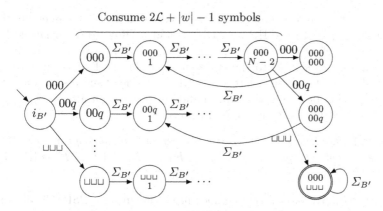

Fig. 2. The DFA gadget B' identifies invalid outcomes. Remark that B' reads three symbols at the start and discard $N - 2$ symbols until it reaches the next configuration. Then it reads additional three symbols to validate the configurations according to the transition δ of M. If it is an *invalid* outcome, then B' accepts the input. This process repeats until B' finds an invalid sequence or reaches the end of the given input.

- $Q_B = Q_{B'} \cup \{i_B\} \cup \{\emptyset^k \mid 1 \leq k \leq 2\mathcal{L} + |w|\}$
- $\Sigma_B = Q \cup \Gamma \cup \{\#\} \cup \{\#_k \mid 1 \leq k \leq 2\mathcal{L} + |w| + 1\}$.
- δ_B is a union of the following three sets.
  1. $\{(i_B, \#_1, i_{B'})\} \cup \{(i_B, \#_{k+1}, \emptyset^k \mid 1 \leq k \leq 2\mathcal{L} + |w|\}$
  2. $\{(\emptyset^1, \gamma, i_{B'}) \mid \gamma \in \Gamma\} \cup \{(\emptyset^{k+1}, \gamma, \emptyset^k) \mid 1 \leq k < 2\mathcal{L} + |w|, \gamma \in \Gamma\}$
  3. $\delta_{B'}$

The symbol $\#_k$ denotes that the initial symbol to compare is the kth one and directs B' to compare every configuration subsequence of kth to $k+2$nd symbols. The transition function δ_B enables to verify every position of given computations using the gadget B'.

Then, given a computation C of M, T contains at least one of $\#_i C$ for any i if $C \in S$ is invalid. Otherwise, T does not contain any of $\#_i C$. Therefore, IIP of S and T being true guarantees that M rejects w. $\square$

Combining Lemmas 7 and 8, we establish the following result for the IIP of regular languages represented by either a DFA or an NFA.

Theorem 9. *When two languages S and T are given by FAs (either DFA or NFA), IIP(S, T) is PSPACE-complete.*

We move to the IIP for the context-free language case, starting with the IIP between a context-free language and a regular language. Since the IIP of S and T is equivalent to $S \subseteq \text{infix}(T)$ by definition, the following result is immediate from the fact that the inclusion problem of an NPDA and an NFA is EXPTIME-complete [10].

Lemma 10. *Let S be a context-free language given by an NPDA and T be a regular language given by an NFA. Then, IIP(S, T) is in EXPTIME.*

On the other hand, Lemma 11 shows a reduction from a membership test on alternating TM [8] to the IIP of a DPDA and a DFA. An alternating TM M is a Turing machine with two types (existential and universal) of states. An existential state is similar to the states in an ordinary TM; when M is on an existential state, it is sufficient to verify that any one of successive computation has an accepting path. On the other hand, when M is on a universal state, M must verify that all successive computations have an accepting path.

Lemma 11. *Let S be a deterministic context-free language given by a DPDA A and T be a regular language given by a DFA B. Then, $IIP(S, T)$ is EXPTIME-hard.*

Proof (Sketch). We reduce the membership problem for a PSPACE alternating TM M on an input w. Note that APSPACE is the class of languages recognized by PSPACE alternating TMs, and is equivalent to EXPTIME [8]. Let us assume that

1. M alternates existential and universal configurations,
2. a universal configuration, unless it is final, has exactly two next existential configurations and
3. the start state q_0 is existential.

Then, we can write a computation in the form of

$$C_0 \# C_1^R \$_1 \overbrace{C_2 \# C_3^R \$_1 \cdots \$_2}^{\text{The first successor of } C_1} \cdots \$_2 \overbrace{C_2' \# C_3'^R \$_1 \cdots \$_2}^{\text{The second successor of } C_1} \cdots \%,$$

where $C_0 = q_0 w$. Since states alternate over the computation, we observe that existential configurations (e.g., C_0) are written in order and universal configurations (e.g., C_1) are written in reversed order. Note that, if we ignore the last % symbol, the form is recursive, where $\$_1$'s and $\$_2$'s define nested pairs. For a non-final universal configuration C_1, $\$_1$ is followed by the first successive configuration (C_2) of C_1 until its $\$_2$ counterpart appears, which is then followed by the second successive configuration C_2' of C_1. Since all non-final universal configurations (e.g., C_1) have exactly two successive configurations by the second assumption above, every such configuration has its corresponding $\$_1$ and $\$_2$.

It is easy to confirm that a subsequent computation sequence is always split by # or $\$_1$. For example, $C_0 \# C_1^R \$_1 C_2 \$_2 C_2' \%$ denotes two valid moves C_0, C_1, C_2 and C_0, C_1, C_2' of M on w, where C_1 is a universal configuration. Also, similar to Lemma 8, we fix the length of each configuration to $N = 2 \cdot l(|w|) + |w| + 1$ and let $\mathcal{L} = l(|w|)$.

Now we design a DPDA A that checks the following five conditions:

1. Each configuration contains only one state symbol $q \in Q$ and $2\mathcal{L} + |w|$ tape symbols.
2. C_0 is a start configuration $\sqcup^{\mathcal{L}} q_0 w \sqcup^{\mathcal{L}}$.
3. The second transitions of a universal configuration, for example, from C_1^R to C_2', are all valid.

4. The separator symbols (#, $\$_1$, $\$_2$) and the end marker symbol % are placed correctly.
5. Every last configuration (preceding $\$_2$ or %) is an accepting configuration and the others are neither accepting nor rejecting.

The new DPDA A verifies the first, second and fifth conditions without using a stack, and checks the other two conditions (3, 4) simultaneously using a stack. Note that since A only validates the transitions from a universal configuration to its second successive configuration, S is a superset of all valid accepting computations for M on w. A target DFA B identifies a prefix $\#_i$ (similar to Lemma 8) followed by an invalid computation, which satisfies one of the following conditions.

1. An existential transition (e.g., C_0 to C_1^R) is invalid.
2. The first transitions of a universal configuration (e.g., C_1^R to C_2) is invalid.

Since T has only invalid computations, if IIP holds, then every computation that S contains must be invalid. Thus, $w \notin L(M)$. Since APSPACE and EXPTIME are equivalent, this concludes the proof.   □

Lemmas 10 and 11 provide a tight complexity bound on the IIP of a context-free language (in either DPDA or NPDA) and a regular language (in either DFA or NFA). Thus, the IIP of a context-free language and a regular language is EXPTIME-complete.

Theorem 12. *When S is given by either a DPDA or an NPDA, and T is given by either a DFA or an NFA, IIP(S, T) is EXPTIME-complete.*

Theorem 12 ensures that the IIP of a context-free language S and a regular language T is decidable. It turns out that the problem is not symmetric; when S is regular and T is context-free, the IIP becomes undecidable. When S is given by a DFA and T is given by a DPDA, we show that the IIP of S and T can determine the emptiness of a linear-bounded automaton (LBA), which is undecidable [21].

Theorem 13. *When S is given by a DFA A and T is given by a DPDA B, IIP(S, T) is undecidable.*

Theorem 13 leads to the undecidability of IIP of context-free languages.

Corollary 14. *When two languages S and T are context-free (given by NPDAs) or deterministic context-free (given by DPDAs), IIP(S, T) is undecidable.*

5   Conclusions

There have been several researches on the language inclusion problem [5,6,10, 11]. We have introduced another language inclusion problem based on the infix set of all the strings in one language. The problem is motivated from an observation on how attackers produce new malicious patterns from known malicious patterns to damage a target system in practice. For instance, an attacker may modify a known malicious pattern for the ReDoS attack to avoid the current detection system by adding additional information before and after the pattern, which can still cause the ReDoS attack. Here the malicious pattern S is being an infix of the new pattern T; namely $S \in \text{infix}(T)$.

The IIP between two languages S and T is to determine whether or not S is included in the set of all infixes of T. We have studied the problem for different language classes including finite, regular and context-free cases for S and T. We have established different decidability results for the considered cases including the undecidability result. We have demonstrated that the problem is co-NP-complete when one of S and T is finite, and PSPACE-complete when both S, T are regular. Moreover, we proved that the problem is EXPTIME-complete when S is context-free and T is regular. Finally, we have obtained the undecidability result when T is context-free and S is either regular or context-free.

For future work, we can consider different classes of languages given by visibly pushdown automata [1,6] or various counter machines [14,15]. For example, when S is given by a DFA and T is given by a $\text{DCM}(1,1)$, $\text{IIP}(S,T)$ is undecidable [14], whereas when S is given by an NPDA augmented with reversal-bounded counters and T is given by a deterministic one counter machine that makes only 1 turn on the counter, the problem becomes decidable [15].

Acknowledgments. We would like to thank Oscar Ibarra and Ian McQuillan for their comments that relate our problem to the infix-density problem. This research was supported by the NRF grant funded by MIST (NRF-2020R1A4A3079947). The first two authors contributed equally to this work.

References

1. Alur, R., Madhusudan, P.: Visibly pushdown languages. In: Proceedings of the 36th Annual ACM Symposium on Theory of Computing, pp. 202–211 (2004)
2. Arora, S., Barak, B.: Computational Complexity - A Modern Approach. Cambridge University Press, Cambridge (2009)
3. Berglund, M., Drewes, F., van der Merwe, B.: Analyzing catastrophic backtracking behavior in practical regular expression matching. In: Proceedings of the 14th International Conference on Automata and Formal Languages, pp. 109–123 (2014)
4. Berglund, M., van der Merwe, B.: Regular expressions with backreferences re-examined. In: Proceedings of the Prague Stringology Conference 2017, pp. 30–41 (2017)
5. Bousquet, N., Löding, C.: Equivalence and inclusion problem for strongly unambiguous Büchi automata. In: Proceedings of the 4th International Conference on Language and Automata Theory and Applications, pp. 118–129 (2010)
6. Bruyère, V., Ducobu, M., Gauwin, O.: Visibly pushdown automata: universality and inclusion via antichains. In: Proceedings of the 7th International Conference on Language and Automata Theory and Applications, pp. 190–201 (2013)
7. Câmpeanu, C., Moreira, N., Reis, R.: Distinguishability operations and closures. Fundamenta Informaticae **148**(3–4), 243–266 (2016)
8. Chandra, A.K., Stockmeyer, L.J.: Alternation. In: Proceedings of the 17th Annual Symposium on Foundations of Computer Science, pp. 98–108 (1974)

9. Chapman, C., Stolee, K.T.: Exploring regular expression usage and context in Python. In: Proceedings of the 25th International Symposium on Software Testing and Analysis, pp. 282–293 (2016)
10. Clemente, L.: On the complexity of the universality and inclusion problems for unambiguous context-free grammars. In: Proceedings of the 8th International Workshop on Verification and Program Transformation and 7th Workshop on Horn Clauses for Verification and Synthesis, pp. 29–43 (2020)
11. Clemente, L., Mayr, R.: Efficient reduction of nondeterministic automata with application to language inclusion testing. Log. Methods Comput. Sci. **15**(1), 12:1-12:73 (2019)
12. Davis, J.C., Coghlan, C.A., Servant, F., Lee, D.: The impact of regular expression denial of service (ReDoS) in practice: an empirical study at the ecosystem scale. In: Proceedings of the 26th ACM Joint Meeting on European Software Engineering Conference and Symposium on the Foundations of Software Engineering, pp. 246–256 (2018)
13. Davis, J.C., Michael IV, L.G., Coghlan, C.A., Servant, F., Lee, D.: Why aren't regular expressions a lingua franca? An empirical study on the re-use and portability of regular expressions. In: Proceedings of the 27th ACM Joint Meeting on European Software Engineering Conference and Symposium on the Foundations of Software Engineering, pp. 443–454 (2019)
14. Eremondi, J., Ibaraa, O.H., McQuillan, I.: On the density of context-free and counter languages. Int. J. Found. Comput. Sci. **29**(02), 233–250 (2018)
15. Eremondi, J., Ibarra, O.H., McQuillan, I.: Deletion operations on deterministic families of automata. Inf. Comput. **256**, 237–252 (2017)
16. Gao, Y., Moreira, N., Reis, R., Yu, S.: A survey on operational state complexity. J. Autom. Lang. Combin. **21**(4), 251–310 (2017)
17. Kirrage, J., Rathnayake, A., Thielecke, H.: Static analysis for regular expression denial-of-service attacks. In: Proceedings of the 7th International Conference on Network and System Security, pp. 135–148 (2013)
18. McNaughton, R., Yamada, H.: Regular expressions and state graphs for automata. IRE Trans. Electron. Comput. EC. **9**(1), 39–47 (1960)
19. Pribavkina, E.V., Rodaro, E.: State complexity of prefix, suffix, bifix and infix operators on regular languages. In: Proceedings of the 14th International Conference on Developments in Language Theory, pp. 376–386 (2010)
20. Pribavkina, E.V., Rodaro, E.: State complexity of code operators. Int. J. Found. Comput. Sci. **22**(7), 1669–1681 (2011)
21. Sipser, M.: Introduction to the Theory of Computation, 3rd edn. Cengage Learning, Boston (2013)
22. Spencer, H.: A regular-expression matcher. In: Software Solutions in C, pp. 35–71. Academic Press Professional, Inc. (1994)
23. Thompson, K.: Programming techniques: regular expression search algorithm. Commun. ACM **11**(6), 419–422 (1968)
24. Wood, D.: Theory of Computation. Harper & Row, New York (1987)
25. Wüstholz, V., Olivo, O., Heule, M.J.H., Dillig, I.: Static detection of DoS vulnerabilities in programs that use regular expressions. In: Proceedings of the 23rd International Conference on Tools and Algorithms for the Construction and Analysis of Systems, Part II, pp. 3–20 (2017)

Column Representation of Sturmian Words in Cellular Automata

Francesco Dolce[1($\boxtimes$)] and Pierre-Adrien Tahay[2($\boxtimes$)]

[1] FIT, Czech Technical University in Prague, Prague, Czech Republic
dolcefra@fit.cvut.cz
[2] FNSPE, Czech Technical University in Prague, Prague, Czech Republic
pierre.adrien.tahay@cvut.cz

Abstract. We prove that, given a Sturmian word $\mathbf{w}$ with quadratic slope, it is possible to construct a one-dimensional cellular automaton such that $\mathbf{w}$ is represented in a chosen column in its space-time diagram. Our proof is constructive and use the continued fraction expansion of the slope of the word.

Keywords: Sturmian words · Cellular automata · Quadratic numbers · Continued fraction expansion

1 Introduction

Sturmian words are infinite words over a binary alphabet that have exactly $n + 1$ factors of length n for each non-negative n. Their origin can be traced back to the astronomer J. Bernoulli III. Their first in-depth study is by Morse and Hedlund [17]. Many combinatorial properties were described in the paper by Coven and Hedlund [5]. Sturmian words are one of the most studied topics in combinatorics on words. They can be defined in different ways and have various interpretations in several domains, including combinatorics and discrete geometry. A possible way to describe them is by using mechanical words.

In this paper we consider characteristic Sturmian words with quadratic slope (see Sect. 2 for precise definitions). In particular, given such a Sturmian word $\mathbf{w}$ with continued fraction of its slope $\alpha = [0, 1 + b_1, b_2, \ldots, b_m, \overline{a_1, a_2, \ldots, a_k}]$ and corresponding directive sequence $\Delta = (b_1, b_2, \cdots, b_m, (a_1, a_2, \cdots, a_k)^\omega)$, we have $\mathbf{w} = \lim_{n \to \infty} w_n$, where $w_0 = \mathtt{a}, w_1 = \mathtt{b}, w_n = w_{n-1}^{b_n} w_{n-2}$ for $1 \le n \le m$ and $w_{m+n} = w_{m+n-1}^{a_{(n \bmod k)}} w_{m+n-2}$ for $n > 0$ (where $a_0 := a_k$). We use such a characterisation to construct a machine, called cellular automaton, that will "print" us the infinite word $\mathbf{w}$. A cellular automaton is a dynamical system defined by an infinite string of symbols over an alphabet and a map, called local rule, that transforms every symbol of the string according to its neighbourhood (see Sect. 3 for a

The research received funding from the Ministry of Education, Youth and Sports of the Czech Republic through the projects CZ.02.01.01/0.0/0.0/16_019/0000765 and CZ.02.1.01/0.0/0.0/16_019/0000778.

© Springer Nature Switzerland AG 2022
V. Diekert and M. Volkov (Eds.): DLT 2022, LNCS 13257, pp. 127–138, 2022.
https://doi.org/10.1007/978-3-031-05578-2_10

more precise definition). A classical example of a 2-dimensional cellular automaton is given by 1970 Conway's Game of Life. In spite of its very simple definition, the Game of Life has some quite remarkable properties. Indeed, Rendell proved that starting from it, it is possible to simulate any Turing machine [18]. In this paper we consider one-dimensional cellular automata. The initial configuration will be given by an infinite string over a (finite) alphabet.

Our main result is the following one.

Theorem 1. *A Sturmian word with quadratic slope can be represented as a column in the space-time diagram of a one-dimensional cellular automaton.*

The task of representing a sequence over a finite alphabet in the space-time diagram of a cellular automaton is a non-trivial and still not entirely explored topic. One of the first results on the subject is the construction of the characteristic sequence of primes numbers, done by Fischer in 1965, using a cellular automaton with more than 30,000 states [8]. In 1997, Korec gives another construction with only 11 states [13]. In 1999, Mazoyer and Terrier establish several geometric constructions of increasing functions, which they call Fischer constructible [16]. A very interesting result is given by Rowland and Yassawi in 2015 [19]: they give a complete characterisation of the construction of q-automatic sequence, with q a power of a prime number, in the columns of the space-time diagram of linear cellular automata. Finally, in 2018, Marcovici, Stoll and Tahay construct different non-automatic sequences, such as the characteristic sequences of polynomials (squares, cubes, etc.) and the Fibonacci word [15]. Our main result in this paper can be seen as an extension of the construction obtained by Marcovici, Stoll and Tahay for this last infinite word. While in [15] the authors use *ad hoc* properties of Fibonacci numbers and Fibonacci finite words, in this article we consider the development of the continued fraction associated a Sturmian word with quadratic slope to define a new algorithm. Such an algorithm could even be generalised to larger families of infinite words (see Sect. 6).

2   Preliminaries

In this section we recall some basic definitions on finite and infinite words. For all undefined terms we refer to [14]. We denote the set of integers, of positive integers and of non-negative integers respectively by $\mathbb{Z}$, $\mathbb{Z}^+$ and $\mathbb{N}$, while $\mathbb{Q}$ and $\mathbb{R}$ denote the set of rational and of real numbers.

2.1   Words

An *alphabet* Σ is a (finite) set of symbols called *letters*. The set of *finite words* Σ^* over Σ is the free monoid having neutral element the *empty word* ε. We also denote by Σ^+ the free semigroup over Σ, e.g., $\Sigma^+ = \Sigma^* \setminus \{\varepsilon\}$. The *length* of a word $w = a_0 a_1 \cdots a_{n-1}$, with $a_i \in \Sigma$, is the non-negative integer $|w| = n$. The length of ε is considered to be 0. When it is possible to write $w = pus$, with $p, u, s \in \Sigma^*$, we call p (resp., s, u) a *prefix* (resp., *suffix*, *factor*) of w.

An *infinite word* over Σ is a sequence $\mathbf{w} = a_0 a_1 a_2 \cdots$, with $a_i \in \Sigma$ for all i. Similarly to finite words we can define the set of infinite words $\Sigma^{\mathbb{N}}$ and extend in a natural way to $\Sigma^* \cup \Sigma^{\mathbb{N}}$ the notions of prefix, suffix and factor. An infinite word $\mathbf{w}$ is *eventually periodic* if $\mathbf{w} = uv^{\omega} = uvvv\cdots$. When $u = \varepsilon$ we say that $\mathbf{w}$ is *purely periodic*. An infinite word that is not eventually periodic is called *aperiodic*. The *factor complexity* of an infinite word $\mathbf{w}$ is the mapping $C_{\mathbf{w}} : \mathbb{N} \to \mathbb{N}$ defined by $C_{\mathbf{w}}(n) = \#\{u \mid u \text{ is a factor of } \mathbf{w} \text{ and } |u| = n\}$.

Aperiodic infinite words with the lowest possible factor complexity, i.e., such that $C_{\mathbf{w}}(n) = n+1$ for all $n \in \mathbb{N}$, are called *Sturmian words* (for other equivalent definitions see [1]). It follows from the definition that all Sturmian words are defined over a binary alphabet, e.g., $\{\mathsf{a}, \mathsf{b}\}$. If both sequences $\mathsf{a}\mathbf{w}$ and $\mathsf{b}\mathbf{w}$ are Sturmian, we call $\mathbf{w}$ a *characteristic Sturmian word*. The family of Sturmian words coincide with the family of irrational mechanical words, as well as the family of binary balanced aperiodic words (for more on balanced words see, for instance, [1,6,10]). In particular, characteristic Sturmian words correspond to balanced irrational mechanical words with intercept equal to the slope [14]. We denote by $\mathbf{c}_{\alpha}$ the unique characteristic Sturmian word with slope (and intercept) α. Since two Sturmian words with the same slope (but different intercept) share the same set of factors, we will focus only on the study of characteristic Sturmian words. Note that every mechanical word with rational slope is a purely periodic word.

Example 1. Let us consider the well-known *Fibonacci word* $\mathbf{f} = \mathsf{abaababaab}\cdots$. It is defined as the fixed point of the morphism sending $\mathsf{a} \mapsto \mathsf{ab}$ and $\mathsf{b} \mapsto \mathsf{a}$. The word $\mathbf{f}$ is the characteristic Sturmian word $\mathbf{c}_{1/\varphi^2}$, where $\varphi = \frac{1+\sqrt{5}}{2}$.

2.2   Continued Fraction Expansion

Let θ be a real number. A continued fraction expansion of θ is defined as $[c_0, c_1, c_2, \ldots]$ whenever

$$\theta = c_0 + \cfrac{1}{c_1 + \cfrac{1}{c_2 + \ddots}}$$

with $c_0 \in \mathbb{Z}$ and $c_i \in \mathbb{Z}^+$ for every positive i. It is known that if $\theta \in \mathbb{Q}$ then there exist exactly two continued fraction expansion of θ and they are both finite. On the other hand every positive irrational number corresponds to a unique infinite continued fraction expansion with $c_0 \in \mathbb{N}$ and $c_i \in \mathbb{Z}^+$ for all $i \geq 1$. Note that, if $c_0 = 0$ then $0 \leq \theta \leq 1$. If θ is a quadratic irrational, then its continued fraction expansion is eventually periodic, that is it will be of the form

$$\theta = [b_0, \ldots, b_m, \overline{a_1, \ldots, a_k}] = [b_0, \ldots, b_m, a_1, \ldots, a_k, a_1, \ldots, a_k, \ldots].$$

Example 2. The golden ratio $\varphi = \frac{1+\sqrt{5}}{2}$ is a quadratic irrational number. Its continued fraction expansion is $\varphi = [\overline{1}] = [1, 1, 1, \ldots]$ The continued fraction expansions of e and π are $[2, 1, 2, 1, 1, 4, 1, 1, 6, \ldots]$ and $[3, 7, 15, 1, 292, 1, 1, 1, 2, \ldots]$ respectively (sequences A003417 and A001203 in the OEIS [20]).

2.3   Standard Sequences and Directed Sequences

Let $\Delta = (d_n)_{n \geq 1}$ be an integer sequence with $d_1 \in \mathbb{N}$ and $d_n \in \mathbb{Z}^+$ for every positive integer n. The *standard sequence* associated to Δ is the sequence of finite words $(w_n)_{n \geq -1}$ defined by

$$w_{-1} = \mathsf{b}, \quad w_0 = \mathsf{a}, \quad w_n = w_{n-1}^{d_n} w_{n-2} \quad \text{for every } n \geq 1.$$

The sequence Δ is also called the *directive sequence* of $(w_n)_{n \geq -1}$. Note that if $d_1 > 0$, every w_n starts with a. Otherwise, w_n starts with b for every $n \neq 0$. Let us consider the infinite word $\mathbf{w} = \lim_{n \to \infty} w_n$. Such an infinite word is well defined and w_n is a prefix of $\mathbf{w}$ for every positive n. We say that Δ (resp. $(w_n)_{n \geq -1}$) is the *directive sequence* (resp. the *standard sequence*) of $\mathbf{w}$.

Example 3. Let us consider the directive sequence $\Delta = (1^\omega) = (1, 1, 1, \ldots)$. The associated standard sequence is $f_{-1} = \mathsf{b}$, $f_0 = \mathsf{a}$, and $f_n = f_{n-1} f_{n-2}$ for every $n \geq 1$. It is known that the Fibonacci word $\mathbf{f}$ defined in Example 1 can be obtained as the limit $\mathbf{f} = \lim_{n \to \infty} f_n$.

Proposition 1 [14]. *Let α be an irrational number with $0 < \alpha < 1$ having continued fraction expansion $\alpha = [0, d_1 + 1, d_2, d_3, \ldots]$, and let $(w_n)_{n \geq -1}$ be the standard sequence associated to $(d_1, d_2, d_3, \ldots)$. Then $\mathbf{c}_\alpha = \lim_{n \to \infty} w_n$.*

Note that mechanical words are defined for $0 \leq \alpha \leq 1$. It is possible to generalise such definition to every $\alpha \in \mathbb{R}$ [14, Remark 2.1.12]: the fraction expansion in the statement of Proposition 1 would be $\alpha = [c_0, d_1 + 1, d_2, d_3, \ldots]$ with $c_0 \in \mathbb{Z}$, but the standard associated sequence would not change.

Example 4. As seen in Example 3, the directive sequence of the Fibonacci word $\mathbf{f}$ is (1^ω). The continued fraction expansion of the corresponding irrational slope (see also Example 1) is $\frac{1}{\varphi^2} = [0, 2, \overline{1}]$.

Example 5. Let us consider the characteristic Sturmian word $\mathbf{v}$ having associated directive sequence $\Delta = ((1, 2)^\omega)$. We have $\mathbf{v} = \lim_{n \to \infty} v_n$, where the standard sequence $(v_n)_{n \geq -1}$ is defined by $v_{-1} = \mathsf{b}$, $v_0 = \mathsf{a}$, $v_{2n+1} = v_{2n} v_{2n-1}$ and $v_{2n} = v_{2n-1}^2 v_{2n-2}$ for every $n \in \mathbb{N}$. Since $\frac{3-\sqrt{3}}{3} = [0, 2, \overline{2, 1}]$, we have, according to Proposition 1, $\mathbf{v} = \mathbf{c}_{(3-\sqrt{3})/3}$.

In the next sections we will also need the lengths of the prefixes of a characteristic Sturmian sequence. Given a standard sequence $(w_n)_{n \geq -1}$ we define for every integer $n \geq -1$ the number $W_n = |w_n|$.

Example 6. Let $(f_n)_{n \geq -1}$, and $(v_n)_{n \geq -1}$ be the standard sequences defined in Examples 3, and 5. Let $F_n = |f_n|$ and $V_n = |v_n|$. For every $n \in \mathbb{N}$ we have the relations $F_n = F_{n-1} + F_{n-2}$, $V_{2n+1} = V_{2n} + V_{2n-1}$ and $V_{2n} = 2V_{2n-1} + V_{2n-2}$.

3   Cellular Automata

In the following we use the terminology developed by Mazoyer and Terrier in [16] and Marcovici, Stoll and Tahay in [15].

Definition 1. *A one-dimensional* cellular automaton *(CA) is a dynamical system* $(\mathcal{A}^{\mathbb{Z}}, T)$, *where* $\mathcal{A}$ *is a finite set, and where the map* $T : \mathcal{A}^{\mathbb{Z}} \to \mathcal{A}^{\mathbb{Z}}$ *is defined by a local rule which acts uniformly and synchronously on the configuration space. More precisely, there exists an integer* $r \in \mathbb{N}$ *called the* radius *of the CA, and a local rule* $\tau : \mathcal{A}^{2r+1} \to \mathcal{A}$ *such that for every* $\mathbf{x} = (x_k)_{k \in \mathbb{Z}}$ *and for every* $k \in \mathbb{Z}$, *we have* $T(\mathbf{x})_k = \tau((x_{k+i})_{-r \leq i \leq r})$.

When the set $\mathcal{A}$ is understood, we will call *cellular automaton* just the map T. The elements of $\mathcal{A}^{\mathbb{Z}}$ are called *configurations*. By the Curtis-Hedlund-Lyndon Theorem [9], a map $T : \mathcal{A}^{\mathbb{Z}} \to \mathcal{A}^{\mathbb{Z}}$ is a CA if and only if it is continuous with respect to the product topology and it commutes with the shift map σ defined by $\sigma(\mathbf{x})_k = x_{k-1}$, for every configuration $\mathbf{x} = (x_k)_{k \in \mathbb{Z}}$ and every $k \in \mathbb{Z}$. Let $0 \in \mathcal{A}$ and $T : \mathcal{A}^{\mathbb{Z}} \to \mathcal{A}^{\mathbb{Z}}$ a cellular automaton. We say that T is 0-*quiescent* if $T(0^{\mathbb{Z}}) = 0^{\mathbb{Z}} = \cdots 000 \cdots$. A configuration $\mathbf{x} = (x_k)_{k \in \mathbb{Z}}$ is called *finite* if the set $\{k \in \mathbb{Z} : x_k \neq 0\}$ is finite. A cellular automaton can be visualized by using a space-time diagram which is a 2-dimensional grid where each cell contains an element of the set $\mathcal{A}$ and is represented by a space coordinate and a time coordinate.

Let us consider the set

$$\mathcal{S} = \left\{ (T^n(\mathbf{x})_0)_{n \geq 0} \in \mathcal{A}^{\mathbb{N}} : T \text{ is a 0-quiescent CA on } \mathcal{A}^{\mathbb{Z}} \text{ and } \mathbf{x} \text{ is finite} \right\}.$$

In other words, $\mathcal{S}$ is the set of sequences of $\mathcal{A}^{\mathbb{N}}$ that can occur as the first column (and thus as any column) in the space-time diagram of some one-dimensional 0-quiescent CA, starting from a finite initial configuration. This set corresponds to the set of *Fischer's produced* sequences in [16].

In a space-time diagram it is also possible to "transmit information" through *signals*, that is to connect two cells (m, n) and $(m', n+t)$ through a monotonous path; we call *slope* of the signal the number $\frac{t}{m'-m}$ (see [16, Definitions 3 and 4] for a formal definition). When $m = m'$, we call such a signal a *vertical signal* or a signal of *infinite* slope. For the sake of simplicity, we usually represent a signal as a straight line between the cells (m, n) and $(m', n + t)$. Signals are usually "porous", i.e., they do not interact between each other. In some case, however, we also need to consider "concrete" signals. In particular, let us define two distinct kinds of walls. We say that a wall is of type (i) whenever a given signal hitting the wall *bounces* from the cell just above, i.e., when a given signal of slope d arrives in a cell (ℓ, t), then such a signal dies and a new signal of slope $-d$ starts from the cell $(\ell, t + 1)$. We say that a wall is of type (ii) whenever a given signal hitting the wall bounces from the same cell, i.e., when a given signal of slope d arrives in a cell (ℓ, t), then such a signal dies and a new signal of slope $-d$ starts from the same cell (ℓ, t). We usually represent a wall of type (i) as a line inside the column $\ell + 1$ when the signals comes from the left (resp. $\ell - 1$ when the signal comes from the right), and a wall of type (ii) as a rectangle containing the cells in the column ℓ (see Fig. 1).

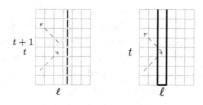

Fig. 1. Walls of type (i) (on the left) and of type (ii) (on the right).

When two signals meet, we can *mark* the cell at the intersection, i.e., assign to it a value from the set $\mathcal{A}$, and define new signals starting from it.

4   Construction of Numbers

To prove our main result we proceed in two steps. First let us construct a CA recognising the lengths W_n of the prefixes w_n of our Sturmian word $\mathbf{w}$.

Let $X \subset \mathbb{Z}$. Let us denote by $\mathbb{1}_X$ the *characteristic function* of X, that is the map $\mathbb{1}_X : \mathbb{Z} \to \{0, 1\}$ defined by $\mathbb{1}_X(x) = 1$ iff $x \in X$.

Proposition 2 [16]. *Let $(S_n)_{n \geq 0}$ be an integer sequence defined by $S_{n+p} = \sum_{i=0}^{p-1} a_i S_{n+i}$, where $p, a_i \in \mathbb{N}$. Then $\mathbb{1}_{\{S_n\}_{n \geq 0}} \in \mathcal{S}$.*

Mazoyer and Terrier give an explicit method to build this sequences in a column of a CA. We propose here a different construction for a particular case that will be necessary for representing a Sturmian word of quadratic slope.

Proposition 3. *Let $(d_n)_{n \geq 1}$ be an eventually periodic integer sequence with $d_1 \in \mathbb{N}$ and $d_i \in \mathbb{Z}^+$ for every $i \geq 2$. Let $(S_n)_{n \geq 0}$ be the integer sequence defined by $S_n = d_n S_{n-1} + S_{n-2}$ for every $n \geq 0$, with $S_{-1}, S_0 \in \mathbb{Z}^+$. Then $\mathbb{1}_{\{S_n\}_{n \geq 0}} \in \mathcal{S}$.*

Proof (Sketch). Since the sequence $(d_n)_{n \geq 1}$ is eventually periodic, then there exist $m \in \mathbb{N}$, $k \in \mathbb{Z}^+$ and $b_1, \ldots, b_m, a_1, \ldots a_k \in \mathbb{N}$ such that $(d_n)_{n \geq 1} = (b_1, \ldots, b_m, (a_1, \ldots, a_k)^\omega)$. Note that we can consider the first S_m rows of the cellular automaton as initial conditions, i.e., we can start the construction from the row of rank S_m. Let $n \geq 1$ be an integer. We are going to consider two distinct cases according to the value of d_{n+1}.

Let us first suppose that $d_{n+1} \neq 1$. Assume that we have already marked the cells $(0, S_n)$, (S_{n-2}, S_n), (S_{n-1}, S_n), $(S_{n-1} + S_{n-2}, S_n)$ and (S_n, S_n). We claim that we can mark the cells $(0, S_{n+1})$, (S_{n-1}, S_{n+1}), (S_n, S_{n+1}), $(S_n + S_{n-1}, S_{n+1})$ and (S_{n+1}, S_{n+1}) In order to do that, we use the relation

$$S_{n+1} = S_n + d_n(d_{n+1} - 1)S_{n-1} + (d_{n+1} - 1)S_{n-2} + S_{n-1}.$$

The idea is to consider intermediate rows, such that their distance is given by the addends in the previous sum. When two signals meet, they died and we can use the cell on the intersection to define other signals. The slope of each signal is determined by the ratio between the difference between the time coordinates and the difference between the space coordinates. For example, the slope of the signal between the cell $(0, S_n)$ and the cell $(S_{n-1}, S_n + d_n(d_{n+1} - 1)S_{n-1})$ is

$$\frac{(S_n + d_n(d_{n+1} - 1)S_{n-1}) - S_n}{S_{n-1} - 0} = d_n(d_{n+1} - 1)$$

When $d_{n+1} = 1$ the construction is different. In this case the three rows S_n, $S_n + d_n(d_{n+1} - 1)S_{n-1}$ and $d_{n+1}S_n$ coincide, as well as the two rows $S_n + S_{n-1}$ and S_{n+1} (resp. the two columns $S_n + S_{n-1}$ and S_{n+1}). We start with the cells $(0, S_n)$, (S_{n-2}, S_n), (S_{n-1}, S_n) and (S_n, S_n) and we claim that we can mark the cells $(0, S_{n+1})$, (S_{n-1}, S_{n+1}), (S_n, S_{n+1}) and (S_{n+1}, S_{n+1}).

The construction of the sequence $(S_n)_{n \geq 0}$ is illustrated in Figs. 2 and 3. Using these figures it is not hard to recover exact definitions of the signals. For sake of simplicity we use the same colour when two signals have the same slope and we represent only the cells on the intersections between two signals.

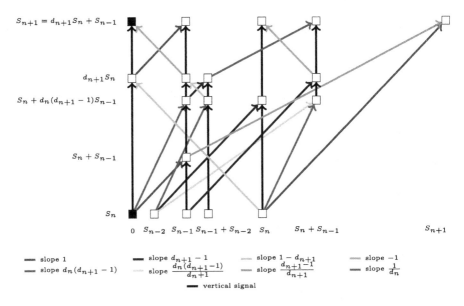

Fig. 2. Construction of the number sequence $(S_n)_{n \geq 0}$ when $d_{n+1} \neq 1$.

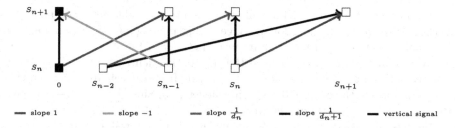

Fig. 3. Construction of the number sequence $(S_n)_{n\geq 0}$ when $d_{n+1} = 1$.

In particular, in both cases we are able to mark the cell $(0, S_n)$. Hence we can mark the sequence $\{S_n\}_{n\geq 0}$ on the column 0. To complete the proof it is enough to put the letter 1 in the cells $(0, S_n)$, for every $n \geq 0$ and the letter 0 in all other cells in the column 0.

Note that the hypothesis of eventual periodicity of the sequence $(d_n)_{n\geq 1}$ in the previous proof is essential to guarantee that the cellular automaton is defined over a finite set $\mathcal{A}$. Indeed, since the signals (and their slope) are periodically repeated, we have $\mathrm{Card}(\mathcal{A}) = \mathcal{O}(k)$, where k is the length of the maximum between the pre-period and the eventual period of $(d_n)_{n\geq 1}$.

Example 7. Let us consider the word **f** defined in Example 1 and the associated numerical sequence $(F_n)_{n\geq 0}$. According to Proposition 3 we have $\mathbb{1}_{\{F_n\}_{n\geq 0}} \in \mathcal{S}$.

5   Construction of Prefixes

In this section we prove our main theorem. In order to do that, we need some preliminary results.

Proposition 4. *Let $\Delta = (d_n)_{n\geq 1}$ be an eventually periodic integer sequence with $d_1 \geq 0$ and $d_n \geq 1$ for every $n > 1$; let $(w_n)_{n\geq -1}$ be the standard sequence associated to Δ defined by $w_{-1} = $ b, $w_0 = $ a, and $w_n = w_{n-1}^{d_n}w_{n-2}$ for every $n \geq 1$. Then $\mathbf{w} = \lim_{n\to\infty} w_n \in \mathcal{S}$.*

In order to prove Proposition 4, we need the following result stating that a letter in a cell of a CA can be recopied in the same column and in any row above. Moreover, we can do it by using only walls and signals of slope 1 and -1.

Lemma 1. *Let $T : \mathcal{A}^{\mathbb{Z}} \to \mathcal{A}^{\mathbb{Z}}$ be a CA, $a \in \mathcal{A}$, and $n, m, t, t' \in \mathbb{Z}$ with $m > 0$, and $t > t' \geq 0$. Suppose that the cells (n, t'), $(n + m, t')$ and (n, t) are marked, the last one with a. Then it is possible to mark the cell $(n, t + m)$ with a.*

Proof. Without loss of generality, suppose that $n = 0$ and $t' = 0$. and so a is in the cell $(0, t)$. Let us prove that we can recopy a into the cell $(0, t + m)$. To mark the wall we are going to use the cell $(m, 0)$. We consider two distinct cases, according to the parity of m.

Let m be odd. We consider a wall of type (i) to the right of the column $\lfloor \frac{m}{2} \rfloor$. Such a wall can be defined, for instance, using two signals of slope 1 and -1 starting respectively at $(0,0)$ and at $(m,0)$ (see left of Fig. 4).

We send a signal of slope 1 from the cell $(0,t)$; when this signal touches the wall we define a new signal of slope -1 (starting from the cell above); when this new signal meets the column 0, we write the letter a. Since $\lfloor \frac{m}{2} \rfloor = \frac{m-1}{2}$, we have that the new a is exactly on the row $t + \frac{m-1}{2} + \frac{m-1}{2} + 1 = t + m$. as wanted.

Suppose now that m is even. We consider this time a wall of type (ii) in the column $\frac{m}{2}$. Such a wall can be defined, for instance, using two signals of slope 1 and -1 starting respectively at $(0,0)$ and at $(m,0)$ (see right of Fig. 4). We send a signal of slope 1 from the cell $(0,t)$; when this signal hit the wall, we define a new signal of slope -1 (starting from the same cell); when this new signal meets the column 0, we write the letter a. The new a is exactly on the row $t + \frac{m}{2} + \frac{m}{2} - 1 + 1 = t + m$, as wanted.

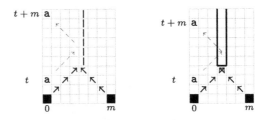

Fig. 4. Recopying of letters.

We can now prove Proposition 4.

Proof (of Proposition 4). Let us denote $S_n = |w_n|$ for all $n \geq -1$. For every $n \geq 1$ we have $S_n = d_n S_{n-1} + S_{n-2}$. Since, for all $n \geq 2$, w_{n-3} is a suffix of w_{n-1}, the word $w_{n-3}w_{n-2}$ is a suffix of w_n. Suppose the word w_n constructed for a given $n \geq 2$ and suppose that the last letter of w_n is in the cell $(0, S_n)$.

Let us first suppose that $d_n \neq 1$. We will show that it is possible to construct the word w_{n+1} with its last letter in the cell $(0, S_{n+1})$. In order to do that, we will use the relation $w_{n+1} = w_n((w_{n-2}^{d_n-1}w_{n-3})^{d_n}w_{n-2})^{d_{n+1}-1}w_{n-2}^{d_n-1}w_{n-3}$. Let us take up the construction of Proposition 3 until the number S_n. Moreover, in the column 0 we mark the cell $(0, S_{n-1} + d_{n-1}(d_n - 1)S_{n-2})$. For this, we define a signal of slope $-(d_{n-1}(d_n - 1))$ from the cell (S_{n-2}, S_{n-1}). This signal meets the vertical signal defined from $(0, S_{n-1})$ in the required cell $(0, S_{n-1} + d_{n-1}(d_n - 1)S_{n-2})$. From this cell we define a signal $\mathbf{P}_1^{(n-1)}$ of slope 1 and from the cells $(S_{n-2}, S_{n-1} + d_{n-1}(d_n - 1)S_{n-2})$, $(S_{n-1}, S_{n-1} + d_{n-1}(d_n - 1)S_{n-2})$ and $(S_{n-1} + S_{n-2}, S_{n-1} + d_{n-1}(d_n - 1)S_{n-2})$ we define three signals $\mathbf{N}_1^{(n-1)}$, $\mathbf{N}_2^{(n-1)}$ and $\mathbf{N}_3^{(n-1)}$ of slope -1. At the intersection of these with the signal $\mathbf{P}_1^{(n-1)}$ we define three walls $\mathbf{M}_1^{(n-1)}$, $\mathbf{M}_2^{(n-1)}$ and $\mathbf{M}_3^{(n-1)}$ in the columns S_{n-2}, S_{n-1} and

$S_{n-1} + S_{n-2}$ respectively. These walls will be of type (i) or of type (ii) according to the parity of the columns where $\mathbf{N}_i^{(n-1)}$ originate. Suppose we have already recopied the words w_{n-3} and w_{n-2} in the suffix of w_n, i.e., we have found the two corresponding heights in the column 0. Now, from these two words, we are going to define signals of slope 1 that will stop against one of the three walls previously defined and send back signals of slope -1 that will recopy the same word in the column 0. Using Lemma 1 we can recopy one by one the letters of w_{n-2} and w_{n-3}. First, we send a signal of slope 1 from each letter of w_{n-2}. When this signal meets the wall $\mathbf{M}_1^{(n-1)}$ we send a signal of slope -1 until the column 0. Since $\mathbf{N}_1^{(n-1)}$ is generated from the column S_{n-2} the letters of w_{n-2} are recopied in the same column but above at distance S_{n-2}. This means that we have recopied a second word w_{n-2} above the first. We repeat this procedure d_{n-1} times and we get the word $w_{n-2}^{d_{n-1}}$ above the w_n already constructed. Next, we send a signal of slope 1 from each letter of w_{n-3} to the wall $\mathbf{M}_3^{(n-1)}$ and, in a similar way, from there we send a signal of slope -1 to the column 0. The letters of w_{n-3} are recopied in the same column but above at distance $S_{n-1} + S_{n-2} = d_{n-1}S_{n-2} + S_{n-3} + S_{n-2}$. Thus, we have recopied a word w_{n-3} above the word $w_{n-3}w_{n-2}w_{n-2}^{d_{n-1}}$, where the subword $w_{n-3}w_{n-2}$ corresponds to a suffix of w_n. So far we have constructed the word $w_n(w_{n-2}^{d_{n-1}}w_{n-3})$ on the column 0. For the next step we need to use also the word w_{n-2} in the suffix of w_n. This time, we send signals of slope 1 from cells in column 0 to the wall $\mathbf{M}_3^{(n-1)}$ and from there, signals of slope -1 to the column 0. Therefore, the letters of w_{n-2} are recopied in the same column but above at distance $S_{n-1} + S_{n-2} = (d_{n-1} + 1)S_{n-2} + S_{n-3}$. Hence we obtain the word $w_n(w_{n-2}^{d_{n-1}}w_{n-3})w_{n-2}$ in the column 0. By using the wall $\mathbf{M}_1^{(n-1)}$ the word $w_n((w_{n-2}^{d_{n-1}}w_{n-3})w_{n-2}^{d_{n-1}}$ can be obtained. From every letter of the word w_{n-3} we send signals of slope 1 to the wall $\mathbf{M}_2^{(n-1)}$ and, from there, signals of slope -1 back to column 0. The letters of w_{n-3} are recopied in the same column but above at distance $S_{n-1} = d_{n-1}S_{n-2} + S_{n-3}$. Hence, we obtain the word $w_n(w_{n-2}^{d_{n-1}}w_{n-3})w_{n-2}^{d_{n-1}}w_{n-3}$. Similarly, it is easy to obtain the word $w_n(w_{n-2}^{d_{n-1}}w_{n-3})^{d_n} = w_n(w_{n-2}^{d_{n-1}}w_{n-3})^{d_n-1}(w_{n-2}w_{n-2}^{d_{n-1}-1}w_{n-3})$. Following the same idea, we use the wall $\mathbf{M}_2^{(n-1)}$ to recopy w_{n-2} in the column 0 but $S_{n-1} = (d_{n-1} - 1)S_{n-2} + S_{n-3} + S_{n-2}$ cells above and so, we obtain the word $w_n(w_{n-2}^{d_{n-1}}w_{n-3})^{d_n}w_{n-2}$. Since the suffix of this word is $w_{n-3}w_{n-2}$ the previous steps can be applied again to obtain the word $w_n((w_{n-2}^{d_{n-1}}w_{n-3})^{d_n}w_{n-2})^{d_{n+1}-1}$ which also has the suffix $w_{n-3}w_{n-2}$. Thus we easily obtain the word $w_n((w_{n-2}^{d_{n-1}}w_{n-3})^{d_n}w_{n-2})^{d_{n+1}-1}w_{n-2}^{d_{n-1}}w_{n-3} = w_{n+1}$.

Let us now consider the case $d_n = 1$. Here we have $S_n = S_{n-1} + S_{n-2}$ and the cell $(S_{n-1} + S_{n-2}, S_{n-1} + d_{n-1}(d_n - 1)S_{n-2}) = (S_n, S_{n-1})$ is not marked. Therefore the wall $\mathbf{M}_3^{(n-1)}$ can no longer be defined as before. This time, we define a wall $\mathbf{M}_4^{(n)}$ at the intersection of a signal $\mathbf{P}_2^{(n)}$ a slope 1 starting from $(0, S_n)$ and a signal $\mathbf{N}_4^{(n)}$ of slope -1 starting from (S_n, S_n). The suffix $w_{n-2}w_{n-3}$ of w_n is below the signals used to define the wall $\mathbf{M}_4^{(n)}$, therefore, we have to construct the word $w_{n-3}w_{n-2}$ in $w_n\big(w_{n-2}^{d_{n-1}}w_{n-3}w_{n-2}\big)^{d_{n+1}-1}w_{n-2}^{d_{n-1}}w_{n-3} = w_{n+1}$ in another way. We have to copy the letters of the word $w_{n-3}w_{n-2}$ in the column 0 above at distance $S_{n-3} + S_{n-2} + d_{n-1}S_{n-2} = S_{n-1} + S_{n-2} = S_n$. In order to do that, we define a wall $\mathbf{M_5^{(n-1)}}$ of type (i) or (ii) (according to the parity) by intersecting a signal $\mathbf{P_3^{(n-1)}}$ of slope 1 from (S_{n-2}, S_{n-1}) and a signal $\mathbf{N_2^{(n-1)}}$ of slope -1 from (S_{n-1}, S_{n-1}) defined as in the previous case. Such a wall is in the column $S_{n-2} + \left\lfloor \frac{S_{n-1}-S_{n-2}}{2} \right\rfloor = S_n - S_{n-1} + \left\lfloor S_{n-1} - \frac{S_n}{2} \right\rfloor = \left\lfloor \frac{S_n}{2} \right\rfloor$.

We send a signal of slope 1 from each letter of $w_{n-3}w_{n-2}$ to the wall $\mathbf{M_5^{(n-1)}}$ and from there we send a signal of slope -1 until the column 0. Therefore, we have constructed the word $w_n w_{n-2}^{d_{n-1}} w_{n-3}w_{n-2}$. The rest of the construction is the same as before, with $\mathbf{M_4^{(n)}}$ playing the role of the wall $\mathbf{M_3^{(n-1)}}$ for the rows above the word $w_{n-3}w_{n-2}$.

Note that the signals $\mathbf{P_i^{(n)}}$ can also be used to destroy the walls $\mathbf{M_i^{(n-1)}}$ previously constructed. Formally, when a signal $\mathbf{P_i^{(n)}}$ meets a wall $\mathbf{M_i^{(n-1)}}$, the last one is destroyed and the signal $\mathbf{P_i^{(n)}}$ continues its move.

Example 8. The Fibonacci word $\mathbf{f}$ defined in Example 1 is in $\mathcal{S}$.

We can now prove our main result.

Proof (of Theorem 1). Let $\mathbf{w}$ be a Sturmian word of quadratic slope, $\alpha = [0, 1 + b_1, b_2, \ldots, b_m, \overline{a_1, a_2, \ldots, a_k}]$ the continued fraction expansion of its slope, and $(w_n)_{n \geq -1}$ the standard sequence associated to the eventually periodic integer sequence $\Delta = (b_1, \cdots, b_m, (a_1, \cdots, a_k)^\omega)$ so that Δ is an eventually periodic integer sequence. Following Proposition 1, we have $\mathbf{w} = \lim_{n \to \infty} w_n$. Using Propositions 3 and 4, it is clear that $\mathbf{w} \in \mathcal{S}$.

6   Conclusions

To prove our results we used the continued fraction expansion associated with $\mathbf{w}$. However, to use Proposition 4 it is enough to know how to decompose each prefix w_n in compositions of powers of smaller prefixes. A different approach could be to use the morphisms

$$G = \begin{cases} a \mapsto a \\ b \mapsto ab \end{cases} \qquad \text{and} \qquad D = \begin{cases} a \mapsto ba \\ b \mapsto b \end{cases}.$$

Indeed, it is known that for every standard Sturmian sequence $\mathbf{w}$ there exist a unique sequence of words $\big(\mathbf{w}^{(i)}\big)_i$ and an infinite sequence $(\psi_i)_i \in \{G, D\}^{\mathbb{N}}$ of morphisms such that $\mathbf{w} = \lim_{n \to \infty} \psi_0 \psi_1 \ldots \psi_n(w^{(n)})$ (see, for instance, [12]). Using these notions, and the strictly related notion of $\mathcal{S}$-adicity, we think it is possible to generalise our results to larger classes of words and languages, namely Arnoux-Rauzy words and dendric words (see, e.g., [2,3,7]).

Blanchard and Kůrka [4] consider a larger family of languages that can be recognised by a non-deterministic Turing machine. This family contains the languages corresponding to quadratic numbers but also the ones corresponding to Hurwitz numbers, such as e (a Hurwitz number is an irrational such that its continued fraction expansion is a polynomial mixture [11]). An interesting question is whether it is possible to generalise our construction to such a family as well.

References

1. Balková, Ľ, Pelantová, E., Starosta, Š: Sturmian jungle (or garden?) on multiliteral alphabets. RAIRO - Theor. Inf. Appl. **44**, 443–470 (2010)
2. Berthé, V., et al.: Acyclic, connected and tree sets. Monatshefte für Mathematik **176**(4), 521–550 (2014). https://doi.org/10.1007/s00605-014-0721-4
3. Berthé, V., Dolce, F., Durand, F., Leroy, J., Perrin, D.: Rigidity and substitutive tree words. Int. J. Found. Comp. S. **29**, 705–720 (2018)
4. Blanchard, F., Kůrka, P.: Language complexity of rotations and Sturmian sequences. Theoret. Comput. Sci. **209**, 179–193 (1998)
5. Coven, E.M., Hedlund, G.A.: Sequences with minimal block growth. Math. Sys. Theory **7**, 138–153 (1973)
6. Dolce, F., Dvořáková, L., Pelantová, E.: On balanced sequences and their asymptotic critical exponent. In: Leporati, A., Martín-Vide, C., Shapira, D., Zandron, C. (eds.) LATA 2021. LNCS, vol. 12638, pp. 293–304. Springer, Cham (2021). https://doi.org/10.1007/978-3-030-68195-1_23
7. Dolce, F., Perrin, D.: Eventually dendric shift spaces. Ergod. Theor. Dyn. Syst. **41**, 2023–2048 (2021)
8. Fischer, P.C.: Generation of primes by a one-dimensional real-time iterative array. J. Assoc. Comput. Mach. **12**, 388–394 (1965)
9. Hedlund, G.A.: Endomorphisms and automorphisms of the shift dynamical system. Math. Sys. Theory **3**, 320–375 (1969)
10. Hubert, P.: Suites équilibrées. Theoret. Comput. Sci. **242**, 91–108 (2000)
11. Hurwitz, A.: Über die Bedingungen, unter welchen eine Gleichung nur Wurzeln mit negativen reellen Theilen besitzt. Math. Ann. **46**, 273–3284 (1895)
12. Justin, J., Pirillo, G.: Episturmian words and episturmian morphisms. Theoret. Comput. Sci. **276**, 281–313 (2002)
13. Korec, I.: Real-time generation of primes by a one-dimensional cellular automaton with 11 states. In: Mathematical Foundations of Computer Science 1997, vol. 1295, pp. 358–367. Berlin, Heidelberg (1997)
14. Lothaire, M.: Algebraic combinatorics on words. Encyclopedia of Mathematics and its Applications, vol. 90. Cambridge University Press, Cambridge (2002)
15. Marcovici, I., Stoll, T., Tahay, P.-A.: Construction of some nonautomatic sequences by cellular automata. Lecture Notes in Comput. Sci. **10875**, 113–126 (2018)
16. Mazoyer, J., Terrier, V.: Signals in one-dimensional cellular automata. Theoret. Comput. Sci. **217**, 53–80 (1999)
17. Morse, M., Hedlund, G.A.: Symbolic dynamics II. Sturmian Trajectories. Am. J. Math. **62**, 1–42 (1940)
18. Rendell, P.: Turing universality of the game of life. In: Adamatzky, A. (eds.) Collision-Based Computing, ECC, vol. 18. Springer, Cham (2002). https://doi.org/10.1007/978-3-319-19842-2
19. Rowland, E., Yassawi, R.: A characterization of p-automatic sequences as columns of linear cellular automata. Adv. in Appl. Math. **63**, 68–89 (2015)
20. Sloane, N.J.A.: On-line Encyclopedia of Integer Sequences. https://oeis.org/

Logarithmic Equal-Letter Runs for BWT of Purely Morphic Words

Andrea Frosini[1], Ilaria Mancini[2], Simone Rinaldi[2], Giuseppe Romana[3], and Marinella Sciortino[3(⊠)]

[1] Universitá di Firenze, Florence, Italy
`andrea.frosini@unifi.it`
[2] Universitá di Siena, Siena, Italy
`ilaria.mancini@student.siena.it`, `simone.rinaldi@unisi.it`
[3] Universitá di Palermo, Palermo, Italy
`{giuseppe.romana01,marinella.sciortino}@unipa.it`

Abstract. In this paper we study the number r_{bwt} of equal-letter runs produced by the *Burrows-Wheeler transform* (BWT) when it is applied to *purely morphic finite words*, which are words generated by iterating prolongable morphisms. Such a parameter r_{bwt} is very significant since it provides a measure of the performances of the BWT, in terms of both compressibility and indexing. In particular, we prove that, when BWT is applied to whichever purely morphic finite word on a binary alphabet, r_{bwt} is $\mathcal{O}(\log n)$, where n is the length of the word. Moreover, we prove that r_{bwt} is $\Theta(\log n)$ for the binary words generated by a large class of prolongable binary morphisms. These bounds are proved by providing some new structural properties of the *bispecial circular factors* of such words.

Keywords: Burrows-Wheeler Transform · Equal-letter runs · Morphisms · Bispecial circular factors

1 Introduction

The Burrows-Wheeler Transform (BWT) is a reversible transformation that produces a permutation of the text given in input, according to the lexicographical order of its cyclic rotations. It was introduced in 1994 in the field of Data Compression [3] and it still represents the main component of some of the most known lossless text compression tools [24] as well as of compressed indexes [9]. BWT is used as pre-processing of memoryless compressors, causing the *boosting* of their performance. The key motivation for this fact is that BWT is likely to create equal-letter runs (clusters) that are longer than the clusters of the original text. In other words, if we denote with $r(w)$ the number of equal-letter runs in the word w, the number $r_{\mathsf{bwt}}(w)$ of equal-letter runs produced by BWT applied to w often becomes lower than $r(w)$. It is important to note that the performance in terms of both space and time of text compressors and compressed indexing data structures applied on a text w can be evaluated by using $r_{\mathsf{bwt}}(w)$ [11].

© Springer Nature Switzerland AG 2022
V. Diekert and M. Volkov (Eds.): DLT 2022, LNCS 13257, pp. 139–151, 2022.
https://doi.org/10.1007/978-3-031-05578-2_11

An upper bound on the number of clusters produced by BWT has been provided in [13], in particular, it has been proved that $r_{\mathsf{bwt}}(w) = \mathcal{O}(z(w)\log^2 n)$ where $z(w)$ is the number of phrases in the $LZ77$ factorization of w and n is the length of w. The ratio between $r_{\mathsf{bwt}}(w)$ and the number of clusters in the BWT of the reverse of w has been studied in [12]. A recent comparative survey illustrating the properties of $r_{\mathsf{bwt}}(w)$ and other repetitiveness measures can be found in [18]. In particular, in this survey the measure γ, which is the size of the smallest string attractor for the sequence [14], and the measure δ, which is defined from the factor complexity function [5], are also considered. From a combinatorial point of view, the parameter r_{bwt} has been studied in order to obtain more information about the combinatorial complexity of a word from the number of clusters produced by applying the BWT. In particular, great attention has been given to the characterization of the words for which the BWT produces the minimal number of clusters [8,17,22,26]. A first combinatorial investigation of the BWT clustering effect has been given in [15,16] in which the BWT-clustering ratio $\rho(w) = \frac{r_{\mathsf{bwt}}(w)}{r(w)}$ has been studied. In particular, it has been proved in [16] that $\rho(w) \leq 2$ and infinite families of words for which ρ assumes its maximum value have been shown. In [15] the behavior of ρ is studied for two very well known families of words, namely Sturmian words and de Brujin binary words.

This paper is focused on investigating the behaviour of BWT when applied to finite words obtained by iterating a morphism. Morphisms are well-known objects in the field of combinatorics on words and they represent a powerful and natural tool to define repetitive sequences. Studying the compressibility of repetitive sequences is an issue that is raising great interest. The morphisms, combined with macro-schemes, have been used to define other mechanisms to generate repetitive sequences, called NU-systems [19]. We consider the morphisms φ that admit a fixed point (denoted by $\varphi^\infty(a)$) starting from a given character $a \in A$, i.e. $\varphi^\infty(a) = \lim_{i\to\infty} \varphi^i(a)$. Such morphisms are called *prolongable* on a. In [25] the measure γ is computed for the prefixes of infinite words that are fixed points of some morphisms. Moreover, a complete characterization of the Lempel-Ziv complexity z for the prefixes of fixed points of prolongable morphisms has been given in [6]. In this paper we analyse the number r_{bwt} of equal-letter runs and the BWT-clustering ratio ρ when BWT is applied to the *purely morphic finite words*, i.e. the words $\varphi^i(a)$ generated by iterating a morphism φ prolongable on a. In [2], the parameter ρ has been computed for the families of finite words generated by some morphisms. In all cases considered in [2], the BWT efficiently clusters since the value of ρ is much lower than 1. In the one-page abstract appeared in [10] we extended such results by providing some new upper bounds on $r_{\mathsf{bwt}}(\varphi^i(a))$ depending on the factor complexity of the fixed point $\varphi^\infty(a)$ for a large class of morphisms.

In this paper, we formalize the notion of BWT-*highly compressible* morphism by evaluating, for a given morphism φ, whether BWT-clustering ratio ρ on the words $\varphi^i(a)$ tends towards zero, when i goes to infinity. We introduce the notion of *run-bounded* morphism and we identify some classes of BWT-highly compressible morphisms, by proving that r_{bwt} is $\mathcal{O}(\log n)$ for words of length n generated by primitive morphisms.

Furthermore, we give some combinatorial properties of several classes of binary morphisms and we improve the results announced in [10] in the case of binary purely morphic words. In fact, in [10] we provided the upper bound $\mathcal{O}(\log n)$ for the parameter r_{bwt} only for one class of binary morphism. Here, we prove that such an upper bound holds for every purely morphic finite word on a binary alphabet. A consequence of these results is that all binary prolongable morphisms, except a few cases, are BWT-highly compressible. Finally, we prove that $r_{\mathsf{bwt}}(w)$ is $\Theta(\log n)$ for the binary finite words w generated by a large class of morphisms. Such bounds are obtained by using a close relation between $r_{\mathsf{bwt}}(w)$ and the combinatorial notion of a bispecial circular factor of w and by providing some new structural properties of the bispecial circular factors of infinite families of finite binary words generated by prolongable morphisms.

2 Preliminaries

Let $A = \{a_1, a_2, \ldots, a_k\}$ be a finite ordered alphabet with $a_1 < a_2 < \ldots < a_k$, where $<$ denotes the standard lexicographic order. We assume that $|A| \geq 2$. The set of words over the alphabet A is denoted by A^*. A finite word $w = w_1 w_2 \cdots w_n \in A^*$ is a finite sequence of letters from A. The length of w, denoted $|w|$, is the number n of its letters, $|w|_a$ denotes the number of occurrences of the letter a in w. An infinite word $x = x_1 x_2 x_3 \ldots$ is a non-ending sequence of elements of the alphabet A.

Given an infinite or finite word x, we say that a word u is a *factor* of x if $x = vuy$ for some words v and y. The word u is a *prefix* (resp. *suffix*) of x if $x = uy$ (resp. $x = yu$) for some word y. A factor u of x is *left special* (*right special*) if there exist $a, b \in A$ with $a \neq b$ such that both au and bu (ua and ub) are factors of x. A factor u is *bispecial* if it is both left and right special. We denote by $f_x(k)$ the number of distinct factors of x having length k. The function f_x is called *factor complexity* of x.

We say that a finite word w has a *period* $p > 0$ if $w_i = w_{i+p}$ for each $i \leq |w| - p$. It is easy to see that each integer $p \geq |w|$ is a period of w. The smallest of all periods is called *minimum period* of w. The notion of period can be also given for infinite words. We say that an infinite word is *ultimately periodic* with period $p > 0$ if exists $K \geq 1$ such that $w_i = w_{i+p}$ for each $i \geq K$. Moreover, if this condition holds for every $i \in \mathbb{N}$, w is said *periodic* (with period p). An infinite word x is *aperiodic* if it is not ultimately periodic.

Given two finite words $w, z \in A^*$, we say that w is a cyclic rotation of z, or equivalently w and z are *conjugate*, if $w = uv$ and $z = vu$, where $u, v \in A^*$. Conjugacy between words is an equivalence relation over A^*. We say that a finite word u is a *circular factor* of w if u is a factor of a conjugate of w. For instance, aa is a circular factor of $abbba$, but it is not a factor. We denote by $\mathcal{C}(w)$ the set of circular factors of a word w. If we denote by $c_w(k)$ the number of distinct circular factors of w having length k, it is easy to see that $f_w(k) \leq c_w(k)$, for each

$k \geq 1$. Note that the notions of left special, right special and bispecial factors can be extended to circular factors. In particular, we say that u is a *bispecial circular factor* of a word w if there exist $a, b \in A$, with $a \neq b$, and $a', b' \in A$, with $a' \neq b'$, such that both aua' and bub' are circular factors of w. We denote by $BS(w)$ the set of bispecial circular factors of w.

The *Burrows-Wheeler Transform (BWT)* is a reversible transformation introduced in the context of Data Compression [3]. Given a word $w \in A^*$, the BWT produces a permutation of w which is obtained by concatenating the last letter of the lexicographically sorted cyclic rotations of w, and we denote it with $\mathtt{bwt}(w)$. Note that $\mathtt{bwt}(w) = \mathtt{bwt}(v)$ if and only if w and v are conjugate.

The *run-length encoding* of a word w, denoted by $\mathtt{rle}(w)$, is a sequence of pairs (w_i, l_i) with $w_i \in A$ and $l_i > 0$, such that $w = w_1^{l_1} w_2^{l_2} \cdots w_r^{l_r}$ and $w_i \neq w_{i+1}$. We denote by $r(w) = |\mathtt{rle}(w)|$, i.e. the number r of equal-letter runs in w. We denote by $r_{\mathtt{bwt}}(w) = r(\mathtt{bwt}(w))$ the number of equal-letter runs in $\mathtt{bwt}(w)$. The *BWT-clustering ratio* $\rho(w) = \frac{r_{\mathtt{bwt}}(w)}{r(w)}$ is a measure introduced to evaluate how much the number of equal-letter runs varies when BWT is applied [16].

Morphisms are fundamental tools of formal languages and a very crucial notion in combinatorics on words. They represent a very interesting way to generate an infinite family of words. Let A and Σ be alphabets. A *morphism* is a map φ from A^* to Σ^* that obeys the identity $\varphi(uv) = \varphi(u)\varphi(v)$ for all words $u, v \in A^*$. By definition, a morphism can be described by just specifying the images of the letters of A. Examples of very well known morphisms are the *Thue-Morse morphism* τ, defined as $\tau(a) = ab$ and $\tau(b) = ba$, and the *Fibonacci morphism* θ, defined as $\theta(a) = ab$ and $\theta(b) = a$. A morphism φ is *primitive* if there exists a positive integer k such that, for every pair of characters $a, b \in A$, the character a occurs in $\varphi^k(b)$. Both τ and θ are primitive morphisms. A morphism is called *non-erasing* if $|\varphi(a)| \geq 1$, for each $a \in A$. In the following we consider only non-erasing morphisms.

Morphisms can be classified by the length of images of letters. If there is a constant k such that $|\varphi(a)| = k$ for all $a \in A$ then we say that φ is k-*uniform* (or just uniform, if k is clear from the context). For instance, the Thue-Morse morphism τ is 2-uniform. The *growth function* of a morphism φ with respect to a letter $a \in A$ and an iteration i is defined by $\varphi_a(i) = |\varphi^i(a)|$. A letter a is said to be *growing* for φ if $\lim_{i \to \infty} \varphi_a(i) = +\infty$, otherwise it is *bounded*. A morphism φ is growing if each letter of the alphabet is growing for φ. For growing morphisms it holds that, for every $a \in A$, $\varphi_a(i) = \Theta(i^{e_a} p_a^i)$, for some $e_a \geq 0$ and $p_a > 1$. Another classification of morphisms is according to its growth function on distinct letters [20]. A growing morphism φ is called *quasi-uniform* if $\varphi_a(i) = \Theta(p^i)$ for every $a \in A$ and some $p > 0$; φ is called *polynomially divergent* if for every $a \in A$ it holds that $\varphi_a(i) = \Theta(i^e p^i)$ for some $p > 1$ and exist $a, b \in A$ such that $e_a \neq e_b \geq 0$; φ is called *exponentially divergent* if exist $a, b \in A$ such that $\varphi_a(i) = \Theta(i^{e_a} p_a^i)$ and $\varphi_b(i) = \Theta(i^{e_b} p_b^i)$, for some $e_a, e_b \geq 0$ and $p_a \neq p_b > 1$.

A morphism is called *prolongable* on a letter $a \in A$ if $\varphi(a) = au$ with $u \in A^+$. Then, for $i \geq 1$, $\varphi^i(a) = au\varphi(u) \cdots \varphi^{i-1}(u)$. In this case, the infinite family of finite words $\{a, \varphi(a), \ldots, \varphi^i(a), \ldots\}$ are prefixes of a unique infinite word denoted by $\varphi^\infty(a)$, that is called *purely morphic word* or *word generated by φ*. In this paper, we assume $\varphi(a) \notin \{a\}^+$. Examples of infinite words generated by a morphism are the *Thue-Morse word* $t = abbabaabbaababba\ldots$ generated by the Thue-Morse morphism τ and the *Fibonacci word* $f = abaababaabaab\ldots$ generated by the Fibonacci morphism θ. More generally, an infinite word is called *morphic* if is generated by applying a *coding* (a 1-uniform morphism from A to a possibly different alphabet Σ) to a purely morphic word.

3 BWT-Highly Compressible Morphisms

In this section, we focus on the morphisms that generate finite words w on which the Burrows-Wheeler transform leads to a significant reduction of the number of equal-letter runs. In particular, we show that some upper bounds depending on the factor complexity of the fixed point of the morphism can be derived.

Definition 1. *A morphism φ prolongable on $a \in A$ is BWT-highly compressible if $\limsup_{i \to \infty} \rho(\varphi^i(a)) = 0$, where ρ is the BWT-clustering ratio.*

The factor complexity of purely morphic words has been studied [20].

Theorem 2 [20]. *Let $x = \varphi^\infty(a)$ be an infinite aperiodic word and let f_x be its factor complexity.*

1. *If φ is growing, then $f_x(n)$ is $\Theta(n)$, $\Theta(n \log \log n)$ or $\Theta(n \log n)$ if φ is quasi-uniform, polynomially divergent or exponentially divergent, respectively*
2. *Let φ be not-growing and let B be the set of its bounded letters*
   (a) if x has arbitrarily large factors of B^* then $f_x(n) = \Theta(n^2)$
   (b) if the factors of B^* in x have bounded length then $f_x(n)$ can be any of $\Theta(n)$, $\Theta(n \log \log n)$ or $\Theta(n \log n)$.

The following theorem summarizes some results in [7] regarding the factor complexity of some particular classes of morphisms. Both Fibonacci morphism θ and Thue-Morse morphism τ are included in these classes.

Theorem 3. *Let $x = \varphi^\infty(a)$ be an aperiodic infinite word. If φ is uniform or primitive, then $f_x(n) = \Theta(n)$.*

The following two examples provide a *BWT*-highly compressible and a not *BWT*-highly compressible morphism, respectively.

Example 4 ($\Theta(n \log \log n)$ factor complexity). Let us consider the binary morphism φ defined as $\varphi(a) = abab$ and $\varphi(b) = bb$. In this case $\varphi_a(i) = (i+1)2^i$ and $\varphi_b(i) = 2^i$. Moreover, $r(\varphi^i(a)) = 2^{i+1}$ and $r_{\text{bwt}}(\varphi^i(a)) = 2i$, for $i > 2$. Hence, φ is *BWT*-highly compressible.

Example 5 ($\Theta(n \log n)$ factor complexity). Let us consider the morphism ψ defined as $\psi(a) = abc$, $\psi(b) = bb$ and $\psi(c) = ccc$. One can verify that $x = \psi^\infty(a) = abcb^2c^3b^4c^9 \ldots$ and $\psi_a(i+1) = \psi_a(i) + 2^i + 3^i$. Moreover, $\rho(\psi^i(a)) = \frac{r_{bwt}(\psi^i(a))}{r(\psi^i(a))} = \frac{4i}{2i+1} > 1$ for $i > 2$. Hence, ψ is not BWT-highly compressible.

The following proposition gives an upper bound on the value $r_{bwt}(\varphi^i(a))$, for some classes of morphisms prolongable on a. Such bounds depend on the factor complexity of the infinite word generated by φ.

Proposition 6. *Let $x = \varphi^\infty(a)$ be an infinite word. Then the following bounds for $r_{bwt}(\varphi^i(a))$, $i \geq 1$, hold:*

1. *if $f_x(n) = \Theta(1)$ then $r_{bwt}(\varphi^i(a)) = \Theta(1)$;*
2. *if $f_x(n) = \Theta(n)$ then $r_{bwt}(\varphi^i(a)) = \mathcal{O}(i)$;*
3. *if $f_x(n) = \Theta(n \log \log n)$ then $r_{bwt}(\varphi^i(a)) = \mathcal{O}(i \log i \log \log i)$;*
4. *if $f_x(n) = \Theta(n \log n)$ then $r_{bwt}(\varphi^i(a)) = \mathcal{O}(i^2 \log i)$.*

Proof. (Sketch) Bound 1 follows from the inequalities $r_{bwt}(v) \leq r_{bwt}(uv^k) \leq r_{bwt}(v) + 2|u|$, for every $u, v \in A^*$ and $k > 0$. The other bounds can be derived by using an upper bound proved in [13] and from the fact that the growth of every morphism is in $\mathcal{O}(\rho_a^i)$, for some $\rho_a > 1$ [23]. □

Note that the upper bounds 2, 3 and 4 of Proposition 6 have been enunciated in [10], and the upper bound 2 extends some known results. In fact, as shown in [2], for the i-th Thue-Morse word $\tau^i(a)$, it holds that $r_{bwt}(\tau^i(a)) = \Theta(i)$. We also remark that, since $n = |\tau^i(a)| = 2^i$, $r_{bwt}(\tau^i(a)) = \Theta(\log n)$. However, the lower bounds can be quite different. In fact, for the i-th Fibonacci word $\theta^i(a)$ it holds that $r_{bwt}(\theta^i(a)) = \Theta(1)$, as shown in [17]. In the next section we show that, in case of binary alphabet, lower and upper bounds can be derived for some classes of morphisms.

In the next example we show a class of morphisms φ_k, over an alphabet of size k, such that $r_{bwt}(\varphi_k^i(a)) = \Theta(n^{\frac{1}{k-1}})$, where $n = |\varphi_k^i(a)|$.

Example 7 ($\Theta(n^2)$ factor complexity). Let us consider the morphism

$$
\varphi_k : \quad
\begin{aligned}
a_1 &\mapsto a_1 a_2 \\
a_2 &\mapsto a_2 a_3 \\
&\ldots \\
a_{k-1} &\mapsto a_{k-1} a_k \\
a_k &\mapsto a_k
\end{aligned}
$$

One can verify that $n = |\varphi_k^i(a)| = \Theta(i^{k-1})$ [6]. Moreover, $r(\varphi_k^i(a)) = \Theta(i^{k-2})$ and $r_{bwt}(\varphi_k^i(a)) = \Theta(i) = \Theta(n^{\frac{1}{k-1}})$. Hence, φ_k is BWT-highly compressible for every $k > 3$.

Here we introduce the notion of *run-bounded* morphism in order to identify some classes of BWT-highly compressible morphisms.

Definition 8. *Let φ be a morphism such that $r(\varphi^i(a)) \leq K$, for every $i > 0$, for some $a \in A$ and $K > 0$. Then we say that φ is run-bounded on a.*

The following two propositions give bounds for $r(\varphi^i(a))$. Note that, since φ is prolongable, $r(\varphi^i(a))$ is not decreasing.

Proposition 9. *Let φ be a morphism prolongable on $a \in A$ and let $\mathcal{R} = \{b \in A \mid \varphi$ is run-bounded on $b\}$. If $\varphi(a) = au$ with $u \in \mathcal{R}^+$ then $r(\varphi^i(a)) = \mathcal{O}(i)$.*

Proof. Recall that $\varphi^i(a) = au\varphi(u)\varphi^2(u)\cdots\varphi^{i-1}(u)$. Since φ is run-bounded on every symbol in u, then $r(\varphi^i(u)) \leq K \cdot |u|$, for some $K > 0$ and every $i > 0$. Hence, we have that $r(\varphi^i(a)) \leq 1 + \sum_{j=0}^{i-1} r(\varphi^j(u)) \leq (K \cdot |u|) \cdot i = \mathcal{O}(i)$.   □

Proposition 10. *Let φ be a morphism prolongable on $a \in A$ such that $r(\varphi(a)) \geq 2$. If exists $t > 0$ such that a occurs at least twice in $\varphi^t(a)$, then the growth of $r(\varphi^i(a))$ is exponential.*

The following corollary can be proved by using Propositions 6 and 10.

Corollary 11. *Let φ be a primitive morphism and prolongable on a and let $n = |\varphi^i(a)|$, $i \geq 1$. It holds that $r_{bwt}(\varphi^i(a)) = \mathcal{O}(\log n)$ and $\lim_{i \to \infty} \rho(\varphi^i(a)) = 0$, therefore φ is BWT-highly compressible.*

Corollary 11 confirms what could be also deduced from results in [2] and [17], namely that the Thue-Morse morphism τ and the Fibonacci morphism θ are *BWT*-highly compressible, since both are primitive. The following example shows that, unlike primitive morphisms, there exist uniform morphisms on generic alphabets that are not *BWT*-highly compressible. The situation becomes different if binary alphabets are considered, as shown in the next section.

Example 12. Let us consider the 3-uniform morphism η defined as $\eta(a) = abc$, $\eta(b) = bbb$ and $\eta(c) = ccc$. It is easy to verify that in this case $r(\eta^i(a)) = r(abcbbbccc \cdots b^{3^{i-1}} c^{3^{i-1}}) = 2i + 1$ and $r_{bwt}(\eta^i(a))) = 4i$, for every $i \geq 2$. Hence, $\rho(\eta^i(a)) = 2 - \frac{2}{2i+1}$ and it is not *BWT*-highly compressible.

4   Upper and Lower Bounds for r_{bwt}

In this section we show that lower and upper bounds for r_{bwt} of a word w over a generic alphabet can be derived by considering the number of extensions of some bispecial circular factors of w. By focusing on binary words over the ordered alphabet $A = \{a, b\}$ generated by a binary morphism μ prolongable on a, we give some new structural properties of their circular bispecial factors. Such results allow us to derive logarithmic lower and upper bounds for some classes of binary morphisms. Furthermore, we prove that for all the binary morphisms μ prolongable on a, except few cases, $\lim_{i \to \infty} \rho(\mu^i(a)) = 0$. Hence they are *BWT*-highly compressible. Note that such results are independent of the order between the letters in A.

Let u be a circular factor of a given word w over a generic alphabet A. Inspired by the notation in [4], we denote by $e_r(u) = |\{x \in A \mid ux \in \mathcal{C}(w)\}| - 1$ the *number of right circular extensions of u in w* minus 1, and by $e_\ell(u) = |\{x \in A \mid xu \in \mathcal{C}(w)\}| - 1$ the *number of left circular extensions of u in w* minus 1. The bispecial circular factors of w can be classified according to the number of their extensions. In particular, a circular factor u is *strictly bispecial* if $|\mathcal{C}(w) \cap AuA| = (e_r(u) + 1)(e_\ell(u) + 1)$, u is *weakly bispecial* if $|\mathcal{C}(w) \cap AuA| = \max\{e_r(u), e_\ell(u)\} + 1$. We denote by $SBS(w)$ and $WBS(w)$ the set of strictly and weakly bispecial circular factors of w, respectively. The following lemma holds. The upper bound is already known in terms of the size of the compact directed acyclic word graph [1].

Lemma 13. *Let w be a word over the alphabet A. Then,*

$$\sum_{u \in WBS(w)} \min\{e_l(u), e_r(u)\} + 1 \le r_{bwt}(w) \le \sum_{u \in BS(w)} e_r(u) + 1.$$

From now on, we suppose that $A = \{a, b\}$. Given a binary morphism μ, the notation $\mu \equiv (\alpha, \beta)$ means that $\mu(a) = \alpha$ and $\mu(b) = \beta$.

4.1   Combinatorial Structure of Binary Morphisms

In this subsection we give a combinatorial characterization of α and β, for several classes of binary morphisms $\mu \equiv (\alpha, \beta)$, with $\alpha \notin \{a\}^+$. Such a characterization, which depends on the factor complexity of the fixed point $\mu^\infty(a)$, is used in the next subsection to derive lower and upper bound for r_{bwt}.

The following proposition consider ultimately periodic purely morphic words and can be proved by using the proof of [21, Corollary 3].

Proposition 14. *Let $x = \mu^\infty(a)$ be an infinite binary ultimately periodic word, where $\mu \equiv (\alpha, \beta)$. Then, one of the following cases occurs:*

1. $\alpha = \eta^\ell$ and $\beta = \eta^t$, for some $\eta \in \{a, b\}^*$ and some $\ell, t \ge 1$;
2. $\alpha = ab^k$ and $\beta = b^\ell$, for some $k \ge 1$, $\ell \ge 1$;
3. $\alpha = (ab)^p a$ and $\beta = (ba)^q b$ for some $p, q \ge 1$;
4. $\alpha = (ab^p)^q a$ and $\beta = b$, for some $p \ge 1$, $q \ge 1$.

Let $R_i = \{q_1 < q_2 < \ldots < q_h\}$ be the set of non-negative integers such that $ab^{q_j}a$ is a circular factor in $\mu^i(a)$, for some $1 \le j \le h$. Denoted by $n_a = |\alpha|_a$, it is easy to verify that $|R_1| \le n_a$, since $\mu(a) = ab^{t_1}ab^{t_2}a \cdots ab^{t_{n_a}}$ with $t_j \ge 0$ for every $1 \le j \le n_a$. As shown later, the size of the sets R_i can be used to derive tighter bounds for r_{bwt} on words generated by binary morphisms. The two following lemmas give a combinatorial characterization of the sets R_i for every non-primitive binary morphism.

Lemma 15. *Let $\mu = (\alpha, \beta)$ be a non-growing morphism prolongable on a and let $x = \mu^\infty(a)$ be its fixed point.*

1. If $\alpha = auba^k$, $\beta = b$, for some $u \in A^*$, and $k \geq 1$ then $R_i = R_1$ for every $i \geq 1$;
2. If $\alpha = auab^k$, $\beta = b$, for some $u \in A^*$, and $k \geq 1$ then $R_i = \bigcup_{h=1}^{i} \bigcup_{j=1}^{n_a-1} \{t_j + (h-1)k\} \cup \{ik\}$ for every $i \geq 1$.

Lemma 16. Let μ be a growing non-primitive binary morphism. If $\mu^\infty(a)$ is ultimately periodic, then $\mu \equiv (ab^k, b^\ell)$, $k \geq 1, \ell > 1$ and $R_i = k \sum_{j=0}^{i-1} \ell^j$, $i \geq 1$.

Otherwise, it holds that $\mu \equiv (auab^k, b^\ell)$, $k \geq 0$, $\ell > 1$, $u \in A^*$ and $R_i = \bigcup_{h=1}^{i} \bigcup_{j=1}^{n_a-1} \{\frac{\ell^{h-1}((\ell-1)t_j+k)-k}{\ell-1}\} \cup \{k\frac{\ell^i-1}{\ell-1}\}$, for every $i \geq 1$.

The following two propositions give a combinatorial characterization of non-primitive morphisms generating aperiodic words.

Proposition 17. Let $\mu = (\alpha, \beta)$ be a non-growing morphism prolongable on a and let $x = \mu^\infty(a)$ be its aperiodic fixed point. Then, one of the following cases must occur:

1. $\alpha = auba^k$, $\beta = b$, for some $u \in A^*$ and $k \geq 1$, and $f_x(n) = \Theta(n)$;
2. $\alpha = auab^k$, $\beta = b$, for every $u \in A^*$ and $k \geq 1$, and $f_x(n) = \Theta(n^2)$.

Remark 18. Note that the morphism $\mu \equiv (auab^k, b)$, $k \geq 1$ always generates a fixed point with quadratic factor complexity for some $u \in A^*$. If $\mu \equiv (auba^k, b)$, $k \geq 1$, then we have to distinguish two cases: if $k > 1$, then μ generates a fixed point with linear factor complexity for some $u \in A^*$; if $k = 1$, by Proposition 14 (case 4), $u \neq (b^p a)^q b^{p-1}$, $p, q \geq 1$, otherwise the fixed point is ultimately periodic.

Proposition 19. Let $\mu = (\alpha, \beta)$ be a growing non-primitive morphism prolongable on a and let $x = \mu^\infty(a)$ be its aperiodic fixed point. Then, $\mu \equiv (av, b^\ell)$, for some $\ell \geq 2$ and $v \in A^+$ such that $|v|_a, |v|_b \geq 1$. Moreover, let $n_a = |av|_a$. Then, it holds that:

1. $n_a < \ell$ iff $f_x(n) = \Theta(n)$;
2. $n_a = \ell$ iff $f_x(n) = \Theta(n \log \log n)$;
3. $n_a > \ell$ iff $f_x(n) = \Theta(n \log n)$.

The following proposition shows that for every binary non-primitive morphism, except for the case of non-growing morphisms with linear factor complexity of the fixed point (case 1 of Lemma 15), the size of the set R_i of non-negative integers j such that $ab^j a$ is a circular factor of $\mu^i(a)$ grows linearly with i.

Proposition 20. Let $\mu \equiv (auab^k, b^\ell)$ a binary morphism with $k \geq 0$, $\ell \geq 1$ and $k + \ell > 1$ and aperiodic fixed point. Then $|R_i| = \Theta(i)$.

4.2   Logarithmic Bounds for r_{bwt} in Case of Binary Morphisms

In this subsection we prove that if w is a finite word of length n generated by iterating a binary morphism, then $r_{\mathrm{bwt}}(w) = \mathcal{O}(\log n)$. Moreover, we identify some classes of binary morphisms for which $\Omega(\log n)$ is a lower bound for r_{bwt}. From Proposition 14, one can easily derive that in case of ultimately periodic words r_{bwt} is $\Theta(1)$. In case of a primitive morphism μ the upper bound $\mathcal{O}(\log n)$ can be deduced by Proposition 6 and Theorem 3, by using the fact that, in this case, $\mu_i(a)$ is exponential [23]. Hence, here we can suppose that the morphism $\mu \equiv (\alpha, \beta)$ is not primitive, with $\alpha \notin \{a\}^+$, and $\mu^\infty(a)$ is aperiodic.

The following lemmas give a structural characterization of the bispecial circular factors of the words generated by iterating a binary morphism. In particular, Lemma 21 shows how to construct bispecial circular factors of $\mu^{i+1}(a)$ starting from the bispecial circular factors of $\mu^i(a)$, $i \geq 1$. In Lemma 22 we prove that all bispecial circular factors can be constructed by starting from the bispecial circular factors of a finite set of words depending of the images of μ on the letters of the alphabet.

Lemma 21. *Let $\mu \equiv (auab^k, b^\ell)$ be a binary morphism, for some $k \geq 0, \ell \geq 1$. If v is a circular bispecial factor of $\mu^i(a)$, $i \geq 1$, then $w = b^k \mu(v)$ is a bispecial circular factor of $\mu^{i+1}(a)$.*

Lemma 22. *Let $\mu \equiv (\alpha, \beta) = (auab^k, b^\ell)$ for some $k \geq 0$, $\ell \geq 1$, let m be the length of the longest equal-letter run of b's that occurs in $auab^k$, and let $M = \max\{\lfloor \frac{m-(\ell+1)k}{\ell^2} \rfloor, 0\}$. Then, every circular bispecial factor w of $\mu^{i+1}(a)$, $i \geq 1$, either appears as a circular factor in $\bigcup_{j=0}^{M}\{\mu(\alpha)\mu(\beta)^j\mu(\alpha)\}$ or $w = b^k\mu(v)$, for some circular bispecial factor v in $\mu^i(a)$ (or $w = b^h$, for some $h \geq 1$, when $k \geq 0$ and $\ell > 1$).*

Theorem 23. *Let $\mu \equiv (auab^k, b^\ell)$ be a non-primitive morphism with $k \geq 0$, $\ell \geq 1$ with aperiodic fixed point. Then $r_{bwt}(\mu^i(a)) = \mathcal{O}(i)$.*

Proof. (Sketch) The case $k = 0$ and $\ell = 1$ follows from Proposition 17 and Proposition 6. In case of binary alphabet, Lemma 13 implies that, for every $i \geq 1$, $r_{\mathrm{bwt}}(\mu^i(a)) \leq |BS(\mu^i(a))| + 1$. If $k \geq 1$ and $\ell = 1$, then $|BS(\mu^i(a))| = \mathcal{O}(i)$ by using Lemma 22. If $k \geq 0$ and $\ell > 1$, $|BS(\mu^i(a))|$ grows exponentially since BS contains the subset $BS_b(\mu^i(a)) = \{b^h \mid b^h, b^{h+1} \in \mathcal{C}(\mu^i(a))\}$ by Lemma 22. However, only the elements of a subset of $BS_b(\mu^i(a))$ size at most $2|R_i|$ produce an increase of 1 for $r_{\mathrm{bwt}}(\mu^i(a))$. The thesis follows from Proposition 20. □

Theorem 24. *Let $\mu \equiv (auab^k, b^\ell)$ be a non-primitive morphism with $k \geq 0$, $\ell \geq 1$ and $x = \mu^\infty(a)$ is aperiodic. Then $r_{bwt}(\mu^i(a)) = \Omega(|R_i|)$. Moreover, when μ is not growing with $f_x(n) = \Theta(n^2)$ or μ is growing, then $r_{bwt}(\mu^i(a)) = \Omega(i)$.*

Proof. (Sketch) Let $R_i = \{q_1 < q_2 < \ldots < q_{|R_i|}\}$ the set of non-negative integers such that $ab^{q_j}a \in \mathcal{C}(\mu^i(a))$. For every $q_j \in R_i$, we can consider the block X_j of lexicographically sorted conjugates starting with $b^{q_j}a$. Among the corresponding characters in $\mathtt{bwt}(\mu^i(a))$, at least one occurrence of the letter a is included. For every $0 \le j \le \frac{\lceil |R_i| - 1\rceil}{2}$, let us consider the blocks X_{2j+1} and X_{2j+3}. Since $q_{2j+1} < q_{2j+3} - 1$, there are at least $|X_{2j+3}|$ lexicographically sorted conjugates starting with $b^{q_{2j+3}-1}a$ and ending with b. The second part of the thesis is proved by using Propositions 17, 19 and 20.   □

Remark 25. Note that Theorem 23 and Theorem 24 also hold when $b < a$.

The following lemma and corollary allow us to states that, for morphisms focused in this subsection, $i = \Theta(\log n)$, where $n = \mu_a(i) = |\mu^i(a)|$.

Lemma 26. *Let $\mu \equiv (\alpha, \beta)$ a binary morphism prolongable on a. Let $n_a = |\alpha|_a, n_b = |\alpha|_b, m_a = |\beta|_a, m_b = |\beta|_b$ and $\alpha_i = |\mu^{i-1}(a)|_a, \beta_i = |\mu^{i-1}(a)|_b$. It holds that:*

$$r(\mu^i(a)) \ge \alpha_i r(\alpha) + \beta_i r(\beta) - |\mu^{i-1}(a)| + 1$$

with

$$\alpha_i = n_a \alpha_{i-1} + m_a n_b \alpha_{i-2} + m_a m_b n_b \alpha_{i-3} + m_a m_b^2 n_b \alpha_{i-4} + \cdots + m_a m_b^{i-3} n_b$$
$$\beta_i = m_b \beta_{i-1} + m_a n_b \beta_{i-2} + m_a n_a n_b \beta_{i-3} + m_a n_a^2 n_b \beta_{i-4} + \cdots + m_a n_a^{i-4} n_b \beta_2 + n_a^{i-2} n_b.$$

From Lemma 26 and Proposition 10, the following corollary follows.

Corollary 27. *Let $\mu = (\alpha, \beta)$ a binary morphism prolongable on a. Then, the growth of $\mu_a(i)$ is exponential except when $\alpha = ab^p$, with $p \ge 1$, and $\beta = b$, where $\mu_a(i) = \Theta(i)$.*

The goal of the following result is to evaluate the BWT-clustering ratio of the finite words generated by iterating a binary morphism. The proof can be derived from Lemma 26, Proposition 14, Corollary 11, and Theorem 23.

Theorem 28. *Let $\mu \equiv (\alpha, \beta)$ be a binary morphism prolongable on a such that $\mu \not\equiv (ab^m, b^n)$ for every $m \ge 1$, $n \ge 1$. Then $\lim_{i \to \infty} \rho(\mu^i(a)) = 0$, consequently μ is BWT-highly compressible.*

5   Conclusions and Further Work

In this paper, we have studied the number $r_{\mathtt{bwt}}(w)$ of equal-letter runs produced by the BWT, when $w = \mu^i(a)$ of length n is the binary word generated after the i-th iteration of a morphism μ prolongable on the letter a with an aperiodic fixed point $x = \mu^\infty(a)$. We have proved that $r_{\mathtt{bwt}}(w)$ is $\Theta(\log n)$ when μ is a non-primitive growing morphism or a non-primitive not-growing morphism such that $f_x(n) = \Theta(n^2)$. The problem of characterizing the primitive morphisms such that $r_{\mathtt{bwt}}(w)$ is $\Omega(\log n)$ is still open. This could allow a tight lower bound to be deduced even for non-primitive not growing morphisms such that $f_x(n) = \Theta(n)$. Moreover, we are interested to extend these bounds also for purely morphic finite words on larger alphabets, and also for generic morphic finite words.

References

1. Belazzougui, D., Cunial, F., Gagie, T., Prezza, N., Raffinot, M.: Composite repetition-aware data structures. In: Cicalese, F., Porat, E., Vaccaro, U. (eds.) CPM 2015. LNCS, vol. 9133, pp. 26–39. Springer, Cham (2015). https://doi.org/10.1007/978-3-319-19929-0_3
2. Brlek, S., Frosini, A., Mancini, I., Pergola, E., Rinaldi, S.: Burrows-Wheeler transform of words defined by morphisms. In: Colbourn, C.J., Grossi, R., Pisanti, N. (eds.) IWOCA 2019. LNCS, vol. 11638, pp. 393–404. Springer, Cham (2019). https://doi.org/10.1007/978-3-030-25005-8_32
3. Burrows, M., Wheeler, D.J.: A block-sorting lossless data compression algorithm. Technical report, DIGITAL System Research Center (1994)
4. Cassaigne, J.: Complexity and special factors. (complexité et facteurs spéciaux.). Bull. Belgian Math. Soc. - Simon Stevin 4(1), 67–88 (1997)
5. Christiansen, A.R., Ettienne, M.B., Kociumaka, T., Navarro, G., Prezza, N.: Optimal-time dictionary-compressed indexes. ACM Trans. Algorithms 17(1), 8:1–8:39 (2021)
6. Constantinescu, S., Ilie, L.: The Lempel-Ziv complexity of fixed points of morphisms. SIAM J. Discret. Math. 21(2), 466–481 (2007)
7. Ehrenfeucht, A., Lee, K.P., Rozenberg, G.: Subword complexities of various classes of deterministic developmental languages without interactions. Theor. Comput. Sci. 1(1), 59–75 (1975)
8. Ferenczi, S., Zamboni, L.Q.: Clustering words and interval exchanges. J. Integer Seq. 16(2), Article 13.2.1 (2013)
9. Ferragina, P., Manzini, G.: Indexing compressed text. J. ACM 52, 552–581 (2005)
10. Frosini, A., Mancini, I., Rinaldi, S., Romana, G., Sciortino, M.: Burrows-Wheeler transform on purely morphic words. In: DCC, pp. 452–452. IEEE (2022)
11. Gagie, T., Navarro, G., Prezza, N.: Fully functional suffix trees and optimal text searching in BWT-runs bounded space. J. ACM 67(1), 2:1–2:54 (2020)
12. Giuliani, S., Inenaga, S., Lipták, Z., Prezza, N., Sciortino, M., Toffanello, A.: Novel results on the number of runs of the Burrows-Wheeler transform. In: Bureš, T., et al. (eds.) SOFSEM 2021. LNCS, vol. 12607, pp. 249–262. Springer, Cham (2021). https://doi.org/10.1007/978-3-030-67731-2_18
13. Kempa, D., Kociumaka, T.: Resolution of the Burrows-Wheeler transform conjecture. In: FOCS, pp. 1002–1013. IEEE (2020)
14. Kempa, D., Prezza, N.: At the roots of dictionary compression: string attractors. In: STOC. pp. 827–840. ACM (2018)
15. Mantaci, S., Restivo, A., Rosone, G., Sciortino, M.: Burrows-Wheeler transform and Run-Length Enconding. In: Brlek, S., Dolce, F., Reutenauer, C., Vandomme, É. (eds.) WORDS 2017. LNCS, vol. 10432, pp. 228–239. Springer, Cham (2017). https://doi.org/10.1007/978-3-319-66396-8_21
16. Mantaci, S., Restivo, A., Rosone, G., Sciortino, M., Versari, L.: Measuring the clustering effect of BWT via RLE. Theoret. Comput. Sci. 698, 79–87 (2017)
17. Mantaci, S., Restivo, A., Sciortino, M.: Burrows-Wheeler transform and Sturmian words. Inform. Process. Lett. 86, 241–246 (2003)
18. Navarro, G.: Indexing highly repetitive string collections, part I: repetitiveness measures. ACM Comput. Surv. 54(2), 29:1–29:31 (2021)
19. Navarro, G., Urbina, C.: On stricter reachable repetitiveness measures. In: Lecroq, T., Touzet, H. (eds.) SPIRE 2021. LNCS, vol. 12944, pp. 193–206. Springer, Cham (2021). https://doi.org/10.1007/978-3-030-86692-1_16

20. Pansiot, J.: Complexité des facteurs des mots infinis engendrés par morphimes itérés. In: ICALP. Lecture Notes Computer Science, vol. 172, pp. 380–389. Springer (1984)
21. Pansiot, J.J.: Decidability of periodicity for infinite words. RAIRO - Theor. Inform. Appl. **20**(1), 43–46 (1986)
22. Restivo, A., Rosone, G.: Burrows-Wheeler transform and palindromic richness. Theoret. Comput. Sci. **410**(30–32), 3018–3026 (2009)
23. Rozenberg, G., Salomaa, A.: The Mathematical Theory of L Systems. Elsevier Science (1980)
24. Seward, J.: The bzip2 home page (2006). http://www.bzip.org
25. Shaeffer, L., Shallit, J.: String attractors for automatic sequences. CoRR abs/2012.06840 (2020)
26. Simpson, J., Puglisi, S.J.: Words with simple Burrows-Wheeler transforms. Electron. J. Combin. **15** (article R83) (2008)

On Perfect Coverings of Two-Dimensional Grids

Elias Heikkilä[1], Pyry Herva[2(✉)], and Jarkko Kari[2]

[1] Nordic Semiconductor, Aalto University, Espoo, Finland
elias.heikkila@aalto.fi
[2] Department of Mathematics and Statistics, University of Turku, 20014 Turku,
Finland
{pysahe,jkari}@utu.fi

Abstract. We study perfect multiple coverings in translation invariant
graphs with vertex set $\mathbb{Z}^2$ using an algebraic approach. In this approach
we consider any such covering as a two-dimensional binary configuration which we then express as a two-variate formal power series. Using
known results, we conclude that any perfect multiple covering has a
non-trivial periodizer, that is, there exists a non-zero polynomial whose
formal product with the power series presenting the covering is a two-
periodic configuration. If a non-trivial periodizer has line polynomial
factors in at most one direction, then the configuration is known to be
periodic. Using this result we find many setups where perfect multiple
coverings of infinite grids are necessarily periodic. We also consider some
algorithmic questions on finding perfect multiple coverings.

Keywords: Perfect multiple coverings · Two-dimensional
configurations · Laurent polynomials · Formal power series · Periodicity

1 Introduction and Preliminaries

A perfect multiple covering in a graph is a set of vertices, a code, such that
the number of codewords in the neighborhood of an arbitrary vertex depends
only on whether the vertex is in the code or not. In this paper we study these
codes on translation invariant graphs with the vertex set $\mathbb{Z}^2$. We present codes
as two-dimensional binary configurations and observe that the perfect covering
condition provides an algebraic condition that can be treated with the algebraic
tools developed in [8]. We focus on periodic codes and, in particular, study setups
where all codes are necessarily periodic. The approach we take was initially
mentioned in an example in the survey [6] by the third author, and considered
in the Master's thesis [5] by the first author.

We start by giving the basic definitions, presenting the aforementioned algebraic approach and stating some past results relevant to us. In Sect. 2 we describe
an algorithm to find the line polynomial factors of any given (Laurent) polynomial. In Sect. 3 we formally define the perfect multiple coverings in graphs and

V. Diekert and M. Volkov (Eds.): DLT 2022, LNCS 13257, pp. 152–163, 2022.
https://doi.org/10.1007/978-3-031-05578-2_12

prove some periodicity results concerning them. We give new algebraic proofs of some known results concerning perfect multiple coverings on the infinite square grid and on the triangular grid [1,12], and provide a new result on the forced periodicity of such coverings on the king grid. Furthermore, we generalize the definition of perfect coverings for two-dimensional binary configurations with respect to different neighborhoods and covering constants. In Sect. 4 we consider some algorithmic questions concerning perfect coverings. Using a standard argument by H. Wang we show that under certain constraints it is algorithmically decidable to determine whether there exist any perfect coverings with given neighborhood and given covering constants.

Configurations, Periodicity, Finite Patterns and Subshifts

A d-dimensional *configuration* is a coloring of the infinite grid $\mathbb{Z}^d$ using finitely many colors, that is, an element of $\mathcal{A}^{\mathbb{Z}^d}$ which we call the d-dimensional *configuration space* where $\mathcal{A}$ is some finite alphabet. For a configuration c we let $c_{\mathbf{u}} = c(\mathbf{u})$ to be the symbol or color that c has in cell $\mathbf{u}$. The *translation* $\tau^{\mathbf{t}}$ by a vector $\mathbf{t} \in \mathbb{Z}^d$ shifts a configuration c such that $\tau^{\mathbf{t}}(c)_{\mathbf{u}} = c_{\mathbf{u}-\mathbf{t}}$ for all $\mathbf{u} \in \mathbb{Z}^d$. A configuration c is $\mathbf{t}$-*periodic* if $\tau^{\mathbf{t}}(c) = c$ and c is *periodic* if c is $\mathbf{t}$-periodic for some non-zero $\mathbf{t} \in \mathbb{Z}^d$. We also say that a configuration c is *periodic in direction* $\mathbf{v} \in \mathbb{Z}^d \setminus \{\mathbf{0}\}$ if c is $k\mathbf{v}$-periodic for some $k \in \mathbb{Q}$. A d-dimensional configuration c is *strongly periodic* if it has d linearly independent vectors of periodicity. Strongly periodic configurations are then periodic in all directions. Two-dimensional strongly periodic configurations are called *two-periodic*.

A finite *pattern* is an assignment of symbols on some finite shape $D \subseteq \mathbb{Z}^d$, that is, an element of $\mathcal{A}^D$ where $\mathcal{A}$ is some fixed alphabet. In particular, the finite patterns in $\mathcal{A}^D$ are called D-*patterns*. Let us denote by $\mathcal{A}^*$ the set of all finite patterns over alphabet $\mathcal{A}$ where the dimension d is known from the context. A finite pattern $p \in \mathcal{A}^D$ *appears* in a configuration $c \in \mathcal{A}^{\mathbb{Z}^d}$ if $\tau^{\mathbf{t}}(c)|_D = p$ for some $\mathbf{t} \in \mathbb{Z}^d$. A configuration c *contains* the pattern p if it appears in c. For a fixed shape D, the set of all D-patterns that appear in c is the set $\mathcal{L}_D(c) = \{\tau^{\mathbf{t}}(c)|_D \mid \mathbf{t} \in \mathbb{Z}^d\}$ and the set of all finite patterns in c is denoted by $\mathcal{L}(c)$ which we call the *language of* c. For a set $\mathcal{S} \subseteq \mathcal{A}^{\mathbb{Z}^d}$ of configurations we define $\mathcal{L}_D(\mathcal{S})$ and $\mathcal{L}(\mathcal{S})$ as the unions of $\mathcal{L}_D(c)$ and $\mathcal{L}(c)$ over all $c \in \mathcal{S}$, respectively.

Let us review some basic concepts of symbolic dynamics we need. For a reference see *e.g.* [3,10,11]. The configuration space $\mathcal{A}^{\mathbb{Z}^d}$ can be made a compact topological space by endowing $\mathcal{A}$ with the discrete topology and considering the product topology it induces on $\mathcal{A}^{\mathbb{Z}^d}$ – the *prodiscrete topology*. This topology is induced by a metric where two configurations are close if they agree on a large area around the origin. Thus $\mathcal{A}^{\mathbb{Z}^d}$ is a compact metric space.

A subset $\mathcal{S} \subseteq \mathcal{A}^{\mathbb{Z}^d}$ of the configuration space is a *subshift* if it is topologically closed and translation-invariant meaning that if $c \in \mathcal{S}$ then for any $\mathbf{t} \in \mathbb{Z}^d$ also $\tau^{\mathbf{t}}(c) \in \mathcal{S}$. Equivalently we can define subshifts using forbidden patterns: Given a set $F \subseteq \mathcal{A}^*$ of *forbidden* finite patterns, the set

$$X_F = \{c \in \mathcal{A}^{\mathbb{Z}^d} \mid \mathcal{L}(c) \cap F = \emptyset\}$$

of configurations that avoid all forbidden patterns is a subshift, and every sub-shift is obtained by forbidding some set of finite patterns. If $F \subseteq \mathcal{A}^*$ is finite then we say that X_F is a *subshift of finite type* (SFT).

The *orbit* of a configuration c is the set $\mathcal{O}(c) = \{\tau^{\mathbf{t}}(c) \mid \mathbf{t} \in \mathbb{Z}^d\}$ of its every translate. The *orbit closure* $\overline{\mathcal{O}(c)}$ is the topological closure of its orbit under the prodiscrete topology. The orbit closure of a configuration c is the smallest subshift that contains c. It consists of all configurations c' such that $\mathcal{L}(c') \subseteq \mathcal{L}(c)$.

The Algebraic Approach

To present a configuration $c \in \mathcal{A}^{\mathbb{Z}^d}$ algebraically we make the assumption that $\mathcal{A} \subseteq \mathbb{Z}$. Then we identify the configuration c with the formal power series

$$c(X) = \sum_{\mathbf{u} \in \mathbb{Z}^d} c_{\mathbf{u}} X^{\mathbf{u}}$$

over d variables $x_1, \ldots, x_d$ where we have denoted $X = (x_1, \ldots, x_d)$ and $X^{\mathbf{u}} = x_1^{u_1} \cdots x_d^{u_d}$ for any $\mathbf{u} = (u_1, \ldots, u_d) \in \mathbb{Z}^d$. For $d = 2$ we usually denote $X = (x, y)$. More generally we study the set of all formal power series $\mathbb{C}[[X^{\pm 1}]] = \mathbb{C}[[x_1^{\pm 1}, \ldots, x_d^{\pm 1}]]$ over d variables $x_1, \ldots, x_d$ with complex coefficients. A power series is *finitary* if it has only finitely many different coefficients and *integral* if its coefficients are all integers. Thus we identify configurations with finitary and integral power series.

We also use Laurent polynomials which we call from now on simply polyno-mials. We use the term "proper" when we talk about proper (*i.e.*, non-Laurent) polynomials. Let us denote by $\mathbb{C}[X^{\pm 1}] = \mathbb{C}[x_1^{\pm 1}, \ldots, x_d^{\pm 1}]$ the set of all (Laurent) polynomials over d variables $x_1, \ldots, x_d$ with complex coefficients, which is the *Laurent polynomial ring*. We say that two polynomials have no common factors if all of their common factors are units and that they have a common factor if they have a non-unit common factor.

A product of a polynomial and a power series is well defined. We say that a polynomial $f = f(X)$ *annihilates* (or is an annihilator of) a power series $c = c(X)$ if $fc = 0$, that is, if their product is the zero power series. We say that a formal power series $c = c(X)$ is *periodic* if it is annihilated by a *difference polynomial* $X^{\mathbf{t}} - 1$ where $\mathbf{t}$ is non-zero. Note that this definition is consistent with the definition of periodicity of configurations defined above. Indeed if $c = c(X)$ is a configuration then multiplying it by a monomial $X^{\mathbf{t}}$ produces the translated configuration $\tau^{\mathbf{t}}(c)$ and hence c is $\mathbf{t}$-periodic if and only if $c = \tau^{\mathbf{t}}(c) = X^{\mathbf{t}}c$, which is equivalent to $(X^{\mathbf{t}} - 1)c = 0$. So it is natural to study the *annihilator ideal*

$$\mathrm{Ann}(c) = \{f \in \mathbb{C}[X^{\pm 1}] \mid fc = 0\}$$

of a power series $c \in \mathbb{C}[[X^{\pm 1}]]$, which indeed is an ideal of the Laurent polynomial ring. Hence the question whether a configuration (or any formal power series) is periodic is equivalent to asking whether its annihilator ideal contains a difference polynomial. Another useful ideal that we study is the *periodizer ideal*

$$\mathrm{Per}(c) = \{f \in \mathbb{C}[X^{\pm 1}] \mid fc \text{ is strongly periodic}\}.$$

Note that clearly Ann(c) is a subset of Per(c). Note also that a configuration c has a non-trivial (= non-zero) annihilator if and only if it has a non-trivial periodizer. The following theorem states that if a configuration has a non-trivial periodizer then it has in fact an annihilator of a particular simple form – a product of difference polynomials.

Theorem 1 [8]. *Let c be a configuration in any dimension that has a non-trivial periodizer. Then there exist pairwise linearly independent $\mathbf{t}_1, \ldots, \mathbf{t}_m$ with $m \geq 1$ such that*

$$(X^{\mathbf{t}_1} - 1) \cdots (X^{\mathbf{t}_m} - 1) \in Ann(c).$$

Line Polynomials

The *support* of a power series $c = \sum_{\mathbf{u} \in \mathbb{Z}^d} c_{\mathbf{u}} X^{\mathbf{u}}$ is the set $\text{supp}(c) = \{\mathbf{u} \in \mathbb{Z}^d \mid c_{\mathbf{u}} \neq 0\}$. Thus a polynomial is a power series with a finite support. A *line polynomial* is a polynomial whose support contains at least two points and the points of the support lie on a unique line. In other words, a polynomial f is a line polynomial if it is not a monomial and there exist vectors $\mathbf{u}, \mathbf{v} \in \mathbb{Z}^d$ such that $\text{supp}(f) \subseteq \mathbf{u} + \mathbb{Q}\mathbf{v}$. In this case we say that f is a line polynomial in direction $\mathbf{v}$. We say that non-zero vectors $\mathbf{v}, \mathbf{v}' \in \mathbb{Z}^d$ are *parallel* if $\mathbf{v}' \in \mathbb{Q}\mathbf{v}$, and clearly then a line polynomial in direction $\mathbf{v}$ is also a line polynomial in any parallel direction. A vector $\mathbf{v} \in \mathbb{Z}^d$ is *primitive* if its components are pairwise relatively prime. If $\mathbf{v}$ is primitive then $\mathbb{Q}\mathbf{v} \cap \mathbb{Z}^d = \mathbb{Z}\mathbf{v}$. For any non-zero $\mathbf{v} \in \mathbb{Z}^d$ there exists a parallel primitive vector $\mathbf{v}' \in \mathbb{Z}^d$. It follows that we may assume the vector $\mathbf{v}$ in the definition of a line polynomial f to be primitive so that $\text{supp}(f) \subseteq \mathbf{u} + \mathbb{Z}\mathbf{v}$. In the following our preferred presentations of directions are in terms of primitive vectors.

Any line polynomial ϕ in a (primitive) direction $\mathbf{v}$ can be written uniquely in the form

$$\phi = X^{\mathbf{u}}(a_0 + a_1 X^{\mathbf{v}} + \ldots + a_n X^{n\mathbf{v}}) = X^{\mathbf{u}}(a_0 + a_1 t + \ldots + a_n t^n)$$

where $\mathbf{u} \in \mathbb{Z}^d, n \geq 1, a_0 \neq 0, a_n \neq 0$ and $t = X^{\mathbf{v}}$. Let us call the single variable proper polynomial $a_0 + a_1 t + \ldots + a_n t^n \in \mathbb{C}[t]$ the *normal form* of ϕ. Moreover, for a monomial $a X^{\mathbf{u}}$ we define its normal form to be a. Thus two line polynomials in the direction $\mathbf{v}$ have the same normal form if and only if they are the same polynomial up to multiplication by $X^{\mathbf{u}}$, for some $\mathbf{u} \in \mathbb{Z}^d$.

Difference polynomials are line polynomials and hence the annihilator provided by Theorem 1 is a product of line polynomials. Annihilation by a difference polynomial means periodicity. More generally, annihilation of a configuration c by a line polynomial in a primitive direction $\mathbf{v}$ can be understood as the annihilation of the one-dimensional $\mathbf{v}$-*fibers* $\sum_{k \in \mathbb{Z}} c_{\mathbf{u}+k\mathbf{v}} X^{\mathbf{u}+k\mathbf{v}}$ of c in direction $\mathbf{v}$, and since annihilation in the one-dimensional setting implies periodicity we conclude that a configuration is periodic if and only if it is annihilated by a line polynomial. It is known that if c has a periodizer with line polynomial factors in at most one direction then c is periodic:

Theorem 2 [9]. *Let c be a two-dimensional configuration and $f \in Per(c)$. Then the following conditions hold.*

- *If f does not have any line polynomial factors then c is two-periodic.*
- *If all line polynomial factors of f are in the same direction then c is periodic in this direction.*

Proof Sketch. The periodizer ideal $Per(c)$ is a principal ideal generated by a polynomial $g = \phi_1 \cdots \phi_m$ where $\phi_1, \ldots, \phi_m$ are line polynomials in pairwise non-parallel directions [9]. Because $f \in Per(c)$ we know that g divides f. If f does not have any line polynomial factors then $g = 1$ and thus $c = gc$ is two-periodic. If f has line polynomial factors and they are in the same primitive direction $\mathbf{v}$ then g is a line polynomial in this direction. Since gc is two-periodic it is annihilated by $(X^{k\mathbf{v}} - 1)$ for some $k \in \mathbb{Z}$. Then the configuration c is annihilated by the line polynomial $(X^{k\mathbf{v}} - 1)g$ in direction $\mathbf{v}$. We conclude that c is periodic in direction $\mathbf{v}$.   □

2   Line Polynomial Factors

The open and closed *discrete half planes* determined by a non-zero vector $\mathbf{v} \in \mathbb{Z}^2$ are the sets $H_{\mathbf{v}} = \{\mathbf{u} \in \mathbb{Z}^2 \mid \langle \mathbf{u}, \mathbf{v}^\perp \rangle > 0\}$ and $\overline{H}_{\mathbf{v}} = \{\mathbf{u} \in \mathbb{Z}^2 \mid \langle \mathbf{u}, \mathbf{v}^\perp \rangle \geq 0\}$, respectively, where $\mathbf{v}^\perp = (v_2, -v_1)$ is orthogonal to $\mathbf{v} = (v_1, v_2)$. Let us also denote by $l_{\mathbf{v}} = \overline{H}_{\mathbf{v}} \setminus H_{\mathbf{v}}$ the discrete line parallel to $\mathbf{v}$ that goes through the origin. In other words, the half plane determined by $\mathbf{v}$ is the half plane "to the right" of the line $l_{\mathbf{v}}$ when moving along the line in the direction of $\mathbf{v}$. We say that a finite set $D \subseteq \mathbb{Z}^2$ has an *outer edge* in direction $\mathbf{v}$ if there exists a vector $\mathbf{t} \in \mathbb{Z}^2$ such that $D \subseteq \overline{H}_{\mathbf{v}} + \mathbf{t}$ and $|D \cap (l_{\mathbf{v}} + \mathbf{t})| \geq 2$. We then call $D \cap (l_{\mathbf{v}} + \mathbf{t})$ an outer edge of D in direction $\mathbf{v}$. An outer edge corresponding to $\mathbf{v}$ means that the convex hull of D has an edge in direction $\mathbf{v}$ in the clockwise orientation around D.

If D does not have an outer edge in direction $\mathbf{v}$ then there exists a vector $\mathbf{t} \in \mathbb{Z}^2$ such that $D \subseteq \overline{H}_{\mathbf{v}} + \mathbf{t}$ and $|D \cap (l_{\mathbf{v}} + \mathbf{t})| = 1$ and then we say that D has a vertex in direction $\mathbf{v}$ and we call $D \cap (l_{\mathbf{v}} + \mathbf{t})$ a vertex of D in direction $\mathbf{v}$. We say that a polynomial f has an outer edge or a vertex in direction $\mathbf{v}$ if its support has an outer edge or a vertex in direction $\mathbf{v}$, respectively. Note that every finite shape D has either an edge or a vertex in any non-zero direction. Note also that in this context directions $\mathbf{v}$ and $-\mathbf{v}$ are not the same: a shape may have an outer edge in direction $\mathbf{v}$ but no outer edge in direction $-\mathbf{v}$. The following lemma shows that a polynomial can have line polynomial factors only in the directions of its outer edges.

Lemma 1 [7]. *Let f be a non-zero polynomial with a line polynomial factor in direction $\mathbf{v}$. Then f has outer edges in directions $\mathbf{v}$ and $-\mathbf{v}$.*

Let $\mathbf{v} \in \mathbb{Z}^2 \setminus \{\mathbf{0}\}$ be any non-zero primitive vector and let $f = \sum f_{\mathbf{u}} X^{\mathbf{u}}$ be a polynomial. Recall that $\mathbf{v}$-*fibers* of f are the polynomials $\sum_{k \in \mathbb{Z}} f_{\mathbf{u}+k\mathbf{v}} X^{\mathbf{u}+k\mathbf{v}}$

for $\mathbf{u} \in \mathbb{Z}^2$. Thus a non-zero $\mathbf{v}$-fiber of a polynomial is either a line polynomial or a monomial. Let us denote by $\mathcal{F}_{\mathbf{v}}(f)$ the set of different normal forms of all non-zero $\mathbf{v}$-fibers of a polynomial f, which is thus a finite set. The following simple example illustrates the concept of fibers and their normal forms.

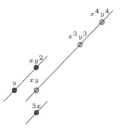

Fig. 1. The support of $f = 3x + y + xy^2 + xy + x^3y^3 + x^4y^4$ and its different $(1,1)$-fibers.

Example 1. Let us determine the set $\mathcal{F}_{\mathbf{v}}(f)$ for $f = f(X) = f(x,y) = 3x + y + xy^2 + xy + x^3y^3 + x^4y^4$ and $\mathbf{v} = (1,1)$. By grouping the terms we can write

$$f = 3x + y(1+xy) + xy(1+x^2y^2+x^3y^3) = X^{(1,0)} \cdot 3 + X^{(0,1)}(1+t) + X^{(1,1)}(1+t^2+t^3)$$

where $t = X^{(1,1)} = xy$. Hence $\mathcal{F}_{\mathbf{v}}(f) = \{3, 1+t, 1+t^2+t^3\}$. See Fig. 1 for a pictorial illustration. □

As noticed in the example above, polynomials are linear combinations of their fibers: for any polynomial f and any non-zero primitive vector $\mathbf{v}$ we can write

$$f = X^{\mathbf{u}_1}\psi_1 + \ldots + X^{\mathbf{u}_n}\psi_n$$

for some $\mathbf{u}_1, \ldots, \mathbf{u}_n \in \mathbb{Z}^2$ where $\psi_1, \ldots, \psi_n \in \mathcal{F}_{\mathbf{v}}(f)$. We use this in the proof of the next theorem.

Theorem 3. *A polynomial f has a line polynomial factor in direction $\mathbf{v}$ if and only if the polynomials in $\mathcal{F}_{\mathbf{v}}(f)$ have a common factor.*

Proof. For any line polynomial ϕ in direction $\mathbf{v}$, and for any polynomial g, the $\mathbf{v}$-fibers of the product ϕg have a common factor ϕ. In other words, if a polynomial f has a line polynomial factor ϕ in direction $\mathbf{v}$ then the polynomials in $\mathcal{F}_{\mathbf{v}}(f)$ have the normal form of ϕ as a common factor.

For the converse direction, assume that the polynomials in $\mathcal{F}_{\mathbf{v}}(f)$ have a common factor ϕ which is thus a line polynomial in direction $\mathbf{v}$. Then there exist vectors $\mathbf{u}_1, \ldots, \mathbf{u}_n \in \mathbb{Z}^d$ and polynomials $\phi\psi_1, \ldots, \phi\psi_n \in \mathcal{F}_{\mathbf{v}}(f)$ such that

$$f = X^{\mathbf{u}_1}\phi\psi_1 + \ldots + X^{\mathbf{u}_n}\phi\psi_n.$$

Hence ϕ is a line polynomial factor of f in direction $\mathbf{v}$. □

Note that Lemma 1 actually follows immediately from Theorem 3: A vertex instead of an outer edge in direction $\mathbf{v}$ or $-\mathbf{v}$ provides a non-zero monomial $\mathbf{v}$-fiber, which implies that the polynomials in $\mathcal{F}_{\mathbf{v}}(f)$ have no common factors.

Thus to find out the line polynomial factors of f we first need to find out the possible directions of the line polynomials, that is, the directions of the (finitely many) outer edges of f, and then we need to check for which of these possible directions $\mathbf{v}$ the polynomials in $\mathcal{F}_{\mathbf{v}}(f)$ have a common factor. There are clearly algorithms to find the outer edges of a given polynomial and to determine whether finitely many line polynomials have a common factor. If such a factor exists then f has a line polynomial factor in this direction by Theorem 3. Thus we have proved the following theorem.

Theorem 4. *There is an algorithm to find the line polynomial factors of a given (Laurent) polynomial.*

3   Perfect Coverings

In this paper a *graph* is a tuple $G = (V, E)$ where V is the (possibly infinite) vertex set of G and $E \subseteq \{\{u, v\} \mid u, v \in V, u \neq v\}$ is the edge set of G. Thus the graphs we consider are *simple* and *undirected*. We also assume that all vertices have only finitely many neighbors in the graph. For a graph $G = (V, E)$ we call any subset $S \subseteq V$ of the vertex set a *code* in G. The distance $d(u, v)$ of two vertices $u, v \in V$ is the length of a shortest path between them. The *(closed) r-neighborhood* of a vertex $u \in V$ is the set $N_r(u) = \{v \in V \mid d(v, u) \leq r\}$, that is, the ball of radius r centered at u. Let us now give the definition of the family of codes we consider.

Definition 1. *Let $G = (V, E)$ be a graph. A code $S \subseteq V$ is an (r, b, a)-covering in G for non-negative integers b and a if the r-neighborhood of every vertex in S contains exactly b elements of S and the r-neighborhood of every vertex not in S contains exactly a elements of S, that is, if for every $u \in V$*

$$|N_r(u) \cap S| = \begin{cases} b \ \text{if } u \in S \\ a \ \text{if } u \notin S \end{cases}.$$

By a *perfect (multiple) covering* we mean any (r, b, a)-covering.

3.1   Infinite Grids

An *infinite grid* is a translation invariant graph with the vertex set $\mathbb{Z}^2$. In other words, in infinite grids $N_r(\mathbf{u}) = \mathbf{u} + N_r(\mathbf{0})$ for all $\mathbf{u} \in \mathbb{Z}^2$. The *square grid* is the graph $(\mathbb{Z}^2, E_S)$ with $E_S = \{\{\mathbf{u}, \mathbf{v}\} \mid \mathbf{u} - \mathbf{v} \in \{(\pm 1, 0), (0, \pm 1)\}\}$, the *king grid* is the graph $(\mathbb{Z}^2, E_K)$ with $E_K = \{\{\mathbf{u}, \mathbf{v}\} \mid \mathbf{u} - \mathbf{v} \in \{(\pm 1, 0), (0, \pm 1), (\pm 1, \pm 1)\}\}$ and the *triangular grid* is the graph $(\mathbb{Z}^2, E_T)$ with $E_T = \{\{\mathbf{u}, \mathbf{v}\} \mid \mathbf{u} - \mathbf{v} \in \{(\pm 1, 0), (0, \pm 1), (1, 1), (-1, -1)\}\}$. See Fig. 2 for the 1-neighborhoods of a vertex in these graphs. A code $S \subseteq \mathbb{Z}^2$ is periodic if $S = S + \mathbf{t}$ for some non-zero $\mathbf{t} \in \mathbb{Z}^2$. It is two-periodic if $S = S + \mathbf{t}_1$ and $S = S + \mathbf{t}_2$ where $\mathbf{t}_1$ and $\mathbf{t}_2$ are linearly independent. The following result is by Axenovich.

(a) The square grid   (b) The king grid (c) The triangular grid

Fig. 2. The 1-neighborhoods of the black vertex in (a) the square grid, (b) the king grid, and (c) the triangular grid.

Theorem 5 [1]. *If $b - a \neq 1$ then any $(1, b, a)$-covering in the square grid is two-periodic.*

A code $S \subseteq \mathbb{Z}^2$ in any infinite grid can be presented as a configuration $c \in \{0, 1\}^{\mathbb{Z}^2}$ which is defined such that $c_{\mathbf{u}} = 1$ if $\mathbf{u} \in S$ and $c_{\mathbf{u}} = 0$ if $\mathbf{u} \notin S$. The positioning of the codewords in the r-neighborhood of any vertex $\mathbf{u} \in \mathbb{Z}^2$ is then presented as a finite pattern $c|_{\mathbf{u}+N_r(\mathbf{0})}$.

Definition 2. *A configuration $c \in \{0, 1\}^{\mathbb{Z}^2}$ is a (D, b, a)-covering for a finite shape $D \subseteq \mathbb{Z}^2$ (the neighborhood) and non-negative integers b and a (the covering constants) if for all $\mathbf{u} \in \mathbb{Z}^2$ the pattern $c|_{\mathbf{u}+D}$ contains exactly b symbols 1 if $c_{\mathbf{u}} = 1$ and exactly a symbols 1 if $c_{\mathbf{u}} = 0$.*

We call also any (D, b, a)-covering perfect and hence a perfect covering is either a code in a graph or a two-dimensional binary configuration.

Definitions 1 and 2 are consistent in infinite grids: a code S in an infinite grid G is an (r, b, a)-covering if and only if the configuration $c \in \{0, 1\}^{\mathbb{Z}^2}$ presenting S is a (D, b, a)-covering where D is the r-neighborhood of $\mathbf{0}$ in G. For a set $D \subseteq \mathbb{Z}^2$ we define its *characteristic polynomial* to be $f_D(X) = \sum_{\mathbf{u} \in D} X^{-\mathbf{u}}$. Let us denote by $\mathbb{1}(X)$ the constant power series $\sum_{\mathbf{u} \in \mathbb{Z}^2} X^{\mathbf{u}}$. If c is a (D, b, a)-covering then from the definition we get that $f_D(X)c(X) = (b - a)c(X) + a\mathbb{1}(X)$ which is equivalent to $(f_D(X) - (b - a))c(X) = a\mathbb{1}(X)$. Thus if c is a (D, b, a)-covering then $f_D(X) - (b - a) \in \mathrm{Per}(c)$. Using our formulation we get a simple proof for Theorem 5:

Reformulation of Theorem 5. *Let D be the 1-neighborhood of $\mathbf{0}$ in the square grid and assume that $b - a \neq 1$. Then every (D, b, a)-covering is two-periodic.*

Proof. Let c be an arbitrary (D, b, a)-covering. We show that $g = f_D - (b - a) = x^{-1} + y^{-1} + 1 - (b - a) + x + y \in \mathrm{Per}(c)$ has no line polynomial factors. Then c is two-periodic by Theorem 2. The outer edges of g are in directions $(1, 1), (-1, -1), (1, -1)$ and $(-1, 1)$ and hence by Lemma 1 any line polynomial factor of g is either in direction $(1, 1)$ or $(1, -1)$. For $\mathbf{v} \in \{(1, 1), (1, -1)\}$ we have $\mathcal{F}_{\mathbf{v}}(g) = \{1 + t, 1 - (b - a)\}$. See Fig. 3 for an illustration. Since $1 - (b - a)$ is a non-trivial monomial, by Theorem 3 the periodizer $g \in \mathrm{Per}(c)$ has no line polynomial factors. □

The following result was already proved in a more general form in [12]. We give a short proof using our algebraic approach.

Theorem 6 [12]. *Let $r \geq 2$ and let D be the r-neighborhood of $\mathbf{0}$ in the square grid. Then every (D, b, a)-covering is two-periodic. In other words, all (r, b, a)-coverings in the square grid are two-periodic for all $r \geq 2$.*

Proof. Let c be an arbitrary (D, b, a)-covering. Again, by Theorem 2, it is enough to show that $g = f_D - (b - a) \in \mathrm{Per}(c)$ has no line polynomial factors. By Lemma 1 any line polynomial factor of g has direction $(1, 1)$ or $(1, -1)$. So assume that $\mathbf{v} \in \{(1, 1), (1, -1)\}$. We have $\phi_1 = 1 + t + \ldots + t^r \in \mathcal{F}_{\mathbf{v}}(g)$ and $\phi_2 = 1 + t + \ldots + t^{r-1} \in \mathcal{F}_{\mathbf{v}}(g)$. See Fig. 3 for an illustration in the case $r = 2$. Since $\phi_1 - \phi_2 = t^r$, the polynomials ϕ_1 and ϕ_2 have no common factors, and hence by Theorem 3 the periodizer g has no line polynomial factors. $\square$

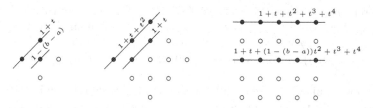

Fig. 3. The constellation on the left illustrates the proof of Theorem 5, the constellation on the center illustrates the proof of Theorem 6 with $r = 2$ and the constellation on the right illustrates the proof of Theorem 7 with $r = 2$.

If $a \neq b$ then for all $r \geq 1$ any (r, b, a)-covering in the king grid is two-periodic:

Theorem 7. *Let $r \geq 1$ be arbitrary and let D be the r-neighborhood of $\mathbf{0}$ in the king grid and assume that $a \neq b$. Then any (D, b, a)-covering is two-periodic. In other words, all (r, b, a)-coverings in the king grid are two-periodic whenever $a \neq b$.*

Proof. Let c be an arbitrary (D, b, a)-covering. By Theorem 2 it is sufficient to show that $g = f_D - (b - a)$ has no line polynomial factors. The outer edges of g are in directions $(1, 0), (-1, 0), (0, 1)$ and $(0, -1)$. Hence by Lemma 1 any line polynomial factor of g has direction $(1, 0)$ or $(0, 1)$. Let $\mathbf{v} \in \{(1, 0), (0, 1)\}$. We have $\phi_1 = 1 + t + \ldots + t^{r-1} + (1 - (b - a))t^r + t^{r+1} + \ldots + t^{2r} \in \mathcal{F}_{\mathbf{v}}(g)$ and $\phi_2 = 1 + t + \ldots + t^{2r} \in \mathcal{F}_{\mathbf{v}}(g)$. See Fig. 3 for an illustration with $r = 2$. Since $\phi_2 - \phi_1 = (b - a)t^r$ is a non-trivial monomial, ϕ_1 and ϕ_2 have no common factors. Thus g has no line polynomial factors by Theorem 3. $\square$

Similarly as in the square grid we can give simple proofs for known results from [12] concerning forced periodicity in the triangular grid:

Theorem 8 [12]. *Let D be the 1-neighborhood of $\mathbf{0}$ in the triangular grid and assume that $b - a \neq -1$. Then every (D, b, a)-covering in the triangular grid is two-periodic. In other words, all $(1, b, a)$-coverings in the triangular grid are two-periodic whenever $b - a \neq -1$.*

Theorem 9 [12]. *Let $r \geq 2$ and let D be the r-neighborhood of $\mathbf{0}$ in the triangular grid. Then every (D, b, a)-covering is two-periodic. In other words, all (r, b, a)-coverings in the triangular grid are two-periodic for $r \geq 2$.*

3.2   General Convex Neighborhoods

A shape $D \subseteq \mathbb{Z}^2$ is *convex* if it is the intersection $D = \text{conv}(D) \cap \mathbb{Z}^2$ where $\text{conv}(D) \subseteq \mathbb{R}^2$ is the real convex hull of D.

Let $D \subseteq \mathbb{Z}^2$ be a finite convex shape. Any (D, b, a)-covering has a periodizer $g = f_D - (b - a)$. As earlier, we study whether g has any line polynomial factors. For any $\mathbf{v} \neq 0$ the set $\mathcal{F}_\mathbf{v}(f_D)$ contains only polynomials $\phi_n = 1 + \ldots + t^{n-1}$ for different $n \geq 1$ since D is convex: if D contains two points then D contains every point between them. Thus $\mathcal{F}_\mathbf{v}(g)$ contains only polynomials ϕ_n for different $n \geq 1$ and, if $b - a \neq 0$, also a polynomial $\phi_{n_0} - (b - a)t^{m_0}$ for some $n_0 \geq 1$ such that $\phi_{n_0} \in \mathcal{F}_\mathbf{v}(f_D)$ and for some $m_0 \geq 0$. If $b - a = 0$ then $g = f_D$ and thus $\mathcal{F}_\mathbf{v}(g) = \mathcal{F}_\mathbf{v}(f_D)$.

Two polynomials ϕ_m and ϕ_n have a common factor if and only if $\gcd(m, n) > 1$. More generally, the polynomials $\phi_{n_1}, \ldots, \phi_{n_r}$ have a common factor if and only if $d = \gcd(n_1, \ldots, n_r) > 1$ and, in fact, their greatest common factor is the dth *cyclotomic polynomial*

$$\prod_{\substack{1 \leq k \leq d \\ \gcd(k, d) = 1}} (t - e^{i \cdot \frac{2\pi k}{d}}).$$

Let us introduce the following notation. For any polynomial f, we denote by $\mathcal{F}'_\mathbf{v}(f)$ the set of normal forms of the non-zero fibers $\sum_{k \in \mathbb{Z}} f_{\mathbf{u}+k\mathbf{v}} X^{\mathbf{u}+k\mathbf{v}}$ for all $\mathbf{u} \notin \mathbb{Z}\mathbf{v}$. In other words, we exclude the fiber through the origin. Let us also denote $\text{fib}_\mathbf{v}(f)$ for the normal form of the fiber $\sum_{k \in \mathbb{Z}} f_{k\mathbf{v}} X^{k\mathbf{v}}$ through the origin. Thus $\mathcal{F}_\mathbf{v}(f) = \mathcal{F}'_\mathbf{v}(f) \cup \{\text{fib}_\mathbf{v}(f)\}$.

Applying Theorems 2 and 3 we have the following theorem that gives sufficient conditions for every (D, b, a)-covering to be periodic for a finite and convex D. The first part of the theorem was also mentioned in [4] in a more general form.

Theorem 10. *Let D be a finite convex shape, $g = f_D - (b - a)$ and let E be the set of the outer edge directions of g.*

- *Assume that $b - a = 0$. For any $\mathbf{v} \in E$ denote $d_\mathbf{v} = \gcd(n_1, \ldots, n_r)$ where $\mathcal{F}_\mathbf{v}(g) = \{\phi_{n_1}, \ldots, \phi_{n_r}\}$. If $d_\mathbf{v} = 1$ holds for all $\mathbf{v} \in E$ then every (D, b, a)-covering is two-periodic. If $d_\mathbf{v} = 1$ holds for all but some parallel $\mathbf{v} \in E$ then every (D, b, a)-covering is periodic.*

– *Assume that $b - a \neq 0$. For any $\mathbf{v} \in E$ denote $d_{\mathbf{v}} = \gcd(n_1, \ldots, n_r)$ where $\mathcal{F}'_{\mathbf{v}}(g) = \{\phi_{n_1}, \ldots, \phi_{n_r}\}$. If the $d_{\mathbf{v}}$'th cyclotomic polynomial and $\mathit{fib}_{\mathbf{v}}(g)$ have no common factors for any $\mathbf{v} \in E$ then every (D, b, a)-covering is two-periodic. If the condition holds for all but some parallel $\mathbf{v} \in E$ then every (D, b, a)-covering is periodic. (Note that the condition is satisfied, in particular, if $d_{\mathbf{v}} = 1$.)*

4   Algorithmic Aspects

All coverings are periodic, in particular, if there are no coverings at all! It is useful to be able to detect such trivial cases.

The set

$$\mathcal{S}(D, b, a) = \{c \in \{0, 1\}^{\mathbb{Z}^2} \mid (f_D - (b - a))c = a\mathbb{1}(X)\}$$

of all (D, b, a)-coverings is an SFT for any given finite shape D and non-negative integers b and a. Hence the question whether there exist any (D, b, a)-coverings for given neighborhood D and covering constants b and a is equivalent to the question whether the SFT $\mathcal{S} = \mathcal{S}(D, b, a)$ is non-empty. The question of emptiness of a given SFT is in general undecidable, but if the SFT is known to be not aperiodic then the problem becomes decidable. In particular, if $g = f_D - (b - a)$ has line polynomial factors in at most one direction then this question is decidable:

Theorem 11. *Let finite $D \subseteq \mathbb{Z}^2$ and non-negative integers b and a be given such that the polynomial $g = f_D - (b - a)$ has line polynomial factors in at most one parallel direction. Then there exists an algorithm to determine whether there exist any (D, b, a)-coverings.*

Proof. Let $\mathcal{S} = \mathcal{S}(D, b, a)$ be the SFT of all (D, b, a)-coverings. Since g has line polynomial factors in at most one direction, by Theorem 2 every element of $\mathcal{S}$ is periodic. Any two-dimensional SFT that contains periodic configurations contains also two-periodic configurations, so $\mathcal{S}$ is either empty or contains a two-periodic configuration. By a standard argumentation by H. Wang [13] there exist semi-algorithms to determine whether a given SFT is empty and whether a given SFT contains a two-periodic configuration. Running these two semi-algorithms in parallel gives us an algorithm to test whether $\mathcal{S} \neq \emptyset$.   □

One may also want to design a perfect (D, b, a)-covering for given D, b and a. This can be effectively done under the assumptions of Theorem 11: As we have seen, if $\mathcal{S} = \mathcal{S}(D, b, a)$ is non-empty it contains a two-periodic configuration. For any two-periodic configuration c it is easy to check if c contains a forbidden pattern. By enumerating two-periodic configurations one-by-one one is guaranteed to find eventually one that is in $\mathcal{S}$.

If the polynomial g has no line polynomial factors then the following stronger result holds:

Theorem 12. *If the polynomial $g = f_D - (b - a)$ has no line polynomial factors for given finite shape $D \subseteq \mathbb{Z}^2$ and non-negative integers b and a then the SFT $\mathcal{S} = \mathcal{S}(D, b, a)$ is finite. One can then effectively construct all the finitely many elements of $\mathcal{S}$.*

The proof of the first part of above theorem relies on the fact that a two-dimensional subshift is finite if and only if it contains only two-periodic cofigurations [2]. If g has no line polynomial factors then every configuration it periodizes (including every configuration in $\mathcal{S}$) is two-periodic by Theorem 2, and hence $\mathcal{S}$ is finite. The "moreover" part of the theorem, *i.e.*, the fact that one can effectively produce all the finitely many elements of $\mathcal{S}$ holds generally for finite SFTs.

References

1. Axenovich, M.A.: On multiple coverings of the infinite rectangular grid with balls of constant radius. Disc. Math. **268**(1), 31–48 (2003)
2. Ballier, A., Durand, B., Jeandal, E.: Structural aspects of tilings. In: Albers, S., Weil, P. (eds.) 25th International Symposium on Theoretical Aspects of Computer Science. Leibniz International Proceedings in Informatics (LIPIcs), vol. 1, pp. 61–72. Schloss Dagstuhl-Leibniz-Zentrum fuer Informatik, Dagstuhl, Germany (2008)
3. Ceccherini-Silberstein, T., Coornaert, M.: Cellular Automata and Groups. Springer Monographs in Mathematics, Springer, Heidelberg (2010). https://doi.org/10.1007/978-3-642-14034-1
4. Geravker, N., Puzynina, S.A.: Abelian Nivat's conjecture for non-rectangular patterns (2021). arXiv:2111.04690
5. Heikkilä, E.: Algebrallinen näkökulma peittokoodeihin. Master's thesis University of Turku (2020)
6. Kari, J.: Low-complexity tilings of the plane. In: Hospodár, M., Jirásková, G., Konstantinidis, S. (eds.) DCFS 2019. LNCS, vol. 11612, pp. 35–45. Springer, Cham (2019). https://doi.org/10.1007/978-3-030-23247-4_2
7. Kari, J., Moutot, E.: Nivat's conjecture and pattern complexity in algebraic subshifts. Theor. Comput. Sci. **777**, 379–386 (2019)
8. Kari, J., Szabados, M.: An algebraic geometric approach to nivat's conjecture. In: Halldórsson, M.M., Iwama, K., Kobayashi, N., Speckmann, B. (eds.) ICALP 2015. LNCS, vol. 9135, pp. 273–285. Springer, Heidelberg (2015). https://doi.org/10.1007/978-3-662-47666-6_22
9. Kari, J., Szabados, M.: An algebraic geometric approach to Nivat's conjecture. Inf. Comput. **271**, 104481 (2020)
10. Kůrka, P.: Topological and Symbolic Dynamics. Collection SMF, Société Mathématique de France (2003)
11. Lind, D., Marcus, B.: An Introduction to Symbolic Dynamics and Coding. Cambridge University Press, Cambridge (1995)
12. Puzynina, S.A.: On periodicity of generalized two-dimensional infinite words. Inf. Comput. **207**(11), 1315–1328 (2009)
13. Wang, H.: Proving theorems by pattern recognition - II. Bell Syst. Tech. J. **40**(1), 1–41 (1961). https://doi.org/10.1002/j.1538-7305.1961.tb03975.x

Automata-Theoretical Regularity Characterizations for the Iterated Shuffle on Commutative Regular Languages

Stefan Hoffmann$^{(\boxtimes)}$ (iD)

Informatikwissenschaften, FB IV, Universität Trier, Trier, Germany
`hoffmanns@informatik.uni-trier.de`

Abstract. We present new automata-theoretical characterizations for the regularity of the iterated shuffle on commutative regular languages. Using these characterizations we show that, for a fixed alphabet, it is tractable to decide whether the iterated shuffle of a regular commutative language is itself regular when the input language is given by a deterministic automaton. Additionally, we introduce two new subclasses of commutative regular languages, called Type I and Type II languages, on which the iterated shuffle is regularity-preserving and show that the iterated shuffle of a commutative language is a Type I language if it is regular. Additionally, we establish various closure properties and show that we can decide if a language given by deterministic automaton is in one of these classes in polynomial time.

Keywords: Finite automata · Commutative languages · Closure properties · Iterated shuffle · Shuffle closure · Regularity-preserving operations

1 Introduction

The shuffle and iterated shuffle have been introduced and studied to understand, or specify, the semantics of parallel programs. This was undertaken (seemingly) independently by Campbell and Habermann [4], by Mazurkiewicz [17] and by Shaw [22]. They introduced *flow expressions*, which allow for sequential operators (catenation and iterated catenation) as well as for parallel operators (shuffle and iterated shuffle) to specify sequential and parallel execution traces.

The shuffle operation as a binary operation is regularity-preserving on all regular languages, see [1,3,10,11,21] for further work related to the binary shuffle in formal language theory.

However, already the iterated shuffle of very simple languages (like $\{ab, ba\}$, whose iterated shuffle equals the set of words with an equal number of a's and b's) can give non-regular languages. Hence, it is interesting to know, and to identify, quite rich classes for which this operation is regularity-preserving.

For further connections on regularity conditions and closure properties, in particular for the star-free languages, see the recent survey [20].

© Springer Nature Switzerland AG 2022
V. Diekert and M. Volkov (Eds.): DLT 2022, LNCS 13257, pp. 164–176, 2022.
https://doi.org/10.1007/978-3-031-05578-2_13

Overview and Contribution. In Sect. 2 we introduce various notions and results that we need. Then, in Sects. 3 and 4, we introduce the Type I and the Type II languages respectively. For languages from both classes, we show that the iterated shuffle is always regular. Type I languages were already investigated in [12] (without calling them Type I languages). We show additional closure properties for the Type I language and prove that when the iterated shuffle of a commutative language is regular, then it is a Type I language. The Type II languages are an extension of the Type I languages. We give automata-theoretical characterizations for both classes of languages.

In Sect. 5 we derive, using a result by Imreh, Ito & Katsura [14], two automata-theoretical characterizations for the regularity of the iterated shuffle of a commutative regular language. One is a condition on the final states, the other refines this (essentially, it says that we can interchange a universal quantifier for final states with an existential quantifier for certain letters). We give a few corollaries and state an even easier characterization for a binary alphabet.

Section 6 uses the results from the previous sections to show that it is possible to decide in polynomial time, for a fixed alphabet, whether the iterated shuffle of a commutative regular language is regular when the input is given by a deterministic automaton. Also, our results imply polynomial time procedures to decide if a given regular language is a Type I or Type II language when the input is given by a deterministic automaton.

2   Preliminaries

General Notions. Let Σ be a finite set of symbols called an *alphabet*. The set Σ^* denotes the set of all finite sequences, i.e., of all *words*. The finite sequence of length zero, or the *empty word*, is denoted by ε. We set $\Sigma^+ = \Sigma^* \setminus \{\varepsilon\}$. For a given word $w \in \Sigma^*$ we denote by $|w|$ its length, and for $a \in \Sigma$ by $|w|_a$ the number of occurrences of the symbol a in w. A *language* is a subset of Σ^*. For $w \in \Sigma^*$, we set $\text{alph}(u) = \{a \in \Sigma \mid |u|_a > 0\}$ and $\text{alph}(L) = \bigcup_{u \in L} \text{alph}(u)$ for $L \subseteq \Sigma^*$. If $L \subseteq \Sigma^*$ and $u \in \Sigma^*$, then the *left quotient* is the languages $u^{-1}L = \{v \in \Sigma^* \mid uv \in L\}$ and the *right quotient* is $Lu^{-1} = \{v \in \Sigma^* \mid vu \in L\}$.

Given two integers $a, b \geq 0$, we denote by $\gcd(a, b)$ and by $\text{lcm}(a, b)$, their *greatest common divisor* and their *least common multiple*, respectively.

We assume the reader to have some basic knowledge in formal language theory, as contained, e.g., in [13, 16]. For instance, we make use of regular expressions to describe languages and the operators Kleene star, Kleene plus and concatenation.

Let $\Gamma \subseteq \Sigma$. Then, we define *projection homomorphisms* $\pi_\Gamma : \Sigma^* \to \Gamma^*$ onto Γ^* by $\pi_\Gamma(x) = x$ for $x \in \Gamma$, $\pi_\Gamma(x) = \varepsilon$ for $x \notin \Gamma$ and $\pi_\Gamma(x_1 x_2 \cdots x_n) = \pi_\Gamma(x_1)\pi_\Gamma(x_2) \cdots \pi_\Gamma(x_n)$ for $x_1, x_2, \ldots, x_n \in \Sigma$.

By $\mathbb{N}_0 = \{0, 1, 2, \ldots\}$, we denote the set of natural numbers, including zero.

A quintuple $\mathcal{A} = (\Sigma, Q, \delta, q_0, F)$ is a *finite complete deterministic automaton (DFA)*, where $\delta : Q \times \Sigma \to Q$ is a *totally defined transition function*, Q a *finite set of states*, $q_0 \in Q$ the *start state* and $F \subseteq Q$ the *set of final states*. As we are not concerned with other models, we will simply call them *automata*. The transition function $\delta : Q \times \Sigma \to Q$ can be extended to a transition function on words $\delta^* : Q \times \Sigma^* \to Q$ by setting $\delta^*(q, \varepsilon) = q$ and $\delta^*(q, wa) := \delta(\delta^*(q, w), a)$ for $q \in Q$, $a \in \Sigma$ and $w \in \Sigma^*$. In the remainder, we drop the distinction between both functions and will also denote this extension by δ. The language *recognized* (or *accepted*) by an automaton $\mathcal{A} = (\Sigma, Q, \delta, q_0, F)$ is $L(\mathcal{A}) = \{w \in \Sigma^* \mid \delta(q_0, w) \in F\}$. A language $L \subseteq \Sigma^*$ is called *regular* if $L = L(\mathcal{A})$ for some finite automaton $\mathcal{A}$.

If $S \subseteq Q$, $q \in Q$, $u \in \Sigma^*$ and $L \subseteq \Sigma^*$, we set $\delta(S, u) = \{\delta(q, u) \mid q \in S\}$ and $\delta(q, L) = \{\delta(q, v) \mid v \in L\}$.

The following number-theoretical result from [18] will be needed.

Lemma 1. *Suppose a, b are positive integers. Then each number of the form $ax + by$, with $x, y \geq 0$, is a multiple of $\gcd(a, b)$ and the largest multiple of $\gcd(a, b)$ that cannot be represented as $ax + by$, with $x, y \geq 0$, is $\operatorname{lcm}(a, b) - (a + b)$.*

Commutative Languages and the Shuffle Operation. For a given word $w \in \Sigma^*$, we define $\operatorname{perm}(w) := \{u \in \Sigma^* \mid \forall a \in \Sigma : |u|_a = |w|_a\}$. If $L \subseteq \Sigma^*$, then we set $\operatorname{perm}(L) := \bigcup_{w \in L} \operatorname{perm}(w)$. A language is called *commutative*, if $\operatorname{perm}(L) = L$. Let $\Sigma = \{a_1, \ldots, a_k\}$. The *Parikh mapping* is $\psi : \Sigma^* \to \mathbb{N}_0^k$ given by $\psi(u) = (|u|_{a_1}, \ldots, |u|_{a_k})$ for $u \in \Sigma^*$. We have $\operatorname{perm}(L) = \psi^{-1}(\psi(L))$. Every commutative $L \subseteq \Sigma^*$ can be identified with its *Parikh image* $\psi(L)$. We make use of this identification occasionally (for example when depicting commutative languages graphically, like in Fig. 1 or 2) without special mentioning.

An automaton $\mathcal{A} = (\Sigma, Q, \delta, q_0, F)$ is called *commutative*, if for all $q \in Q$ and $a, b \in \Sigma$ we have $\delta(q, ab) = \delta(q, ba)$. A commutative automaton recognizes a commutative language. Furthermore, if $L \subseteq \Sigma^*$ is commutative and regular, then the automaton with the least number of states recognizing L is commutative (see [6, Proof of Theorem 15] for a proof).

The *shuffle operation*, denoted by $\sqcup\!\sqcup$, is defined by

$$u \sqcup\!\sqcup v = \{w \in \Sigma^* \mid w = x_1 y_1 x_2 y_2 \cdots x_n y_n \text{ for some words}$$

$$x_1, \ldots, x_n, y_1, \ldots, y_n \in \Sigma^* \text{ such that } u = x_1 x_2 \cdots x_n \text{ and } v = y_1 y_2 \cdots y_n\},$$

for $u, v \in \Sigma^*$ and $L_1 \sqcup\!\sqcup L_2 := \bigcup_{x \in L_1, y \in L_2} (x \sqcup\!\sqcup y)$ for $L_1, L_2 \subseteq \Sigma^*$.

In writing formulas without brackets, we suppose that the shuffle operation binds stronger than the set operations, and the concatenation operator has the strongest binding.

If $L_1, \ldots, L_n \subseteq \Sigma^*$, we set $\bigsqcup\!\bigsqcup_{i=1}^n L_i = L_1 \sqcup\!\sqcup \ldots \sqcup\!\sqcup L_n$. The *iterated shuffle* of $L \subseteq \Sigma^*$ is $L^{\sqcup\!\sqcup,*} = \bigcup_{n \geq 1} \bigsqcup\!\bigsqcup_{i=1}^n L \cup \{\varepsilon\}$. Note that $\emptyset^{\sqcup\!\sqcup,*} = \{\varepsilon\}$.

Theorem 2 (Fernau et al. [7]). *Let $U, V, W \subseteq \Sigma^*$ and $V_i \subseteq \Sigma^*$, $i \in I$, for an arbitrary index set I. Then,*

1. $U \sqcup\!\sqcup V = V \sqcup\!\sqcup U$ *(commutative law)*;
2. $(U \sqcup\!\sqcup V) \sqcup\!\sqcup W = U \sqcup\!\sqcup (V \sqcup\!\sqcup W)$ *(associative law)*;
3. $U \sqcup\!\sqcup \left(\bigcup_{i \in I} V_i\right) = \bigcup_{i \in I} (U \sqcup\!\sqcup V_i)$ *(distributive over arbitrary unions)*;
4. $(U^{\sqcup\!\sqcup,*})^{\sqcup\!\sqcup,*} = U^{\sqcup\!\sqcup,*}$;

5. $(U \cup V)^{\sqcup\!\sqcup,*} = U^{\sqcup\!\sqcup,*} \sqcup\!\sqcup V^{\sqcup\!\sqcup,*}$;
6. $(U \sqcup\!\sqcup V^{\sqcup\!\sqcup,*})^{\sqcup\!\sqcup,*} = (U \sqcup\!\sqcup (U \cup V)^{\sqcup\!\sqcup,*}) \cup \{\varepsilon\}$.

We also write $\bigsqcup\!\sqcup_{a \in \Gamma} L_a$ to mean $L_{a_1} \sqcup\!\sqcup \ldots \sqcup\!\sqcup L_{a_k}$ for some ordering $\Sigma = \{a_1, \ldots, a_k\}$ with $k = |\Sigma|$. Note that by Theorem 2 (commutative law) this is well-defined and does not depend on the chosen ordering.

A commutative language $L \subseteq \Sigma^*$ is *periodic* if there exists $v \in \mathbb{N}_0^\Sigma$ (we identify $\mathbb{N}_0^\Sigma$, the set of functions $\Sigma \to \mathbb{N}_0$, with the set of points $\mathbb{N}_0^{|\Sigma|}$ without specifically introducing an ordering on Σ) and numbers $c_a \geq 0$ for $a \in \Sigma$ such that

$$\psi(L) = \left\{ p \in \mathbb{N}_0^\Sigma \mid \forall a \in \Sigma \, \exists n_a \in \mathbb{N}_0 : p = v + \sum_{a \in \Sigma} n_a \cdot c_a \cdot \psi(a) \right\}. \quad (1)$$

These languages were introduced in [5] and it was shown that every regular commutative language is a finite union of periodic languages.

In fact, a variant of this result already goes back to Ginsburg & Spanier [8, Lemma 1.2].

For reference, we state the following simple folklore result.

Lemma 3. *If $L \subseteq \{a\}^*$ is recognizable by a DFA with a single final state, then L is empty, or $L = \{a^n\}$ for some $n \geq 0$ or $L = a^n(a^p)^*$ for some $n \geq 0$, $p > 0$.*

With Lemma 3 we can easily prove the next characterization of periodic languages (as far as we know, this was not stated elsewhere).

Proposition 4. *The language $L \subseteq \Sigma^*$ is periodic if and only if $L = \bigsqcup\!\sqcup_{a \in \Sigma} L_a$ where each $L_a \subseteq a^*$ is non-empty and recognizable by a DFA with a single final state.*

In [14,15], the iterated shuffle on commutative regular languages was investigated and the following regularity condition stated [15, Theorem 4.6.1].

Theorem 5 (Imreh, Ito & Katsura [14,15]). *Let $L \subseteq \Sigma^*$ be commutative and regular. Then $L^{\sqcup\!\sqcup,*}$ is regular if and only if for all $u \in L \setminus \{\varepsilon\}$ and all $a \in \mathrm{alph}(u)$ with $a^+ \cap L = \emptyset$ we have $a^i \Sigma^* \cap (L \cap \mathrm{alph}(u)^*) \neq \emptyset$ for every positive integer i.*

3   Type I Languages

Referring to $v \in \mathbb{N}_0^\Sigma$ from Eq. (1) for periodic languages, write $v = (v_a)_{a \in \Sigma}$ (or $v = (v_1, \ldots, v_k)$ if $\Sigma = \{a_1, \ldots, a_k\}$). Then a periodic language is called *diagonal periodic* if $v_a > 0$ implies $c_a > 0$ for each $a \in \Sigma$. These languages were introduced in [12]. We define a *Type I language* as a finite union of diagonal periodic languages. These languages are obviously regular and commutative. See Fig. 1 for an example of the Parikh image of a Type I language and its iterated shuffle.

Without denoting them as Type I languages, finite unions of diagonal periodic languages were investigated in [12] and the following results were obtained.

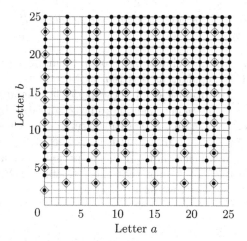

Fig. 1. The Parikh image of the Type I language $bb(bbb)^* \cup aaa(aaaa)^* \shuffle bbb(bbbb)^*$ (red diamonds) and its iterated shuffle (black dots), which is regular. (Color figure online)

1. The diagonal periodic languages have the form $\{\varepsilon\}$ or $\bigsqcup_{a \in \Gamma} a^{k_a}(a^{p_a})^*$ for $\Gamma \subseteq \Sigma$ with numbers $k_a \geq 0$ and $p_a > 0$ for $a \in \Gamma$.
2. The iterated shuffle of a Type I language is a Type I language.
3. The binary shuffle of two Type I languages yields a Type I language.
4. The class of Type I languages equals the positive Boolean algebra (i.e., a class of language closed under union and intersection) generated by languages of the form Γ^*, Γ^+, $\{u \in \Sigma^+ \mid |u|_a \geq n\}$ and $\{u \in \Sigma^* \mid |u|_a \equiv k \pmod{n}\}$ for $\Gamma \subseteq \Sigma$, $0 \leq k < n$.

Also, it was observed that commutative group languages and languages contained in $\mathbf{Com}^+(\Sigma^*)$, the positive variety of languages recognized by commutative ordered semigroups (see [19] for formal definitions of these notions), are Type I languages.

Here, we state additional closure properties, give an automata-theoretical characterization and prove that if the iterated shuffle of any commutative language is regular, then it is a Type I language.

Using that $\pi_\Gamma(\bigsqcup_{a \in \Sigma} L_a) = \bigsqcup_{a \in \Gamma} L_a$ for languages $L_a \subseteq \{a\}^*$, we can additionally conclude that Type I languages are closed under projection.

Proposition 6. *If $L \subseteq \Sigma^*$ is a Type I language and $\Gamma \subseteq \Sigma$, then $\pi_\Gamma(L)$ is a Type I language.*

Type I languages are special because when the iterated shuffle of a commutative language is regular, then it is a Type I language. The basic proof idea is that we can always append a word and repeat it infinitely often without leaving the iterated shuffle. In particular, by doing sufficiently often so we can "eliminate" parts that are not diagonal periodic.

Theorem 7. *Let $L \subseteq \Sigma^*$ be any commutative language. If $L^{\shuffle,*}$ is regular, then it is a Type I language.*

Proof (sketch). The iterated shuffle $L^{\shuffle,*}$ of a commutative language is commutative. So, if it is regular, as written in Sect. 2 it is a finite union of periodic languages. By Proposition 4 these periodic languages have the form $\bigsqcup_{a \in \Sigma} L_a$ with $L_a \subseteq a^*$ being regular and recognizable by automata with a single final state. By Lemma 3, we can assume (the case $L_a = \emptyset$ can be excluded) $|L_a| = 1$ or $L_a = a^{k_a}(a^{p_a})^*$ with $p_a > 0$. Now, if we take such a part $U \subseteq L^{\shuffle,*}$ of the union which is not a diagonal periodic language, we can write $U = \bigsqcup_{a \in \Sigma} a^{k_a}(a^{p_a})^* \subseteq L^{\shuffle,*}$ such that there exists $b \in \Sigma$ with $k_b > 0$ and $p_b = 0$. As $\bigsqcup_{a \in \Sigma} a^{k_a} \subseteq L^{\shuffle,*}$, we have

$$U \shuffle \left(\bigsqcup_{a \in \Sigma} a^{k_a} \right)^{\shuffle,*} = \left(\bigsqcup_{a \in \Sigma \setminus \{b\}} a^{k_a}(a^{p_a})^*(a^{k_a})^* \right) \shuffle b^{k_b}(b^{k_b})^* \subseteq L^{\shuffle,*}.$$

Let $a \in \Sigma \setminus \{b\}$. Now if $k_a = p_a = 0$, then $a^{k_a}(a^{p_a})^*(a^{k_a})^* = \{\varepsilon\}$ and if precisely one of k_a or p_a is non-zero, it is a language of the form $a^{k_a}(a^p)^*$ with $p \in \{k_a, p_a\}$ being non-zero. So, only the case that k_a and p_a are both non-zero is left. In that case, by Lemma 1, the language $(a^{p_a})^*(a^{k_a})^*$ equals $F \cup a^{\mathrm{lcm}(p_a,k_a)-(p_a+k_a)}(a^{\gcd(p_a,k_a)})^+$ (note that the plus operator was used, and, in fact, $\mathrm{lcm}(p_a, k_a) - (p_a + k_a)$ is the largest multiple of $\gcd(p_a, k_a)$ not of the form $up_a + vk_a$) with $F \subseteq \{\varepsilon, a, a^2, \dots, a^{\mathrm{lcm}(p_a,k_a)-(p_a+k_a)-\gcd(p_a,k_a)}\}$. However, as the words in $(a^{p_a})^*(a^{k_a})^*$ have length divisible by $\gcd(p_a, k_a)$, we have $F \cup a^{\mathrm{lcm}(p_a,k_a)-(p_a+k_a)}(a^{\gcd(p_a,k_a)})^+ = F \cdot (a^{\mathrm{lcm}(p_a,k_a)})^* \cup a^{\mathrm{lcm}(p_a,k_a)-(p_a+k_a)}(a^{\gcd(p_a,k_a)})^+$.

Using that shuffle and concatenation distribute over union, we can write the language $U \shuffle \left(\bigsqcup_{a \in \Sigma} a^{k_a} \right)^{\shuffle,*}$ as a finite union of periodic languages that are in fact diagonal periodic languages. Repeating this procedure for other parts $U \subseteq L^{\shuffle,*}$ that are not diagonal periodic languages, we can write $L^{\shuffle,*}$ as a finite union of diagonal periodic languages. Hence it is a Type I language. □

Next, we give a characterization in terms of automata.

Theorem 8. *Let $L \subseteq \Sigma^*$. Then L is a Type I language if and only if there exists a commutative automaton $\mathcal{A} = (\Sigma, Q, \delta, q_0, F)$ with $L = L(\mathcal{A})$ such that for each $u \in L(\mathcal{A})$ and $a \in \mathrm{alph}(u)$ there exists a final state $q = \delta(q_0, ua^n) \in F$ for some $n > 0$ such that[1] $q = \delta(q, a^n)$.*

Note that $\{a\} \cup aaaa(aa)^*$ is not a Type I language and this language shows that the divisibility condition in Theorem 8 is necessary, as it fulfills the condition with an $n \geq 0$ and $p > 0$ such that n is not divisible by p (for example for $u = a$ we have $n = 3$ and $p = 2$).

---

[1] As noted by a reviewer, this is a more compact way of expressing the fact that there exist $n \geq 0$ and $p > 0$ such that $q = \delta(q_0 u a^n)$, $q = \delta(q, a^p)$ and n is divisible by p.

Let $\mathcal{A} = (\Sigma, Q, \delta, q_0, F)$. It can be shown that the condition from Theorem 8 has to hold true for the minimal automaton of the commutative language. This directly yields that if $\mathcal{A}$ with $F = \{q_f\}$ is the minimal automaton for $L(\mathcal{A})$, then $L(\mathcal{A})$ is a Type I language if and only if $q \in \delta(q, a^+)$ for each $a \in \mathrm{alph}(L(\mathcal{A}))$.

Let $\mathcal{A} = (\Sigma, Q, \delta, q_0, F)$ be an automaton and $u \in \Sigma^*$. Then $u^{-1}L(\mathcal{A}) = L((\Sigma, Q, \delta, \delta(q_0, u), F))$. Hence, Theorem 8 directly yields the next closure property. Note that for commutative languages $L \subseteq \Sigma^*$ we have $u^{-1}L = Lu^{-1}$.

Proposition 9. *If $L \subseteq \Sigma^*$ is a Type I language and $u \in \Sigma^*$, then $u^{-1}L$ and Lu^{-1} are Type I languages.*

Lastly in this section, we point out that the set of words of sufficient length from a regular commutative language forms a Type I language. More specifically, if L is accepted by an n-state commutative automaton and $u \in L$ with $|u|_a \geq n$, then it induces a loop in the automaton and by commuativity, a loop labelled by a occurs at the state at which the word ends. This can be used to construct an automaton as in Theorem 8.

Proposition 10. *Let $\mathcal{A} = (\Sigma, Q, \delta, q_0, F)$ be a commutative automaton. Then $L(\mathcal{A}) \cap \{u \in \Sigma^* \mid \forall a \in \mathrm{alph}(u) : |u|_a \geq |Q|\}$ is a Type I language.*

4   Type II Languages

Here, we introduce Type II languages, which properly contain the Type I languages, and show that the iterated shuffle of a Type II language yields a Type I language.

Definition 11. *We call a commutative language a Type II language, if it is a finite union of languages of the form*

$$\bigsqcup_{a \in \Gamma} a^{k_a}(a^{p_a})^* \tag{2}$$

with $\Gamma \subseteq \Sigma$ and which fulfills the condition that if there exists $a \in \Gamma$ such that $p_a = 0$ and $k_a > 0$, then $k_b = 0$ for all $b \in \Gamma \setminus \{a\}$.

Actually, the choice of a subalphabet Γ is not necessary in the previous definition, as we can set $k_a = p_a = 0$ for unused letters. We use this form without reference to a subalphabet occasionally.

Note that every Type I language is also a Type II language.

Type II languages are clearly regular (more specifically, languages having the form stated in Eq. (2) are periodic).

Proposition 12. *Every Type II language is regular.*

Example 1. The shuffle $\{a\} \sqcup \{b\}$ of the two Type II languages $\{a\}$ and $\{b\}$ is neither a Type I nor a Type II language. Hence, contrary to Type I languages, Type II languages are not closed under the binary shuffle operation.

For Type II languages, the iterated shuffle yields a Type I language. In particular, it is also regularity-preserving on this class.

Theorem 13. *The iterated shuffle of a Type II language is a Type I language.*

The question arises how Type II languages relate to the regularity of the iterated shuffle in general.

Proposition 14. *The iterated shuffle of a periodic language is regular if and only if it has the form as stated in Eq. (2), which is equivalent for periodic languages to being a Type II language.*

Remark 1. However, in general Type II languages do not characterize those languages that have regular iterated shuffle. For general languages, the situation is more complicated. For example for $L = \{a, b, ab, ba\}$ we have $L^{\sqcup,*} = \{a, b\}^*$, but L is not a Type II language.

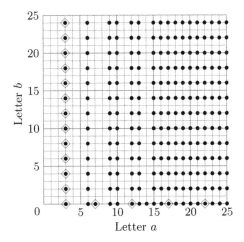

Fig. 2. The Parikh image of the Type II language $aaa \sqcup (bb)^* \cup aaaaaaa \sqcup (aaaaa)^*$ (red diamonds) and its iterated shuffle (black dots), which is regular. (Color figure online)

Lastly in this section, we give an automata-theoretical characterization of Type II languages.

Theorem 15. *Let $L \subseteq \Sigma^*$. Then L is a Type II language if and only if there exists a commutative automaton $\mathcal{A} = (\Sigma, Q, \delta, q_0, F)$ with $L = L(\mathcal{A})$ such that when $u \in L(\mathcal{A})$, at least one of the following two conditions is true:*

1. *for each $a \in \mathrm{alph}(u)$ there exists $n > 0$ such that $q = \delta(q_0, ua^n) \in F$ and $\delta(q, a^n) = q$ (observe that this is the same condition as in Theorem 8),*
2. *there exists $\Delta_u \subseteq \Sigma$ with $|\Delta_u| \leq 1$ such that for each $a \in \Sigma \setminus \Delta_u$ and $v \in \Delta_u^*$ with $|v|_b = |u|_b$ for each $b \in \Delta_u$ we have $\delta(q_0, va^{i \cdot |u|_a}) \in F$ for all $i \geq 0$.*

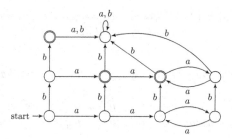

Fig. 3. Example automaton such that the iterated shuffle of its accepted language is regular according to Theorem 16.

5   Automata-Theoretical Characterizations

Here, in Theorems 16 and 18 we derive two conditions on commutative automata that characterize the regularity of the iterated shuffle of the recognized language, state some corollaries and an easy condition in Theorem 19, stating that only the existence of a single final state contained in a cycle for certain letters is equivalent to regularity of the iterated shuffle.

Theorem 16. *Let* $\mathcal{A} = (\Sigma, Q, \delta, q_0, F)$ *be commutative and* $L = L(\mathcal{A})$. *Then* $L^{\sqcup,*}$ *is regular if and only if for each* $a \in \Sigma$ *with* $a^+ \cap L = \emptyset$ *and* $u \in \Sigma^* a \Sigma^* \cap L$ *there exists* $q \in \delta(q_0, \mathrm{alph}(u)^*) \cap F$ *such that* $\delta(q, a^p) = q$ *for some* $p > 0$.

Proof. Note that L is a commutative language. First, suppose $L^{\sqcup,*}$ is regular and we have a letter $a \in \Sigma$ with $a^+ \cap L = \emptyset$ and a word $u \in \Sigma^* a \Sigma^* \cap L$. By Theorem 5, for each $i > 0$ there exists $u_i \in a^i \Sigma^* \cap (L \cap \mathrm{alph}(u)^*)$. Consider the states $q_i = \delta(q_0, u_i) \in F$. As the state set is finite, by the pigenhole principle, there exists a single state $q \in F$ such that $q = \delta(q_0, u_i)$ for infinitely many indices $i > 0$. Using finiteness again on the prefixes a^i of the infinitely many u_i with $\delta(q_0, u_i) = q$, there exist $0 < r < s$ such that we can write $u_r = a^r v_r$ and $u_s = a^s v_s$ with $v_r, v_s \in \Sigma^*$, $q = \delta(q_0, u_r) = \delta(q_0, u_s)$ and $\delta(q_0, a^r) = \delta(q_0, a^s)$. By commutativity, $\delta(q, a^{s-r}) = \delta(q_0, a^r v_r a^{s-r}) = \delta(q_0, a^r a^{s-r} v_r) = \delta(q_0, a^s v_r) = \delta(\delta(q_0, a^s), v_r) = \delta(\delta(q_0, a^r), v_r) = \delta(q_0, a^r v_r) = q$. Hence with $p = s - r > 0$ one implication is shown.

Conversely, suppose the condition of the statement holds true for L. Let $a \in \Sigma$ with $a^+ \cap L = \emptyset$ and $u \in \Sigma^* a \Sigma^* \cap L$. By assumption there exists $q \in F$, $v \in \mathrm{alph}(u)^*$ and $p > 0$ such that $q = \delta(q_0, v)$ and $\delta(q, a^p) = q$. Hence $v(a^p)^* \subseteq L \cap \mathrm{alph}(u)^*$. By commutativity $(a^p)^* v \subseteq L$. Hence, the condition from Theorem 5 is fulfilled and $L^{\sqcup,*}$ is regular. □

Using Theorem 16, we next give some more concrete classes of automata that recognize language whose iterated shuffle is regular. Recall that $q \in \delta(q, a^+)$ is equivalent to the fact that there exists $p > 0$ such that $q = \delta(q, a^p)$, i.e., the state is contained in a cycle for the letter a in the automaton.

Corollary 17. *Let $\mathcal{A} = (\Sigma, Q, \delta, q_0, F)$ be a commutative automaton. The iterated shuffle $L(\mathcal{A})^{\sqcup,*}$ is regular when $\mathcal{A}$ fulfills one of the following additional properties:*

1. *$\mathcal{A}$ has the property that for each $q \in F$ and $a \in \Sigma$ we have $q \in \delta(q_0, a^+)$ or $q \in \delta(q, a^+)$.*
2. *$\mathcal{A}$ has the property that for each $q \in F$ and $a \in \Sigma$ we have $q \in \delta(q, a^+)$.*
3. *$\mathcal{A}$ is a permutation automaton, i.e., the map $q \mapsto \delta(q, a)$ is a permutation for each $a \in \Sigma$.*

Note that Theorem 16 only applies to regular languages. For example, the iterated shuffle of the non-regular language $aaaaa \sqcup (bb)^* \cup a \sqcup b \sqcup \mathrm{perm}((ab)^*)$ is regular.

Next, using commutativity, we can strengthen Theorem 16 in the sense that a single final state works for certain letters (instead of a different final state for each letter as in Theorem 16).

Theorem 18. *Let $\mathcal{A} = (\Sigma, Q, \delta, q_0, F)$ be a the minimal automaton for a commutative language $L \subseteq \Sigma^*$. Set $\Gamma = \{a \in \Sigma \mid a^+ \cap L = \emptyset\}$. Then $L^{\sqcup,*}$ is regular if and only if for each $u \in L$ there exists $q \in \delta(q_0, \mathrm{alph}(u)^*) \cap F$ such that $\delta(q, a^{p_a}) = q$ for numbers $p_a > 0$ and each $a \in \Gamma \cap \mathrm{alph}(u)$.*

Observe that in Theorem 18 we only require that the final state is in a cycle for letters from $\Gamma \cap \mathrm{alph}(u)$. For example, in Fig. 3 for $u = bb$ we have $\Gamma \cap \mathrm{alph}(u) = \emptyset$, and hence the condition is vacuously fulfilled for the state at which this word words ends, which is contained in no cycle at all.

For a binary alphabet, Theorem 18 yields even simpler conditions.

Theorem 19. *Let Σ be a binary alphabet and $L \subseteq \Sigma^*$. Set $\Gamma = \{a \in \Sigma \mid a^+ \cap L = \emptyset\}$. Let $L \subseteq \Sigma^*$ be commutative and regular. Then, the following are equivalent:*

1. *$L^{\sqcup,*}$ is regular*
2. *in the minimal automaton $\mathcal{A} = (\Sigma, Q, \delta, q_0, F)$ for L there exists a final state q reachable from the start state such that $q \in \delta(q, a^+)$ for every $a \in \Gamma \cap \mathrm{alph}(L)$,*
3. *there exists $N > 0$ such that $U = \pi_\Gamma(L) \cap \bigcap_{a \in \Gamma} \{u \in \Sigma^* \mid |u|_a \geq N\}$ is a Type I language with $\mathrm{alph}(U) = \Gamma$.*

In Theorem 19 we can choose N to be the number of states of a recognizing automaton for $L \subseteq \Sigma^*$. Furthermore, observe that the condition $\mathrm{alph}(U) = \Gamma$ is important, as for example for $L = \{ab, ba\}$ and $N = 3$ we have $U = \emptyset$, which is a Type I language, but $L^{\sqcup,*}$ is not regular.

Remark 2. $|\Sigma| = 2$ is crucial in Theorem 19. Let $L = a \sqcup b \cup aa^* \sqcup bb^* \sqcup cc^*$. Then $L^{\sqcup,*}$ is not regular, but the last condition of Theorem 19 is true.

6   Decision Problems

In [8,9] it was shown that for regular $L \subseteq \Sigma^*$, it is decidable if perm(L) is regular. As perm$(L)^{\sqcup,*}$ = perm(L^*), also the regularity of the iterated shuffle on commutative regular languages is decidable. As noted in [15], Theorem 5 (and a related theorem) also yields this result. However, it is not immediately clear that this can be done in polynomial time. In fact, the proof in [8] uses Presburger arithmetic, and deciding the truth of Presburger formulae is complete for the class of language recognized by alternating Turing machines using at most double exponential time and a linear number of alternations [2]. So, in total, the proof from [8] only yields containment in double exponential space. By a statement given in [7, Theorem 45] it follows that for a regular language given by a regular expression over a binary alphabet it is co-NP-hard to decide if the commutative closure is regular.

With Theorem 16 we can in fact decide regularity of the iterated shuffle of a commutative regular language in polynomial time when the input is given as a deterministic automaton and the alphabet is fixed in advance, i.e., not part of the input. Let $\Sigma = \{a_1, \ldots, a_k\}$. The algorithm iterates through all words of the form $a_1^{i_1} \cdots a_k^{i_k}$ where $i_1 + \ldots + i_k$ is at most the number of states n of the input automaton. This is essentially the bottleneck, yielding a running time in $O(n^k)$. We do not know if a polynomial-time algorithm is possible when the alphabet is allowed to vary as part of the input.

Theorem 20. *Fix Σ. Given a regular commutative language in Σ^* by a DFA we can decide in polynomial time whether the iterated shuffle of this language is regular.*

For the introduced classes of languages Theorems 8 and 15 yield decidability in polynomial time if the language is given by a deterministic automaton.

Theorem 21. *Fix Σ. Given a regular language by a DFA we can decide in polynomial time whether it is a Type I language, a Type II language or neither.*

7   Conclusion

We have given automata-theoretical characterizations for the regularity of the iterated shuffle of commutative regular languages and identified two wide subclasses of commutative regular languages for which the iterated shuffle is regularity-preserving. We investigated the relation of these classes to common operations and the iterated shuffle. Then we used our results to derive the tractability of related decision problems when the input is given by deterministic automata for a fixed alphabet. We do not know what is the computational complexity when the alphabet is allowed to vary as part of the input. Furthermore, when the input is given by non-deterministic automata (see, for example, the textbook [13] for their definition) or regular expressions, if these decision problems are tractable as well or not is an open problem.

Acknowledgement. I thank the anonymous reviewers for careful reading and helping me identifying some unclear formulations and typos throughout the text.

References

1. Almeida, J., Ésik, Z., Pin, J.: Commutative positive varieties of languages. Acta Cybern. **23**(1), 91–111 (2017)
2. Berman, L.: The complexitiy of logical theories. Theor. Comput. Sci. **11**, 71–77 (1980)
3. Berstel, J., Boasson, L., Carton, O., Pin, J., Restivo, A.: The expressive power of the shuffle product. Inf. Comput. **208**(11), 1258–1272 (2010)
4. Campbell, R.H., Habermann, A.N.: The specification of process synchronization by path expressions. In: Gelenbe, E., Kaiser, C. (eds.) OS 1974. LNCS, vol. 16, pp. 89–102. Springer, Heidelberg (1974). https://doi.org/10.1007/BFb0029355
5. Ehrenfeucht, A., Haussler, D., Rozenberg, G.: On regularity of context-free languages. Theor. Comput. Sci **27**, 311–332 (1983)
6. Fernau, H., Hoffmann, S.: Extensions to minimal synchronizing words. J. Autom. Lang. Combin. **24**(2–4), 287–307 (2019)
7. Fernau, H., Paramasivan, M., Schmid, M.L., Vorel, V.: Characterization and complexity results on jumping finite automata. Theor. Comput. Sci **679**, 31–52 (2017)
8. Ginsburg, S., Spanier, E.H.: Bounded regular sets. Proceedings of the American Mathematical Society **17**, 1043–1049 (1966)
9. Gohon, P.: An algorithm to decide whether a rational subset of n∧ k is recognizable. Theor. Comput. Sci. **41**, 51–59 (1985)
10. Gómez, A.C., Pin, J.: Shuffle on positive varieties of languages. Theor. Comput. Sci. **312**(2–3), 433–461 (2004)
11. Hoffmann, S.: Commutative regular languages – properties and state complexity. In: Ćirić, M., Droste, M., Pin, J.É. (eds.) CAI 2019. LNCS, vol. 11545, pp. 151–163. Springer, Cham (2019). https://doi.org/10.1007/978-3-030-21363-3_13
12. Hoffmann, S.: Regularity conditions for iterated shuffle on commutative regular languages. In: Maneth, S. (ed.) CIAA 2021. LNCS, vol. 12803, pp. 27–38. Springer, Cham (2021). https://doi.org/10.1007/978-3-030-79121-6_3
13. Hopcroft, J.E., Ullman, J.D.: Introduction to Automata Theory, Languages, and Computation. Addison-Wesley Publishing Company, Boston (1979)
14. Imreh, B., Ito, M., Katsura, M.: On shuffle closure of commutative regular languages. In: Bridges, D.S., Calude, C.S., Gibbons, J., Reeves, S., Witten, I.H. (eds.) DMTCS 1996, pp. 276–288. Springer-Verlag, Singapore (1996)
15. Ito, M.: Algebraic Theory of Automata and Languages. World Scientific, Singapore (2004)
16. Kozen, D.: Automata and Computability. Undergraduate Texts in Computer Science, Springer, Heidelberg (1997). https://doi.org/10.1007/978-3-642-85706-5
17. Mazurkiewicz, A.: Parallel recursive program schemes. In: Bečvář, J. (ed.) MFCS 1975. LNCS, vol. 32, pp. 75–87. Springer, Heidelberg (1975). https://doi.org/10.1007/3-540-07389-2_183
18. Pighizzini, G., Shallit, J.: Unary language operations, state complexity and Jacobsthal's function. Int. J. Found. Comput. Sci. **13**(1), 145–159 (2002)
19. Pin, J.: Syntactic semigroups. In: Rozenberg, G., Salomaa, A. (eds.) Handbook of Formal Languages, vol. 1, pp. 679–746. Springer, Heidelberg (1997). https://doi.org/10.1007/978-3-642-59136-5_10

20. Pin, J.É.: How to prove that a language is regular or star-free? In: Leporati, A., Martín-Vide, C., Shapira, D., Zandron, C. (eds.) LATA 2020. LNCS, vol. 12038, pp. 68–88. Springer, Cham (2020). https://doi.org/10.1007/978-3-030-40608-0_5
21. Restivo, A.: The shuffle product: new research directions. In: Dediu, A.-H., Formenti, E., Martín-Vide, C., Truthe, B. (eds.) LATA 2015. LNCS, vol. 8977, pp. 70–81. Springer, Cham (2015). https://doi.org/10.1007/978-3-319-15579-1_5
22. Shaw, A.C.: Software descriptions with flow expressions. IEEE Trans. Softw. Eng. **4**, 242–254 (1978)

On the Complexity of Decision Problems for Counter Machines with Applications to Coding Theory

Oscar H. Ibarra[1] and Ian McQuillan[2(✉)]

[1] Department of Computer Science, University of California,
Santa Barbara, CA 93106, USA
ibarra@cs.ucsb.edu
[2] Department of Computer Science, University of Saskatchewan,
Saskatoon, SK S7N 5A9, Canada
mcquillan@cs.usask.ca

Abstract. We study the computational complexity of several decision problems (including the emptiness, disjointness, finiteness, and containment problems) for various restrictions of two-way reversal-bounded multicounter machines (2NCM). We then apply the results to some problems in coding theory. We examine generalizations of various types of codes with marginal errors; for example, a language L is k-infix-free ($k \geq 0$) if there is no non-empty string y in L that is an infix of more than k strings in $L - \{y\}$. This allows for bounded error versus standard infix-free languages. We show that it is PSPACE-complete to decide, given k and a 2NCM M whose input is finite-crossing, whether $L(M)$ is not k-infix-free. It follows that the problem is also PSPACE-complete for one-way nondeterministic and deterministic finite automata (even for the two-way models), answering an open question in [12]. We also look at the complexity of the problem for restricted models of 2NCM and for other types of codes, and improve/generalize some previous results.

1 Introduction

When introducing a new machine model, it is common to study several standard decision problems, determine whether they are decidable, and find their computational complexity. Frequently studied decision problems for a class of machines $\mathcal{M}$ include:

- *emptiness problem for $\mathcal{M}$*: Given a machine $M \in \mathcal{M}$, is $L(M) = \emptyset$?
- *disjointness problem for $\mathcal{M}$*: Given machines $M_1, M_2 \in \mathcal{M}$, is $L(M_1) \cap L(M_2) = \emptyset$?
- *containment problem for $\mathcal{M}$*: Given machines $M_1, M_2 \in \mathcal{M}$, is $L(M_1) \subseteq L(M_2)$?

The research of I. McQuillan was supported, in part, by Natural Sciences and Engineering Research Council of Canada.

V. Diekert and M. Volkov (Eds.): DLT 2022, LNCS 13257, pp. 177–188, 2022.
https://doi.org/10.1007/978-3-031-05578-2_14

– *equivalence problem for* $\mathcal{M}$: Given machines $M_1, M_2 \in \mathcal{M}$, is $L(M_1) = L(M_2)$?
– *finiteness problem for* $\mathcal{M}$: Given machines $M \in \mathcal{M}$, is $L(M)$ finite?

The *non-emptiness problem* is the negation of the emptiness problem and similarly for the other problems (the negation of the finiteness problem is called the infiniteness problem). It is sometimes more natural in terms of computational complexity to analyze the negations.

Machine models with a one- or two-way input tape plus one or more data stores such as stacks [1], pushdowns, queues, and counters [4], are frequently studied. However, even one-way machines with two counters (or pushdowns, stacks, or one queue) have the same power as Turing machines [4,14] making all of the decision problems undecidable. However, restrictions lead to lesser computational capacity. A machine with one or more counters is called r-reversal-bounded if, during any accepting computation, each counter makes at most r changes between non-decreasing and non-increasing the size of the counters; and it is reversal-bounded if it is r-reversal-bounded for some r. It is known that one-way nondeterministic reversal-bounded multicounter machines (NCM) have lesser power than Turing machines, and it has decidable emptiness, finiteness, and disjointness problems [7]. While containment and equivalence are undecidable, they are decidable for deterministic machines (DCM) [7]. With two-way inputs, it is known that two-way deterministic reversal-bounded multicounter machines (2DCM) have the same power as Turing machines [7], making all the problems undecidable. Again, restrictions make some problems decidable. Such a machine is called c-crossing if every input that is accepted has an accepting computation in which the input head crosses the boundary between any two adjacent cells at most c times; and it is finite-crossing if it is c-crossing for some c. Note that 1-crossing corresponds to one-way machines. It is known that any two-way nondeterministic finite-crossing reversal-bounded multicounter machine can be converted to an equivalent one-way machine [7]. While this is not the case for deterministic machines, the more general two-way deterministic finite-crossing reversal-bounded multicounter machines still have decidable containment and equivalence problems making them one of the most general classes of languages where this is the case.

Here, we summarize existing—and determine new—computational complexity results for decision problems on one-way and two-way finite-crossing reversal-bounded multicounter machines. We study the problem depending upon whether each of the number of crossings, counters, and counter reversals are fixed or not. At one end of the spectrum, if all are allowed to vary then the non-emptiness problem is PSPACE-complete for both nondeterministic and deterministic machines, and if all are fixed, then they are in P.

We then explore applications to coding theory. A code property or relation can be used to define a family of languages. For example, the infix relation (sometimes called subword relation) can be used to define infix-free languages—a language L is infix-free if $x, y \in L$, and x is an infix of y implies $x = y$; similarly with prefix- and suffix-free languages. In [11,12], Ko et al. generalized these notions to allow for marginal errors in codes. They introduced the notion of k-infix-free (resp., k-prefix-free, k-suffix-free) languages. A language L is k-infix-free if there is no non-empty string y in L that is an infix of more than k strings in $L - \{y\}$. The special cases, k-prefix-free language and k-suffix-free language, are defined similarly. Note that when $k = 0$, the definition reduces to the well-studied concepts [10,15] of infix-free (resp., prefix-free, suffix-free) languages. It was shown in [12] that it is PSPACE-complete to decide, given k and an NFA M, whether $L(M)$ is not k-prefix-free (resp., k-suffix-free). It was also shown that it is PSPACE-hard to decide whether $L(M)$ is not k-infix-free, but left open the question of whether the problem is in PSPACE. When k is fixed, it was remarked in [12] that the problem can be shown to be in PSPACE.

In this paper, we show that the k-infix-problem is in PSPACE (thus, PSPACE-complete) for finite-crossing 2NCM and for two-way NFA. In contrast, if k and the number of crossings are fixed, then the problem is NP-complete. Furthermore, when k, the number of crossings, counters, and reversals are fixed, the k-infix-problem is in P. We also investigate the complexity of the problem for other restricted models of 2NCM and improve/generalize some previous results. All omitted proofs will appear in an extended journal version.

2   Preliminaries

We assume a familiarity with the basics of automata and formal language theory. This includes the basic definitions of one-way deterministic finite automata (DFA), one-way nondeterministic finite automata (NFA), two-way deterministic finite automata (2DFA), two-way nondeterministic finite automata (2NFA), deterministic Turing machines (DTM), and nondeterministic Turing machines (NTM). All two-way machines are assumed to have left and right end-markers $(\triangleright, \triangleleft)$ around the input. We also omit the definition of generalized sequential machines [4], which we assume to be one-way, nondeterministic, and to have final states. We will assume familiarity with the basics of complexity theory, the complexity classes P, NP, and PSPACE, and with the concepts of hardness and completeness for a complexity class [4].

Let Σ be a finite alphabet. Then Σ^* (Σ^+) is the set of all words (resp. non-empty words) over Σ which includes the empty word λ. A language L with respect to Σ is any subset of Σ^*. Given L, the complement of L with respect to Σ, $\overline{L} = \Sigma^* - L$. Given a word $w \in \Sigma^*$, if $w = xyz$ for some strings, $x, y, z \in \Sigma^*$, then we say x is a prefix of w, y is an infix of w, and z is a suffix of w.

We will only describe reversal-bounded multicounter machines informally and refer to [7] for formal definitions. A two-way m-counter machine M is a 2NFA where the input is surrounded by left and right end-markers $\triangleright$ and $\triangleleft$, which is augmented with m counters. On each move, each counter can be kept the same or incremented/decremented by 1, and tested for zero. The counters are r-reversal-bounded if, during any accepting computation, each counter makes at most r changes between non-decreasing and non-increasing mode (and vice versa). A two-way m-counter machine is c-crossing if it has the property that every input that is accepted has an accepting computation in which the input head crosses the boundary between any two adjacent cells at most c times. A machine is finite-crossing if it is c-crossing for some given c, and reversal-bounded if it is r-reversal-bounded for some r.

We will use the following notations in the paper:

- NCM(m, r): one-way m-counter machines where all counters are r-reversal-bounded.
- NCM(m): $\bigcup_{r>0}$ NCM(m, r).
- NCM: $\bigcup_{m>0, r>0}$ NCM(m, r).
- 2NCM: two-way machines with multiple reversal-bounded counters.
- 2NCM(m) and 2NCM(m, r) are defined similarly.
- 2NCM(m, r, c): c-crossing two-way m-counter machines where all counters are r-reversal-bounded.

In the deterministic versions of the above models, 'N' is replaced by 'D'. So, for example, DCM and DCM(m) denote the deterministic versions of NCM and NCM(m), respectively. It is evident that e.g. NCM$(0, 0)$ corresponds to NFA, and similarly with other models. Also, since every 2NFA is finite-crossing, it follows that $\bigcup_{c>0}$ NCM$(0, 0, c) = 2$NFA, and similarly for 2DFA.

3   Complexity of Decision Problems for Restrictions of 2NCM

We first recall some known results.

Lemma 1 [7]. *We can construct, given a* 2NCM(m, r, c) *(resp.* 2DCM(m, r, c)) M *of size* n, *a machine* 2NCM$(m\lceil(r+1)/2\rceil, 1, c)$ *(resp.* 2DCM$(m\lceil(r+1)/2\rceil, 1, c)$) M' *such that* $L(M) = L(M')$ *and* M' *is of size at most* $n\lceil(r+1)/2\rceil$, *in time polynomial in the size of* M.

The following result is from [2] (with additional details in [6]):

Lemma 2 (Theorem 5 [2]). *We can effectively construct, given a* 2NCM$(m, 1, c)$ M, *an* NCM$(2m, 1)$ M' *such that* $L(M') = L(M)$. *If* M *has size* n, *then* M' *has size* $n' \leq dn^c$ *for some constant* d. *Moreover* M' *can be constructed in time polynomial in* n', *hence exponential in* n.

Proof. We briefly describe the construction in [6] of the NCM$(2m, 1)$ M' from the 2NCM$(m, 1, c)$ M, as it is informative in the proof of Proposition 5. On an input $w = a_0 a_1 \cdots a_l$ to M (a_0 and a_l are the left and right end-markers), M' simulates M by keeping track of the ordered sequence, R_i, of transition rules M uses when its input head moves right from a_{i-1} to a_i, or left from a_i to a_{i-1}. Note that there are at most c such transition rules in R_i. So, in simulating an accepting computation of M, M' need only nondeterministically guess a sequence, $R_1 R_2 \cdots R_l$, of ordered transition rules that corresponds to the desired accepting computation of M, noting that M' can check (from the specification of M) whether R_{i+1} is a valid successor of R_i. All counter increases are simulated by the first set of m counters and all counter decreases by the next set of m counters, which are verified to be the same at the end. We refer the reader to [6] for the details of the construction of M' from M. Note that M' can 'stay' on a symbol an unbounded number of moves before moving off the symbol. This is handled in a rather tricky way and is described in [6]. The number of possible distinct such ordered transition rules R_i is t^c, where t is the number of transition rules of M. It is also clear that M' can be constructed in time polynomial in its size n'. Note that n' is exponential in the size n of M. □

Lemma 3 (Lemma 2 [2]). *Let M be an $\mathsf{NCM}(m,1)$ with size n. Let t be the number of transition rules of M (note that $t \leq n$). Then $L(M) \neq \emptyset$ if and only if M accepts some input within time (i.e., number of moves and also maximum counter value) $(mt)^{d'm} \leq (mn)^{d'm}$ for some constant d'.*

We state the following while pointing out the complexity.

Lemma 4 (Lemma 1 [6], Lemma 3.2 [7]). *We can effectively construct in polynomial time, given $M \in 2\mathsf{DCM}(m,1,c)$, an equivalent $M' \in 2\mathsf{DCM}(m+1,1,c+2)$ which always halts. Further, if $M \in \mathsf{DCM}(m,1)$, then $M' \in \mathsf{DCM}(m,1)$.*

It is therefore clear that given a finite-crossing 2DCM M, another machine M' can be constructed in polynomial time accepting $\overline{L(M)}$.

3.1   The Non-emptiness Problem

We characterize the complexity of non-emptiness for finite-crossing 2DCM and 2NCM as follows. We include both deterministic and nondeterministic machines in order to highlight problems where there are differences.

Proposition 5. *The non-emptiness problems for both classes $2\mathsf{DCM}(m,r,c)$ and $2\mathsf{NCM}(m,r,c)$ are PSPACE-complete. This is also true when $m = r = 0$, i.e. for 2DFA and 2NFA.*

Proof. First we show that the problem is in PSPACE. Let M be a $2\mathsf{NCM}(m,r,c)$ of size n, and we assume $r = 1$ by Lemma 1. Note that m and c are taken into account in the size n. From Lemma 2, if we were to construct the $\mathsf{NCM}(2m,1)$ M' equivalent to M, it would be of size at most dn^c for some constant d. Then from Lemma 3, $L(M') \neq \emptyset$ if and only if it accepts some input within time

$$t = (2mdn^c)^{2md'},$$

for some constants d, d'. Hence, we would need to decide if M' accepts some string with a maximum value in each of the $2m$ counters during the computation is at most t. Thus the counters can be simulated by an NTM M'' using space $\log t$, which is polynomial in n, m, and c. The idea is for the NTM M'', when given a specification of M as input, to simulate the one-way equivalent of M (i.e. M') indirectly by guessing the sequence of crossing sequences $R_1, \ldots, R_l$, and the one-way input to M' symbol-by-symbol (so there is no need to write the symbol since it is one-way unlike had it directly simulated the two way M). Then M'' uses the specification of M to verify that the guessed $R_1, \ldots, R_l$ are correct. So, the NTM M'' guesses an input string w symbol-by-symbol and uses polynomial space to simulate the sequence of ordered transition rules $R_1 R_2 \cdots R_l$ simulating the computation of the 2NCM M as described in the proof of Lemma 2, noting that an R_i need only space $d''c$ for some constant d''. It follows that the non-emptiness of $L(M)$ is in PSPACE.

That the problem is PSPACE-hard follows from the fact that deciding, given a list of c DFA's, whether they have a non-empty intersection (hence, a 2DFA making $2c$ passes on the input tape) is already PSPACE-hard [13].   □

For the case of NCM or 2NCM where the number of input crossings if fixed, we have a lower complexity:

Proposition 6. *For fixed c, the non-emptiness problems for DCM(m, r, c) and NCM(m, r, c) are NP-complete. This is also true for $(c = 1, m = 2)$ and $(c = 1, r = 1)$ over unary alphabets.*

Proof. The NP-hardness was shown in [2] for DCM(2) over a unary alphabet; hence for $c = 1, m = 2$ over unary alphabets (and for $c = r = 1$ by Lemma 1).

From the construction in the proof of Lemma 2, when c is fixed, a NCM$(2m, 1)$ M' can be built simulating the c-crossing 2NCM$(m, 1, c)$ in time polynomial in the size of M. In [3], it was shown that the non-emptiness problem for NPCM (i.e., NPDA with 1-reversal-bounded counters hence also for r-reversal-bounded counters by Lemma 1) is in NP. Hence, non-emptiness for NCM is in NP.   □

The following result was also shown in [2].

Proposition 7 (Theorem 6, [2]). *Let m, r, c be a fixed integers. Then the non-emptiness problems for 2DCM(m, r, c) and 2NCM(m, r, c) are in P.*

The status of the non-emptiness problem for two-way NCM based on their restrictions is summarized as follows.

Proposition 8. *The following are true:*

1. *The non-emptiness problems for 2DCM, 2NCM, 2DCM$(2, r)$, 2NCM$(2, r)$, 2NCM$(m, 1)$, and 2DCM$(m, 1)$ are undecidable.*
2. *The non-emptiness problems for both 2DCM(m, r, c) and 2NCM(m, r, c) are PSPACE-complete. This is also true for $m = r = 0$; i.e. for 2NFA and 2DFA.*
3. *For fixed c, the non-emptiness problems for 2DCM(m, r, c) and 2NCM(m, r, c) are NP-complete. This is also true for both $(m = 2, c = 1)$ (i.e. DCM$(2, r)$) and $r = c = 1$ (i.e. DCM$(m, 1)$), even over unary languages.*
4. *For fixed c, m, r, the non-emptiness problems for both 2DCM(m, r, c) and 2NCM(m, r, c) are in P.*
5. *The non-emptiness problems for both 2DCM$(1, r, 1)$ and 2NCM$(1, r, 1)$ (i.e. DCM$(1, r)$ and NCM$(1, r)$) are in P.*

We note that the non-emptiness for 2DCM(1) is decidable [8] but the complexity is open; for 2NCM(1), decidability is open.

3.2   Other Decision Problems

Next, we examine the non-disjointment problem, which parallels non-emptiness in terms of complexity.

Proposition 9. *The following are true:*

1. *The non-disjointment problems for* 2DCM, 2NCM, 2DCM$(2, r)$, 2NCM$(2, r)$, 2NCM$(m, 1)$, *and* 2DCM$(m, 1)$ *are undecidable.*
2. *The non-disjointment problems for* 2DCM(m, r, c) *and* 2NCM(m, r, c) *are* PSPACE-*complete. This is also true for* $m = r = 0$; *i.e. for* 2DFA *and* 2NFA.
3. *For fixed* c, *the non-disjointment problems for both families* 2DCM(m, r, c) *and* 2NCM(m, r, c) *are* NP-*complete. This is also true for* $(m = 2, c = 1)$ *(i.e.* DCM$(2, r)$*) and* $r = c = 1$ *(i.e.* DCM$(m, 1)$*), even over unary languages.*
4. *For fixed* c, m, r, *the non-disjointment problems for both* 2DCM(m, r, c) *and* 2NCM(m, r, c) *are in* P.

For the non-containment, non-equivalence, and non-universe problems, there are differences between deterministic and nondeterministic machines.

Proposition 10. *The following are true:*

1. *The non-containment, non-equivalence, and non-universe problems for all of* 2DCM, 2DCM$(2, r)$, 2DCM$(m, 1)$, *and* NCM$(1, 1)$ *are undecidable.*
2. *The non-containment, non-equivalence, and non-universe problems for the family* 2DCM(m, r, c) *are* PSPACE-*complete. This is also true for* $m = r = 0$; *i.e. for* 2DFA.
3. *For fixed* c, *the non-containmment non-equivalence, and non-universe problems for* 2DCM(m, r, c) *are* NP-*complete. This is also true for* $(m = 2, c = 1)$ *(i.e.* DCM$(2, r)$*) and* $r = c = 1$ *(i.e.* DCM$(m, 1)$*), even over unary languages.*
4. *For fixed* c, m, r, *the non-containment, non-equivalence, and non-universe problems for* 2DCM(m, r, c) *are in* P.

Next, we study the complexity of the infiniteness problem. In this case, hardness is relatively straightforward using the non-emptiness problem.

Proposition 11. *The following are true:*

1. *The infiniteness problems for all of* 2DCM, 2NCM, 2DCM$(2, r)$, 2NCM$(2, r)$, 2NCM$(m, 1)$, *and* 2DCM$(m, 1)$ *are undecidable.*
2. *The infiniteness problems for* 2DCM(m, r, c) *and* 2NCM(m, r, c) *are* PSPACE-*hard. This is also true for* $m = r = 0$.
3. *For fixed* c, *the infiniteness problems for* 2DCM(m, r, c) *and* 2NCM(m, r, c) *are* NP-*complete. This is also true for both* $(m = 2, c = 1)$ *(i.e.* DCM$(2, r)$*) and* $r = c = 1$ *(i.e.* DCM$(m, 1)$*), even over unary languages.*

While we know completeness for the case where crossings are fixed, when they can vary, we do not know whether the infiniteness problem for 2NCM(m, r, c) is in PSPACE. However, we can prove that it is in exponential space.

Proposition 12. *The infiniteness problems for both families* 2DCM(m, r, c) *and* 2NCM(m, r, c) *are in exponential space.*

There are still two open problems. First, is the infiniteness problem for finite-crossing 2NCM in PSPACE, thereby making it PSPACE-complete? Also, if c, m, r are all fixed, is it in P (we only know they are in NP by part 3)?

The above proposition is in contrast to c-crossing 2NFA for fixed c, since such an 2NFA can be converted to an equivalent NFA in polynomial time, and infiniteness of NFA is in P; hence, infiniteness of c-crossing 2NFA for fixed c is in P. When c is not fixed (i.e. for 2NFA generally), it is already known that the problem is PSPACE-hard [5], but the authors could not find a result that shows infiniteness is in PSPACE. The following shows that infiniteness is indeed in PSPACE.

Proposition 13. *The infiniteness problems for* 2DFA *and* 2NFA *are* PSPACE-*complete.*

This is in contrast to 2DCM(2), where the emptiness and infiniteness problems are undecidable [7].

4   Complexity of k-Infix-Freeness of Languages

Let $k \geq 0$. A language L is k-infix-free if there is no string $y \neq \lambda$ in L that is an infix of more than k strings in $L - \{y\}$. A language L is k-prefix-free (k-suffix-free) if there is no string $y \neq \lambda$ in L that is a prefix (suffix resp.) of more than k strings in $L - \{y\}$.

The following was shown in [12]:

Proposition 14 *([12]).*

1. *Deciding, given $k \geq 0$ and DFA M, whether $L(M)$ is not k-prefix-free is in P.*
2. *Deciding, given $k \geq 0$ and NFA M, whether $L(M)$ is not k-prefix-free is PSPACE-complete.*
3. *Deciding, given $k \geq 0$ and NFA (or DFA) M, whether $L(M)$ is not k-suffix-free is PSPACE-complete.*
4. *Deciding, given $k \geq 0$ and NFA (or DFA) M, whether $L(M)$ is not k-infix-free is PSPACE-hard.*

It was left open in [12], whether k-infix-freeness for NFA is in PSPACE, though it was mentioned that for fixed k, the problem can be shown to be in PSPACE.

The following result answers the open problem in the affirmative, even for the more general model of M being a finite-crossing 2NCM:

Lemma 15. *Deciding, given $k \geq 0$ and a 2NCM(m, r, c) (or 2DCM(m, r, c)) M, whether $L(M)$ is not k-infix-free (resp. k-suffix-free) is PSPACE-complete. Also, deciding, given $k \geq 0$ and a 2NCM(m, r, c) M, whether $L(M)$ is not k-prefix-free is PSPACE-complete. These remain true when $c = 1, m = r = 0$.*

Proof. First, we show that the problem is in PSPACE for 2NCM(m, r, c) and k-infix-freeness. Let $k \geq 0$ and L be a language accepted by an 2NCM(m, r, c) M with input alphabet Σ. (Note that the input to M has end-markers which we do not explicitly show.) Let $\$, \#$ be new symbols not in Σ. Define the language

$$L' = \{y\$x_1\#z_1\$ \cdots \$x_{k+1}\#z_{k+1} \mid y \in L, y \neq \lambda, x_i, z_i \in \Sigma^* \text{ such that } x_i z_i \neq \lambda,$$
$$x_i y z_i \in L, x_i y z_i \neq x_j y z_j \text{ for } i \neq j\}.$$

Clearly, L is not k-infix-free if and only if $L' \neq \emptyset$. We construct another finite-crossing 2NCM M' to accept L'. (M' has left and right end markers on the input.) M' when given an input, does the following:

1. M' checks that the input has the correct format. This requires no counter and only 2 crosses on the input.
2. M' simulates M and checks that y is in L. This requires m r-reversal-bounded counters (to simulate the counters of M) and make at most c crosses on the input.
3. For $1 \leq i \leq k+1$, M' moves its input head at the beginning of x_i. It simulates M on $x_i y z_i$. To do this, it starts reading x_i, and whenever it crosses from reading x_i to y (which happens at most c times), M' moves to the left end-marker and continues the simulation reading y. The case is similar when simulating the cross of y moving left to x_i, y moving right to z_i, and moving left from z_i to y. This process requires m counters that are $((k+2)r)$-reversal-bounded and $(k + 2)c$ crosses on the input.
4. If $k > 0$, M' does the following: For $1 \leq i \leq k$, M' checks that $x_i y z_i \neq x_j y z_j$ for $i + 1 \leq j \leq k + 1$. We describe how M' can do this using a 1-reversal-bounded counter, say C. M' nondeterministically moves its head to some position p within $x_i y z_i$ while recording the distance of p from the left end-marker in counter C and the symbol, s_i, in that position in the state. (Note that since $x_i y z_i$ is not contiguous, if M' nondeterministically selects a position beyond x_i, M' needs to move its head to the left end marker and scan segment y and resume incrementing C and then return to the beginning of z_i.) Then M' moves its head to the beginning of x_j and starts decrementing counter C to zero to locate the symbol s_j corresponding to position p within $x_j y z_j$ and verifies that $s_i \neq s_j$.

The entire process above requires $k + (k - 1) + \cdots + 1 = k(k+1)/2$ 1-reversal-bounded counters and at most $4k(k + 1)/2 = 2k(k + 1)$ crosses on the input.

Thus M' needs $m' = 2m + k(k + 1)/2$ counters that are $r' = ((k+2)r)$-reversal-bounded and $c' = 2 + c + (k+2)c + 2k(k+1) = 2k^2 + 2k + 2 + c(k+3)$ crossings. Now let n be the size of M. (Note that the number m of 1-reversal-bounded counters of M is already taken into account in n.) Clearly, M' has size at most $p(kn)$ for some effectively computable low-order polynomial $p(.)$, independent of k and n. (Note also that in step 4, because M' does not need to remember all the guessed symbols s_i, s_j at the same time but it can operate one at a time, it does not cause an exponential blowup in states).

Then M' is of size $p(kn)$, is c'-crossing (which is polynomial in k and c) and has m' counters (which is polynomial in m and k) that are r'-reversal-bounded (that is polynomial in k and r). It follows from Proposition 5 that non-emptiness of $L(M')$ is in PSPACE. Hence, deciding if $L(M)$ is not k-infix-free is in PSPACE. It is straightforward to modify this proof for k-suffix-freeness and k-prefix-freeness.

That the problems are PSPACE-hard (even without counters) was already shown in Proposition 14. $\square$

Corollary 16. *Deciding, given $k \geq 0$ and an NFA (or 2NFA, DFA, or 2DFA) M, whether $L(M)$ is not k-infix-index is PSPACE-complete.*

The above corollary answers an open question in [12] of whether or not k-infix-freeness of NFA languages is in PSPACE. Not only do we extend this for finite-crossing 2NCM but also for 2NFA.

In [12], it was mentioned that it could be shown that for a fixed k, deciding, given an NFA M, whether $L(M)$ is not k-infix-free is in PSPACE. In fact, this problem is in P, as this is a special case of the next result.

Lemma 17. *Let m, r, c, and k be fixed integers. Deciding, given a 2NCM(m, r, c) M, whether $L(M)$ is not k-infix-free (resp. k-suffix-free, k-prefix-free) is in P.*

Altogether we can prove the following:

Proposition 18. *The following are true:*

1. *Deciding, given k, m, r, c and a 2NCM(m, r, c) (or 2DCM(m, r, c)) M, whether $L(M)$ is not k-infix-free (resp. suffix-free) is PSPACE-complete. This is also true for k-prefix-free with 2NCM(m, r, c). These remain true when c, m, r are fixed with $c = 1, m = r = 0$, and for 2NFA and 2DFA.*
2. *Let k, m, r be fixed. Deciding, given c and an NCM(m, r, c) (or DCM(m, r, c)) M, whether $L(M)$ is not k-infix-free (resp. k suffix-free, k-prefix-free) is PSPACE-complete. Further, this is true when $k = m = r = 0$, and for 2NFA and 2DFA.*
3. *Let k, c be fixed. Deciding, given m, r and a 2NCM(m, r, c) (or 2DCM(m, r, c)) M, whether $L(M)$ is not k-infix-free (resp. k-suffix-free, k-prefix-free) is NP-complete. In particular, this is true when $(k = 0, m = 2, c = 1)$, or $(k = 0, r = c = 1)$.*
4. *Let m, r, c, k be fixed. Deciding, given an 2NCM(m, r, c) M, whether $L(M)$ is not k-infix-free (resp. k-suffix-free, k-prefix-free) is in P.*

The only open cases not resolved is for determining the complexity of whether a 2DCM(m, r, c) is not k-prefix-free when k is allowed to vary but c is fixed. We can resolve this when c is at least two, and partially resolve it when $c = 1$.

Proposition 19. *The following are true:*

1. *Let $c \geq 2, m, r$ be fixed. Deciding, given k and $M \in$ 2DCM(m, r, c), whether $L(M)$ is not k-prefix-free is PSPACE-complete. This remains true when $c = 2$ and $m = r = 0$ (i.e. given a 2DFA that makes one cross on the input).*

2. *Let* $c = 1$. *Deciding, given* $k \geq 0$ *and* $m \in 2\mathsf{DCM}(m,r,c)$ *(i.e.* $M \in$ *$\mathsf{DCM}(m,r)$), whether* M *is not* k-*prefix-free is* NP-*hard. This remains true when* $k = 0$ *and either* $m = 2$ *or* $r = 1$.

It is open, whether the non-k-prefix-freeness problem for DCM is in NP (although we know it is in PSPACE), or whether the problem is PSPACE-hard. Also, it is open to determine the complexity of testing if a (one-way) $\mathsf{DCM}(m,r)$ is not k-prefix-free when k is allowed to vary but m,r are fixed. We have not been able to completely solve these problems, but they are in PSPACE by Proposition 18 (1). We also know that it is NP-hard if at least one of m,r varies, but it is open if it is in NP.

5 Generalizations of k-Infix-Freeness

The infix relation was used to define infix-free languages and k-infix-free languages. This notion can be generalized to allow for repeating occurrences of an infix up to some bounded number of times. The authors are unaware of this exact definition in the literature.

Definition 20. *A string* y *is an order* n *repeated-infix of a string* w *if* $w = x_1 y x_2 \cdots x_n y x_{n+1}$ *for some strings* $x_1, ..., x_{n+1}$. *Let* $n \geq 1$ *and* $k \geq 0$. *A language* L *is* k-*infix*$^{(n)}$-*free if there is no string* $y \neq \lambda$ *in* L *that is an order* n *repeated-infix of more than* k *strings in* $L - \{y\}$.

The constructions in the proof of Lemma 15 can be generalized to this case.

Proposition 21. *Deciding, given* $n, c \geq 1$, $k, m, r \geq 0$ *and a* $2\mathsf{NCM}(m,r,c)$ *(or* $2\mathsf{DCM}(m,r,c)$) M, *whether* $L(M)$ *is not* k-*infix*$^{(n)}$-*free is* PSPACE-*complete.*

It is worth mentioning that this definition of order n repeated infix allowed the occurrences of the repeated string y within w to be non-adjacent. This type of operation has been defined and studied more frequently where the repetitions of y are adjacent [9]. Proposition 21 would hold with this definition as well.

Next we consider another generalization. Let $k \geq 0$ and L be a language. We say that y is not k-infix-free in L if y is in L and there are more than k strings of the form xyz (for some x, z) in $L - \{y\}$. Using this, we define L to be infinitely not k-infix-free if there are infinitely many strings y that are not k-infix-free in L. The special cases, infinitely not k-prefix-free and infinitely not k-suffix-free are defined similarly.)

Proposition 22. *It is decidable, given* $k, m, r \geq 0, c \geq 1$ *and a* $2\mathsf{NCM}(m,r,c)$ *(or* $2\mathsf{DCM}(m,r,c)$) M, *whether* $L(M)$ *is infinitely not* k-*infix-free.*

It is open whether the problem above is in PSPACE. However, we can show that it is PSPACE-hard, even for special cases:

Proposition 23

1. *It is* PSPACE-*hard to decide, given* $k \geq 0$ *and an* NFA M, *whether* $L(M)$ *is infinitely not* k-*prefix-free.*
2. *It is* PSPACE-*hard to decide, given* $k \geq 0$ *and a* DFA M, *whether* $L(M)$ *is infinitely not* k-*suffix-free.*

References

1. Ginsburg, S., Greibach, S., Harrison, M.: Stack automata and compiling. J. ACM **14**(1), 172–201 (1967)
2. Gurari, E.M., Ibarra, O.H.: The complexity of decision problems for finite-turn multicounter machines. J. Comput. Syst. Sci. **22**(2), 220–229 (1981)
3. Hague, M., Lin, A.W.: Model checking recursive programs with numeric data types. In: Gopalakrishnan, G., Qadeer, S. (eds.) CAV 2011. LNCS, vol. 6806, pp. 743–759. Springer, Heidelberg (2011). https://doi.org/10.1007/978-3-642-22110-1_60
4. Hopcroft, J.E., Ullman, J.D.: Introduction to Automata Theory, Languages, and Computation. Addison-Wesley, Reading (1979)
5. Hunt, H.B.: On the time and tape complexity of languages i. In: Proceedings of the Fifth Annual ACM Symposium on Theory of Computing, STOC 1973, pp. 10–19 (1973)
6. Ibarra, O., Yen, H.: On the containment and equivalence problems for two-way transducers. Theor. Comput. Sci. **429**, 155–163 (2012)
7. Ibarra, O.H.: Reversal-bounded multicounter machines and their decision problems. J. ACM **25**(1), 116–133 (1978)
8. Ibarra, O.H., Jiang, T., Tran, N., Wang, H.: New decidability results concerning two-way counter machines. SIAM J. Comput. **23**(1), 123–137 (1995)
9. Ito, M., Kari, L., Kincaid, Z., Seki, S.: Duplication in DNA sequences. In: Ito, M., Toyama, M. (eds.) DLT 2008. LNCS, vol. 5257, pp. 419–430. Springer, Heidelberg (2008). https://doi.org/10.1007/978-3-540-85780-8_33
10. Rozenberg, G., Salomaa, A. (eds.): Handbook of Formal Languages. Springer, Heidelberg (1997). https://doi.org/10.1007/978-3-642-59136-5
11. Han, Y.-S., Ko, S.-K., Salomaa, K.: Generalizations of code languages with marginal errors. In: Potapov, I. (ed.) DLT 2015. LNCS, vol. 9168, pp. 264–275. Springer, Cham (2015). https://doi.org/10.1007/978-3-319-21500-6_21
12. Ko, S.K., Han, Y.S., Salomaa, K.: Generalizations of code languages with marginal errors. Int. J. Found. Comput. Sci. **32**, 509–529 (2021)
13. Kozen, D.: Lower bounds for natural proof systems. In: Proceedings of the 18th Symposium on the Foundations of Computer Science, pp. 254–266. IEEE (1977)
14. Minsky, M.L.: Recursive unsolvability of Post's problem of "tag" and other topics in theory of Turing Machines. Ann. Math. **74**(3), 437–455 (1961)
15. Shyr, H.J.: Free Monoids and Languages, 3rd edn. Hon Min Book Company, Taichung (2001)

Visit-Bounded Stack Automata

Jozef Jirásek[✉] and Ian McQuillan

Department of Computer Science, University of Saskatchewan,
Saskatoon, SK S7N 5A9, Canada
jirasek.jozef@usask.ca, mcquillan@cs.usask.ca

Abstract. An automaton is *k-visit-bounded* if during any computation its work tape head visits each tape cell at most k times. In this paper we consider stack automata which are k-visit-bounded for some integer k. This restriction resets the visits when popping (unlike similarly defined Turing machine restrictions) which we show allows the model to accept a proper superset of context-free languages and also a proper superset of languages of visit-bounded Turing machines. We study two variants of visit-bounded stack automata: one where only instructions that move the stack head downwards increase the number of visits of the destination cell, and another where any transition increases the number of visits. We prove that the two types of automata recognize the same languages. We then show that all languages recognized by visit-bounded stack automata are effectively semilinear, and hence are letter-equivalent to regular languages, which can be used to show other properties.

Keywords: Stack automata · Visit-bounded automata · Semilinear languages

1 Introduction

When introducing a machine model or a grammar system, one of the most useful properties is that of semilinearity. The idea of a language being semilinear is defined formally in Sect. 2, but equivalently, a language is semilinear if and only if it has the same Parikh image as some regular language [6]. In particular, when this property is effective for a machine model $\mathcal{M}$, there is a procedure to construct a letter-equivalent finite automaton from any such machine. It is well-known due to Parikh that the context-free languages have this property [12]. When this property is effective along with effective closure under homomorphism, inverse homomorphism, and intersection with regular languages (the full trio properties), it immediately implies several useful properties.

1. It provides a procedure to decide emptiness, finiteness, and membership [8].
2. The class can be augmented by reversal-bounded counters and the resulting class is still semilinear [5]—more generally, the smallest full trio (or even full AFL) containing the languages accepted by $\mathcal{M}$ that is also closed under intersection with one-way nondeterministic reversal-bounded multicounter machines [8] is also semilinear. The resulting family has the positive decidable properties of (1).

V. Diekert and M. Volkov (Eds.): DLT 2022, LNCS 13257, pp. 189–200, 2022.
https://doi.org/10.1007/978-3-031-05578-2_15

3. All bounded languages accepted by $\mathcal{M}$ are so-called *bounded semilinear languages* [9], and they can all be accepted by a deterministic machine model, one-way deterministic reversal-bounded multicounter machines [9], where we can decide containment and equivalence of two machines.
4. Properties related to counting functions and slenderness (having at most k strings of each length) can be decided [10].

It is also one of the key properties of a class of grammars being mildly context-sensitive [11], which was developed to encompass the properties that are important for computational linguistics.

Stack automata are a generalization of pushdown automata with the ability to push and pop at the top of the stack, and an added ability to read the contents of the stack in a two-way read-only fashion [2]. They are quite powerful however and can accept non-semilinear languages [1,3]. Checking stack automata are stack automata that cannot pop, and cannot push after reading from the stack. Here, we consider a restriction on stack automata. Given a subset E of the stack instructions (push, pop, stay, move left, or right), a machine is k-visit$_E$-bounded if, during any computation, its stack head visits each tape cell while performing an instruction of E at most k times; and it is visit$_E$-bounded if it is k-visit$_E$-bounded for some k. We omit E if it contains all instructions.

Importantly in this definition, when a cell is popped, the count towards this bound disappears with it, and any new symbols pushed start with a count of zero. This makes the definition in some ways more general than had we defined Turing machines with a visit-bounded worktape. This type of model was studied by Greibach [4], who studied one-way input with a single Turing machine work tape which it can edit (precisely, Greibach defines the machines to be preloaded with a string from a language family such as the regular languages—but as we are restricting our study to regular languages, this preloading does not affect the capacity). Greibach showed that the languages accepted by finite-visit Turing machines are a semilinear subset of the checking stack languages.

Here we show that a stack language is visit-bounded if and only if it is visit$_E$-bounded where E only contains an instruction to move left. We then show that the family of languages accepted by visit-bounded stack automata only contain semilinear languages, in contrast to stack automata generally. Furthermore, they form a language family properly between the context-free and stack languages.

Lastly, we show that the class of languages of Turing machines with a finite-visit (or finite-crossing) restriction (and a one-way input tape) is properly contained in the class of languages of finite-visit stack automata (as the former does not contain all context-free languages), demonstrating the power of our model while still preserving semilinearity. This makes the family useful towards showing that other families are semilinear.

2   Preliminaries

We refer to [6,7] for an introduction to automata and formal language theory. An *alphabet* Σ is a finite set of *symbols*. A *string* over Σ is a finite sequence of symbols from Σ. The set of all strings over Σ, including the empty string λ, is denoted by Σ^*. A *language* is a subset of Σ^*.

Let w be a string over $\Sigma = \{a_1, a_2, \dots, a_n\}$. The *length* of w, denoted by $|w|$, is the number of characters in w, with $|\lambda| = 0$. For $a \in \Sigma$, the number of occurrences of the character a in the string w is denoted by $|w|_a$. The *Parikh image* of a string w, denoted $\Psi(w)$, is the vector $(|w|_{a_1}, |w|_{a_2}, \dots, |w|_{a_n})$. We note that two strings have the same Parikh image if one is a permutation of the other. For a language $L \subseteq \Sigma^*$, let $\Psi(L) = \{\Psi(w) \mid w \in L\}$. Two languages L_1 and L_2 are *letter-equivalent* if $\Psi(L_1) = \Psi(L_2)$. Equivalently, every string in L_1 is a permutation of some string in L_2, and vice versa.

A subset Q of $\mathbb{N}^m$ (m-tuples) is a *linear set* if there exist $\vec{v_0}, \vec{v_1}, \dots, \vec{v_r} \in \mathbb{N}^m$ such that $Q = \{\vec{v_0} + i_1\vec{v_1} + \dots + i_r\vec{v_r} \mid i_1, \dots, i_r \in \mathbb{N}\}$. We call $\vec{v_0}$ the constant and $\vec{v_1}, \dots, \vec{v_r}$ the periods. A finite union of linear sets is a *semilinear set*. A language $L \subseteq \Sigma^*$ is semilinear if $\Psi(L)$ is a semilinear set. It is known that a language L is semilinear if and only if there exists a regular language L' with $\Psi(L) = \Psi(L')$ [6]. For a family of languages accepted by a class of machines $\mathcal{M}$, we say that the family is *effectively semilinear* if there is an algorithm to always determine the constant and the periods for each linear set (or equivalently the letter-equivalent finite automaton). The following is a classical result in automata theory.

Theorem 1 (Parikh's Theorem [12]). *Let L be a context-free language. Then $\Psi(L)$ is a semilinear set.*

Let NFA be the class of nondeterministic finite automata and NPDA be the class of nondeterministic pushdown automata. Given a class of machines $\mathcal{M}$, let $\mathcal{L}(\mathcal{M})$ be the family of languages accepted by $\mathcal{M}$.

2.1 Stack Automata

A nondeterministic one-way *stack automaton* is a 6-tuple $M = (Q, \Sigma, \Gamma, \delta, q_0, F)$, where:

- Q is the finite set of states,
- Σ and Γ are the input and work tape alphabets;
- Let $I = \{\mathsf{S}, \mathsf{L}, \mathsf{R}, \mathsf{push}(x), \mathsf{pop} \mid x \in \Gamma\}$ be the *instruction set*, then:
- $\delta \subseteq Q \times (\Sigma \cup \{\lambda\}) \times (\Gamma \cup \{\triangleright\}) \times Q \times I$ is the transition relation,
- $q_0 \in Q$ is the initial state, and
- $F \subseteq Q$ is the set of final states.

The special symbol $\triangleright$ denotes the left end of the work tape, which is identified with the bottom of the stack.

We will define the contents of the stack slightly differently (but equivalently) from previous definitions in order to better capture the new restrictions. The work tape shall be represented as a series of pairs (x, i), denoting individual tape cells, where $x \in \Gamma \cup \{\triangleright\}$ is the symbol written in this cell, and $i \in \mathbb{N}$ is the number of times the automaton has visited this cell. Note that the transition function of the automaton only has access to the symbols written on the tape, and the automaton can not inspect the visit counters of the cells.

A *configuration* of the automaton M is a triple (q, w, γ), where:

- $q \in Q$ is the current state,
- $w \in \Sigma^*$ is the input that is still to be read,
- $\gamma \in (\{\triangleright\} \times \mathbb{N})(\Gamma \times \mathbb{N})^* \lrcorner (\Gamma \times \mathbb{N})^*$ is the current content of the work tape. The special symbol $\lrcorner$ denotes the position of the tape head, which is scanning the cell immediately preceding this symbol.

Now let $E \subseteq \{\mathsf{S}, \mathsf{L}, \mathsf{R}, \mathsf{push}, \mathsf{pop}\}$ be a set of *expensive* instructions. These are the instructions that are counted as visits to the tape cell. The automaton performs all instructions on the work tape as usual for stack automata. When an expensive instruction is performed, the number of visits of the tape cell under the head after the instruction is completed is increased by one.

We define the *move relation* $\vdash$ between configurations of M using a set of expensive instructions E as follows: For $\iota \in \{\mathsf{S}, \mathsf{L}, \mathsf{R}, \mathsf{push}, \mathsf{pop}\}$, let the *cost* of ι be $c(\iota) = 1$ if $\iota \in E$, and $c(\iota) = 0$ if $\iota \notin E$. Then:

- $(p, aw, \alpha(x, i) \lrcorner \beta) \vdash (q, w, \alpha(x, i + c(\mathsf{S})) \lrcorner \beta)$
  if $(p, a, x, q, \mathsf{S}) \in \delta$,
- $(p, aw, \alpha(x, i)(y, j) \lrcorner \beta) \vdash (q, w, \alpha(x, i + c(\mathsf{L})) \lrcorner (y, j) \beta)$
  if $(p, a, x, q, \mathsf{L}) \in \delta$,
- $(p, aw, \alpha(x, i) \lrcorner (y, j) \beta) \vdash (q, w, \alpha(x, i)(y, j + c(\mathsf{R})) \lrcorner \beta)$
  if $(p, a, x, q, \mathsf{R}) \in \delta$,
- $(p, aw, \alpha(x, i) \lrcorner) \vdash (q, w, \alpha(x, i)(y, c(\mathsf{push})) \lrcorner)$
  if $(p, a, x, q, \mathsf{push}(y)) \in \delta$, and
- $(p, aw, \alpha(x, i)(y, j) \lrcorner) \vdash (q, w, \alpha(x, i + c(\mathsf{pop})) \lrcorner)$
  if $(p, a, y, q, \mathsf{pop}) \in \delta$;

where $p, q \in Q$, $a \in \Sigma \cup \{\lambda\}$, $w \in \Sigma^*$, $x \in \Gamma \cup \{\triangleright\}$, $y \in \Gamma$, $i, j \in \mathbb{N}$, $\alpha \in \{\lambda\} \cup ((\triangleright \times \mathbb{N})(\Gamma \times \mathbb{N})^*)$, $\beta \in (\Gamma \times \mathbb{N})^*$, and the work tape string on both sides of the relation is well-formed (in particular, $x = \triangleright$ if and only if $\alpha = \lambda$). Let $\vdash^*$ denote the reflexive and transitive closure of $\vdash$.

A *computation* of a stack automaton M on a string $w \in \Sigma^*$ is a sequence of configurations $c_0 \vdash c_1 \vdash \cdots \vdash c_n$, where $c_0 = (q_0, w, (\triangleright, 0) \lrcorner)$, and $c_n = (q_n, \lambda, \gamma_n)$. If $q_n \in F$, this computation is *accepting*. The automaton M *accepts* a string w if there exists an accepting computation of M on w. The *language accepted by* M, denoted by $L(M)$, is the set of all strings from Σ^* that M accepts.

Let SA be the class of all stack automata. A stack automaton is called a *non-erasing stack automaton* if it uses no pop instructions. A non-erasing stack automaton is called a *checking stack automaton* if it cannot push again after either a L or R instruction. The class of non-erasing stack automata is denoted by NESA, and checking stack automata by CSA.

For an integer k and a set of expensive instructions E, we say that a computation of a stack automaton M is k-$visit_E$-*bounded*, if the number of visits of every cell in every configuration in this computation is less than or equal to k. We say that M is k-$visit_E$-*bounded* if for every string $w \in L(M)$ the automaton M has a k-visit$_E$-bounded accepting computation on w. Finally, M is $visit_E$-*bounded* if there is a finite $k \in \mathbb{N}$ such that M is k-visit$_E$-bounded. If we leave off the subscript E, it is assumed that $E = \{\mathsf{S}, \mathsf{L}, \mathsf{R}, \mathsf{push}, \mathsf{pop}\}$. Let $\mathsf{VISIT}_E(k)$ be the class of k-visit$_E$-bounded, VISIT_E be all visit$_E$-bounded machines, and again we leave off the subscript E if $E = \{\mathsf{S}, \mathsf{L}, \mathsf{R}, \mathsf{push}, \mathsf{pop}\}$. It is immediate that $\mathcal{L}(\mathsf{SA}) = \mathcal{L}(\mathsf{VISIT}_\emptyset)$.

Note the important distinguishing feature of the stack automaton model which sets it apart from known visit-bounded Turing machine models: whenever a tape cell is popped from the top of the stack, the number of visits of that cell is reset. Whenever a new cell is pushed to the top of the stack, this new cell begins with a visit count of 0 (or 1, if push is an expensive instruction). This allows a visit-bounded stack automaton to perform some computations that an analogous visit-bounded Turing machine could not.

3 Visit-Bounded Automata

As we have seen in definitions in Sect. 2, the notion of a visit-bounded stack automaton is dependent on the choice of the set of expensive instructions E which increase the visit counters of tape cells. To begin, we consider two expensive instruction sets: $E = \{\mathsf{L}\}$, and $E = \{\mathsf{S}, \mathsf{L}, \mathsf{R}, \mathsf{push}, \mathsf{pop}\}$. In the first case, only L instructions increase the visit counters. In the second case, all instructions increase the visit counters.

Example 2. Let $M = (\{q_0\}, \{a\}, \{\}, \{(q_0, a, \triangleright, q_0, \mathsf{S})\}, q_0, \{q_0\})$ be a stack automaton. This simple automaton scans its input consisting of a number of symbols a, while the work tape head rests on the bottom of the stack marker.

Observe that M is visit$_{\{\mathsf{L}\}}$-bounded, as it never performs an L instruction, and thus the number of visits of the only used tape cell never increases above 0. On the other hand, M is not visit-bounded, as the S instructions in the only computation of M on string a^k increases the visit counter of the tape cell to k.

Every visit-bounded automaton is also visit$_{\{\mathsf{L}\}}$-bounded. Indeed, the number of visits to a cell can not increase if we only consider a limited subset of expensive instructions. Perhaps surprisingly, as we will show in Theorem 3, the converse is also true if we only consider languages accepted by the automaton. For any visit$_{\{\mathsf{L}\}}$-bounded automaton A, we can construct a visit-bounded automaton B with $L(B) = L(A)$. Therefore, limiting the usage of any instruction other than L does not reduce the descriptive power of the automaton model.

Theorem 3. *Let $A = (Q, \Sigma, \Gamma, \delta, q_0, F)$ be a visit$_{\{\mathsf{L}\}}$-bounded stack automaton. Then there exists a visit-bounded stack automaton B such that $L(B) = L(A)$. Hence, $\mathcal{L}(\mathsf{VISIT}) = \mathcal{L}(\mathsf{VISIT}_{\{\mathsf{L}\}})$.*

Proof. Let A be visit$_{\{\mathsf{L}\}}$-bounded, *i.e.*, A visits every tape cell using the L instruction at most k times. We prove the theorem by describing a construction of the automaton B. The basic idea of the construction is that B emulates a computation of A, but every symbol on the work tape of A shall be represented by multiple copies of the same symbol on the work tape of B. Instructions of A operating on a specific tape cell will be distributed among the copies of this cell by B in such a way that every copy is only visited a fixed number of times. By careful counting we show that any computation of A can be emulated by B in such a way that the number of visits to every cell of B on any instruction can be bounded as a function of k. This means that there is a constant ℓ which depends on k such that B is ℓ-visit-bounded, $i.e.$, B is visit-bounded.

The detailed construction appeared in Appendix A of the submitted paper. □

As a consequence of Theorem 3, the classes of languages accepted by visit$_{\{L\}}$-bounded and visit-bounded automata are identical.

We can also observe the following result for context-free languages:

Corollary 4. *For all $E \subseteq \{S, L, R, push, pop\}$, $\mathcal{L}(\mathsf{NPDA}) \subsetneq \mathcal{L}(\mathsf{VISIT}_E)$.*

Proof. A pushdown automaton can be seen as a stack automaton which never uses the L and R instructions. This automaton is trivially visit$_{\{L\}}$-bounded, and by Theorem 3 its language can be accepted by some visit-bounded stack automaton. Strictness can be seen using $\{a^n b^n c^n \mid n > 0\}$. □

We conclude this section with a comparison to Turing machines. Consider nondeterministic Turing machines with a one-way read-only input and a single work tape. If there is a bound on the number of changes of direction on the work tape (reversal-bounded), we denote these machines by TMRB; if there is a bound on the number of times the boundary of each pair of adjacent cells is crossed (finite-crossing), we denote these machines by TMFC; and if there is a bound on the number of visits to each cell (finite-visit), we denoted these by TMFV. Greibach studies these machines [4] where the work tape is preloaded with regular languages (or other families but we do not consider others), and the work tape is confined to the preloaded space. This preloading does not impact the languages accepted however as shown in the proof of the following, along with a comparison to visit-bounded stack automata.

Proposition 5. $\mathcal{L}(\mathsf{TMRB}) \subsetneq \mathcal{L}(\mathsf{TMFC}) = \mathcal{L}(\mathsf{TMFV}) \subsetneq \mathcal{L}(\mathsf{VISIT})$.

Proof. First we will argue that preloading these Turing machines with regular languages does note affect the languages accepted. Indeed, preloading can be simulated by guessing and writing a preloaded string and then simulating. In the other direction, a new dummy symbol B can be introduced, and the machine can be preloaded with B^*, the machine then guesses some start position and simulates using B as the blank symbol. It will only accept if it is preloaded with a string that is longer than the number of cells visited and it guesses the correct start position. Greibach shows that $\mathcal{L}(\mathsf{TMRB}) \subsetneq \mathcal{L}(\mathsf{TMFC}) = \mathcal{L}(\mathsf{TMFV})$ in Theorems 2.15 and 3.12. To show that $\mathcal{L}(\mathsf{TMFV}) \subseteq \mathcal{L}(\mathsf{VISIT})$, in Lemma 4.21 of [4], Greibach shows that every $L \in \mathcal{L}(\mathsf{TMFV})$ can be accepted by a Turing machine preset with a regular language where the machine does not ever change the work tape contents, and every accepting computation is k-visit-bounded. Such a machine M can be accepted by a k-visit-bounded stack automaton by first guessing the stack contents, and then simulating. The inclusion is strict as noted in the proof of Theorem 4.26 [4] as the context-free Dyck language cannot be accepted by a TMFV.   □

4   Semilinearity

The main result of this section is to prove that the language accepted by any visit-bounded stack automaton is semilinear. To prove this, we give a procedure that, given a visit-bounded stack automaton M, constructs a pushdown automaton P, such that $L(P)$ and $L(M)$ are letter-equivalent. Specifically, we show that for any string $w \in L(M)$, the automaton P can accept some permutation of w, and vice versa. It is known that languages of pushdown automata are semilinear, and semilinearity is preserved under letter-equivalence, hence this proves the main result.

Theorem 6. *Let* $M = (Q, \Sigma, \Gamma, \delta, q_0, F)$ *be a visit-bounded stack automaton. Then the language accepted by* M *is effectively semilinear.*

Proof. Let M be k-visit-bounded for an integer k. Further, assume that the automaton ends its computation with an empty stack. If it does not, this can be achieved by deleting the entire content of the stack before accepting, which adds at most one visit to every tape cell.

The central concept used for the proof is the *visit history* of a tape cell. Using the analogy of a physical work tape with paper cells, every time the automaton makes a move, it records the transition it has just used (a 5-tuple (p, a, x, q, ι)) on both the cell it left and the cell it entered. If the transition used an S instruction, those two refer to the same cell. In this way, since every cell is visited at most k times before being destroyed, the visit history of every cell contains at most $2k$ entries: k for transitions which were used to enter the cell, and another k for transitions which were used to leave. We shall refer to the i-th entering transition as $t_{\text{in}}[i]$ and the i-th leaving transition as $t_{\text{out}}[i]$. Also note that throughout the computation of the machine every transition used is recorded exactly twice: once in the cell it begins in, and once in the cell it ends in. This connection links the visit histories of all cells into a linked list-like structure which records the entire computation of M. Since every transition contains the input symbol being read (if any), following these links allows us to see the string being accepted.

The main idea is to construct a pushdown automaton P, which emulates the **push** and **pop** instructions in some computation of M, while nondeterministically guessing the entire history of every cell pushed on the stack. As long as P can ensure the integrity of links between every pair of adjacent cells, the entire linked list can be followed to reconstruct a computation of M, including the L, R, and S instructions. Then if P also reads all input symbols corresponding to every t_{in} transition in all histories, it accepts a permutation of the string accepted by M in this computation.

An important fact affecting the construction of P is that cells on the work tape of M can be erased and replaced by another cell. Therefore, not all of the transitions in the history of one cell need to correspond to transitions in the history of one adjacent cell. Some transitions could connect to a cell that had been in that place but was previously erased, and some transitions might connect to a cell that will be in that place in the future, after the currently following cell is erased. Therefore, the representation of every cell in P will additionally carry a *completed transition counter*, an index ctc in the range $1 \leq ctc \leq k$, which indicates how many transitions in the history of the current cell have already been matched with corresponding transitions in the histories of adjacent cells.

We can now describe the construction of the pushdown automaton P.

Definition 7. *A* history card *is a* $(2k + 2)$-tuple $(x, ctc, t_{in}[1], \ldots, t_{in}[k], t_{out}[1], \ldots, t_{out}[k])$, *where:*

- *$x \in (\Gamma \cup \{\triangleright\})$ is the stack symbol written on the tape cell,*
- *$1 \leq ctc \leq k$ is the completed transition counter,*
- *$t_{in}[i] \in (\delta \cup \{\emptyset\})$, for $1 \leq i \leq k$, are the transitions ending in this cell, and*
- *$t_{out}[j] \in (\delta \cup \{\emptyset\})$, for $1 \leq j \leq k$, are the transitions originating in this cell.*

Not all possible history cards can appear in some computation of M. We impose several consistency constraints on the history cards that P can use, to ensure that the information on each card is filled in properly and does not contradict itself.

Definition 8. *A history card is* internally consistent, *if all the following hold:*

- *$t_{in}[i] \neq \emptyset \iff t_{out}[i] \neq \emptyset$ for all $1 \leq i \leq k$. If there is an incoming transition, there has to be a corresponding outgoing transition.*
- *If $t_{in}[i] = \emptyset$, then also $t_{in}[i + 1] = \emptyset$. Similarly if $t_{out}[i] = \emptyset$, then also $t_{out}[i + 1] = \emptyset$. This holds for all $1 \leq i < k$. Transitions are always stored in a contiguous block of indices starting from the beginning of the card.*
- *The transition $t_{in}[1]$ performs the* push(x) *instruction, where x is the symbol stored on this card. The last non-empty $t_{out}[i]$ performs the* pop *instruction. No other t_{in} transitions are* push *and no other t_{out} transitions are* pop *instructions. The history of a cell begins when it is pushed and ends when it is popped from the stack. Each of these events can only happen once in the lifetime of the cell.*
  - *The exception to the three rules above is a card with $x = \triangleright$. This card represents the bottom of the stack of M, and here the computation of M begins and ends. Therefore $t_{in}[1] = \emptyset$, there is exactly one i such that $t_{in}[i] \neq \emptyset$ and $t_{out}[i] = \emptyset$, no t_{in} is a* push *or R instruction, and no t_{out} is a* pop *or L instruction.*
- *The work tape symbol read in every t_{out} transition is the symbol x on this card.*

Denote by H the set of all internally consistent history cards. Note that $|H| \leq (|\Gamma| + 1)k(|\delta| + 1)^{2k}$. The set H shall be the working alphabet of the pushdown automaton P. An example of history cards and links between them corresponding to a computation of M is shown in Fig. 1. The links are not explicitly stored but will be implied.

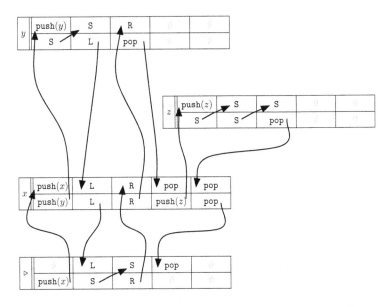

Fig. 1. Histories of tape cells after executing the following sequence of instructions: $\mathsf{push}(x), \mathsf{push}(y), \mathsf{S}, \mathsf{L}, \mathsf{L}, \mathsf{S}, \mathsf{R}, \mathsf{R}, \mathsf{pop}, \mathsf{push}(z), \mathsf{S}, \mathsf{S}, \mathsf{pop}, \mathsf{pop}$. Only the instructions used in the transitions are shown, states and symbols read are omitted. Transitions t_{in} shown in the top row, and t_{out} in the bottom row. Arrows show links between history cards formed by pairs of identical transitions.

Now we describe an algorithm used by P to simulate a computation of M. This algorithm employs two subprocedures: the first one advances the completed transition counter on a card step by step, verifying that the transitions on the card can link together to form a continuous computation, until either a **push** or a **pop** transition, or the end of the computation is reached. The facts that need to be verified are that S instructions on this card link to each other, and that every outgoing L instruction is followed by an incoming R instruction.

The second procedure takes two history cards as input and attempts to link together transitions between them. An outgoing **push** instruction on the bottom card has to link to the first incoming instruction on the top card. Every outgoing R instruction on the bottom card has to link to an incoming R instruction on the top card, and every outgoing L instruction on the top card has to link to an incoming L instruction on the bottom card. Finally, the last transition of the top card, performing a **pop** instruction, has to link to an incoming **pop** transition on the bottom card.

The complete description of both procedures appeared in Appendix B of the submitted paper. The important fact is that since there are only finitely many different history cards, the pushdown automaton itself does not have to perform either of these procedures. The results for all possible inputs can be encoded into its transition function.

A description of the algorithm performed by P is in Algorithm 4.1.

1 Nondeterministically choose a history card containing the symbol ▷. Push this
   card on the stack.
2 Read all input symbols that are read in any incoming transition on this card.
3 **while** *There is a history card on the stack* **do**
4   Advance the *ctc* of the card on top of the stack, verifying the consistency of
     instruction links, until either a **push** or a **pop** instruction, or the end of the
     computation is encountered.
5   **if** *The transition encountered performs a* **push** *instruction* **then**
6   Nondeterministically choose a history card containing the symbol being
        pushed.
7   Verify that the chosen card can be matched to the card currently on top
        of the stack.
8   **if** *The cards can be matched together* **then**
9   Move the *ctc* of the card on top of the stack to the incoming **pop**
          instruction corresponding to the removal of the cell represented by
          the new card.
10   Push the newly chosen card on top of the stack, initializing its *ctc* to
          1.
11   Read all input symbols that are read in any incoming transition on
          the new card.
12   **else**
13   Halt the computation and reject.
14   **end**
15   **else if** *The transition encountered performs the* **pop** *instruction, or the end
     of the computation is encountered* **then**
16   Erase the top card from the stack.
17   **else if** *A transition on the card can not be linked properly* **then**
18   Halt the computation and reject.
19   **end**
20 **end**
21 Finish the computation and accept.

Algorithm 4.1: The algorithm performed by the pushdown automaton P
emulating a computation of a visit-bounded stack automaton.

If the computation of P succeeds, this means that all transitions in all the
history cards used can be linked together to form one possible contiguous computation of M. Further, P reads every symbol that is read by every instruction
in this computation, just not necessarily in the same order as M. However,
this means that the string read by P is a permutation of the string that is
read by the corresponding computation of M. Therefore, the language of P is
letter-equivalent to the language of M. Finally, since the languages of pushdown
automata are semilinear, and semilinearity is preserved under letter-equivalence,
this means that the language of M is semilinear as well.   □

5   Other Expensive Instruction Sets

We have considered automata models with $E = \{L\}$ and $E = \{S, L, R, push, pop\}$. We can ask whether models with other expensive instruction sets also describe the same class of languages.

It is possible to show that $\text{visit}_{\{R\}}$-bounded automata accept the same class of languages as $\text{visit}_{\{L\}}$-bounded automata. The proof uses similar ideas as in the construction in the proof of Theorem 3, though we do not include it here. Adding the S instruction to a set of expensive instructions does not change the class of languages accepted, as every S instruction can be replaced by a pair of R and L instructions, or push and pop instructions when operating on top of the stack. Therefore it is always possible to construct an equivalent automaton which never uses the S instruction.

Hence, if we consider expensive instruction sets E containing either L or R, any visit-bounded automaton is also visit_E bounded for such E. Therefore all models with such an expensive instruction set accept the same class of languages.

Making expensive instructions exactly the push instructions has no effect on the languages accepted (i.e. it accepts all stack languages), as any cell can only be pushed on the stack once. We only include the push instruction as a possible expensive instruction for completeness.

Finally, we shall see that a model with $E = \{pop\}$ also accepts all stack automaton languages. Using a procedure similar to the one in the construction of automaton B in Theorem 3 we can clone symbols on the stack and replace every pop transition by a sequence $pop - pop - push$, such that every cell is visited at most twice by pop instructions.

These results can be summarized as follows. The hierarchy is depicted in Fig. 2.

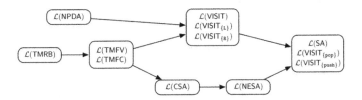

Fig. 2. The language families listed are related such that the families that are equal are written together in a box, inclusions are shown with an arrow that are proper in every case, and no lines connecting them indicate that they are incomparable.

Theorem 9. *The hierarchy shown in Fig. 2 is correct.*

Proof. That $\mathcal{L}(\mathsf{TMRB}) \subsetneq \mathcal{L}(\mathsf{TMFC}) = \mathcal{L}(\mathsf{TMFV}) \subsetneq \mathcal{L}(\mathsf{VISIT})$ is shown in Proposition 5. That $\mathcal{L}(\mathsf{TMFV}) \subsetneq \mathcal{L}(\mathsf{CSA})$ is shown in [4]. That $\mathcal{L}(\mathsf{CSA}) \subsetneq \mathcal{L}(\mathsf{NESA}) \subsetneq \mathcal{L}(\mathsf{SA})$ is well known [3]. That $\mathcal{L}(\mathsf{NPDA}) \subsetneq \mathcal{L}(\mathsf{VISIT})$ was shown in Corollary 4. That $\mathcal{L}(\mathsf{VISIT}) = \mathcal{L}(\mathsf{VISIT}_{\{L\}})$ is from Theorem 3, and the equality with $\mathcal{L}(\mathsf{VISIT}_{\{R\}})$ is mentioned above. The equality of $\mathcal{L}(\mathsf{SA})$ with $\mathcal{L}(\mathsf{VISIT}_{\{pop\}})$ and

$\mathcal{L}(\mathsf{VISIT}_{\{push\}})$ is also mentioned above. The proper inclusion of $\mathcal{L}(\mathsf{VISIT})$ in $\mathcal{L}(\mathsf{SA})$ follows from Theorem 6 since stack automata can accept non-semilinear languages [3]. Also, $\mathcal{L}(\mathsf{CSA})$ contains languages not accepted by $\mathcal{L}(\mathsf{VISIT})$ since $\mathcal{L}(\mathsf{CSA})$ contains non-semilinear languages [3]. Also, it is known that $\mathcal{L}(\mathsf{NESA})$ does not contain all context-free languages [4]. □

Hence, all the families above that are semilinear are contained in $\mathcal{L}(\mathsf{VISIT})$, making it the most powerful such family.

References

1. Ginsburg, S., Greibach, S., Harrison, M.: One-way stack automata. J. ACM **14**(2), 389–418 (1967)
2. Ginsburg, S., Greibach, S., Harrison, M.: Stack automata and compiling. J. ACM **14**(1), 172–201 (1967)
3. Greibach, S.: Checking automata and one-way stack languages. J. Comput. Syst. Sci. **3**(2), 196–217 (1969)
4. Greibach, S.A.: One way finite visit automata. Theor. Comput. Sci. **6**, 175–221 (1978)
5. Harju, T., Ibarra, O., Karhumäki, J., Salomaa, A.: Some decision problems concerning semilinearity and commutation. J. Comput. Syst. Sci. **65**(2), 278–294 (2002)
6. Harrison, M.: Introduction to Formal Language Theory. Addison-Wesley, Reading (1978)
7. Hopcroft, J.E., Ullman, J.D.: Introduction to Automata Theory, Languages, and Computation. Addison-Wesley, Reading (1979)
8. Ibarra, O., McQuillan, I.: Semilinearity of families of languages. Int. J. Found. Comput. Sci. **31**(8), 1179–1198 (2020)
9. Ibarra, O.H., McQuillan, I.: On families of full trios containing counter machine languages. Theor. Comput. Sci. **799**, 71–93 (2019)
10. Ibarra, O.H., McQuillan, I., Ravikumar, B.: On counting functions and slenderness of languages. Theor. Comput. Sci. **777**, 356–378 (2019)
11. Joshi, A.K.: Tree adjoining grammars: how much context-sensitivity is required to provide reasonable structural descriptions? In: Natural Language Parsing, pp. 206–250. Cambridge University Press, Cambridge (1985)
12. Parikh, R.: On context-free languages. J. ACM **13**(4), 570–581 (1966)

Well Quasi-Orders Arising from Finite Ordered Semigroups

Ondřej Klíma[(⊠)] and Jonatan Kolegar

Department of Mathematics and Statistics, Masaryk University,
Kotlářská 2, 611 37 Brno, Czech Republic
{klima,kolegar}@math.muni.cz

Abstract. In 1985, Bucher, Ehrenfeucht and Haussler studied derivation relations associated with a given set of context-free rules. Their research motivated a question regarding homomorphisms from the semigroup of all words onto a finite ordered semigroup. The question is which of these homomorphisms induce a well quasi-order on the set of all words. We show that this problem is decidable and the answer does not depend on the homomorphism, but it is a property of the ordered semigroup.

Keywords: Finite semigroups · Well quasi-orders · Unavoidable words

1 Introduction

The notion of well quasi-order (*wqo*) is a well-established tool in mathematics and in many areas of theoretical computer science that was rediscovered by many authors (see [8] by Kruskal). A comprehensive overview of the applications of the notion in theory of formal languages and combinatorics on words can be found in the book [10] by de Luca and Varricchio or in the survey paper [2] by D'Alessandro and Varricchio. Since our contribution belongs to formal language theory, we recall the central notion of *wqo* directly for the set of all words over a finite alphabet A. A quasi-order $\leq$ on a set A^* is a *wqo* if it has no infinite antichains and no decreasing infinite chains (the latter property is often called *well-foundedness*). There are several equivalent conditions for the notion (see, *e.g.*, [10, Theorem 6.1.1]); among them we recall the following: for every infinite sequence of words $w_1, w_2, \ldots$ there exist integers $0 < i < j$ such that $w_i \leq w_j$. We point out that the important property which makes the notion of *wqo* a useful tool is that every subset of A^* upper closed with respect to a monotone *wqo* $\leq$ is a regular language (see [10, Theorem 6.3.1]).

The first example of a *wqo* in the area of formal languages was given by Higman [5]. We mention the simplest consequence of the general statement, namely the result that the *embedding* relation $\trianglelefteq$ on A^* is a *wqo*. The embedding relation $\trianglelefteq$ is often called subword ordering, because a word u *embeds* in a word

The research was supported by Grant 19-12790S of the Grant Agency of the Czech Republic.

V. Diekert and M. Volkov (Eds.): DLT 2022, LNCS 13257, pp. 201–212, 2022.
https://doi.org/10.1007/978-3-031-05578-2_16

v if u is a scattered subword of v, *i.e.*, $u \trianglelefteq v$ if there are factorizations of the same length $u = a_1 \ldots a_k$ and $v = v_1 \ldots v_k$ such that, for all $i \in \{1, \ldots, k\}$, we have $a_i \in A$, $v_i \in A^+$, and a_i appears in v_i.

The considered notion of embedding relation can be modified by requiring different conditions on the factorizations. For example, if the alphabet A is quasi-ordered by $\preceq$, then we may replace the condition that a_i appears in v_i by the condition that there is a letter $b \in A$ such that $a_i \preceq b$ and b appears in v_i. In this way we obtain a quasi-order considered by Higman where he extended his result to infinite alphabets. Another variant is the following *gap embedding* considered by Schütte and Simpson in [14] for an alphabet equipped with a linear order $\sqsubseteq$: the defining condition that a_i appears in v_i is replaced by the condition that the letter a_i is the last letter in v_i and it is the least letter in v_i with respect to $\sqsubseteq$. Notice that yet another modification of the gap embedding is the *priority embedding* in [4] by Haase, Schmitz, and Schnoebelen.

Our paper concentrates on an application of *wqos* motivated by the work of Bucher, Ehrenfeucht and Haussler [1], which leads to a purely algebraic question in the realm of ordered semigroups. Notice that the topic is nicely overviewed in the recent survey paper [13] by Pin.

Before we introduce the primary question, we briefly recall the role of ordered semigroups in the algebraic theory of regular languages. At first, when we talk about an ordered semigroup $(S, \cdot, \leq)$, we assume that the partial order $\leq$ is monotone (or *stable*), *i.e.*, compatible with the multiplication $\cdot$ in the sense that, for arbitrary $x, y, s \in S$, the inequality $x \leq y$ implies both $s \cdot x \leq s \cdot y$ and $x \cdot s \leq y \cdot s$. The finite ordered semigroups are used to recognize regular languages similarly to unordered semigroups – see, *e.g.*, the fundamental survey on the algebraic theory of regular languages [12] by Pin. The modification is natural as the syntactic semigroup of a regular language is implicitly ordered in the following way. In the syntactic congruence of the regular language, words are related if they have the same set of contexts. Then one may also compare these sets of contexts by the inclusion relation; this comparison gives the syntactic quasi-order and consequently the partial order on the syntactic semigroup of the considered regular language. Let us note that in the literature the syntactic quasi-order is not always defined in this way, as sometimes the dual quasi-order is considered instead, *e.g.*, in [12].

The starting point in the study of well quasi-orders in [1] was a research by Ehrenfeucht, Haussler, and Rozenberg [3] concerning certain rewriting systems preserving regularity, where a well quasi-order plays a role of a sufficient condition guaranteeing the required property of the rewriting system. Particular attention in that research is paid to rewriting systems R with rules of the form $a \to u$ with a being a letter and u being a word. For such a rewriting system, several conditions equivalent to the fact that the derivation relation $\overset{*}{\Rightarrow}_R$ is a well quasi-order are stated in [1]. For example, one of the equivalent conditions is that the set $L = \{aua \mid a \in A, u \in A^*, a \overset{*}{\Rightarrow}_R aua\}$ is unavoidable (in the sense that every infinite word over the alphabet A contains a finite factor from the language L). Unfortunately, they did not give algorithms to test the conditions.

They also showed that every derivation relation that is a *wqo* originates from a rewriting system induced by a semigroup homomorphism $\sigma : A^+ \to S$ onto a finite ordered semigroup $(S, \cdot, \leq)$ by the following formula:

$$R_\sigma = \{a \to u \mid a \in A, u \in A^+, \sigma(a) \leq \sigma(u)\}.$$

The open question is to characterize homomorphisms σ such that $\overset{*}{\Rightarrow}_{R_\sigma}$ is a well quasi-order. Another research goal is a characterization of finite ordered semigroups S such that the relation $\overset{*}{\Rightarrow}_{R_\sigma}$ is a *wqo* for every alphabet A and every homomorphism $\sigma : A^+ \to S$. For the purpose of this paper we call these ordered semigroups *congenial*.

Let us note that the examples of *wqos* mentioned earlier also fit to the introduced scheme of relations arising from a homomorphism onto a finite ordered semigroup. Indeed, for an alphabet A we may consider an ordered semigroup $(P(A), \cup, \subseteq)$ consisting of non-empty subsets of A equipped with the operation of union, and ordered by the inclusion relation. Then for the homomorphism $\sigma : A^+ \to P(A)$, where $\sigma(a) = \{a\}$ for $a \in A$, the relation $\overset{*}{\Rightarrow}_{R_\sigma}$ coincides with the embedding relation $\trianglelefteq$.[1] For the gap embedding, one may construct the semigroup $A \times A$ (ordered by equality), where the multiplication is given by $(a, b) \cdot (c, d) = (min(a, c), d)$, where min is taken with respect to $\sqsubseteq$ (and the homomorphism $\sigma : A^+ \to A \times A$ is the extension of the diagonal mapping).

Up to our knowledge, and also according to the survey paper [13], there is just one significant contribution to the mentioned open questions. Namely, in the paper [9] by Kunc, the questions are solved for the semigroups ordered by the equality relation. It is stated in [9] (implicitly contained in the proof of Theorem 10) that the property depends only on the semigroup S, not on the actual homomorphism σ. The congenial semigroups ordered by the equality relation are characterized as finite chains of finite simple semigroups. One of the equivalent characterizations of this transparent structural property is the following condition which can be checked in polynomial time: for every $s, t \in S$, we have $(s \cdot t)^\omega \cdot s = s$ or $t \cdot (s \cdot t)^\omega = t$, where $(s \cdot t)^\omega$ is the power of $s \cdot t$ that is idempotent.

Our research aims to give an analogous characterization in the general case; however, we do not fulfill that program yet, and our contribution brings tentative results. As the main result, we show that the problem of whether a homomorphism induces a well quasi-order is decidable. The proof is an almost straightforward application of the mentioned characterization in [1]. We show that the mentioned side result from [9] holds in the full generality: the property is indeed a property of an ordered semigroup and does not depend on the homomorphism. Next, we give some necessary and sufficient conditions ensuring congeniality.

---

[1] For a variant of Higman's Lemma where the alphabet A is equipped with the quasi-order $\preceq$, we take for the ordered semigroup S the subsemigroup of $P(A)$ consisting of all downward closed subsets of A with respect to the considered quasi-order $\preceq$. Note that for an infinite alphabet the constructed ordered semigroup is not finite and therefore it is not our focus.

Due to the space constraints, some proofs are omitted. They are available in the full version of the paper [7].

2   Preliminaries

We briefly recall basic notions and fix notation used in the paper. When we talk about ordered semigroup or semigroup, we write simply S instead of formal notation $(S, \cdot, \leq)$ and $(S, \cdot)$. Throughout the paper we work with finite semigroups with the exception of the free monoid A^* (the free semigroup A^+) formed by (non-empty) words over an alphabet A. We use the symbol ε for the empty word. An element e in a semigroup S is called *idempotent* if $e \cdot e = e$. For all $s \in S$, the set $\{s^n \mid n \in \mathbb{N}\}$ contains exactly one idempotent, which is denoted s^ω. We put $s^{\omega+1} = s^\omega \cdot s$ which equals (by definition) to $s \cdot s^\omega$. By S^1 we mean the monoid $S \cup \{1\}$ with a new neutral element 1 added when S is not a monoid and $S^1 = S$ otherwise. We denote $\mathrm{eval}_S \colon S^+ \to S$ the evaluation homomorphism from the free semigroup over S defined by the rule $\mathrm{eval}_S(s) = s$ for all $s \in S$. Here, an element $w \in S^+$ is a word $w = s_1 s_2 \ldots s_k$, where $s_i \in S$ for $i \in \{1, \ldots, k\}$, and for such w we have $\mathrm{eval}_S(w) = \mathrm{eval}_S(s_1 s_2 \ldots s_k) = s_1 \cdot s_2 \cdots s_k$.

Furthermore, we use the Green relations, a basic notion in the theory of semigroups (see [6] by Howie). For the reader's convenience we recall that an *ideal* of a semigroup S is a non-empty subset $I \subseteq S$ such that for all $t \in I$ and $s \in S$ we have $t \cdot s \in I$ and $s \cdot t \in I$. The ideal generated by an element $s \in S$ is equal to $S^1 s S^1 = \{x \cdot s \cdot y \mid x, y \in S^1\}$. Then the Green relation $\mathcal{J}$ is defined by the rule $s \mathcal{J} t \iff S^1 s S^1 = S^1 t S^1$. We say that a semigroup is *simple* if it has no proper ideal, *i.e.*, if all elements of the semigroup are $\mathcal{J}$-equivalent.

The following lemma is well known (see, *e.g.*, [11, Theorem 1.11] by Pin).

Lemma 1. *Let S be a finite semigroup. There exists $n \in \mathbb{N}$ such that for every sequence of elements $s_1, \ldots, s_n \in S$ there are indices $i, j \in \{1, \ldots, n\}, i \leq j$ for which the product $s_i \cdot s_{i+1} \cdots s_j$ is an idempotent.*

For words $w, x, y, z \in A^*$ such that $w = xyz$, we say that x is a *prefix*, y is a *factor* and z is a *suffix*. We say that the prefix x of w is proper if $|x| < |w|$, where $|w|$ is the length of the word w. A language $L \subseteq A^*$ is unavoidable if an arbitrary infinite word u over A has a factor v in L. We denote the set of all infinite words over A by A^∞.

A *quasi-order* $\leq$ on a set X is a reflexive and transitive binary relation. It is called a *well quasi-order (wqo)* if for an infinite sequence $(x_n)_{n \in \mathbb{N}}$ of elements of X there exist indices $m, n \in \mathbb{N}$ such that $m < n$ and $x_m \leq x_n$. Many equivalent defining conditions are known (see, *e.g.*, [10, Theorem 6.1.1]).

The next definition is partially motivated by results from [1]. We prefer to follow the formalism and notation used in [9].

Definition 2. *Let $\sigma \colon A^+ \to S$ be a homomorphism onto a finite ordered semigroup. We denote $\leq_\sigma$ a quasi-order on A^* defined by setting $u \leq_\sigma v$ if and only if there exist factorizations $u = a_1 \ldots a_n$ and $v = v_1 \ldots v_n$, such that for all $i \in \{1, \ldots, n\}$ we have $a_i \in A, v_i \in A^+$ and $\sigma(a_i) \leq \sigma(v_i)$.*

We refer to the list of inequalities $\sigma(a_j) \leq \sigma(v_j)$ for $j \in \{1, \ldots, n\}$ as to the *proof of* $u \leq_\sigma v$, and we can use the proof to form other inequalities. We say that $a_i \ldots a_j \leq_\sigma v_i \ldots v_j$ is the *consequence of the proof given by the factor* $a_i \ldots a_j$. Note that $u \leq_\sigma v$ implies $\sigma(u) \leq \sigma(v)$ and either $u = v = \varepsilon$ or $u, v \in A^+$. Finally, it is clear that $\leq_\sigma$ is a stable quasi-order on A^*.

The following interpretation of the results from [1] needs some other observations [1, Sect. 3] concerning rewriting systems of the form mentioned in the introduction (in [1] called *OS schemes*): If the relation $\overset{*}{\Rightarrow}_R$ with rules of the form $a \to u, a \in A, u \in A^+$ is a *wqo*, then there is a homomorphism $\sigma : A^+ \to S$ onto a finite ordered semigroup such that the relations $\overset{*}{\Rightarrow}_R$ and $\leq_\sigma$ coincide. And vice versa, if $\leq_\sigma$ is a *wqo*, then there is a system R with the property $\overset{*}{\Rightarrow}_R = \leq_\sigma$.

Proposition 3 [1]. *Let $\sigma : A^+ \to S$ be a homomorphism onto a finite ordered semigroup. Then the following conditions are equivalent:*

1. *The relation $\leq_\sigma$ is a well quasi-order on A^*.*
2. *The language $L_\sigma = \{awa \mid a \in A, w \in A^*, a \leq_\sigma awa\}$ is unavoidable over A.*
3. *The language $\{aw \mid a \in A, w \in A^*, a \leq_\sigma aw\}$ is unavoidable over A.*

We say that an ordered semigroup S is *congenial* if for every homomorphism $\sigma : A^+ \to S$ the corresponding relation $\leq_\sigma$ is a well quasi-order. The class of congenial semigroups is denoted $\mathcal{C}$.

We finish this section with a basic observation that it is enough to consider the case of the homomorphism eval_S when a congeniality of S is tested. We establish the following auxiliary lemma with the proof essentially same as to the unordered case (see [9, Theorem 10, (iii)$\Longrightarrow$(i)]).

Lemma 4. *Let $\sigma : A^+ \to S$ be a homomorphism to an ordered semigroup S such that $\leq_\sigma$ is a wqo. Let B be an alphabet, $\alpha : B^+ \to A^+$ be a homomorphism of free semigroups such that $\alpha(B) \subseteq A$, and $\varphi = \sigma \circ \alpha$. Then the quasi-order $\leq_\varphi$ is a wqo.* $\square$

The following is a direct consequence of Lemma 4 for $A = S$ and $\sigma = \mathrm{eval}_S$.

Lemma 5. *A semigroup S is congenial if and only if $\leq_{\mathrm{eval}_S}$ is a wqo.* $\square$

3   What Makes an Ordered Semigroup Congenial

Due to [9, Lemma 2], we know that a semigroup S ordered by equality is congenial if and only if for every $s, t \in S$ either $s = (s \cdot t)^\omega \cdot s$ or $t = t \cdot (s \cdot t)^\omega$. The natural generalization of this condition for an ordered semigroup S is

$$\forall s, t \in S: s \leq (s \cdot t)^\omega \cdot s \text{ or } t \leq t \cdot (s \cdot t)^\omega. \tag{1}$$

We show that this condition is indeed necessary. Note that a semigroup satisfying (1) also satisfies $s \leq s^{\omega+1}$ as it is just the condition (1) with $s = t$.

Proposition 6. *Every congenial semigroup satisfies the condition (1).* $\square$

The following example shows that the condition (1) is not a sufficient condition. It indicates that the ordered situation is more complicated.

Example 7. Denote $\mathcal{F}_{LRB}(3)$ the free left-regular band (*i.e.*, semigroup satisfying the identities $xyx = xy$ and $x^2 = x$) over three generators a, b, c. The semigroup has 15 elements represented by words listed in Fig. 1, where the order is depicted. For the product of a pair of elements, we simply concatenate the words and then

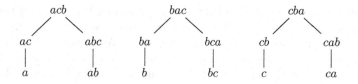

Fig. 1. The order $\leq$ of $\mathcal{F}_{LRB}(3)$.

omit the second occurrence of each letter if it occurs. It is a routine to check that $\mathcal{F}_{LRB}(3)$ satisfies the condition (1).

Now we take $\sigma \colon \{a, b, c\}^+ \to \mathcal{F}_{LRB}(3)$ where $\sigma(a) = a, \sigma(b) = b, \sigma(c) = c$. For the language L_σ given by Proposition 3, we see that $u \in L_\sigma$ if and only if $\sigma(u) \in \{a, ac, acb, b, ba, bac, c, cb, cba\}$. The periodical infinite word generated by abc, that is $(abc)^\infty = abcabc\ldots$, has no factor in the language L_σ since $\sigma((abc)^n a) = abc$ (for $n \in \mathbb{N}$) and similarly for factors starting and ending with b, resp. c. This means that L_σ is avoidable and $\mathcal{F}_{LRB}(3) \notin C$. Therefore, the condition (1) is not the characterization of the class of congenial semigroups.

We also add an example of ordered semigroup which is not completely regular.

Example 8. We consider two ordered versions B_2^+ and B_2^- of the Brandt semigroup B_2. The semigroup is generated by two elements a and b satisfying $a^2 = 0$, $b^2 = 0$, $aba = a$, and $bab = b$. The semigroup has five elements a, b, ab, ba, and 0, where $ab, ba, 0$ are idempotents. The orders are given in Fig. 2. In both cases we consider a homomorphism $\sigma \colon \{a, b\}^+ \to B_2$, where $\sigma(a) = a$, $\sigma(b) = b$.

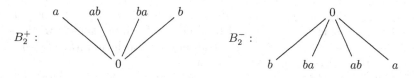

Fig. 2. Orders of B_2^+ and B_2^-.

Firstly, we deal with B_2^+. Taking the sequence of words $(a^i)_{i \in \mathbb{N}}$, we get an infinite antichain with respect to $\leq_\sigma$ showing that $\leq_\sigma$ is not a *wqo* and thus

$B_2^+ \notin C$. For the ordered semigroup B_2^-, we see that $a \leq a^2 = 0, b \leq b^2 = 0$, and $a = aba$, and so $a^2, b^2, aba \in L_\sigma$ for L_σ from Proposition 3. The language $\{a^2, aba, b^2\}$ is unavoidable, which implies $B_2^- \in C$.

Motivated by the previous examples and basic observations, we show the first sufficient condition ensuring the congeniality.

Proposition 9. *Let S be a finite ordered semigroup satisfying the inequality* $x \leq x \cdot (y \cdot x)^\omega$. *Then S is congenial.*

Proof. Let $\sigma : A^+ \to S$ be an arbitrary homomorphism. We show that the set L_σ from Proposition 3 is unavoidable. Let $v \in A^\infty$. Since the alphabet is finite, some letter $a \in A$ has infinitely many occurrences in v. We consider the factorization $v = w_0 a w_1 a w_2 a w_3 \dots$, where the words w_i do not contain the letter a. We take the sequence $s_1 = \sigma(w_1 a), s_2 = \sigma(w_2 a), \dots$ and use Lemma 1 to show that there exist indices i, j such that $\sigma(w_i a \dots w_j a)$ is an idempotent. If we denote $x = \sigma(a)$, $y = \sigma(w_i a \dots w_j)$, then we get $y \cdot x = (y \cdot x)^\omega$ and $x \leq x \cdot (y \cdot x)^\omega = x \cdot y \cdot x$. Therefore $a \leq_\sigma a w_i a \dots w_j a$ and the infinite word v has a factor in L_σ.   □

4   Effective Characterization of the Class C

In order to check the condition in Proposition 3, we introduce some technical notation. Let A be an alphabet, and $w \in A^+$ be a word. Then we write $f(w)$ for the first letter in w, *i.e.*, the letter a such that $w \in aA^*$. Dually, $\ell(w)$ means the last letter in the word w. Moreover, we denote the set of all factors and suffixes of a given word in a usual way with an exception that we do not consider letters as factors and suffixes here:

$$Fac(w) = \{u \in A^+ \setminus A \mid \exists p, q \in A^*, w = puq\}, \quad \text{and}$$

$$Suf(w) = \{u \in A^+ \setminus A \mid \exists p \in A^*, w = pu\}.$$

Now, let $\sigma : A^+ \to S$ be a homomorphism onto a finite ordered semigroup. We introduce the main technical notation: for $w \in A^+$ we put

$$Fac_\sigma(w) = \{(\sigma(u), f(u), \ell(u)) \in S \times A \times A \mid u \in Fac(w)\}, \quad \text{and}$$

$$Suf_\sigma(w) = \{(\sigma(u), f(u), \ell(u)) \in S \times A \times A \mid u \in Suf(w)\}.$$

Notice that $Fac_\sigma(w) = Suf_\sigma(w) = Fac(w) = Suf(w) = \emptyset$ whenever $w \in A$. Furthermore, $Suf_\sigma(w) \subseteq S \times A \times \{\ell(w)\}$ for every word w, that is a useful property motivating the following definition. A non-empty subset M of the set $S \times A \times A$ is called *coherent* if there is a letter $a \in A$ such that $M \subseteq S \times A \times \{a\}$; if such a letter exists, we denote it by $\ell(M)$.

Clearly, $w \in A^+$ does not avoid $L_\sigma = \{ava \mid a \in A, v \in A^*, a \leq_\sigma ava\}$ if and only if there exist $w', v, w'' \in A^*$ such that $w = w'avaw''$ and $\sigma(a) \leq \sigma(ava)$. The latter condition is equivalent to $(\sigma(ava), a, a) \in Fac_\sigma(w)$ with $\sigma(a) \leq \sigma(ava)$. In other words, a word $w \in A^+$ avoids the set L_σ if and only if $Fac_\sigma(w)$ is disjoint with the set $F = \{(s, a, a) \in S \times A \times A \mid \sigma(a) \leq s\}$. Now, we are ready to formulate a direct consequence of Proposition 3.

Lemma 10. *Let $\sigma : A^+ \to S$ be a homomorphism onto a finite ordered semi-group S. Then the relation $\leq_\sigma$ is a well quasi-order if and only if the set $\{w \in A^+ \setminus A \mid Fac_\sigma(w) \cap F = \emptyset\}$ is finite.* ☐

To test whether the considered set is finite, we use that every $Fac_\sigma(w)$ is a subset of $S \times A \times A$, and therefore there are only finitely many of them. In fact, we compute all possible $Suf_\sigma(w)$ disjoint with F instead of computing all $Fac_\sigma(w)$. It is enough as $Fac_\sigma(w)$ is a union of all $Suf_\sigma(u)$ where u is a prefix of w. Naturally, we compute sets $Suf_\sigma(u)$ recursively, since $Suf_\sigma(wa)$ can be determined by $Suf_\sigma(w)$ in the following way. Informally speaking, we add a at the end of all elements of $Suf_\sigma(w)$ and evaluate the suffix $\ell(Suf_\sigma(w))a$ of wa of length two. Therefore, we see the sets $Suf_\sigma(w)$ as states of the following finite deterministic incomplete automaton $\mathcal{A}_\sigma$ over the alphabet A. Notice that the automaton does not have final states.

We put $\mathcal{A}_\sigma = (Q, A, \delta, \iota)$ where $Q = \{\iota\} \uplus \bar{A} \uplus \mathcal{P}$, $\bar{A} = \{\bar{a} \mid a \in A\}$, and $\mathcal{P} = \{M \subseteq S \times A \times A \mid M \neq \emptyset, M \cap F = \emptyset, M \text{ coherent}\}$. For a given set $M \in \mathcal{P}$ and a letter $a \in A$ we define

$$M * a = \{(s \cdot \sigma(a), b, a) \mid (s, b, c) \in M\} \cup \{(\sigma(\ell(M)a), \ell(M), a)\}.$$

Similarly, for $\bar{b} \in \bar{A}$ we put $\bar{b} * a = \{(\sigma(ba), b, a))\}$. Furthermore, we define the partial transition function $\delta : Q \times A \to Q$ by $\delta(\iota, a) = \bar{a}$ for the initial state ι, and for $q \in Q \setminus \{\iota\}$ we put $\delta(q, a) = q * a$ if $q * a \in \mathcal{P}$. Note that the condition $q * a \in \mathcal{P}$ is equivalent to $q * a \cap F = \emptyset$ since $q * a$ is always non-empty and coherent. In particular, we have $\ell(q*a) = a$. As usual, the partial function δ can be extended to the partial function $\delta : Q \times A^+ \to Q$, which is denoted by δ too.

The following lemma summarises the properties of the previous constructions, with an obvious proof by an induction with respect to the length of words.

Lemma 11. *Let $\sigma : A^+ \to S$ be a homomorphism onto a finite ordered semi-group S, and $\mathcal{A}_\sigma$ be the automaton defined as above. For every word $w \in A^+ \setminus A$, the state $\delta(\iota, w)$ is defined in $\mathcal{A}_\sigma$ if and only if $Fac_\sigma(w) \cap F = \emptyset$. Moreover, if $\delta(\iota, w)$ is defined, then $\delta(\iota, w) = Suf_\sigma(w)$.* ☐

Now, we are ready to state the main result. The proof is straightforward consequence of Proposition 3 and the constructions and lemmas in this section.

Theorem 12. *Let $\sigma : A^+ \to S$ be a homomorphism onto a finite ordered semi-group S. Then $\leq_\sigma$ is a wqo if and only if the automaton $\mathcal{A}_\sigma$ does not contain an infinite path starting in the initial state ι.* ☐

The purpose of Theorem 12 is the following statement.

Corollary 13. *Let $\sigma : A^+ \to S$ be a homomorphism onto a finite ordered semi-group S. Then it is decidable whether $\leq_\sigma$ is a well quasi-order.* ☐

Recall that all states in $\mathcal{P}$ are coherent subsets of $S \times A \times A$. Since the automaton $\mathcal{A}_\sigma$ is finite, the existence of an infinite path starting in the initial state ι is equivalent to the existence of a loop reachable from ι. If we assume that there is a loop labeled by u and reachable by v, then we have that $vu^\infty = vuuu\ldots$ avoids L_σ. Hence the periodical infinite word u^∞ avoids L_σ too.

Corollary 14. *Let $\sigma : A^+ \to S$ be a homomorphism onto a finite ordered semi-group S. Then there is an infinite word avoiding L_σ if and only if there is a periodic infinite word u^∞ with that property.* □

The number of states of the automaton $\mathcal{A}_\sigma$ is bounded by $|A| \times 2^{|S| \times |A|} + |A| + 1$, that gives the obvious exponential bound for the time complexity of the algorithm based on Theorem 12.

One may modify the construction of $\mathcal{A}_\sigma$ if the condition (3) from Proposition 3 replaces the Condition (2). This means that Corollary 14 holds if we take the set $\{aw \mid a \in A, w \in A^+, \sigma(a) \le \sigma(aw)\}$ instead of the set L_σ.

5   Other Necessary and Sufficient Conditions

The motivation for this section is to examine whether the condition that $\le_\sigma$ is a *wqo* depends on the homomorphism σ or it is just a property of the semi-group. Therefore, we try to prove necessary conditions from Sect. 3 under the assumption that $\le_\sigma$ is a *wqo*.

Proposition 15. *Let $\sigma : A^+ \to S$ be a homomorphism onto a finite ordered semigroup S such that $\le_\sigma$ is a wqo. Then for every $u \in A^+$ there exists an integer $p > 1$ such that $u \le_\sigma u^p$.*

Proof. We show the statement by induction with respect to the length of the word u. For every $a \in A$ the definition of *wqo* implies that $a^k \le_\sigma a^\ell$ for some integers $k < \ell$, and by the definition of $\le_\sigma$ we have $a \le_\sigma a^p$ for some $p > 1$.

Assume that the statement is true for all words shorter than a given word $u \in A^+ \setminus A$. Similarly to the initial step, we have $u^k \le_\sigma u^\ell$ for some $k < \ell$. We consider the consequences of the proof of $u^k \le_\sigma u^\ell$ given by factors u of u^k. The first non-trivial inequality among these consequences in the order from left to right is of the form $u \le_\sigma u^m v$ where v is a proper prefix of u. Notice that for $v = \varepsilon$ we are done.

For the considered prefix v of u we may also have some inequality of the form $v \le_\sigma u^j w$ with $j \in \mathbb{N}$, in particular the proof of the inequality $u^k \le_\sigma u^\ell$ has such a consequence. We analyze the inequalities of that form for all prefixes of u.

On the set $P = \{v \in A^* \mid v \text{ is prefix of } u\}$ we define the relation $\to$ as follows: $v \to w$ if there is $j \ge 0$ such that $v \le_\sigma u^j w$ and $|v| < |u^j w|$. (Notice that if $w = v$, then $j > 0$.) Since $\le_\sigma$ is a stable quasi-order the relation $\to$ is transitive. As is discussed above, there is $v \in P$, $v \ne u$ such that $u \to v$ and so at least one of the following cases occurs.

Case I: $u \to \varepsilon$. This means $u \le_\sigma u^j$ with $j > 1$ and we are done.

Case II: there is $v \in P$ such that $u \to v \ne \varepsilon$, and there is no w such that $v \to w$. In particular, $v \ne u$. Let $\bar{v}$ be the suffix of u such that $v\bar{v} = u$. Since $v \ne \varepsilon$, we have $|\bar{v}| < |u|$ and by the induction assumption there is p such that $\bar{v} \le_\sigma \bar{v}^p$. If we consider the proof of $u \le_\sigma u^j v$, then the consequence given by the prefix v is trivial equality (by the assumption that there is no inequality of the form $v \le_\sigma u^j w$ with $|v| < |u^j w|$). Then the consequence of the proof given by the suffix $\bar{v}$ is in the form $\bar{v} \leq_\sigma \bar{v}u^{j-1}v$. Now we use this inequality $(p-1)$-times to get $\bar{v}^{p-1} \leq_\sigma (\bar{v}u^{j-1}v)^{p-1} = \bar{v}u^{(p-1)j-1}v$. Then we multiply the former inequality by $\bar{v}$ on the right and we get $\bar{v}^p \leq_\sigma \bar{v}u^{(p-1)j}$. Since we assumed $\bar{v} \leq_\sigma \bar{v}^p$, we also get $\bar{v} \leq_\sigma \bar{v}u^{(p-1)j}$. Finally, we multiply by v on the left and obtain $u \leq_\sigma u^{(p-1)j+1}$.

Case III: there is $v \in P$ such that $u \to v \neq \varepsilon$, and $v \to v$. This means that there is $j > 0$ such that $v \leq_\sigma u^j v$. Now, it is enough to multiply the former inequality by the suffix $\bar{v}$ of u on the right, and we get $u \leq_\sigma u^{j+1}$. $\square$

We try to show that an ordered semigroup S is congenial whenever we have an onto homomorphism $\sigma : A^+ \to S$ determining the wqo $\leq_\sigma$. This means that for every homomorphism $\varphi : B^+ \to S$, the set $L_\varphi = \{bwb \mid b \in B, w \in B^*, b \leq_\varphi bwb\}$ has to be unavoidable. Hence, every periodic infinite word w^∞ must contain a factor from L_φ. In particular, if B contains n letters $b_1, b_2, \ldots, b_n$, then, for the word $w = b_1 b_2 \ldots b_n$, there is an index $i \in \{1, \ldots, n\}$ and an integer $p \in \mathbb{N}$ such that $\varphi(b_i) \leq \varphi(b_i(b_{i+1} \ldots b_i)^p)$. Since the homomorphism σ is onto, we may consider words $w_j \in A^+$ such that $\sigma(w_j) = \varphi(b_j)$. In this setting, we want to show that $\sigma(w_i) \leq \sigma(w_i(w_{i+1} \ldots w_i)^p)$. In fact, we aim on the stronger inequality $w_i \leq_\sigma w_i(w_{i+1} \ldots w_i)^p$. Proposition 15 is a special case of this property for $n = 1$. The following statement fulfills the sketched program.

Theorem 16. *Let S be a finite ordered semigroup. Then the following conditions are equivalent:*

(i) *There exists an alphabet A and an onto homomorphism $\sigma : A^+ \to S$ such that $\leq_\sigma$ is a well quasi-order.*

(ii) *There exists an alphabet A and an onto homomorphism $\sigma : A^+ \to S$ such that, for every $n \in \mathbb{N}$ and $a_1, \ldots, a_n \in A$, there exists $i \in \{1, \ldots, n\}$ and $p \in \mathbb{N}$ such that $a_i \leq_\sigma a_i(a_{i+1} \ldots a_n a_1 \ldots a_i)^p$.*

(iii) *There exists an alphabet A and an onto homomorphism $\sigma : A^+ \to S$ such that, for every $n \in \mathbb{N}$ and $u_1, \ldots, u_n \in A^+$, there exists $i \in \{1, \ldots, n\}$ and $p \in \mathbb{N}$ such that $u_i \leq_\sigma u_i(u_{i+1} \ldots u_n u_1 \ldots u_i)^p$.*

(iv) *There exists an alphabet A and an onto homomorphism $\sigma : A^+ \to S$ such that, for every $n \in \mathbb{N}$ and $u_1, \ldots, u_n \in A^+$, there exists $i \in \{1, \ldots, n\}$ and $p \in \mathbb{N}$ such that $\sigma(u_i) \leq \sigma(u_i(u_{i+1} \ldots u_n u_1 \ldots u_i)^p)$.*

(v) *For every $n \in \mathbb{N}$ and $s_1, \ldots, s_n \in S$, there exists $i \in \{1, \ldots, n\}$ and $p \in \mathbb{N}$ such that $s_i \leq s_i \cdot (s_{i+1} \cdots s_n \cdot s_1 \cdots s_i)^p$.*

(vi) *For every alphabet B and a homomorphism $\varphi : B^+ \to S$ the relation $\leq_\varphi$ is a well quasi-order.*

Proof. We show the implications from top to bottom. The omitted implications are easy to see. In the conditions (i)–(iv), the same pair (A, σ) is employed.

"(i) $\implies$ (ii)": We consider a new alphabet $B = \{b_1, \ldots, b_n\}$ of size n and a homomorphism $\alpha : B^+ \to A^+$ such that $\alpha(b_i) = a_i$ for all $i \in \{1, \ldots, n\}$. We denote the composition $\sigma \circ \alpha$ by φ. By Lemma 4, we know that the relation $\leq_\varphi$ is a wqo. In particular, the infinite word $(b_1 b_2 \ldots b_n)^\infty$ has a factor in $L_\varphi = \{bwb \mid b \in B, w \in B^*, b \leq_\varphi bwb\}$. Therefore, there is $i \in \{1, \ldots, n\}$ and

$p \in \mathbb{N}$ such that $b_i \leq_\varphi b_i(b_{i+1} \ldots b_n b_1 \ldots b_i)^p$. Finally, we get $\sigma(a_i) = \varphi(b_i) \leq \varphi(b_i(b_{i+1} \ldots b_n b_1 \ldots b_i)^p) = \sigma(a_i(a_{i+1} \ldots a_n a_1 \ldots a_i)^p)$.

"(ii) $\implies$ (iii)": We apply the condition (ii) on the word $u = u_1 u_2 \ldots u_n$ which we see as a concatenation of individual letters. So, there is $i \in \{1, \ldots, n\}$, $p \in \mathbb{N}$ and $u_i', u_i'' \in A^*$ such that $u_i = u_i' a u_i''$ and $a \leq_\sigma a(u_i'' u_{i+1} \ldots u_n u_1 \ldots u_{i-1} u_i' a)^p$. If we multiply this inequality by the word u_i' on left and by the word u_i'' on right, we get $u_i \leq_\sigma u_i(u_{i+1} \ldots u_n u_1 \ldots u_{i-1} u_i)^p$.

"(v) $\implies$ (vi)": It follows from Corollary 14. $\qquad\square$

The condition (ii) from Theorem 16 was mentioned in [1] in the setting of rewriting systems, namely it occurs as condition (c) in the concluding section. Also, the condition in Proposition 15 is mentioned there as the condition (b). It is mentioned in [1] without proof that the conditions are equivalent.

The equivalence of the conditions (i) and (vi) in Theorem 16 gives the following result saying that whether the induced quasi-order $\leq_\sigma$ is a *wqo* does not depend on the homomorphism σ and it is a property of the ordered semigroup.

Corollary 17. *Let* $\sigma : A^+ \to S$ *be a homomorphism onto a finite ordered semigroup* S. *Then* $\leq_\sigma$ *is a wqo if and only if the semigroup* S *is congenial.* $\qquad\square$

We get the following characterization of congeniality using the condition (v) of Theorem 16.

Corollary 18. *Let* S *be an ordered semigroup. Then* S *is congenial if and only if for every* $n \in \mathbb{N}$ *and* $s_1, \ldots, s_n \in S$, *there exists* $i \in \{1, \ldots, n\}$ *such that* $s_i \leq s_i \cdot (s_{i+1} \cdots s_n \cdot s_1 \cdots s_i)^\omega$. $\qquad\square$

Unfortunately, it is not possible to bound n in Corollary 18. Indeed, there is a sequence of ordered semigroups S_m such that S_m satisfies the condition in Corollary 18 if $n < m$ and does not satisfy the condition for $n = m$.

6 Conclusion

We have shown that for a homomorphism $\sigma \colon A^+ \to S$ onto a finite ordered semigroup, it is decidable whether $\leq_\sigma$ is a *wqo*. We also proved that the question does not depend on σ, but it is indeed a property of the given ordered semigroup. One may expect more effective or transparent characterization similar to that of the unordered case in [9]. Nevertheless, our observations suggest that such a characterization could be more difficult to obtain.

We conclude with a brief discussion of the applications of our results. We do not see any direct impact of the research to the work done in [1]. On the other hand, in [9], the *wqo* was applied to prove regularity of maximal solutions of very general language equations and inequalities (see also [13]). The theory developed in [9] may be naturaly extended to the ordered case, so our new class of ordered semigroups inducing well quasi-orders may find the application there.

Acknowledgement. We are grateful to the referees for their numerous valuable suggestions which improved the paper, in particular, its introductory part. We also thank to Michal Kunc for inspiring discussions.

References

1. Bucher, W., Ehrenfeucht, A., Haussler, D.: On total regulators generated by derivation relations. Theor. Comput. Sci. **40**, 131–148 (1985). https://doi.org/10.1016/0304-3975(85)90162-8
2. D'Alessandro, F., Varricchio, S.: Well quasi-orders in formal language theory. In: Ito, M., Toyama, M. (eds.) DLT 2008. LNCS, vol. 5257, pp. 84–95. Springer, Heidelberg (2008). https://doi.org/10.1007/978-3-540-85780-8_6
3. Ehrenfeucht, A., Haussler, D., Rozenberg, G.: On regularity of context-free languages. Theor. Comput. Sci. **27**, 311–332 (1983). https://doi.org/10.1016/0304-3975(82)90124-4
4. Haase, C., Schmitz, S., Schnoebelen, P.: The power of priority channel systems. Log. Methods Comput. Sci. **10**(4) (2014). https://doi.org/10.2168/LMCS-10(4:4)2014
5. Higman, G.: Ordering by divisibility in abstract algebras. Proc. Lond. Math. Soc. **s3-2**(1), 326–336 (1952). https://doi.org/10.1112/plms/s3-2.1.326
6. Howie, J.M.: An Introduction to Semigroup Theory. Academic Press, London (1976)
7. Klíma, O., Kolegar, J.: Well quasi-orders arising from finite ordered semigroups (2022). https://arxiv.org/abs/2203.06535
8. Kruskal, J.B.: The theory of well-quasi-ordering: a frequently discovered concept. J. Comb. Theory Ser. A **13**(3), 297–305 (1972). https://doi.org/10.1016/0097-3165(72)90063-5
9. Kunc, M.: Regular solutions of language inequalities and well quasi-orders. Theor. Comput. Sci. **348**, 277–293 (2005). https://doi.org/10.1016/j.tcs.2005.09.018
10. de Luca, A., Varricchi, S.: Finiteness and Regularity in Semigroups and Formal Languages. Springer, Heidelberg (1999). https://doi.org/10.1007/978-3-642-59849-4
11. Pin, J.É.: Varieties of Formal Languages. Foundations of Computer Science. North Oxford Academic, London (1986)
12. Pin, J.-E.: Syntactic semigroups. In: Rozenberg, G., Salomaa, A. (eds.) Handbook of Formal Languages, pp. 679–746. Springer, Heidelberg (1997). https://doi.org/10.1007/978-3-642-59136-5_10
13. Pin, J.É.: How to prove that a language is regular or star-free? In: Leporati, A., Martín-Vide, C., Shapira, D., Zandron, C. (eds.) LATA 2020. LNCS, vol. 12038, pp. 68–88. Springer, Cham (2020). https://doi.org/10.1007/978-3-030-40608-0_5
14. Schütte, K., Simpson, S.G.: Ein in der reinen Zahlentheorie unbeweisbarer Satz über endliche Folgen von natürlichen Zahlen. Arch. Math. Log. **25**(1), 75–89 (1985). https://doi.org/10.1007/BF02007558

The Billaud Conjecture for $|\Sigma| = 4$, and Beyond

Szymon Łopaciuk$^{(\boxtimes)}$ (ID) and Daniel Reidenbach (ID)

Department of Computer Science, Loughborough University,
Loughborough LE11 3TU, UK
{s.p.lopaciuk,d.reidenbach}@lboro.ac.uk

Abstract. The Billaud Conjecture, first stated in 1993, is a fundamental problem on finite words and their heirs, i.e., the words obtained by a projection deleting a single letter. The conjecture states that every morphically primitive word, i.e., a word which is not a fixed point of any non-identity morphism, has at least one morphically primitive heir. In this paper we give the proof of the Conjecture for alphabet size 4, and discuss the potential for generalising our reasoning to larger alphabets. We briefly discuss how other language-theoretic tools relate to the Conjecture, and their suitability for potential generalisations.

Keywords: Billaud Conjecture · Morphic primitivity · Fixed point

1 Introduction

The context of our research is the notion of morphic primitivity of words: a word is morphically primitive if the only morphism for which it is a fixed point is the identity morphism. In 1993, Billaud [1] stated the following conjecture, which is still open to this day, relating to the concept of morphic primitivity of words:

Conjecture 1 (The Billaud Conjecture). There exists at least one letter x in every morphically primitive word w such that the word obtained by deleting all occurrences of x in w is also morphically primitive.

Let us call the word w the *parent*, and the words obtained through a deletion of a letter from w *heirs*. We can consider, as an example, morphically primitive parent *abcbac*: the heir resulting from the deletion of the letter c is *abba*, and it is morphically primitive. Notably, as the Billaud Conjecture is an implication and not a characterisation, while there are morphically imprimitive words such as $(abc)^2$ whose all heirs are morphically imprimitive, there are also words like $a(bc)^2a$, which are morphically imprimitive, yet have morphically primitive heirs (in our example *abba*).

The conjecture has been open for almost three decades and only a few special cases where it holds have so far been established; no counterexamples have been found, and the conjecture is widely believed to be true. Shortly after Billaud

© Springer Nature Switzerland AG 2022
V. Diekert and M. Volkov (Eds.): DLT 2022, LNCS 13257, pp. 213–225, 2022.
https://doi.org/10.1007/978-3-031-05578-2_17

posed his question, Geser [4] noted that in the context of morphic imprimitivity of words we need only to consider idempotent morphisms; an extended version of this statement was proved by Levé and Richomme [10]. It has been shown that the conjecture holds in the following special cases: Nevisi and Reidenbach [13] proved that the conjecture is correct for all words (with three or more different letters) if they contain each letter exactly twice. Restricting the alphabet size, Zimmermann [17] showed that the conjecture holds for the alphabet size of 3. (We note the case of alphabet size of 2 is trivial.) Walter [16] identified and proved some of the cases necessary for showing that the conjecture holds for alphabet size of 4. Levé and Richomme [10], working with the contrapositive of the conjecture, proved that if all morphically imprimitive heirs of a word are fixed points of non-trivial morphisms with exactly one expanding letter each, then the same holds for the parent word; the letter x is expanding if the morphism ϕ for which the word is a fixed point is such that $|\phi(x)| \geqslant 2$. The latter two results form the basis for the completion of our proof for the Billaud Conjecture for alphabet size 4.

Morphic primitivity of words has been studied by Head [7] in the context of L-systems, by Hamm and Shallit [6], and in the context of erasing pattern languages by Reidenbach [14]; all these authors gave alternative characterisations of the concept. The term 'morphic imprimitivity' itself was coined by Reidenbach and Schneider [15], who described a characteristic factorisation of these words. Holub [8] gave a polynomial-time algorithm to find an imprimitivity factorisation of a word (further refined by Matocha and Holub [11], and Kociumaka et al. [9]).

In this paper we show that the Billaud Conjecture holds for the class of words over the alphabet of size 4. We begin by introducing some concepts that are vital to our reasoning in Sect. 3. Prior to giving our proof in Sect. 5, and in order to provide the reader with some intuition for the direction of our reasoning, in Sect. 4, we present a generalised overview, or a 'blueprint', of the steps one can follow in order to attempt to prove the Billaud Conjecture holds for any fixed alphabet size. Moreover, alongside our presentation of the proof technique, we discuss other tools available in the literature that are of relevance to the conjecture, namely the *synchronised shuffle* [3], and *pattern expressions* [2].

Due to space constraints, most of the proofs have been omitted.

2   Preliminaries

We denote the set of positive integers with $\mathbb{N}_+$, and non-negative ones with $\mathbb{N}_0$. Moreover, for any two integers m, n, $m \geqslant n$, let $[\![n, m]\!]$ be the set of all integers i such that $n \leqslant i \leqslant m$. An *alphabet* Σ is an enumerable set of *symbols* or *letters*. As a *word (over an alphabet Σ)* we refer to a finite sequence of elements, symbols, of Σ. The cardinality of a set A shall be denoted by $|A|$, similarly the length of a word w is $|w|$. The *empty word*, denoted by λ, is the special word for which $|\lambda| = 0$. The set of all words over an alphabet Σ is denoted with Σ^*, and of all non-empty words with Σ^+.

We write the *concatenation* of two words u, v as $u \cdot v$ or uv; we extend this to languages: $L \cdot L' := \{uv \mid u \in L, v \in L'\}$. Sometimes, where a set operation is implied, we shall write w instead of $\{w\}$ for a word w, e.g., $w^* = \{w\}^*$. Given a word w and $n \in \mathbb{N}_0$, w^n denotes the n-fold concatenation of the word w. A word v is a *factor of* w, denoted $v \sqsubseteq w$, if there exist words w_1, w_2 satisfying $w = w_1 \cdot v \cdot w_2$; the factor v is a *prefix* if $w_1 = \lambda$, and a *suffix* if $w_2 = \lambda$. We say that a word v is a *circular factor* of w if it is a factor of w, or if there exists a prefix v' and a suffix v'' of w such that $v = v''v'$. The number of occurrences of a factor v in w is denoted by $|w|_v$. Given words u and v, we define their *shuffle product*, $u \sqcup\!\sqcup v$, as the set of all words w such that $w = u_1 v_1 u_2 v_2 \cdots u_n v_n$, where $u = u_1 u_2 \cdots u_n$ and $v = v_1 v_2 \cdots v_n$ for some $n \in \mathbb{N}_+$; we extend this definition to languages, so that $L \sqcup\!\sqcup L' := \{u \sqcup\!\sqcup v \mid u \in L, v \in L'\}$. Let $\mathrm{symb}(w)$ denote the set of all letters $x \sqsubseteq w$. Finally, we define a *regular expression E (over Σ)* and its language $L(E)$ in the usual way, and we use $|$ as the alternative operator.

For any alphabets A and B and all words $w, v \in A^*$, a *(homo-)morphism* $\phi : A^* \to B^*$ is a function that satisfies $\phi(w)\phi(v) = \phi(wv)$. A morphism $\phi : A^* \to A^*$ is *idempotent* if $\phi = \phi \circ \phi$. A word w is a *fixed point* of a morphism ϕ if $\phi(w) = w$. If a morphism ϕ is not idempotent, and there exists a finite fixed point word w of ϕ, then there exists an idempotent morphism ϕ' such that $\phi'(w) = w$ [4]. If X is a set of letters, then we denote with π_X the morphism *deleting all* $x \in X$, i.e., $\pi_X(x) = \lambda$ for all $x \in X$, and $\pi_X(x) = x$ otherwise; we extend this to sets: given a set of words Y, let $\pi_X(Y) := \{\pi_X(w) \mid w \in Y\}$. We call a morphism *trivial* or the *identity* if it maps every letter to itself.

A word w is *morphically primitive* [15] if there is no word w' with $|w'| < |w|$ such that w and w' can be mapped onto each other by morphisms; otherwise w is *morphically imprimitive*. This definition is equivalent to the notion of morphic primitivity used at the beginning of Sect. 1, as we shall explain in due course. Let $\phi : \Sigma^* \to \Sigma^*$ be an idempotent morphism; when discussing morphic imprimitivity we shall restrict ourselves only to idempotent morphisms, unless explicitly stated otherwise. We define the following three sets, similarly to Levé and Richomme [10], which form a partition on $\mathrm{symb}(w)$, in the context of ϕ: the set of *expanding letters*, $E_\phi = \{x \in \Sigma \mid |\phi(x)| \geqslant 2\}$, of *mortal letters*, $M_\phi = \{x \in \Sigma \mid \phi(x) = \lambda\}$, and of *constant letters*, $C_\phi = \{x \in \Sigma \mid \phi(x) = x\}$. An *imprimitivity factorisation f* of w is a tuple $\langle x_1, x_2, \ldots, x_n; v_1, v_2, \ldots, v_n \rangle$ where $v_1, v_2, \ldots, v_n \in \Sigma^+$, $x_1, x_2, \ldots, x_n \in \Sigma$, $n \in \mathbb{N}_+$, such that w can be factorised as $u_0 v_1 u_1 v_2 u_2 \cdots v_n u_n$ for some $u_0, u_1, \ldots, u_n \in \Sigma^*$, and there is a non-trivial morphism ϕ with $\phi(w) = w$ such that for all $i \in \mathbb{N}_+$, $|v_i|_{x_i} = 1$ and $v_i = \phi(x_i)$, and $u_0, u_1, \ldots, u_n \in C_\phi^*$. We shall say that ϕ *determines f*, and define sets E_f, M_f, and C_f to be equal to E_ϕ, M_ϕ, and C_ϕ. As we shall operate with the following concepts interchangeably, we restate the following result by Reidenbach and Schneider [15]. For every word w, the following statements are equivalent: w is morphically imprimitive, w is a fixed point of a non-trivial morphism, and w has an imprimitivity factorisation.

3 Languages of Fixed Points and Their Parents

In this section we shall present a conjecture that will form the basis of our hypothesised proof technique for the Billaud Conjecture for an arbitrary alphabet size, and more specifically the basis of the proof of the Billaud Conjecture for alphabet size 4.

The classical way of thinking about the Billaud Conjecture is centred around the parent word and the operation of deleting letters from said word. In the original statement presented in Conjecture 1, we delete letters from any morphically primitive word, and assert that at least one heir obtained in this manner is morphically primitive as well. The contrapositive statement of the conjecture asserts that the parent word is morphically imprimitive if all of its heirs have this property. Levé and Richomme [10] extend this way of thinking to sets of words. In this section, we present an alternative, yet similar, version of the Billaud Conjecture, which can be summarised as follows. Consider some alphabet Σ and $|\Sigma|$ sets of 'heir candidate' words, and consider all the ways that the words of the 'heir candidate' sets can be combined to create a set of parent words. We conjecture that if every 'heir candidate' set consists only of morphically imprimitive words, the set of parents (if not empty) consists only of morphically imprimitive words as well. We will discuss this conjecture in more detail in due course, but for now we illustrate what we mean with an example:

Example 1. Let $\Sigma := \{a, b, c\}$, and let us define the morphisms $\phi_c : a \mapsto ab, b \mapsto \lambda$, $\phi_b : a \mapsto ac, c \mapsto \lambda$, and $\phi_a : b \mapsto bc, c \mapsto \lambda$. Let W_c, W_b, and W_a be sets of words consisting of all fixed points of ϕ_c, ϕ_b, and ϕ_a, respectively. For example, $ab \in W_c$, and $(ac)^7 \in W_b$. We can also represent these sets with regular expressions: $W_c = L((ab)^*)$, $W_b = L((ac)^*)$, and $W_a = L((bc)^*)$.

We now consider the set W of all words w such that $\pi_x(w) \in W_x$, $x \in \{a, b, c\}$. It can be shown, due to the properties of regular languages that we discuss below, that $W = L((abc)^*)$. We have started with three sets of morphically imprimitive words W_c, W_b, and W_a, and we can see that W, constructed by combining these sets, is also a set of morphically imprimitive words, as every word of the form $(abc)^*$ is a fixed point of, e.g., the morphism $\phi : a \mapsto abc, bc \mapsto \lambda$.

We commence our discussion by analysing the languages of fixed points of morphisms. To this end, we define formally what we mean by this concept:

Definition 1. *Given a morphism ϕ, let F_ϕ be the set of all words w such that $\phi(w) = w$. Moreover, given a set of morphisms Φ, let F_Φ be the union of all F_ϕ with $\phi \in \Phi$. We refer to F_ϕ (or F_Φ) as the* language of fixed points *of ϕ (or Φ).*

In the context of our Example 1 it can be seen that W_c, W_b, and W_a are languages of fixed points of ϕ_c, ϕ_b, and ϕ_a. We first consider the languages of fixed points of a single morphism, and make the following simple observation about the nature of these languages:

Lemma 1. *Let ϕ be an idempotent morphism over the finite alphabet Σ. Then $F_\phi = \{\phi(x) \mid x \in \Sigma\}^*$.*

From the above lemma, it follows in particular that:

Corollary 1. *Let ϕ be an idempotent morphism. Then F_ϕ is regular.*

In Example 1 we note that due to the properties of regular languages, we can deduce that our language W is regular, and in particular we are able to give its structure; we shall now give more detail on why that is in the following example:

Example 2. Let W_c, W_b, and W_a be as in Example 1. Let us recall that $W_c = L((ab)^*) = F_{\phi_c}$. In such a case, the set W_c' of all possible words w satisfying $\pi_c(w) \in W_c$ can be represented as follows $L((ac^*b|c)^*)$, or alternatively as $W_c \shuffle L(c^*)$, as it is a language of repetitions of ab interleaved by an arbitrary number of cs. If we define W_b' and W_a' in an analogous fashion to be the languages of all possible parents of the words of W_b and W_a respectively, then we can see that $W = W_c' \cap W_b' \cap W_a'$, i.e., W is the set of those parents, whose respective heirs are in W_c, W_b, and W_a. It is well-known that regular languages are closed under intersection, and in this case we can represent W as follows: $W = L((abc)^*)$.

Let us now formally define the concept of the set of all possible parents of fixed points of morphisms. For the ease of notation let us denote by Σ_x the set $\Sigma \setminus \{x\}$ for some $x \in \Sigma$.

Definition 2. *Let Σ be an alphabet and let $x \in \Sigma$. Let $\phi : \Sigma_x^* \to \Sigma_x^*$ be an idempotent morphism, and let Φ be a set of idempotent morphisms with the same domain and codomain as ϕ. Then the set of all possible parents of fixed points of ϕ is defined as $P_{\phi,\Sigma} := F_\phi \shuffle x^*$. Similarly, $P_{\Phi,\Sigma} := \bigcup_{\phi \in \Phi} P_{\phi,\Sigma} = F_\Phi \shuffle x^*$.*

In fact, the sets W_c', W_b', and W_a' in Example 2 are equal to $P_{\phi_c,\Sigma}$, $P_{\phi_b,\Sigma}$, and $P_{\phi_a,\Sigma}$, respectively. Let us recall that the class of regular languages is known to be closed under the shuffle operation, as shown by Ginsburg and Spanier [5]. As a consequence of this, henceforth, we shall extend our regular expression notation with the infix operator $\shuffle$.

Moreover, in the context of the above result, it is clear that for any idempotent morphism ϕ and any alphabet Σ the language $P_{\phi,\Sigma}$ is regular. Therefore, due to regular languages being closed under both the intersection and the shuffle operation, we can give a generalisation of our statement in Example 2, i.e., we can conclude that a language W, satisfying $\pi_x(W) = F_{\phi_x}$ with $\phi_x : \Sigma_x^* \to \Sigma_x^*$ for all $x \in \Sigma$, is regular. We leave open the question of deciding, in the general case, whether the regular expression generating W generates only morphically imprimitive words. Nevertheless we note that a suitable characterisation could provide additional insights into the Billaud Conjecture.

We now formally present the conjecture announced earlier in this section.

Conjecture 2. Let $\Sigma = \{a_1, a_2, \ldots, a_n\}$ be an alphabet with $n \geqslant 3$, and let $\Phi_1, \Phi_2, \ldots, \Phi_n$ be sets of idempotent, non-trivial morphisms such that for every $\phi \in \Phi_i$, $i \in [\![1, n]\!]$, we have that $\phi : \pi_{a_i}(\Sigma)^* \to \pi_{a_i}(\Sigma)^*$. Let $L := \bigcap_{i \in [\![1,n]\!]} P_{\Phi_i,\Sigma}$. Then, either $L = \varnothing$ or there exists a set of idempotent, non-trivial morphisms Φ such that $L \subseteq F_\Phi$.

We can show the equivalence of our conjecture and the Billaud Conjecture:

Proposition 1. *Conjecture 2 is equivalent to the Billaud Conjecture.*

From the prior remarks, the implication from our conjecture to the Billaud Conjecture is apparent. In the other direction, we can show that the Billaud Conjecture implies our conjecture by assuming the former, and showing that L, as it is defined above, is a subset of some F_Φ. As all heirs of every word $w \in L$ are morphically imprimitive, by the Billaud Conjecture we have that w is a fixed point of some non-trivial morphism ϕ; then, taking Φ to be the union of all such morphisms ϕ, we have that $L \subseteq P_\Phi$.

4   A 'Blueprint' for a Billaud Conjecture Proof for Fixed Σ

We are now ready to discuss the way of how Conjecture 2 can serve as a template for proving the Billaud Conjecture for a fixed alphabet size. One of the difficulties of applying Conjecture 2 to solve the Billaud Conjecture directly is the inherent necessity of considering sets $\Phi_1, \Phi_2, \ldots$ that can contain an infinite number of morphisms. In fact, in order to prove the Billaud Conjecture, it is sufficient to show that for every Σ, the conjecture holds for the sets $\Phi_1, \Phi_2, \ldots, \Phi_{|\Sigma|}$ each containing *all* possible morphisms. However, we will now describe how, by solving a finite number of subproblems, we can nevertheless show that the Billaud Conjecture holds for an alphabet Σ of a fixed size N.

To start, for every Σ_x let us consider a partition $\boldsymbol{\Phi}_x$ of the class of all idempotent morphisms over Σ_x into a finite number of subsets $\Phi_{x,1}, \Phi_{x,2}, \ldots, \Phi_{x,|\boldsymbol{\Phi}_x|}$. Each of the sets of morphisms $\Phi_{x,i}$, $i \in [\![1, |\boldsymbol{\Phi}_x|]\!]$, corresponds to an equivalence class under the relation that $E_\phi = E_{\phi'}$, $M_\phi = M_{\phi'}$, and $C_\phi = C_{\phi'}$ for any two morphisms $\phi, \phi' \in \Phi_{x,i}$.

Example 3. Let $\Sigma = \{a, b, c\}$. For any non-trivial idempotent morphism $\phi : \Sigma_c^* \to \Sigma_c^*$, we can have the following 2 options for $\langle E_\phi, M_\phi, C_\phi \rangle$: $\langle \{a\}, \{b\}, \varnothing \rangle$ and $\langle \{b\}, \{a\}, \varnothing \rangle$. We note that E_ϕ and M_ϕ cannot be empty for a non-trivial ϕ, hence we only have the two options. Then, we can consider, e.g., that $\Phi_{c,1}$ is the set of all morphisms $\phi : \Sigma_c^* \to \Sigma_c^*$ with $E_\phi = \{a\}$ and $M_\phi = \{b\}$, and $\Phi_{c,2}$ the set of all $\phi : \Sigma_c^* \to \Sigma_c^*$ with $E_\phi = \{b\}$ and $M_\phi = \{a\}$. There are no idempotent, non-trivial morphisms $\phi : \Sigma_c^* \to \Sigma_c^*$ beyond $\Phi_{c,1} \cup \Phi_{c,2}$.

We note that $|\boldsymbol{\Phi}_x|$ is finite, as there is a finite number of ways in which we can partition Σ_x into three sets, even when, as we do, the definition of a partition is relaxed to allow the third set to be empty. Let us now represent with H_Σ the set $\times_{x \in \Sigma} \boldsymbol{\Phi}_x$, whose every element has the structure $\langle \Phi_{a_1, i_1}, \Phi_{a_2, i_2}, \ldots, \Phi_{a_N, i_N} \rangle$ for $\Sigma = \{a_1, a_2, \ldots, a_N\}$ and $i_1, i_2, \ldots, i_N \in \mathbb{N}_+$. This set represents all of the cases of our proof blueprint, and in each of the cases we shall prove Conjecture 2 for $\langle \Phi_1, \Phi_2, \ldots, \Phi_N \rangle \in H_\Sigma$.

Example 4. Let $\Sigma = \{a, b, c\}$. Given a morphism $\phi \in \Phi_{x,i}$, let us also define the following notation: $R_{x,i} := \langle E_\phi, M_\phi, C_\phi \rangle$. We recall from Example 3 that $\Phi_{c,1}$ is the set of all morphisms ϕ with $R_{c,1} = \langle a, b, \varnothing \rangle$ (let us for readability abuse the notation and omit the set braces), $\Phi_{c,2}$ a set of all those with $R_{c,2} = \langle b, a, \varnothing \rangle$.

We can similarly write for $\boldsymbol{\Phi}_b$ that $\{R_{b,1}, R_{b,2}\} = \{\langle a, c, \varnothing \rangle, \langle c, a, \varnothing \rangle\}$, and for $\boldsymbol{\Phi}_a$ that $\{R_{a,1}, R_{a,2}\} = \{\langle b, c, \varnothing \rangle, \langle c, b, \varnothing \rangle\}$. Therefore there are six cases to consider, as $|H_\Sigma| = 6$, and H_Σ consist of all six combinations of the sets of morphisms.

Once we show that, for every case of H_Σ, Conjecture 2 holds, we can conclude that it holds for Σ in general. We shall show why that is, and to that end let us assume to the contrary that our conjecture holds for all the cases of H_Σ, but it does not in general for Σ. Since Conjecture 2 is equivalent to the Billaud Conjecture, let us assume that there is a word $w \in \Sigma^*$, such that each of its heirs $\pi_x(w)$, $x \in \Sigma$, is a fixed point of some non-trivial, idempotent morphism ϕ_x, and such that w is morphically primitive. For every $x \in \Sigma$, there must exist an $i \in \mathbb{N}_+$ such that $\phi_x \in \Phi_{x,i}$, as all $\Phi_{x,j}$, $j \in [\![1, |\boldsymbol{\Phi}_x|]\!]$, together form a partition on a set of all non-trivial, idempotent morphisms of the form $\Sigma_x^* \to \Sigma_x^*$. Therefore, given $\Sigma = \{a_1, a_2, \ldots, a_N\}$, there is a case $\langle \boldsymbol{\Phi}_{a_1}, \boldsymbol{\Phi}_{a_2}, \ldots, \boldsymbol{\Phi}_{a_N} \rangle \in H_\Sigma$ such that $\phi_{a_i} \in \boldsymbol{\Phi}_{a_i}$ for all $i \in [\![1, N]\!]$. Therefore, as Conjecture 2 is assumed to hold in such a case, we can show that w is a morphically imprimitive word. This is a contradiction to our assumption that w was morphically primitive.

We can consult the following partial example to see how one can prove a simple single instance of Conjecture 2:

Example 5. Let $\Sigma = \{a, b, c\}$. Let us also use the following shorthand notation within this example: $[x \mapsto w]$ shall stand for a morphism ϕ such that $\phi(x) := w$ and $\phi(y) := \lambda$ for all $y \in \mathrm{symb}(w) \backslash \{x\}$.

We consider a partial example, namely a case $H := \langle \boldsymbol{\Phi}_a, \boldsymbol{\Phi}_b, \boldsymbol{\Phi}_c \rangle$ for Conjecture 2 where $\boldsymbol{\Phi}_c = \{[a \mapsto ab], [a \mapsto ab^2]\}$, $\boldsymbol{\Phi}_b = \{[c \mapsto ac], [c \mapsto a^2c]\}$, and $\boldsymbol{\Phi}_a = \{[b \mapsto bc], [b \mapsto bc^2]\}$. The set $\boldsymbol{\Phi}_c$ contains a very small subset of morphisms ϕ for which $E_\phi = \{a\}$, $M_\phi = \{b\}$, and $C_\phi = \varnothing$. We can make similar observations about $\boldsymbol{\Phi}_b \subset \Phi_b'$ and $\boldsymbol{\Phi}_c \subset \Phi_c'$, and see that H is a restriction of some $H' \in H_\Sigma$.

We shall now show that $L := \bigcap_{x \in \Sigma} P_{\Phi_x, \Sigma}$ only contains words that are fixed points of non-trivial morphisms. We note, that F_{Φ_c} can be represented as a regular expression $L((ab)^* | (abb)^*)$, and so $P_{\Phi_c, \Sigma} = L(((ab)^* | (abb)^*) \sqcup c^*) = L((ac^*b|c)^* | (ac^*bc^*b|c)^*)$. We can analogously represent the languages $P_{\Phi_b, \Sigma}$ and $P_{\Phi_a, \Sigma}$ and then apply known algorithms (Thompson's, Kleene's [12], and finite automaton intersection) to obtain their intersection, $L = L((abc)^*)$. As explained in Example 1, such L is a subset of, e.g., $F_{[a \mapsto abc]}$, and so Conjecture 2 holds for the case H.

In Example 5 we are still working with regular languages for simplicity. In reality, a set of all morphisms with given letter roles is infinite, and as such, an infinite union of regular languages is no longer regular. As a result, our reasoning needs to be more robust: it is however possible, and indeed this is what we accomplish in our proof of the Billaud Conjecture for $|\Sigma| = 4$.

A question relevant to the conjecture is to find an algorithm that could yield a proof of the Billaud Conjecture given a case $H \in H_\Sigma$. As mentioned before, in general, F_Φ and $P_{\Phi, \Sigma}$ may not be regular for an infinite Φ. If we consider a set Φ of all non-trivial morphisms with specific letter roles, the set F_Φ is recursive, as we can check if a word w satisfies $\phi(w) = w$ for some ϕ, and whether $\phi \in \Phi$. Similarly, P_{Φ, Σ_x} is recursive as we can decide if a word belongs to P_{Φ, Σ_x} by first deleting all of its occurrences of x, and then checking if the result is in F_Φ. Since recursive languages are closed under intersection, $L := \bigcap_{x \in \Sigma} P_{\Phi_x, \Sigma_x}$, where $H = \langle \Phi_x \rangle_{x \in \Sigma}$, is recursive as well. However, in order to prove Conjecture 2 for Σ, we need to show that $L \subseteq F_\Phi$ for some set of non-trivial, idempotent morphisms. Clearly, this is true if L is a subset of the set of all morphically imprimitive words over Σ; however the inclusion of recursive languages is undecidable in general.

Therefore, we need to know more about the classes of languages of fixed points, and their parents, their closure and inclusion properties to be able to answer our question. The set L_Φ for a class of morphisms Φ with specific letter roles can be modelled using the so-called *pattern expressions*, introduced by Câmpeanu and Yu [2], which are a language descriptor more powerful than both regular expressions and patterns. The operation of intersection of languages of parents of fixed points of morphisms can similarly be expressed directly on the languages of fixed points themselves using an operation of *synchronised shuffle*, introduced by De Simone [3]. Due to space constraints we state without proof that, in fact, given two sets of morphisms Φ, Ψ of some case $H \in H_\Sigma$, the synchronised shuffle of F_Φ and F_Ψ is equivalent to $P_{\Phi, \Sigma} \cap P_{\Psi, \Sigma}$. If pattern expression languages were closed under synchronised shuffle, this would constitute a step closer to constructing the described algorithm and solving the Billaud Conjecture; unfortunately, it can be shown that this is not the case in general. Nevertheless, further work related to these concepts, with a particular focus on bettering our understanding of the languages of fixed points, could be valuable for proving the Billaud Conjecture.

5   A Proof for the Billaud Conjecture for $|\Sigma| = 4$

We apply the general reasoning presented in the previous section to show that the Billaud Conjecture holds for $|\Sigma| = 4$. To this end, we begin by presenting some results by others, which are prerequisites for our result. First, however, we recall the following definition by Levé and Richomme [10]:

Definition 3. *Let w be a morphically imprimitive word, and let F be the set of all of its imprimitivity factorisations. Then, let $\mathrm{minCardExp}(w) := \min_{f \in F} |E_f|$ be the minimal number of expanding letters of w.*

The following are the results by others which are vital prerequisites for our completion of the announced proof of the special case of the Billaud Conjecture.

Theorem 1 (Levé and Richomme [10]). *Let w be a word. If for all $x \in \mathrm{symb}(w)$ we have $\mathrm{minCardExp}(\pi_x(w)) = 1$, then w is morphically imprimitive.*

Lemma 2 (Walter [16]). *Let w be a word with $|\mathrm{symb}(w)| = 4$, and let all of the heirs of w be morphically imprimitive. If $\mathrm{minCardExp}(\pi_x(w)) = 2$ for exactly three or exactly four letters $x \in \mathrm{symb}(w)$, then w is morphically imprimitive.*

Any heir of a word w with $\mathrm{symb}(w) = 4$ consists of three letters. Therefore, 1 and 2 are the only valid values of $\mathrm{minCardExp}(\pi_x(w))$ for any $x \in \Sigma$, as at least one letter has to be expanding, and the three letters cannot all be expanding. The above two results show that the Billaud Conjecture holds for the cases when there are 0, 3, or 4 heirs w' of w with $\mathrm{minCardExp}(w') = 2$. Subsequently, we discuss the remaining cases when there are exactly 1 or 2 heirs w' with $\mathrm{minCardExp}(w') = 1$, and present the main theorem.

Before we commence with our main lemma, we present a technical lemma describing a structure of certain factors of certain regular languages, that will be of use in the proof for our main lemma:

Lemma 3. *Let the word $w \in \Sigma^*$ have one of the following structures:*

- *$w \in \{uxu', vyv'\}^*$, where $x, y \in \Sigma$, such that $x \neq y$ and $x, y \not\sqsubseteq uu'vv'$;*
- *$w \in \{uxu'\}^*$, where $x \in \Sigma$ and $x \not\sqsubseteq uu'$, in which case let us assume that $vyv' = \lambda$.*

Then, any circular factor s of w such that $s \in x(\Sigma \backslash x)^*x$ is equal to $s = xu'(vyv')^k ux$ and $|s|_y = k$ where $k \in \mathbb{N}_0$, and $|s|_z = |uu'|_z + k|vv'|_z$ for every letter $z \notin \{x, y\}$.

Finally, we present our main lemma, which complements the previous results by others. As the full proof alone exceeds the page limit, we present a small subset of representative cases, while the full proof can be found in the appendix. Our proof is split into cases corresponding to all relevant $H \in H_\Sigma$, and in each case we show the conjecture holds for all words $w \in \Sigma^*$ whose heirs are fixed points of any morphisms satisfying the case H.

Lemma 4. *Let w be a word with $|\mathrm{symb}(w)| = 4$, and let all heirs of w be morphically imprimitive. If $\mathrm{minCardExp}(\pi_x(w)) = 2$ for exactly one or exactly two letters $x \in \mathrm{symb}(w)$, then w is morphically imprimitive.*

Proof fragment. Let us assume that $\mathrm{symb}(w) = \{a, b, c, d\}$; the letters a, b, c, d are distinct. Let us assume to the contrary that w is morphically primitive.

Throughout the entire reasoning that follows, let us always assume that the variables $i, j, k, \ell, m, n, p, q, r, s, t, \not{p}$ are non-negative integers. These variables may be reused between cases, and should not be assumed to have the same value between different (sub-)cases; conversely, they shall stay constant within the given (sub-)case.

Let us assume w.l.o.g. that $\mathrm{minCardExp}(\pi_d(w)) = 2$. Let us further assume, w.l.o.g. that the expanding letters in $\pi_d(w)$ are a and b. Thus we can write that:

$$\pi_d(w) \in \{c^i a c^j, c^k b c^\ell\}^* \text{ where } i + j, \ k + \ell \geqslant 1. \tag{1}$$

As a consequence of the statement of the lemma, we know that all heirs of w are morphically imprimitive. In particular, that implies that the word $\pi_c(w)$ is morphically imprimitive, and that $\mathrm{minCardExp}(\pi_c(w)) = 2$ or $\mathrm{minCardExp}(\pi_c(w)) = 1$. We present the following overview of the cases for $\pi_c(w)$ and its imprimitivity factorisation f, followed by a discussion of duplicate cases:

E_f	M_f	C_f	Case	E_f	M_f	C_f	Case	E_f	M_f	C_f	Case	E_f	M_f	C_f	Case
a,b	d		Case 1	a	d	b	Case 4	b	d	a	(Case 4)	d	b	a	(Case 6)
a	b,d		Case 2	b	a,d		(Case 2)	d	a,b		Case 5	a,d	b		Case 7
a	b	d	Case 3	b	a	d	(Case 3)	d	a	b	Case 6	b,d	a		(Case 7)

As a result of the assumption about both a, b being expanding in $\pi_d(w)$, the cases for f where a, b are swapped need not be considered separately, as they are symmetric. These cases correspond to the bracketed cases in the table above.

In the full proof we shall examine the cases and further heirs to show for the purpose of contradiction that w cannot be morphically primitive. Firstly, however, we present the following claim, which will prove vital in our subsequent case analysis:

Claim 1. If the number of occurrences of a or the number of occurrences of c is the same for every circular factor v of $\pi_d(w)$ with $v \in b\{a, c\}^*b$, then we have a contradiction to the assumption that $\mathrm{minCardExp}(\pi_d(w)) = 2$.

Proof of Claim 1. Let us factorise the word $\pi_d(w)$ in the following fashion, where $N := |w|_b - 1$:

$$\pi_d(w) = v'bv_1bv_2b\cdots bv_Nbv'', \text{ where } v', v'', v_1, v_2, \ldots, v_N \in \{a, c\}^*.$$

Let us, for ease of reasoning, denote the word $v''v'$ with v_0. The words bv_0b, $bv_1b, \ldots, bv_Nb$ are all the circular factors v of $\pi_d(w)$ as defined in the statement of the claim. Due to the structure of $\pi_d(w)$, described in (1), every word v_J, $J \in [\![0, N]\!]$, necessarily has the following form as a consequence of Lemma 3:

$$v_J = c^\ell (c^i ac^j)^{|v_J|_a} c^k. \tag{2}$$

Therefore, we can express the relation between $|v_J|_a$ and $|v_J|_c$ in any v_J, $J \in [\![0, N]\!]$, as follows:

$$|v_J|_c = k + \ell + |v_J|_a(i + j). \tag{3}$$

If, as per the assumption of our claim, $|v_0|_x = |v_1|_x = \ldots = |v_N|_x$ for $x = a$ or $x = c$, then, due to the linear dependence between $|v_J|_a$ and $|v_J|_c$ described in (3), $|v_0|_x = |v_1|_x = \ldots = |v_N|_x$ for both $x = a$ and $x = c$.

Therefore, due to (2), we have that in fact $v_0 = v_1 = \ldots = v_N$. In particular, since $v_0 = v''v'$, we can give the following representation of $\pi_d(w)$:

$$\pi_d(w) \in \{v'bv''\}^* \text{ and } b \not\sqsubseteq v'v''.$$

Hence, $\pi_d(w)$ has an imprimitivity factorisation f' with $E_{f'} = \{b\}$, which contradicts our assumption that $\mathrm{minCardExp}(\pi_d(w)) = 2$.   □ (Claim 1)

In order to illustrate our reasoning, we now consider a number of representative sub-cases of Case 1 of the full proof.

<u>Case 1.</u> Let $E_f = \{a, b\}$, $M_f = \{d\}$; then we can write:

$$\pi_c(w) \in \{d^n a d^m, d^p b d^q\}^*, \text{ where } n + m, p + q \geqslant 1. \tag{4}$$

Additionally, we assume that $\text{minCardExp}(\pi_c(w)) = 2$, as otherwise $\pi_c(w)$ has an alternative imprimitivity factorisation such that one of the Cases 2–6 applies. Moreover, recall that:

$$\pi_d(w) \in \{c^i a c^j, c^k b c^\ell\}^* \text{ where } i + j, k + \ell \geqslant 1. \tag{1 revisited}$$

Let g be an imprimitivity factorisation of $\pi_a(w)$. If follows from the above structure of $\pi_d(w)$ that $|\pi_a(w)|_b < |\pi_a(w)|_c$ and $|\pi_a(w)|_b < |\pi_a(w)|_d$, and therefore we know that the letter b is not in M_g. Moreover, due to the fact that $\text{minCardExp}(\pi_d(w)) = \text{minCardExp}(\pi_c(w)) = 2$, we can further assume, due to the statement of our lemma where we postulate that there are at most two heirs with said property, that $\text{minCardExp}(\pi_a(w)) = 1$.

Let us summarise all possible letter role combinations in g, and outline the sub-cases of the present case:

E_g	M_g	C_g	Case	E_g	M_g	C_g	Case	E_g	M_g	C_g	Case
b	c,d		Case 1.1	c	b,d		$b \notin M_g$ ↯	d	b,c		$b \notin M_g$ ↯
b	c	d	Case 1.2(a)	c	b	d	$b \notin M_g$ ↯	d	b	c	$b \notin M_g$ ↯
b	d	c	Case 1.2(b)	c	d	b	Case 1.3(a)	d	c	b	Case 1.3(b)

<u>Case 1.1.</u> Let $E_g = \{b\}$ and $M_g = \{c,d\}$; then we can write:

$$\pi_a(w) \in \{u\}^*, \text{ where } u \in \left(c^r \sqcup d^t\right) b \left(c^s \sqcup d^p\right), \text{ and } r+s, t+p \geqslant 1.$$

Due to the above structure of $\pi_a(w)$ the number of cs in every circular factor of $\pi_a(w)$ of the form $b\{c,d\}^* b$ is the same, and equal to $r + s$ (as can be calculated due to Lemma 3). Hence, as $a, c \in \text{symb}(\pi_d(w))$, it follows that $r + s$ is also the number of cs in every circular factor of $\pi_d(w)$ of the form $b\{a,c\}^* b$. Thus, Claim 1 applies, and we have a contradiction to the assumption that $\text{minCardExp}(\pi_d(w)) = 2$.

Case 1.2(a). Let $E_g = \{b\}$, $M_g = \{c\}$, and $C_g = \{d\}$; then we can write:

$$\pi_a(w) \in \{c^r b c^s, d\}^* \text{ where } r + s \geqslant 1.$$

As in the preceding case, due to the above structure of $\pi_a(w)$, the number of cs in every circular factor of $\pi_a(w)$ of the form $b\{c,d\}^* b$ is the same, and equal to $r + s$ (as can be calculated due to Lemma 3). We can apply Claim 1 in the same way as in Case 1.1 to reach a contradiction to the assumption that $\text{minCardExp}(\pi_d(w)) = 2$.

Case 1.3(a). Let $E_g = \{c\}$, $M_g = \{d\}$, and $C_g = \{b\}$; then we can write:

$$\pi_a(w) \in \{d^r c d^s, b\}^* \text{ where } r + s \geqslant 1. \tag{5}$$

Due to Lemma 3 we can write the following equations for any circular factor $v \in b\{a, c, d\}^* b$ of w: $|v|_d = |v|_c (r + s)$ due to (5); $|v|_c = k + \ell + |v|_a (i + j)$ due to (1); and $|v|_d = p + q + |v|_a (n + m)$ due to (4). We can solve the above equations for $|v|_a$ to obtain a value independent of a choice of an occurrence of the factor v in w. Therefore, the number of as in any circular factor of $\pi_d(w)$ of the form $b\{a, c\}^* b$ is the same, and Claim 1 applies as in the preceding cases.

The Remaining Cases. The omitted cases of Case 1 can be analogously shown to hold using Claim 1. With help of Claim 1 and another claim phrased below, we can use a similar analysis to prove the remaining cases.

Claim 2. If every circular factor of $\pi_a(w)$ of the form $d\{b, c\}^* d$ is the same, and if every circular factor of $\pi_c(w)$ of the form $d\{a, b\}^* d$ is the same as well, then the word w is morphically imprimitive.   □

Theorem 1, Lemma 2, and Lemma 4 directly imply our main result:

Theorem 2. *The Billaud Conjecture holds for all words w with $|\mathrm{symb}(w)| = 4$.*

Acknowledgements. We thank the anonymous reviewers for their thorough and helpful comments, and for suggesting a more concise proof of Lemma 4.

References

1. Billaud, M.: A problem with words. Newsgroup 'comp.theory' (1993)
2. Câmpeanu, C., Yu, S.: Pattern expressions and pattern automata. Inf. Process. Lett. **92**(6), 267–274 (2004)
3. De Simone, R.: Languages infinitaires et produit de mixage. Theoret. Comput. Sci. **31**(1–2), 83–100 (1984)
4. Geser, A.: Your 'Problem with Words'. Private communication to M, Billaud (1993)
5. Ginsburg, S., Spanier, E.H.: Mappings of languages by two-tape devices. J. ACM **12**(3), 423–434 (1965)
6. Hamm, D., Shallit, J.: Characterization of finite and one-sided infinite fixed points of morphisms on free monoids. Technical Report CS-99-17, University of Waterloo, Ontario, Canada (1999)
7. Head, T.: Fixed languages and the adult languages of OL schemes. Int. J. Comput. Math. **10**(2), 103–107 (1981)
8. Holub, Š: Polynomial-time algorithm for fixed points of nontrivial morphisms. Discret. Math. **309**(16), 5069–5076 (2009)
9. Kociumaka, T., Radoszewski, J., Rytter, W., Waleń, T.: Linear-time version of Holub's algorithm for morphic imprimitivity testing. Theoret. Comput. Sci. **602**, 7–21 (2015)
10. Levé, F., Richomme, G.: On a conjecture about finite fixed points of morphisms. Theoret. Comput. Sci. **339**(1), 103–128 (2005)
11. Matocha, V., Holub, Š: Complexity of testing morphic primitivity. Kybernetika **49**(2), 216–223 (2013)
12. McNaughton, R., Yamada, H.: Regular expressions and state graphs for automata. IRE Trans. Electron. Comput. **EC-9**(1), 39–47 (1960)

13. Nevisi, H., Reidenbach, D.: Morphic primitivity and alphabet reductions. In: Proceedings of DLT 2021. LNCS, vol. 7410, pp. 440–451. Springer, Berlin (2012)
14. Reidenbach, D.: Discontinuities in pattern inference. Theoret. Comput. Sci. **397**(1), 166–193 (2008)
15. Reidenbach, D., Schneider, J.C.: Morphically primitive words. Theoret. Comput. Sci. **410**(21), 2148–2161 (2009)
16. Walter, T.: Über die Billaudsche Vermutung. Diplomarbeit, Universität Stuttgart, Fakultät Informatik, Elektrotechnik und Informationstechnik, Germany (2011)
17. Zimmermann, P.: A Problem With Words From Michel Billaud. Private communication to M, Billaud (1993)

Weighted Tree Automata
with Constraints

Andreas Maletti and Andreea-Teodora Nász[(✉)]

Faculty of Mathematics and Computer Science, Universität Leipzig,
PO box 100 920, 04009 Leipzig, Germany
{maletti,nasz}@informatik.uni-leipzig.de

Abstract. The HOM problem, which asks whether the image of a regular tree language under a given tree homomorphism is again regular, is known to be decidable [Godoy & Giménez: The HOM problem is decidable. JACM 60(4), 2013]. However, the problem remains open for regular weighted tree languages. It is demonstrated that the main notion used in the unweighted setting, the tree automaton with equality and inequality constraints, can straightforwardly be generalized to the weighted setting and can represent the image of any regular weighted tree language under any nondeleting, nonerasing tree homomorphism. Several closure properties as well as decision problems are also investigated for the weighted tree languages generated by weighted tree automata with constraints.

1 Introduction

Numerous extensions of nondeterministic finite-state string automata have been proposed in the past few decades. On the one hand, the qualitative evaluation of inputs was extended to a quantitative evaluation in the weighted automata of [23]. This development led to the fruitful study of recognizable formal power series [22], which are well-suited for representing factors such as costs, consumption of resources, or time and probabilities related to the processed input. The main algebraic structure for the weight calculations are semirings [16,17], which offer a nice compromise between generality and efficiency of computation (due to their distributivity).

On the other hand, finite-state automata have been generalized to other input structures such as infinite words [21] and trees [4]. Finite-state tree automata were introduced independently in [7,24,25] and they and the tree languages they generate, called regular tree languages, have been intensively studied since their inception [4]. They are successfully utilized in various applications in many diverse areas like natural language processing [18], picture generation [8], and compiler construction [28]. Indeed several applications require the combination of the two mentioned generalizations and a broad range of weighted tree automaton (WTA) models has been studied (cf. [13, Chapter 9] for an overview). It is

Research financially supported by a scholarship awarded to T. Nasz by the Free State of Saxony (Funding no. LAU-R-I-9-2-1021).

V. Diekert and M. Volkov (Eds.): DLT 2022, LNCS 13257, pp. 226–238, 2022.
https://doi.org/10.1007/978-3-031-05578-2_18

well-known that finite-state tree automata cannot ensure that two subtrees (of potentially arbitrary size) are always equal in an accepted tree [14]. An extension proposed in [20] aims to remedy this problem and introduces a tree automaton model that explicitly can require certain subtrees to be equal or unequal. Such models are very useful when investigating transduction models (see [13] for an overview) that can copy subtrees (thus resulting in equal subtrees) and they are the main tool used in the seminal paper [15] that proved that the HOM problem is decidable.

The HOM problem was a long-standing open problem in the theory of tree languages and recently solved in [15]. It asks whether the image of an (effectively presented) regular tree language under a given tree homomorphism is again regular. This is not necessarily the case as tree homomorphisms can create copies of subtrees. Indeed removing this ability from the tree homomorphism, obtaining a linear tree homomorphism, yields that the mentioned image is always regular [14]. In the solution to the HOM problem provided in [15] the image is first represented by a tree automaton with constraints and then it is investigated whether this tree automaton actually generates a regular tree language.

In the weighted setting, the HOM problem is also interesting as it once again provides an answer whether a given homomorphic image of a regular weighted tree language can efficiently be represented. While preservation of regularity has been investigated [3,10–12] also in the weighted setting, the decidability of the HOM problem remains wide open. With the goal of investigating this problem, we introduce weighted tree automata with constraints (WTAc for short) in this contribution. We demonstrate that those WTAc can again represent all homomorphic images of the regular weighted tree languages. Thus, in principle, it only remains to provide a decision procedure for determining whether a given WTAc generates a regular weighted tree language. We approach this task by providing some common closure properties following essentially the steps also taken in [15]. For zero-sum free semirings we can also show that decidability of support emptiness and finiteness are directly inherited from the unweighted case [15].

2   Preliminaries

We denote the set of nonnegative integers by $\mathbb{N}$, and for every $k \in \mathbb{N}$, we let $[k] = \{i \in \mathbb{N} \mid 1 \leq i \leq k\}$. For all sets T and Z let T^Z be the set of all mappings $\varphi \colon Z \to T$, and correspondingly we sometimes write φ_z instead of $\varphi(z)$ for every $\varphi \in T^Z$. The cardinality of Z is denoted by $|Z|$.

A *ranked alphabet* (Σ, rk) is a pair consisting of a finite set Σ and a mapping $\mathrm{rk} \in \mathbb{N}^\Sigma$ that assigns a rank to each symbol of Σ. If there is no risk of confusion, we denote a ranked alphabet (Σ, rk) by Σ. We write $\sigma^{(k)}$ to indicate that $\mathrm{rk}(\sigma) = k$. Moreover, for every $k \in \mathbb{N}$ we let $\Sigma_k = \mathrm{rk}^{-1}(k)$. Let $X = \{x_i \mid i \in \mathbb{N}\}$ be a countable set of (formal) variables. For each $n \in \mathbb{N}$ we let $X_n = \{x_i \mid i \in [n]\}$. Given a ranked alphabet Σ and a set Z, the set $T_\Sigma(Z)$ of Σ" *trees indexed by* Z is the smallest set such that $Z \subseteq T_\Sigma(Z)$

and $\sigma(t_1,\ldots,t_k) \in T_\Sigma(Z)$ for every $k \in \mathbb{N}$, $\sigma \in \Sigma_k$, and $t_1,\ldots,t_k \in T_\Sigma(Z)$. We abbreviate $T_\Sigma(\emptyset)$ simply to T_Σ, and any subset $L \subseteq T_\Sigma$ is called a *tree language*.

Let Σ be a ranked alphabet, Z a set, and $t \in T_\Sigma(Z)$. The set $\mathrm{pos}(t)$ of *positions* of t is inductively defined by $\mathrm{pos}(z) = \{\varepsilon\}$ for all $z \in Z$ and by $\mathrm{pos}(\sigma(t_1,\ldots,t_k)) = \{\varepsilon\} \cup \bigcup_{i\in[k]}\{iw \mid w \in \mathrm{pos}(t_i)\}$ for all $k \in \mathbb{N}$, $\sigma \in \Sigma_k$, and $t_1,\ldots,t_k \in T_\Sigma(Z)$. The size $|t|$ of t is defined as $|t| = |\mathrm{pos}(t)|$. For $w \in \mathrm{pos}(t)$ and $t' \in T_\Sigma(Z)$, the *label* $t(w)$ of t at w, the *subtree* $t|_w$ of t at w, and the *substitution* $t[t']_w$ of t' into t at w are defined by $z(\varepsilon) = z|_\varepsilon = z$ and $z[t']_\varepsilon = t'$ for all $z \in Z$ and for $t = \sigma(t_1,\ldots,t_k)$ by $t(\varepsilon) = \sigma$, $t(iw') = t_i(w')$, $t|_\varepsilon = t$, $t|_{iw'} = t_i|_{w'}$, $t[t']_\varepsilon = t'$, and $t[t']_{iw'} = \sigma(t_1,\ldots,t_{i-1},t_i[t']_{w'},t_{i+1},\ldots,t_k)$ for all $k \in \mathbb{N}$, $\sigma \in \Sigma_k$, $t_1,\ldots,t_k \in T_\Sigma(Z)$, $i \in [k]$, and $w' \in \mathrm{pos}(t_i)$. For all $\sigma \in \Sigma \cup Z$, we let $\mathrm{pos}_\sigma(t) = \{w \in \mathrm{pos}(t) \mid t(w) = \sigma\}$ and $\mathrm{var}(t) = \{x \in X \mid \mathrm{pos}_x(t) \neq \emptyset\}$. Finally, for every $t \in T_\Sigma(Z)$, finite $V \subseteq Z$, and $\theta \in T_\Sigma(Z)^V$, the substitution θ applied to t is written as $t\theta$ and defined by $v\theta = \theta_v$ for every $v \in V$, $z\theta = z$ for every $z \in Z \setminus V$, and $\sigma(t_1,\ldots,t_k)\theta = \sigma(t_1\theta,\ldots,t_k\theta)$ for all $k \in \mathbb{N}$, $\sigma \in \Sigma_k$, and $t_1,\ldots,t_k \in T_\Sigma(Z)$. We also write the substitution $\theta \in T_\Sigma(Z)^V$ as (i) $[v_1 \leftarrow \theta_{v_1},\ldots,v_n \leftarrow \theta_{v_n}]$ if $V = \{v_1,\ldots,v_n\}$ or (ii) $[\theta_{x_1},\ldots,\theta_{x_n}]$ if $V = X_n$.

A *commutative semiring* [16,17] is a tuple $(\mathbb{S},+,\cdot,0,1)$ such that $(\mathbb{S},+,0)$ and $(\mathbb{S},\cdot,1)$ are commutative monoids, $\cdot$ distributes over $+$, and $0 \cdot s = 0$ for all $s \in \mathbb{S}$. Examples include (i) the Boolean semiring $\mathbb{B} = (\{0,1\},\vee,\wedge,0,1)$, (ii) the semiring $\mathbb{N} = (\mathbb{N},+,\cdot,0,1)$, (iii) the tropical semiring $\mathbb{T} = (\mathbb{N} \cup \{\infty\},\min,+,\infty,0)$, and (iv) the arctic semiring $\mathbb{A} = (\mathbb{N} \cup \{-\infty\},\max,+,-\infty,0)$. Given semirings $(\mathbb{S},+,\cdot,0,1)$ and $(\mathbb{T},\oplus,\odot,\bot,\top)$, a *semiring homomorphism* is a mapping $h \in \mathbb{T}^\mathbb{S}$ such that $h(0) = \bot$, $h(1) = \top$, and $h(s_1 + s_2) = h(s_1) \oplus h(s_2)$ as well as $h(s_1 \cdot s_2) = h(s_1) \odot h(s_2)$ for all $s_1,s_2 \in \mathbb{S}$. When there is no risk of confusion, we refer to a semiring $(\mathbb{S},+,\cdot,0,1)$ simply by its carrier set $\mathbb{S}$. A semiring $\mathbb{S}$ is a *ring* if there exists $-1 \in \mathbb{S}$ such that $-1 + 1 = 0$. Let Σ be a ranked alphabet. Any mapping $A \in \mathbb{S}^{T_\Sigma}$ is called a *weighted tree language* over $\mathbb{S}$ and its support is $\mathrm{supp}(A) = \{t \in T_\Sigma \mid A_t \neq 0\}$.

Let Σ and Δ be ranked alphabets and let $h' \in T_\Delta(X)^\Sigma$ be a mapping such that $h'_\sigma \in T_\Delta(X_k)$ for all $k \in \mathbb{N}$ and $\sigma \in \Sigma_k$. We extend h' to $h \in T_\Delta^{T_\Sigma}$ by (i) $h_\alpha = h'_\alpha \in T_\Delta(X_0) = T_\Delta$ for all $\alpha \in \Sigma_0$ and (ii) $h_{\sigma(t_1,\ldots,t_k)} = h'_\sigma[h_{t_1},\ldots,h_{t_k}]$ for all $k \in \mathbb{N}$, $\sigma \in \Sigma_k$, and $t_1,\ldots,t_k \in T_\Sigma$. The mapping h is called the *tree homomorphism induced by* h', and we identify h' and its induced tree homomorphism h. It is *nonerasing* if $h'_\sigma \notin X$ for all $k \in \mathbb{N}$ and $\sigma \in \Sigma_k$, and it is *nondeleting* if $\mathrm{var}(h'_\sigma) = X_k$ for all $k \in \mathbb{N}$ and $\sigma \in \Sigma_k$. Let $h \in T_\Delta^{T_\Sigma}$ be a nonerasing and nondeleting homomorphism. Then h is *input finitary*; i.e., the set $h^{-1}(u)$ is finite for every $u \in T_\Delta$ because $|t| \leq |u|$ for each $t \in h^{-1}(u)$. Additionally, let $A \in \mathbb{S}^{T_\Sigma}$ be a weighted tree language. We define the weighted tree language $h(A) \in \mathbb{S}^{T_\Delta}$ for every $u \in T_\Delta$ by $h(A)_u = \sum_{t\in h^{-1}(u)} A_t$.

3   Weighted Tree Grammars with Constraints

Let us start with the formal definition of our weighted tree grammars. They are a weighted variant of the tree automata with equality and inequality constraints originally introduced in [1,5] (with constraints on direct subtrees). An overview of further developments for these automata can be found in [26]. We essentially use the version recently utilized to solve the HOM problem [15, Definition 4.1].

Definition 1. *A* weighted tree grammar with constraints *(WTGc) is a tuple* $G = (Q, \Sigma, F, P, \mathrm{wt})$ *such that*

- *Q is a finite set of nonterminals and $F \in \mathbb{S}^Q$ assigns final weights,*
- *Σ is a ranked alphabet of input symbols,*
- *P is a finite set of productions of the form (ℓ, q, E, I), where $\ell \in T_\Sigma(Q) \setminus Q$, $q \in Q$, and $E, I \subseteq \mathbb{N}^* \times \mathbb{N}^*$ are finite sets, and*
- *$\mathrm{wt} \in \mathbb{S}^P$ assigns a weight to each production.*   □

In the following, let $G = (Q, \Sigma, F, P, \mathrm{wt})$ be a WTGc. The components of a production $p = (\ell, q, E, I) \in P$ are the left-hand side ℓ, the governing nonterminal q, the set E of equality constraints, and the set I of inequality constraints. Correspondingly, the production p is also written $\ell \xrightarrow{E, I} q$ or even $\ell \xrightarrow{E, I}_{\mathrm{wt}_p} q$ if we want to indicate its weight. Additionally, we simply list an equality constraint $(v, v') \in E$ as $v = v'$ and an inequality constraint $(v, v') \in I$ as $v \neq v'$. A production $\ell \xrightarrow{E, I} q \in P$ is *normalized* if $\ell = \sigma(q_1, \ldots, q_k)$ for some $k \in \mathbb{N}$, $\sigma \in \Sigma_k$, and $q_1, \ldots, q_k \in Q$, and it is *unconstrained* if $E = \emptyset = I$; in this case we also simply write $\ell \to q$. The WTGc G is a *weighted tree automaton with constraints* (WTAc) if all productions $p \in P$ are normalized, and it is a *weighted tree grammar* (WTG) [14] if all productions $p \in P$ are unconstrained. If G is both a WTAc as well as a WTG, then it is a *weighted tree automaton* (WTA) [14]. All these devices have *Boolean final weights* if $F \in \{0, 1\}^Q$. Finally, if we utilize the Boolean semiring $\mathbb{B}$, then we reobtain the unweighted versions and omit the 'W' in the abbreviations and the mapping 'wt' from the tuple.

The semantics for our WTGc G is a slightly non-standard *derivation semantics* when compared to [15, Definitions 4.3 and 4.4]. Let $(v, v') \in \mathbb{N}^* \times \mathbb{N}^*$ and $t \in T_\Sigma$. If $v, v' \in \mathrm{pos}(t)$ and $t|_v = t|_{v'}$, we say that t satisfies (v, v'), otherwise t dissatisfies (v, v'). Let now $C \subseteq \mathbb{N}^* \times \mathbb{N}^*$ be a finite set of constraints. We write $t \models C$ if t satisfies all $(v, v') \in C$, and $t \not\models^\forall C$ if t dissatisfies all $(v, v') \in C$. Universally dissatisfying C is generally stronger than simply not satisfying C.

Definition 2. *A sentential form (for G) is simply a tree of $\xi \in T_\Sigma(Q)$. Given an input tree $t \in T_\Sigma$, sentential forms $\xi, \zeta \in T_\Sigma(Q)$, a production $p = \ell \xrightarrow{E, I} q \in P$, and a position $w \in \mathrm{pos}(\xi)$, we write $\xi \Rightarrow_{G,t}^{p,w} \zeta$ if $\xi|_w = \ell$, $\zeta = \xi[q]_w$, and the constraints E and I are fulfilled on $t|_w$; i.e., $t|_w \models E$ and $t|_w \not\models^\forall I$. A sequence $d = (p_1, w_1) \cdots (p_n, w_n) \in (P \times \mathbb{N}^*)^*$ is a derivation of G for t if there exist $\xi_1, \ldots, \xi_n \in T_\Sigma(Q)$ such that $t \Rightarrow_{G,t}^{p_1, w_1} \xi_1 \Rightarrow_{G,t}^{p_2, w_2} \cdots \Rightarrow_{G,t}^{p_n, w_n} \xi_n$. It is leftmost if additionally $w_1 \prec w_2 \prec \cdots \prec w_n$, where $\preceq$ is the lexicographic order on $\mathbb{N}^*$ in which prefixes are larger, so ε is the largest element.*   □

Note that the sentential forms $\xi_1, \ldots, \xi_n$ are uniquely determined if they exist, and for any derivation d for t there exists a unique permutation of d that is a left-most derivation for t. The derivation d is *complete* if $\xi_n \in Q$, and in that case it is also called a derivation to ξ_n. The set of all complete left-most derivations for t to $q \in Q$ is denoted by $D_G^q(t)$. The WTGc G is *unambiguous* if $\sum_{q\in\text{supp}(F)} |D_G^q(t)| \leq 1$ for every $t \in T_\Sigma$.

Definition 3. *The* weight *of a derivation* $d = (p_1, w_1) \cdots (p_n, w_n)$ *is defined to be* $\text{wt}_G(d) = \prod_{i=1}^{n} \text{wt}(p_i)$. *The weighted tree language generated by* G, *written simply* $G \in \mathbb{S}^{T_\Sigma}$, *is defined for every* $t \in T_\Sigma$ *by*

$$G_t = \sum_{q\in Q,\, d\in D_G^q(t)} F_q \cdot \text{wt}_G(d).$$

□

Two WTGc are *equivalent* if they generate the same weighted tree language. Finally, a weighted tree language is *regular* if it is generated by a WTG, and it is *constraint-regular* if it is generated by a WTGc. Since the weights of productions are multiplied, we can assume that $\text{wt}_p \neq 0$ for all $p \in P$.

Example 4. Consider the WTGc $G = (Q, \Sigma, F, P, \text{wt})$ over $\mathbb{A}$ with $Q = \{q, q'\}$, $\Sigma = \{\alpha^{(0)}, \gamma^{(1)}, \sigma^{(2)}\}$, $F_q = -\infty$, $F_{q'} = 0$, and P and 'wt' given by the productions $p_1 = \alpha \to_0 q$, $p_2 = \gamma(q) \to_1 q$, and $p_3 = \sigma(\gamma(q), q) \xrightarrow{11=2}_1 q'$. The tree $t = \sigma(\gamma(\gamma(\alpha)), \gamma(\alpha))$ has the unique left-most derivation

$$d = (p_1, 111)\,(p_2, 11)\,(p_1, 21)\,(p_2, 2)\,(p_3, \varepsilon)$$

to the nonterminal q'. Overall, we have $\text{supp}(G) = \{\sigma(\gamma^{i+1}(\alpha), \gamma^i(\alpha)) \mid i \in \mathbb{N}\}$ and $G_t = |\text{pos}_\gamma(t)|$ for every $t \in \text{supp}(G)$. □

For the restricted model of WTAc we introduce another semantics, called initial algebra semantics, which is often more convenient in proofs.

Definition 5. *If* G *is a WTAc, then for each* $q \in Q$ *we define* $\text{wt}_G^q \in \mathbb{S}^{T_\Sigma}$ *for every* $t = \sigma(t_1, \ldots, t_k)$ *with* $k \in \mathbb{N}$, $\sigma \in \Sigma_k$, *and* $t_1, \ldots, t_k \in T_\Sigma$ *by*

$$\text{wt}_G^q(t) = \sum_{\substack{p=\sigma(q_1,\ldots,q_k)\xrightarrow{E,I} q \in P \\ t\models E,\, t\not\models I}} \text{wt}_p \cdot \prod_{i=1}^{k} \text{wt}_G^{q_i}(t_i) \ .$$

□

It is a routine matter to verify $\text{wt}_G^q(t) = \sum_{d\in D_G^q(t)} \text{wt}_G(d)$ for every $q \in Q$ and $t \in T_\Sigma$. Indeed as for WTG and WTA [13] also every WTGc can be turned into an equivalent WTAc at the expense of additional nonterminals by decomposing the left-hand sides.

Lemma 6 (cf. [15, Lemma 4.8]). *WTGc and WTAc are equally expressive.*

Proof. Let $G = (Q, \Sigma, F, P, \mathrm{wt})$ be a WTGc with a non-normalized production $p = \sigma(\ell_1, \ldots, \ell_k) \xrightarrow{E,I} q \in P$, let U be an infinite set with $Q \subseteq U$ and let $\varphi \in U^{T_\Sigma(Q)}$ be an injective map such that $\varphi_q = q$ for all $q \in Q$. We define the WTGc $G' = (Q', \Sigma, F', P', \mathrm{wt}')$ such that $Q' = Q \cup \{\varphi_{\ell_1}, \ldots, \varphi_{\ell_k}\}$, $F'_q = F_q$ for all $q \in Q$ and $F'_{q'} = 0$ for all $q' \in Q' \setminus Q$, and

$$P' = (P \setminus \{p\}) \cup \{\sigma(\varphi_{\ell_1}, \ldots, \varphi_{\ell_k}) \xrightarrow{E,I} q\} \cup \{\ell_i \to \varphi_{\ell_i} \mid i \in [k], \ell_i \notin Q\},$$

and for every $p' \in P'$

$$\mathrm{wt}'_{p'} = \begin{cases} \mathrm{wt}_{p'} & \text{if } p' \in P \setminus \{p\} \\ \mathrm{wt}_p & \text{if } p' = \sigma(\varphi_{\ell_1}, \ldots, \varphi_{\ell_k}) \xrightarrow{E,I} q \\ 1 & \text{otherwise.} \end{cases}$$

$\square$

Example 7. Consider the WTGc G of Example 4 and its non-normalized production $p = \sigma(\gamma(q), q) \xrightarrow{11=2}_1 q'$. Applying the construction in the proof of Lemma 6 we replace p by the productions $\sigma(q'', q) \xrightarrow{11=2}_1 q$ and $\gamma(q) \to_0 q''$, where q'' is some new nonterminal. The such obtained WTGc is already a WTAc. $\square$

Another routine normalization turns the final weights into Boolean final weights [2, Lemma 6.1.1]. This is achieved by adding special copies of all nonterminals that terminate the derivation and pre-apply the final weight.

Lemma 8. *WTAc and WTAc with Boolean final weights are equally expressive.*

Let $d \in D_G^q(t)$ be a derivation for some $q \in Q$ and $t \in T_\Sigma$. Since we often argue with the help of such derivations d, it is a nuisance that we might have $\mathrm{wt}_G(d) = 0$. This anomaly can occur even if $\mathrm{wt}_p \neq 0$ for all $p \in P$ due to the presence of zero-divisors, which are elements $s, s' \in \mathbb{S} \setminus \{0\}$ such that $s \cdot s' = 0$. However, we can fortunately avoid such anomalies altogether utilizing a construction of [19] based on Dickson's Lemma [6], which has been lifted to tree automata in [9]. We note that the construction preserves Boolean final weights.

Lemma 9. *For every WTAc G there exists a WTAc $G' = (Q', \Sigma, F', P', \mathrm{wt}')$ that is equivalent and $\mathrm{wt}'_{G'}(d') \neq 0$ for all $q' \in Q'$, $t' \in T_\Sigma$, and $d' \in D_{G'}^{q'}(t')$.*

For zero-sum free semirings [16,17] we obtain that the support $\mathrm{supp}(G)$ of an WTAc can be generated by a TAc. A semiring is *zero-sum free* if $s = 0 = s'$ for every $s, s' \in \mathbb{S}$ such that $s + s' = 0$. Clearly, rings are never zero-sum free, but the mentioned semirings $\mathbb{B}$, $\mathbb{N}$, $\mathbb{T}$, and $\mathbb{A}$ are all zero-sum free.

Corollary 10 (of Lemmata 6 and 9). *If $\mathbb{S}$ is zero-sum free, then $\mathrm{supp}(G)$ is constraint-regular for every WTGc G.*

Proof. We apply Lemmata 6 and 8 to obtain an equivalent WTAc with Boolean final weights and then Lemma 9 to obtain the WTAc $G' = (Q', \Sigma, F', P', \mathrm{wt}')$ with Boolean final weights. As mentioned we can assume that $\mathrm{wt}'_{p'} \neq 0$ for all $p' \in P'$. Let $q' \in \mathrm{supp}(F')$ and $t' \in T_\Sigma$ with $D^{q'}_{G'}(t') \neq \emptyset$. Since $\mathrm{wt}'_{G'}(d') \neq 0$ for every $d' \in D^{q'}_{G'}(t')$ and $s+s' \neq 0$ for all $s, s' \in \mathbb{S} \setminus \{0\}$ due to zero-sum freeness, we obtain $t' \in \mathrm{supp}(G')$. Thus, the existence of a complete derivation for t' to an accepting nonterminal (i.e., one with final weight 1) characterizes whether we have $t' \in \mathrm{supp}(G')$. Consequently, the TAc $(Q', \Sigma, \mathrm{supp}(F'), P')$ generates the tree language $\mathrm{supp}(G')$, which is thus constraint-regular. $\qquad\square$

4   Closure Properties

In this section we investigate several closure properties of the constraint-regular weighted tree languages. We start with the (point-wise) sum, which is given by $(A + A')_t = A_t + A'_t$ for every $t \in T_\Sigma$ and $A, A' \in \mathbb{S}^{T_\Sigma}$. Given WTGc G and G' generating A and A' we can trivially use a disjoint union construction to obtain a WTGc generating $A + A'$. We omit the details.

Proposition 11. *The constraint-regular weighted tree languages (over the same ranked-alphabet) are closed under sums.* $\qquad\square$

The corresponding (point-wise) product is the HADAMARD product, which is given by $(A \cdot A')_t = A_t \cdot A'_t$ for every $t \in T_\Sigma$ and $A, A' \in \mathbb{S}^{T_\Sigma}$. With the help of a standard product construction we show that the constraint-regular weighted tree languages are also closed under HADAMARD product. As preparation we introduce a special normal form. A WTAc $G = (Q, \Sigma, F, P, \mathrm{wt})$ is *constraint-determined* if $E = E'$ and $I = I'$ for all productions $\sigma(q_1, \ldots, q_k) \xrightarrow{E,I} q \in P$ and $\sigma(q_1, \ldots, q_k) \xrightarrow{E',I'} q \in P$. In other words, two productions cannot differ only in the sets of constraints. It is straightforward to turn any WTAc into an equivalent constraint-determined WTAc by introducing additional nonterminals (e.g. annotate the constraints to the state on the right-hand side).

Theorem 12. *The constraint-regular weighted tree languages (over the same ranked alphabet) are closed under HADAMARD product.*

Proof. Let $A, A' \in \mathbb{S}^{T_\Sigma}$ be constraint-regular. Without loss of generality (see Lemma 6) we can assume constraint-determined WTAc $G = (Q, \Sigma, F, P, \mathrm{wt})$ and $G' = (Q', \Sigma, F', P', \mathrm{wt}')$ that generate A and A', respectively. We construct the direct product WTAc $G \times G' = (Q \times Q', \Sigma, F'', P'', \mathrm{wt}'')$ such that $F''_{\langle q, q' \rangle} = F_q \cdot F'_{q'}$ for every $q \in Q$ and $q' \in Q'$ and for every production $p = \sigma(q_1, \ldots, q_k) \xrightarrow{E,I} q \in P$ and production $p' = \sigma(q'_1, \ldots, q'_k) \xrightarrow{E',I'} q' \in P'$ the production

$$p'' = \sigma(\langle q_1, q'_1 \rangle, \ldots, \langle q_k, q'_k \rangle) \xrightarrow{E \cup E', I \cup I'} \langle q, q' \rangle$$

belongs to P'' and its weight is $\mathrm{wt}''_{p''} = \mathrm{wt}_p \cdot \mathrm{wt}'_{p'}$. No other productions belong to P''. The proof that $G \times G' = A \cdot A'$ is a straightforward induction proving $\mathrm{wt}^{\langle q,q' \rangle}_{G \times G'}(t) = \mathrm{wt}^q_G(t) \cdot \mathrm{wt}^{q'}_{G'}(t)$ for all $t \in T_\Sigma$ using the initial algebra semantics. The WTAc G and G' are required to be constraint-determined, so that we can uniquely identify the productions $p \in P$ and $p' \in P'$ that construct a production $p'' \in P''$. □

Example 13. Let $G = (\{q\}, \Sigma, F, P, \mathrm{wt})$ and $G' = (\{z\}, \Sigma, F', P', \mathrm{wt}')$ be WTAc over $\mathbb{A}$ and $\Sigma = \{\alpha^{(0)}, \gamma^{(1)}, \sigma^{(2)}\}$, $F_q = F'_z = 0$, and the productions

$$\alpha \to_0 q \qquad \gamma(q) \to_2 q \qquad \sigma(q,q) \xrightarrow{1=2}_0 q \qquad (P)$$
$$\alpha \to_0 z \qquad \gamma(z) \xrightarrow{11\neq12}_1 z \qquad \sigma(z,z) \to_1 z. \qquad (P')$$

We observe that

$$\mathrm{supp}(G) = \{t \in T_\Sigma \mid \forall w \in \mathrm{pos}_\sigma(t) \colon t|_{w1} = t|_{w2}\}$$
$$\mathrm{supp}(G') = \{t \in T_\Sigma \mid \forall w \in \mathrm{pos}_\gamma(t) \colon \text{ if } t(w1) = \sigma \text{ then } t|_{w11} \neq t|_{w12}\}$$

and $G_t = 2|\mathrm{pos}_\gamma(t)|$ as well as $G'_{t'} = |\mathrm{pos}_\gamma(t')| + |\mathrm{pos}_\sigma(t')|$ for all $t \in \mathrm{supp}(G)$ and $t' \in \mathrm{supp}(G')$. We obtain the WTAc $G \times G' = (\{\langle q, z \rangle\}, \Sigma, F'', P'', \mathrm{wt}'')$ with $F''_{\langle q,z \rangle} = 0$ and the following productions.

$$\alpha \to_0 \langle q, z \rangle \qquad \gamma(\langle q, z \rangle) \xrightarrow{11\neq12}_3 \langle q, z \rangle \qquad \sigma(\langle q, z \rangle, \langle q, z \rangle) \xrightarrow{1=2}_1 \langle q, z \rangle$$

Hence we obtain the equality $(G \times G')_t = 3|\mathrm{pos}_\gamma(t)| + |\mathrm{pos}_\sigma(t)| = G_t \cdot G'_t$ for every tree $t \in \mathrm{supp}(G) \cap \mathrm{supp}(G')$. □

Next, we use an extended version of the classical power set construction to obtain an unambiguous WTAc that keeps track of the reachable nonterminals, but preserves only the homomorphic image of its weight. The unweighted part of the construction mimics a power-set construction and the handling of constraints roughly follows [15, Definition 3.1].

Theorem 14. *Let $h \in \mathbb{T}^\mathbb{S}$ be a semiring homomorphism into a finite semiring $\mathbb{T}$. For every WTAc $G = (Q, \Sigma, F, P, \mathrm{wt})$ over $\mathbb{S}$ there exists an unambiguous WTAc $G' = (\mathbb{T}^Q, \Sigma, F', P', \mathrm{wt}')$ such that for every $t \in T_\Sigma$ and $\varphi \in \mathbb{T}^Q$*

$$\mathrm{wt}^\varphi_{G'}(t) = \begin{cases} 1 & \text{if } \varphi_q = h_{\mathrm{wt}^q_G(t)} \text{ for all } q \in Q \\ 0 & \text{otherwise.} \end{cases}$$

Moreover, $G'_t = h_{G(t)}$ for every $t \in T_\Sigma$.

Proof. Let $\mathcal{C} = \{E \mid \sigma(q_1, \ldots, q_k) \xrightarrow{E,I} q \in P\} \cup \{I \mid \sigma(q_1, \ldots, q_k) \xrightarrow{E,I} q \in P\}$ be the constraints that occur in G. We let $F'_\varphi = \sum_{q \in Q} h_{F(q)} \cdot \varphi_q$ for every $\varphi \in \mathbb{T}^Q$.

For all $k \in \mathbb{N}$, $\sigma \in \Sigma_k$, nonterminals $\varphi^1, \ldots, \varphi^k \in \mathbb{T}^Q$, and constraints $\mathcal{E} \subseteq \mathcal{C}$ we let $p' = \sigma(\varphi^1, \ldots \varphi^k) \xrightarrow{\mathcal{E},\mathcal{I}} \varphi \in P'$, where $\mathcal{I} = \mathcal{C} \setminus \mathcal{E}$ and for every $q \in Q$

$$\varphi_q = \sum_{\substack{p = \sigma(q_1, \ldots, q_k) \xrightarrow{E,I} q \in P \\ E \subseteq \mathcal{E}, I \subseteq \mathcal{I}}} h_{\mathrm{wt}(p)} \cdot \varphi_{q_1}^1 \cdot \ldots \cdot \varphi_{q_k}^k . \tag{1}$$

No additional productions belong to P'. Finally, we set $\mathrm{wt}'_{p'} = 1$ for all $p' \in P'$. In general, the WTAc G' is certainly not deterministic due to the choice of constraints, but G' is unambiguous since the resulting $2^{|\mathcal{C}|}$ rules for each left-hand side have mutually exclusive constraint sets. In fact, for each $t \in T_\Sigma$ there is exactly one left-most complete derivation of G' for t, and it derives to $\varphi \in \mathbb{T}^Q$ such that $\varphi_q = h_{\mathrm{wt}^q_G(t)}$ for every $q \in Q$. The weight of that derivation is 1. These statements are proven inductively. The final statement $G'_t = h_{G(t)}$ is an easy consequence of the previous statements. $\qquad \square$

Example 15. Reconsider the WTAc obtained as the disjoint union of the WTAc G and G' of Example 13 as well as the semiring homomorphism $h \in \mathbb{B}^\mathbb{A}$ given by $h_a = 1$ for all $a \in \mathbb{A} \setminus \{-\infty\}$ and $h_{-\infty} = 0$. The set $\mathcal{C}$ of utilized constraints is $\{(1,2),(11,12)\}$ and we write $\varphi \in \mathbb{B}^Q$ simply as subsets of Q. We obtain the unambiguous WTAc G'' with the following (sensible, i.e. having satisfiable constraints) productions for all $Q', Q'' \subseteq \{q, z\}$, which all have weight 1.

$$\alpha \xrightarrow{1 \neq 2, 11 \neq 12} \{q, z\}$$

$$\gamma(Q') \xrightarrow{1=2, 11=12} Q' \cap \{q\} \qquad\qquad \gamma(Q') \xrightarrow{1 \neq 2, 11=12} Q' \cap \{q\}$$

$$\gamma(Q') \xrightarrow{1=2, 11 \neq 12} Q' \qquad\qquad \gamma(Q') \xrightarrow{1 \neq 2, 11 \neq 12} Q'$$

$$\sigma(Q', Q'') \xrightarrow{1=2, 11=12} Q' \cap Q'' \qquad\qquad \sigma(Q', Q'') \xrightarrow{1 \neq 2, 11=12} Q' \cap Q'' \cap \{z\}$$

$$\sigma(Q', Q'') \xrightarrow{1=2, 11 \neq 12} Q' \cap Q'' \qquad\qquad \sigma(Q', Q'') \xrightarrow{1 \neq 2, 11 \neq 12} Q' \cap Q'' \cap \{z\}$$

Each $t \in T_\Sigma$ has exactly one left-most complete derivation in G''; it derives to Q', where (i) $q \in Q'$ iff $t \in \mathrm{supp}(G)$ and (ii) $z \in Q'$ iff $t \in \mathrm{supp}(G')$. $\qquad \square$

Corollary 16 (of Theorem 14). *Let $\mathbb{S}$ be finite. For every WTAc over $\mathbb{S}$ there exists an equivalent unambiguous WTAc.* $\qquad \square$

Corollary 17 (of Theorem 14). *Let $\mathbb{S}$ be zero-sum free. For every WTAc G over $\mathbb{S}$ there exists an unambiguous TAc generating $\mathrm{supp}(G)$.*

Proof. Utilizing Lemma 8 we can first construct an equivalent WTAc with Boolean final weights. If $\mathbb{S}$ is zero-sum free, then there exists a semiring homomorphism $h \in \mathbb{B}^\mathbb{S}$ by [27]. By Lemma 9 we can assume that each derivation of G has non-zero weight and sums of non-zero elements remain non-zero by zero-sum freeness. Thus we can simply replace the factor $h_{\mathrm{wt}(p)}$ by 1 in (1). The such obtained TAc generates $\mathrm{supp}(G)$. $\qquad \square$

Let $A, A' \in \mathbb{S}^{T_\Sigma}$. It is often useful (see [15, Definition 4.11]) to restrict A to the support of A' but without changing the weights of those trees inside the support. Formally, we define $A|_{\mathrm{supp}(A')} \in \mathbb{S}^{T_\Sigma}$ for every $t \in T_\Sigma$ by $A|_{\mathrm{supp}(A')}(t) = A(t)$ if $t \in \mathrm{supp}(A')$ and $A|_{\mathrm{supp}(A')}(t) = 0$ otherwise. Utilizing the unambiguous WTAc and the HADAMARD product, we can show that $A|_{\mathrm{supp}(A')}$ is constraint-regular if A and A' are constraint-regular and the semiring $\mathbb{S}$ is zero-sum free.

Theorem 18. *Let $\mathbb{S}$ be zero-sum free. For all WTAc G and G' there exists a WTAc H such that $H = G|_{\mathrm{supp}(G')}$.*

Proof. By Corollary 10 the support $\mathrm{supp}(G')$ is constraint-regular. Hence we can obtain an unambiguous WTAc G'' for $\mathrm{supp}(G')$ using Theorem 14. Without loss of generality we assume that both G and G'' are constraint-determined; we note that the normalization preserves unambiguous WTAc. Finally we construct $G \times G''$, which by Theorem 12 generates exactly $G|_{\mathrm{supp}(G')}$ as required. $\square$

5   Towards HOM Problem

The strategy of [15] for deciding the HOM problem first represents the homomorphic image L of the regular tree language with the help of an WTGc G. For deciding whether L is regular, a tree automaton G' simulating the behavior of G up to a certain bounded height is constructed. If $G' = G$, then L is regular. If not, pumping arguments are used to prove that it is impossible to find any TA for L. Overall, they reduce the HOM problem to an equivalence problem.

Towards solving the HOM problem in the weighted case we now proceed similarly. First, we show that WTGc can encode each (well-defined) homomorphic image of a regular weighted tree language. This ability motivated their definition in the unweighted case [15, Proposition 4.6], and it also applies in the weighted case with minor restrictions that just enforce that all obtained sums are finite.

Theorem 19. *Let $G = (Q, \Sigma, F, P, \mathrm{wt})$ be a WTA and $h \in T_\Delta^{T_\Sigma}$ be a nondeleting and nonerasing tree homomorphism. There exists a WTGc G' with $G' = h(G)$.*

Proof. We construct a WTGc G' for $h(G)$ in two stages. First, we construct the WTGc $G'' = (Q \cup \{\bot\}, \Delta \cup \Delta \times P, F'', P'', \mathrm{wt}'')$ such that for every production $p = \sigma(q_1, \ldots, q_k) \to q \in P$ and $h_\sigma = u = \delta(u_1, \ldots, u_n)$,

$$p'' = \Big(\langle \delta, p \rangle (u_1, \ldots, u_n) [\![q_1, \ldots, q_k]\!] \xrightarrow{E, \emptyset} q \Big) \in P'' \quad \text{with} \quad E = \bigcup_{i \in [k]} \mathrm{pos}_{x_i}(u)^2$$

where the substitution $\langle \delta, p \rangle (u_1, \ldots, u_n) [\![q_1, \ldots, q_k]\!]$ replaces for every $i \in [k]$ only the left-most occurrence of x_i in $\langle \delta, p \rangle (u_1, \ldots, u_n)$ by q_i and all other occurrences by $\bot$. Moreover, we let $\mathrm{wt}''_{p''} = \mathrm{wt}_p$. Additionally, we let $p''_\delta = \delta(\bot, \ldots, \bot) \to \bot \in P''$ with $\mathrm{wt}''_{p''_\delta} = 1$ for every $k \in \mathbb{N}$ and $\delta \in \Delta_k \cup \Delta_k \times P$. No other productions are in P''. Finally, we let $F''_q = F_q$ for all $q \in Q$ and $F''_\bot = 0$.

We can now delete the annotation. First we remove all productions to $\bot$ that are labeled with symbols from $\Delta \times P$. Second, we use a deterministic relabeling to remove the second components of labels of $\Delta \times P$. Thus, we overall obtain a WTGc G' (using only equality constraints) such that $G' = h(G)$.

The sole purpose of the annotations is to establish a one-to-one correspondence between the valid runs of G and those of G'', before evaluating the sums to compute $h(G)$. This simplifies the understanding of the correctness of the construction, but is otherwise superfluous and may be omitted for efficiency. $\square$

Let us illustrate the construction on a simple example.

Example 20. Consider the WTA $G = (\{q, q'\}, \Sigma, F, P, \mathrm{wt})$ over the semiring $\mathbb{N}$ with $\Sigma = \{\alpha^{(0)}, \phi^{(1)}, \gamma^{(1)}, \epsilon^{(1)}\}$, $F_q = 0$, $F_{q'} = 1$, and the set of productions and their weights given by $p_1 = \alpha \rightarrow_1 q, p_2 = \gamma(q) \rightarrow_2 q, p_3 = \epsilon(q) \rightarrow_1 q$ and $p_4 = \phi(q) \rightarrow_1 q'$. We have $\mathrm{supp}(G) = \{\phi(t) \mid t \in T_{\Sigma \setminus \{\phi\}}\}$ and $G_t = 2^{|\mathrm{pos}_\gamma(t)|}$ for $t \in \mathrm{supp}(G)$. Consider the ranked alphabet $\Delta = \{\alpha^{(0)}, \gamma^{(1)}, \sigma^{(2)}\}$ and the homomorphism h induced by $h_\alpha = \alpha$, $h_\gamma = h_\epsilon = \gamma(x_1)$, and $h_\phi = \sigma(\gamma(x_1), x_1)$. So $\mathrm{supp}(h(G)) = \{\sigma(\gamma^{n+1}(\alpha), \gamma^n(\alpha)) \mid n \in \mathbb{N}\}$ and $h(G)_t = \sum_{k=0}^n \binom{n}{k} 2^k = 3^n$ for every $t = \sigma(\gamma^{n+1}(\alpha), \gamma^n(\alpha)) \in \mathrm{supp}(h(G))$. A WTGc for $h(G)$ is constructed as follows. First, we let $G'' = (\{q, q', \bot\}, \Delta \cup \Delta \times P, F'', P'', \mathrm{wt}'')$ with $F''_{q'} = 1$, $F''_q = F''_\bot = 0$ and the productions and their weights are given by

$$\langle \alpha, p_1 \rangle \rightarrow_1 q \quad \langle \gamma, p_2 \rangle (q) \rightarrow_2 q \quad \langle \gamma, p_3 \rangle (q) \rightarrow_1 q \quad \langle \sigma, p_4 \rangle (\gamma(q), \bot) \xrightarrow{11=2}_1 q'$$

and $\delta(\bot, \ldots, \bot) \rightarrow_1 \bot$ for all $\delta \in \Delta \cup \Delta \times P$. Next we remove the second component of the labels and add weights of productions that become equal. This applies to the production $\gamma(q) \rightarrow q$, which obtains the sum of the two productions (with annotations p_2 and p_3). So we obtain the WTGc $G' = (\{q, q', \bot\}, \Delta, F'', P', \mathrm{wt}')$ with the following productions for all $\delta \in \Delta$.

$$\alpha \rightarrow_1 q \quad \gamma(q) \rightarrow_3 q \quad \sigma(\gamma(q), \bot) \xrightarrow{11=2}_1 q' \quad \delta(\bot, \ldots, \bot) \rightarrow_1 \bot$$

$\square$

Although for zero-sum free semirings, the support of a regular weighted tree language is again regular, in general, the converse is not true, so we cannot apply the decision procedure from [15] to the support of $h(G)$ in order to decide its regularity. Instead, we hope to extend the unweighted argument in a way that tracks the weights sufficiently close. For this, we prepare two decidability results, which rely mostly on the corresponding results in the unweighted case. To this end, we need to relate our WTGc constructed in Theorem 19 to those used in [15]. This requires that the equality constraints in every production refer to positions that occur in its left-hand side and are labeled by the same nonterminal.

Definition 21. *A WTGc $G = (Q, \Sigma, F, P, \mathrm{wt})$ is classic if $\{v, v'\} \subseteq \mathrm{pos}(\ell)$ and $\ell(v) = \ell(v') \in Q$ for every production $\ell \xrightarrow{E,I} q \in P$ and $(v, v') \in E$.* $\square$

Theorem 22. *Let $\mathbb{S}$ be a zero-sum free semiring, $G = (Q, \Sigma, F, P, \mathrm{wt})$ be a WTA and $h \in T_\Delta^{T_\Sigma}$ be a nondeleting and nonerasing tree homomorphism. Finally, let $A = h(G)$. Emptiness and finiteness of $\mathrm{supp}(A)$ are decidable.*

The proof of Theorem 22 applies the corresponding result for the unweighted case. In short, we use Theorem 19 to represent A by a WTGc for which we drop the weights. The resulting TGc representing supp(A) is then modified into an equivalent, classic one. For this, emptiness and finiteness are decidable by [15].

References

1. Bogaert, B., Tison, S.: Equality and disequality constraints on direct subterms in tree automata. In: Finkel, A., Jantzen, M. (eds.) STACS 1992. LNCS, vol. 577, pp. 159–171. Springer, Heidelberg (1992). https://doi.org/10.1007/3-540-55210-3_181
2. Borchardt, B.: The Theory of Recognizable Tree Series. Ph.D. thesis, Technische Universität Dresden (2005)
3. Bozapalidis, S., Rahonis, G.: On the closure of recognizable tree series under tree homomorphisms. J. Autom. Lang. Comb. **10**(2–3), 185–202 (2005)
4. Comon, H., et al.: Tree automata – Techniques and applications (2007)
5. Comon, H., Jacquemard, F.: Ground reducibility and automata with disequality constraints. In: Enjalbert, P., Mayr, E.W., Wagner, K.W. (eds.) STACS 1994. LNCS, vol. 775, pp. 149–162. Springer, Heidelberg (1994). https://doi.org/10.1007/3-540-57785-8_138
6. Dickson, L.E.: Finiteness of the odd perfect and primitive abundant numbers with n distinct prime factors. Amer. J. Math. **35**(4), 413–422 (1913)
7. Doner, J.: Tree acceptors and some of their applications. J. Comput. System Sci. **4**(5), 406–451 (1970)
8. Drewes, F.: Grammatical Picture Generation: A Tree-Based Approach. Springer, Heidelberg (2006). https://doi.org/10.1007/3-540-32507-7
9. Droste, M., Heusel, D.: The supports of weighted unranked tree automata. Funda. Inform. **136**(1–2), 37–58 (2015)
10. Ésik, Z., Kuich, W.: Formal tree series. J. Autom. Lang. Comb. **8**(2), 219–285 (2003)
11. Fülöp, Z., Maletti, A., Vogler, H.: Preservation of recognizability for synchronous tree substitution grammars. In: Proceedings of the Workshop Applications of Tree Automata in Natural Language Processing, pp. 1–9. ACL (2010)
12. Fülöp, Z., Maletti, A., Vogler, H.: Weighted extended tree transducers. Fundamenta Informaticae **111**(2), 163–202 (2011)
13. Fülöp, Z., Vogler, H.: Weighted tree automata and tree transducers. In: Droste, M., Kuich, W., Vogler, H. (eds.) Handbook of Weighted Automata. Monographs in Theoretical Computer Science. An EATCS Series, pp. 313–403. Springer, Heidelberg (2009). https://doi.org/10.1007/978-3-642-01492-5_9
14. Gécseg, F., Steinby, M.: Tree automata. Technical report 1509.06233, arXiv (2015)
15. Godoy, G., Giménez, O.: The HOM problem is decidable. J. ACM **60**(4), 1–44 (2013)
16. Golan, J.S.: Semirings and Their Applications. Kluwer Academic, Dordrecht (1999)
17. Hebisch, U., Weinert, H.J.: Semirings - Algebraic Theory and Applications in Computer Science. World Scientific, Singapore (1998)
18. Jurafsky, D., Martin, J.H.: Speech and Language Processing, 2nd edn. Prentice Hall, Hoboken (2008)
19. Kirsten, D.: The support of a recognizable series over a zero-sum free, commutative semiring is recognizable. Acta Cybernet. **20**(2), 211–221 (2011)

20. Mongy-Steen, J.: Transformation de noyaux reconnaissables d'arbres. Forêts RATEG. Ph.D. thesis, Université de Lille (1981)
21. Perrin, D.: Recent results on automata and infinite words. In: Chytil, M.P., Koubek, V. (eds.) MFCS 1984. LNCS, vol. 176, pp. 134–148. Springer, Heidelberg (1984). https://doi.org/10.1007/BFb0030294
22. Salomaa, A., Soittola, M.: Automata-Theoretic Aspects of Formal Power Series. Springer, New York (1978). https://doi.org/10.1007/978-1-4612-6264-0
23. Schützenberger, M.P.: On the definition of a family of automata. Inf. Control 4(2–3), 245–270 (1961)
24. Thatcher, J.W.: Characterizing derivation trees of context-free grammars through a generalization of finite automata theory. J. Comput. Syst. Sci. 1(4), 317–322 (1967)
25. Thatcher, J.W., Wright, J.B.: Generalized finite automata theory with an application to a decision problem of second-order logic. Math. Syst. Theory 2(1), 57–81 (1968). https://doi.org/10.1007/BF01691346
26. Tison, S.: Tree automata, (dis-)equality constraints and term rewriting: what's new? In: Proceedings of the 22nd International Conference on Rewriting Techniques and Applications. LIPIcs, vol. 10, pp. 1–2. Schloss Dagstuhl – Leibniz-Zentrum für Informatik (2011)
27. Wang, H.: On characters of semirings. Houston J. Math. 23(3), 391–405 (1997)
28. Wilhelm, R., Seidl, H., Hack, S.: Compiler Design. Springer, Heidelberg (2013). https://doi.org/10.1007/978-3-642-17540-4

Performing Regular Operations with 1-Limited Automata

Giovanni Pighizzini[1], Luca Prigioniero[1(✉)] (ID), and Šimon Sádovský[2]

[1] Dipartimento di Informatica, Università degli Studi di Milano,
via Celoria, 18, 20133 Milan, Italy
{pighizzini,prigioniero}@di.unimi.it
[2] Department of Computer Science, Comenius University,
Mlynská Dolina, 842 48 Bratislava, Slovakia
sadovsky@dcs.fmph.uniba.sk

Abstract. The descriptional complexity of basic operations on regular languages using 1-limited automata, a restricted version of one-tape Turing machines, is investigated. When simulating operations on deterministic finite automata with deterministic 1-limited automata, the sizes of the resulting devices are polynomial in the sizes of the simulated machines. The situation is different when the operations are applied on deterministic 1-limited automata: while for boolean operations the simulations remain polynomial, for product, star, and reversal they cost exponential in size. These bounds are tight.

1 Introduction

It is well known that regular languages are recognized by finite automata and are closed under several language operations. When a class of languages benefits of such strong closure properties, it is quite natural to ask how much these operations cost in terms of size of the description of recognizing devices. In this paper we focus on the complexity of union, intersection, complementation, product, star, and reversal. The costs of these operations on deterministic finite automata (1DFAs) have been widely studied in the literature [6,7,15,16], while the case of two-way finite automata (in both deterministic and nondeterministic version) has also been considered [4,5].

In this paper we study the descriptional complexity of language operations on deterministic 1-limited automata (D1-LAs). *Limited automata* are a kind of single-tape Turing machines with rewriting restrictions, introduced by Hibbard in 1967 [2] and recently reconsidered and deeply investigated (see, e.g., [1,8–12,14]). These devices are two-way finite automata with the extra capability of overwriting the contents of each tape cell only in the first d visits, for a fixed constant $d \geq 0$ (we use the name *d-limited automaton* to explicitly mention

The research of Šimon Sádovský was supported, in part, by Slovak Scientific Grant Agency VEGA (Grant 1/0601/20) and by Comenius University in Bratislava (Grant UK/258/2021).

V. Diekert and M. Volkov (Eds.): DLT 2022, LNCS 13257, pp. 239–250, 2022.
https://doi.org/10.1007/978-3-031-05578-2_19

the constant d). For any fixed $d \geq 2$, d-limited automata have the same power as pushdown automata, namely they accept exactly context-free languages [2], while deterministic 2-limited automata recognize exactly the class of deterministic context-free languages [10]. For $d = 0$ no rewritings are possible, hence the resulting models are two-way finite automata. The computational power does not increase if the rewritings in any cell are restricted *only to the first visit*. In other words, 1-limited automata are no more powerful than finite automata [13]. However, their descriptions can be significantly more succinct. In particular, a double exponential size gap between 1-limited automata and one-way deterministic finite automata has been proved in [9], while exponential size gaps have been proved for the conversions from 1-limited automata into one-way nondeterministic finite automata and from deterministic 1-limited automata into one-way deterministic finite automata.

In the study of the descriptional complexity of language operations given a family of recognizers (*source* devices), the goal is the investigation of the size of the devices (*target* devices) accepting the languages obtained by applying some operations to (the languages accepted by) the source devices. Up to now, in the literature it has been analyzed the size of *target* devices of the same family as the *source* devices. However, the results on the succinctness of the description of 1-limited automata suggested us to propose a different approach. Here, for each operation we study, we first take finite automata as source devices, and we simulate the operations on them with 1-limited automata as target devices. We emphasize that we consider *deterministic machines* only. Therefore, we prove that, despite the capabilities of 1-limited automata of rewriting the cells of the tape during the first visit do not make this model more powerful than finite automata, using these machines as target devices for simulating operations between finite automata yields 1-limited automata more succinct than the equivalent finite automata. In fact, if we consider operations between 1DFAs (as source devices), we are able to create D1-LAs accepting the languages obtained by applying such operations that are smaller than the equivalent 1DFAs obtained by using standard constructions [16]. In particular, while the 1DFAs accepting the languages obtained by applying the operations of reversal, product, and star on the languages accepted by 1DFAs cost exponential, the constructions we provided yield equivalent D1-LAs whose sizes are only polynomial in the sizes of the source 1DFAs.

On the other hand, when considering 1-limited automata as source and target devices, the simulations cost polynomial only in the case of union, intersection, and complementation. In the case of reversal, product, and star, however, we were able to find exponential lower bounds witnessing the fact that there is no smaller automaton than the one obtained by converting the simulated D1-LAs into 1DFAs first (obtaining exponentially larger machines), and then applying the corresponding (polynomial-size) language operation construction for obtaining a D1-LA.

2   Preliminaries

We assume the reader familiar with notions from formal languages and automata theory, in particular with *one-way* and *two-way* deterministic finite automata (1DFAs and 2DFAs for short, respectively). For further details see, e.g., [3]. Given a set S, $\#S$ denotes its cardinality and 2^S the family of all its subsets. Given an alphabet Σ, we denote by $|w|$ the length of a string $w \in \Sigma^*$, by w^R the reversal of w, and by ε the empty string. Given two languages $L, L' \subseteq \Sigma^*$, L^c denotes the *complement* of L, L^* denotes the (*Kleene*) *star* of L, L^R denotes the *reversal of* L, and $L \cdot L'$, $L \cup L'$, and $L \cap L'$ denote the *product* (or *concatenation*), *union*, and *intersection* of L and L', respectively (with the usual meaning).

A *deterministic 1-limited automaton* (D1-LA) is a 2DFA which can rewrite the contents of each tape cell in the first visit only. Formally, it is a tuple $\mathcal{A} = (Q, \Sigma, \Gamma, \delta, q_0, F)$, where Q is a finite set of states, Σ is a finite *input alphabet*, Γ is a finite *working alphabet* such that $\Sigma \cup \{\triangleright, \triangleleft\} \subseteq \Gamma$, $\triangleright, \triangleleft \notin \Sigma$ are two special symbols, called the *left* and the *right end-markers*, and $\delta : Q \times \Gamma \to Q \times \Gamma \times \{-1, +1\}$ is the transition function. At the beginning of the computation, the input is stored onto the tape surrounded by the two end-markers, the left end-marker being at the position zero. Hence, on input w, the right end-marker is on the cell in position $|w| + 1$. The head of the automaton is on cell 1 and the state of the finite control is the *initial state* q_0. In one move, according to the transition function and to the current state, $\mathcal{A}$ reads a symbol from the tape, changes its state, replaces the symbol just read from the tape by a new symbol, and moves its head to one position forward or backward. Furthermore, the head cannot pass the end-markers, except at the end of computation, to accept the input, as explained below. However, replacing symbols is allowed to modify the content of each cell only during the first visit (after that, the contents of the cell is said to be *frozen*), with the exception of the cells containing the end-markers, which are never modified. For technical details see [10]. $\mathcal{A}$ accepts an input w if and only if there is a computation path which starts from the initial state q_0 with the input tape containing w surrounded by the two end-markers and the head on the first input cell, and which ends in a *final state* $q \in F$ after passing the right end-marker. It is an easy observation that one can enforce 1-limited automata to always rewrite each cell in the first visit so that they know whether they are scanning the cell for the first time or not.

The *size* of a machine is given by the total number of symbols used to write down its description. Therefore, the size of deterministic 1-limited automata is bounded by a polynomial in the number of states and of working symbols, namely, it is $\Theta(\#Q \cdot \#\Gamma \cdot \log(\#Q \cdot \#\Gamma))$. In the case of deterministic finite automata, since no writings are allowed and hence the working alphabet is not provided, the size is linear in the number of instructions and states, which is bounded by a polynomial in the number of states and in the number of input symbols, namely, it is $\Theta(\#\Sigma \cdot \#Q \cdot \log(\#Q))$.

3   Product and Kleene Star

We start our investigation by studying the operations of product and star. It is known that the costs for these operations on 1DFAs are exponential due to the need of simulating in a deterministic way the nondeterministic choices used for decomposing the input string. However, we show that, using D1-LAs as simulating machines, the costs reduce to polynomials. Then, we analyze the simulations of these operations when the given machines are D1-LAs. In this case, by studying suitable witness languages, we prove that the costs become exponential.

3.1   Simulations of Operations on 1DFAs

We now describe how to obtain a D1-LA $\mathcal{A} = (Q, \Sigma, \Gamma, \delta, q_0, F)$ accepting the concatenation of the languages accepted by two 1DFAs $\mathcal{A}' = (Q', \Sigma, \delta', q_0', F')$ and $\mathcal{A}'' = (Q'', \Sigma, \delta'', q_0'', F'')$, in such a way that the size of $\mathcal{A}$ is polynomial in the sizes of $\mathcal{A}'$ and $\mathcal{A}''$. Let $n' = \#Q'$, $n'' = \#Q''$, $Q' = \{q_0', q_1', \ldots, q_{n'-1}'\}$, and $Q' = \{q_0'', q_1'', \ldots, q_{n''-1}''\}$.

Let us start by briefly recalling how a 1DFA accepting $\mathcal{L}(\mathcal{A}') \cdot \mathcal{L}(\mathcal{A}'')$ can work. It simulates $\mathcal{A}'$ on the whole input word and, every time a final state is entered, it starts a parallel simulation of the automaton $\mathcal{A}''$ on the remaining input suffix. When the end of the input is reached, if some computation of $\mathcal{A}''$ is in a final state, the 1DFA accepts. Since the simulating 1DFA keeps in its finite control, at the same time, a state of $\mathcal{A}'$ and the set of states reached by all the parallel simulations of $\mathcal{A}''$, its size is $\Theta(n' \cdot 2^{n''})$, which is optimal [16].

In our case the goal is to avoid the exponential blowup in size by exploiting the rewriting capability of 1-LAs. To this end, $\mathcal{A}$ still simulates the behavior of $\mathcal{A}'$ by using a state component of size n', and marks the cells from which the simulations of $\mathcal{A}''$ can start, that are the cells next to the ones $\mathcal{A}'$ enters some accepting state. So the simulation can be executed in a sequential rather than parallel way. Moreover, instead of storing the set of states reached by the simulations of $\mathcal{A}''$ in the finite control, $\mathcal{A}$ encodes and writes it along the tape. This information is then accessed, using the ability of 1-LAs of scanning the tape in a two-way fashion, to start and recover the simulations of $\mathcal{A}''$.

In order to encode the set of states reached by the computations of $\mathcal{A}''$, the tape is logically divided into blocks of n'' cells (possibly with a final shorter block). Thus, the i-th cell of each block is marked with ✔ if the state q_i'' is reached by some simulation of $\mathcal{A}''$ ending in the last cell *before* the block, otherwise it is marked with ✗.

The written information is organized into three tracks. In particular, for each frozen cell:

- The first track contains a copy of the input symbol originally contained in the cell before the rewriting, so that it can be still accessed during the simulations of $\mathcal{A}''$;
- The second track contains a marker indicating whether (✔) or not (✗) the automaton $\mathcal{A}'$ has entered an accepting state *right before* reading the cell,

i.e., by reading the input prefix which ends in the cell immediately to the left. So that for any cell containing ✔ a simulation of $\mathcal{A}''$ can be started;
- The third track contains a marker indicating whether (✔) or not (✗) the corresponding states are reachable by some simulation of $\mathcal{A}''$, as explained above.

To make the storing and the recovering of the information about the simulation of the automaton $\mathcal{A}''$ possible while keeping the cost of the simulation polynomial in the size of the simulated devices, the behavior of the simulating 1-LA will be restricted to virtual windows of length $2n''$ that cover two successive blocks of cells. The *right block* covered by a window contains, in some position, the leftmost cell that has not been overwritten so far, to which we refer as relative *frontier*. We refer to the positions relative to the current window as pairs in $\{0, 1, \ldots, n'' - 1\} \times \{\text{L}, \text{R}\}$, where the pairs whose second element is L (resp., R) denote the left (resp., right) block of the window.

We now present some details on how $\mathcal{A}$ recognizes $\mathcal{L}(\mathcal{A}') \cdot \mathcal{L}(\mathcal{A}'')$. The D1-LA stores in its finite control the position of the frontier in the right block of the window, the relative position of the head within the window, and the state of the automaton $\mathcal{A}'$, which is updated every time the cell at the frontier is read. At the beginning of the computation, the simulated state of the automaton $\mathcal{A}'$ is initialized with q_0', and the relative frontier and the relative position both point at position 0 into the right block of the window.

Let us now show how the 1-LA can overwrite each block, cell by cell, with an encoding of the set of states reached by all computations of $\mathcal{A}''$ at the end of the previous block and how it can mark the cells in which the simulations of $\mathcal{A}''$ start. Let (i, R), $i \in \{0, \ldots, n'' - 1\}$, be the position of the frontier. Before visiting the cell in that position, the 1-LA has to gather the information to write in the leftmost cell that has not been rewritten yet. In particular, it has

1. To check whether the simulated automaton $\mathcal{A}'$ accepts the input scanned so far: This can be easily done by using the state component devoted to the simulation of $\mathcal{A}'$ for simulating a move of $\mathcal{A}'$ on the current input symbol and verify whether it enters a state in F'. In that case, $\mathcal{A}$ will write ✔ on the second track, ✗ otherwise.
2. To check whether the state q_i'' can be reached by some computation of $\mathcal{A}''$ before entering the (first cell of the) right block of the current window: This operation is split into two phases. First, the 1-LA starts (from the initial state q_0'') the computations of $\mathcal{A}''$ from each cell of the left block whose second track contains ✔. Then, it recovers, in turn, the computations of $\mathcal{A}''$ from the states indicated in the third track of the cells of the left block, starting from the leftmost position of the window, i.e., relative position $(0, \text{L})$. If, during these two phases, the computation of $\mathcal{A}''$ reaches the state q_i'' after simulating the transition on the symbol in the last cell of the left block, i.e., relative position $(n'' - 1, \text{L})$, the simulating automaton has to write ✔ in the third track, ✗ otherwise.

After gathering this information, the 1-LA moves the head to the frontier, overwrites the cell, and the frontier is moved to the next cell. When the last cell of the window is overwritten, the window shifted forward of one block (i.e., it is shifted $n'' - 1$ cells to the left), so the right block becomes the left one and the frontier points at position $(0, \text{R})$.

When the machine detects the end of the input, indicated by the right end-marker $\lhd$, it has to check whether some simulation of $\mathcal{A}''$ halts in some accepting state. This can be done with the same approach described in Item 2, but the two procedures of the two phases continue the simulations until the last cell of the input rather than stopping in position $(n'' - 1, \text{L})$. The D1-LA accepts if, during the two phases, some state in F'' is reached at the end of the input or if the simulated state of $\mathcal{A}'$ is final and the initial state of $\mathcal{A}''$ is final as well.

By computing the size of the resulting D1-LA $\mathcal{A}$, we are able to state our result on the acceptance of the product of two regular languages (represented by 1DFAs) by a D1-LA.

Theorem 1. *Let* $\mathcal{A}' = (Q', \Sigma, \delta', q'_0, F')$ *and* $\mathcal{A}'' = (Q'', \Sigma, \delta'', q''_0, F'')$ *be two* 1DFA*s. Then there exists a* D1-LA *accepting* $\mathcal{L}(\mathcal{A}') \cdot \mathcal{L}(\mathcal{A}'')$ *with* $O(\#Q' \#Q''^4)$ *states and* $5\#\Sigma + 2$ *working symbols.*

Let us now turn our attention to the star operation. Let $\mathcal{A} = (Q, \Sigma, \delta, q_I, F)$ be a 1DFA. The D1-LA $\mathcal{N}$ for $\mathcal{L}(\mathcal{A})^*$ can implement an approach similar to the one used for the product, so we now illustrate the main differences.

In this case, the only automaton to be simulated is $\mathcal{A}$. The first simulation is started from the leftmost input cell. $\mathcal{N}$ then starts a new simulation every time a (simulated) final state is entered by some simulated computation of $\mathcal{A}$. If, at the end of the input, some simulation reaches a final state, then the 1-LA accepts.

To implement this strategy, the tape of $\mathcal{N}$ is still organized as for the simulation of the product, i.e., it is logically split into blocks of size $\#Q$ and three tracks are used to store a copy of the input, indicating whether or not some simulation of $\mathcal{A}$ has entered an accepting state on the previous cell, and a marker indicating whether or not the corresponding states are reachable by some simulation of $\mathcal{A}$.

Before entering a new cell, $\mathcal{N}$ first checks whether the prefix already visited is in $\mathcal{L}(\mathcal{A})^*$. This is done by recovering the simulations of $\mathcal{A}$ (from the states encoded on the third track) and starting the new ones (from the cells of the second track marked with $\checkmark$), and checking whether some of them reaches a state in F. After that, $\mathcal{N}$ checks whether the state whose index is equal to the index of the frontier (relative to the block) is reached at the end of the previous block by some simulation. Once this information is computed, the automaton moves the head on the cell at the frontier and overwrites it.

When the right endmarker is reached, $\mathcal{N}$ only needs to check whether some simulated device is in a final state and, in that case, accepts.

Theorem 2. *Let* $\mathcal{A} = (Q, \Sigma, \delta, q_0, F)$ *be a* 1DFA. *Then there exists a* D1-LA *accepting* $\mathcal{L}(\mathcal{A})^*$ *with* $O(\#Q^4)$ *states and* $5\#\Sigma + 2$ *working symbols.*

3.2   Simulations of Operations on D1-LAs

We now focus on the size costs of the operations of product and star on D1-LAs. An immediate approach is to convert the source D1-LAs to 1DFAs, and then to apply the constructions shown in the previous section. Since converting D1-LAs into 1DFAs costs exponential in size [9], this procedure yields exponential-size D1-LAs for the two operations we are considering. Here, we show that this strategy cannot be improved, in fact we prove exponential lower bounds for these operations.

For each integer $k \geq 2$, let us consider the language of the strings obtained by concatenating at least two blocks of length k, in which the first and the last blocks are equal: $L_k = \{w\{a, b\}^{kn} w \mid n \geq 0, w \in \{a, b\}^k\}$.

A D1-LA $\mathcal{A}_k$ may recognize L_k as follows. It first scans the leftmost block w of length k of the input, overwriting each symbol with a marked copy. Then, $\mathcal{A}_k$ repeats a subroutine which overwrites any subsequent block of length k, say x, with some fixed symbol $\sharp$, while checking in the meantime whether x equals w or not. This can be achieved as follows. A boolean variable matched is used to keep track of whether or not the prefixes of x and w compared so far match. At the beginning of the inspection of x, the device assigns true to matched, then it iteratively inspects the symbols of x. Suppose that all the symbols to the left of the j-th symbol of x have been inspected and overwritten by $\sharp$. Before inspecting the j-th symbol of x, first, $\mathcal{A}_k$, with the help of a counter modulo k, moves the head leftward to the position j of w and stores the unmarked scanned symbol σ in its finite control; second, it moves the head rightward until reaching the position j of x, namely, the leftmost position that has not been overwritten so far. At this point, $\mathcal{A}_k$ compares the scanned symbol (i.e., the j-th symbol of x) with σ. If the two symbols differ, the machine assigns false to matched. If, after inspecting a block of length k, $\mathcal{A}_k$ detects that the next symbol is the right endmarker, then it stops the computation, accepting in case matched contains true. Otherwise $\mathcal{A}_k$ repeats the subroutine described above in order to inspect the next block.

It is possible to implement $\mathcal{A}_k$ with a number of states linear in k and 7 working symbols (the input symbols and their marked copies, the endmarkers, and the symbol $\sharp$).

Let us now consider the language L_k^2, namely the product of L_k with itself. In this case, the ability of rewriting the tape cell contents of D1-LAs does not come in handy. This is because, ideally, the D1-LA cannot know in advance where to "split" the input string into two parts belonging to L_k. This idea is confirmed by the proof of the following result:

Theorem 3. *For any integer $k \geq 2$,*

- *There exist two D1-LAs $\mathcal{A}'$ and $\mathcal{A}''$ of size linear in k such that any D1-LA accepting $\mathcal{L}(\mathcal{A}') \cdot \mathcal{L}(\mathcal{A}'')$ needs size at least exponential in k.*
- *There exists a D1-LA $\mathcal{A}$ of size linear in k such that any D1-LA accepting $\mathcal{L}(\mathcal{A})^*$ needs size exponential in k.*

Proof. Let us consider the language L_k. Using the approach described above, it is possible to recognize L_k with a D1-LA of size linear in k.

Let us turn our attention to the language $L_k \cdot L_k = L_k^2$. To give a lower bound for the size required by any 1DFA accepting it, we are now going to describe a set of pairwise distinguishable strings for this language. We remind the reader that two strings x, y are *distinguishable* with respect to a language L when there is a string z such that exactly one of the two strings xz and yz belongs to L. The cardinality of each set of strings which are pairwise distinguishable with respect to L gives a lower bound for the number of states of each 1DFA accepting L.

Let us consider the list $x_1, x_2, \ldots, x_N$, with $N = 2^k$, of all the strings in $\{a, b\}^k$ in some fixed order. For each subset $S \subseteq \{1, 2, \ldots, N\}$, we define a string w_S as follows. Let $S = \{i_1, i_2, \ldots, i_n\}$, $1 \leq i_1 < i_2 < \ldots < i_n \leq N$. We define $w_S = x_{i_1} x_{i_1} x_{i_1} x_{i_2} x_{i_1} x_{i_3} x_{i_1} \cdots x_{i_n} x_{i_1}$ if $S \neq \emptyset$, otherwise $w_\emptyset = \varepsilon$. In other words, if S is nonempty, then w_S is the ordered sequence of factors corresponding to the elements of S interleaved with occurrences of x_{i_1}. In particular, x_{i_1} occurs at the beginning of the sequence and after every factor. Now, consider two sets $S, T \subseteq \{1, 2, \ldots, N\}$, with $S \neq T$. Hence, there is a string $x \in \{a, b\}^k$ contained exactly in one of them. Without loss of generality, assume $x \in S$ and $x \notin T$. We prove that $w_S x \in L_k^2$ and $w_T x \notin L_k^2$. Let $x = x_{i_\ell}$. If $\ell > 1$, then $x_{i_1} x_{i_1} x_{i_1} x_{i_2} x_{i_1} \cdots x_{i_{\ell-1}} x_{i_1} \in L_k$ and $x_{i_\ell} x_{i_1} x_{i_{\ell+1}} x_{i_1} \cdots x_{i_n} x_{i_1} x_{i_\ell} \in L_k$. If $\ell = 1$, then $x_{i_1} x_{i_1} \in L_k$ and $x_{i_1} x_{i_2} x_{i_1} \cdots x_{i_n} x_{i_1} x_{i_1} \in L_k$. Hence, in both cases, $w_S x \in L_k^2$. On the other hand, the string $w_T x$ is not in L_k^2 because x does not occur in any other position of w_T. Actually, for the same reason, $w_T x \notin L_k^*$. This observation easily allows to extend our result to the star operation. Hence x distinguishes w_S and w_T with respect to both the languages L_k^2 and L_k^*. Since there are 2^N subsets of $\{1, 2, \ldots, N\}$, each 1DFA accepting $L_k \cdot L_k$ and each 1DFA accepting L_k^* needs at least 2^{2^k} states. Moreover, since the conversion of D1-LAs into 1DFAs costs exponential [9], each D1-LA accepting $L_k \cdot L_k$ and each D1-LA accepting L_k^* has size at least $2^{O(k)}$.   □

In conclusion, starting from two D1-LAs $\mathcal{A}'$ and $\mathcal{A}''$ accepting the languages L' and L'' (resp., from a D1-LA $\mathcal{A}$ accepting a language L), a D1-LA for $L' \cdot L''$ (resp., L^*) can be obtained by converting $\mathcal{A}'$ and $\mathcal{A}''$ (resp., $\mathcal{A}$) into 1DFAs, and then applying the transformation of Theorem 1 (resp., Theorem 2). These constructions are optimal, in fact we proved that the exponential blowup in size due to the conversion into 1DFAs cannot be avoided.

4   Union, Intersection, and Complementation

4.1   Simulations of Operations on 1DFAs

It is well known that for union, intersection, and complement, the simulations are easier than the ones for product and star. Even if the target machines are 1DFAs, it is possible to obtain polynomial-size simulating devices. For union and intersection, the resulting 1DFA is obtained by simulating in parallel the 1DFAs accepting the two given languages. Hence, it has a number of states which is the product of the number of states of the two given 1DFAs. This cannot be improved in the worst case [16].

If we use a 2DFA as target machine, it can perform the simulation of the first 1DFA during a sweep from left to right, then, when the end of the input is reached, the head is brought at the beginning of the tape and the simulation of the second 1DFA is started. In the case of the union, the 2DFA accepts if the simulation of at least one 1DFA accepts, while, in the case of the intersection, the input is accepted if both the simulated 1DFAs accept. The 2DFAs implementing these simulations only need to store, in their state, the copies of the simulated machines, plus one state used to move backward the head at the end of the first simulation. So the total number of states of the simulating devices is 1 plus the sum of the numbers of states of the two simulated 1DFAs.

From the resulting 2DFAs we can directly obtain equivalent D1-LAs that, during the first sweep, simply overwrite each tape cell with a copy of the symbol it originally contains.

Theorem 4. *Let* $\mathcal{A}' = (Q', \Sigma, \delta', q'_0, F')$ *and* $\mathcal{A}'' = (Q'', \Sigma, \delta'', q''_0, F'')$ *be two* 1DFA*s. Then there exist*

- *a* D1-LA *for the language* $\mathcal{L}(\mathcal{A}') \cup \mathcal{L}(\mathcal{A}'')$ *and*
- *a* D1-LA *for the language* $\mathcal{L}(\mathcal{A}') \cap \mathcal{L}(\mathcal{A}'')$

with $\#Q' + \#Q'' + 1$ *states and* $2\#\Sigma + 2$ *working symbols.*

The D1-LA for the complement can be obtained with a construction analogous to the standard one used for obtaining a 1DFA for complementation, i.e., just by complementing the set of the accepting states.

Theorem 5. *Let* $\mathcal{A} = (Q, \Sigma, \delta, q_0, F)$ *be a* 1DFA*. Then there exists one* D1-LA *with* $\#Q$ *states and* $\#\Sigma + 3$ *working symbols which accepts* $\mathcal{L}(\mathcal{A})^c$.

4.2 Simulations of Operations on D1-LAs

Let us now suppose that source and target machines are D1-LAs. We give constructions based on a result on linear-time simulations of 1-LAs in polynomial size: In [1] it is showed that, given a 1-LA, paying a polynomial growth in size it is possible to obtain an equivalent one that works in linear time. The idea of the construction is similar to the technique used for the simulation of the product of Sect. 3: the simulating device works on a virtual window of fixed size that is shifted along the tape in a one-way manner. Along each window it is stored the information useful to simulate the behavior of the 1-LA on the cells to the left of the window without accessing such portion of the tape anymore. In this way, it is possible to bound the number of visits to each cell (for further details we address the reader to [1, Theorem 1 and Lemma 6]).

Lemma 1. *For each* D1-LA $\mathcal{A} = (Q, \Sigma, \Gamma, \delta, q_0, F)$ *there exists an equivalent* D1-LA $\mathcal{A}'$ *working in linear time with* $O(\#Q^4)$ *states and* $(\#Q + 1) \cdot \#(\Gamma \setminus \Sigma)$ *working symbols.*

For the simulation of union and intersection of the languages accepted by two D1-LAs, the machines are simulated in parallel. In particular, two (possibly different) virtual windows are used and shifted independently. Before entering a new cell, the simulating device computes the information about the windows of the simulated D1-LAs (in this phase, only the cells of the two windows are visited: it is used the window of the first simulated device and then, when the information has been gathered, the window of the second simulated device is used). Then the new cell is entered and the information is written (on two tracks of the tape), together with the symbols written by the simulated devices (on two extra tracks).

When the end of the input is reached, in the case of the union the simulating device accepts if at least one simulation accepts, and in the case of the intersection it accepts if both the simulated devices accept.

Theorem 6. *Let* $\mathcal{A}' = (Q', \Sigma, \Gamma', \delta', q_0', F')$ *and* $\mathcal{A}'' = (Q'', \Sigma, \Gamma', \delta'', q_0'', F'')$ *be two* D1-LA*s,* $n' = \#Q'$, *and* $n'' = \#Q''$. *Then there exist*

- *a* D1-LA *for the language* $\mathcal{L}(\mathcal{A}') \cup \mathcal{L}(\mathcal{A}'')$ *and*
- *a* D1-LA *for the language* $\mathcal{L}(\mathcal{A}') \cap \mathcal{L}(\mathcal{A}'')$

with $O(n'^4 n''^4)$ *states and* $(n'+1)(n''+1)\#(\Gamma' \setminus \Sigma)\#(\Gamma'' \setminus \Sigma)$ *working symbols.*

To accept the complement of the language accepted by a D1-LA $\mathcal{A}$, again Lemma 1 can be used to perform a linear-time (and therefore, halting) simulation of $\mathcal{A}$. The simulating D1-LA accepts if $\mathcal{A}$ enters a loop or if it is not in an accepting state at the end of its computation.

Theorem 7. *Let* $\mathcal{A} = (Q, \Sigma, \delta, q_0, F)$ *be a* D1-LA. *Then there exists a* D1-LA *with* $O(\#Q^4)$ *states and* $(\#Q+1)\#(\Gamma' \setminus \Sigma)$ *working symbols which accepts* $\mathcal{L}(\mathcal{A})^c$.

5   Reversal

The last operation we study is the reversal. Even in this case, the D1-LA for the reversal of the language accepted by a 1DFA $\mathcal{A}$ can be obtained by exploiting just the capability of the simulating machine of scanning the input in a two-way fashion, so, again, we first give our result for 2DFAs. Roughly, starting from the initial state of $\mathcal{A}$ with the head positioned on the last symbol of the input word, it accepts if, simulating the transitions of the 1DFA scanning the input from right to left, enters a final state when the head reaches the left endmarker. This approach yields a 2DFA with a number of states equal to the one of the simulated machine, plus two states for adjusting the position of the head along the tape at the beginning and at the end of the computation.

As a consequence, we are able to construct an equivalent D1-LA that uses the same strategy of the obtained 2DFA, with the only difference that, during the first sweep from left to right, it rewrites on each cell a copy of the symbol it scans.

Theorem 8. *Let $\mathcal{A} = (Q, \Sigma, \delta, q_0, F)$ be a* 1DFA. *Then there exists one* D1-LA *with* $\#Q + 2$ *states and* $2\#\Sigma + 2$ *working symbols which accepts* $\mathcal{L}(\mathcal{A})^R$.

In the case of D1-LAs, the reversal has an exponential cost in size. The exponential upper bound can be obtained by converting the D1-LA into a 1DFA and then applying Theorem 8. A matching exponential lower bound has been proved in [1].

Theorem 9 ([1, Theorem 4]). *For any integer $k \geq 2$, there exists a* D1-LA $\mathcal{A}$ *of size linear in k such that any* D1-LA *accepting* $\mathcal{L}(\mathcal{A})^R$ *needs size exponential in k.*

References

1. Guillon, B., Prigioniero, L.: Linear-time limited automata. Theor. Comput. Sci. **798**, 95–108 (2019)
2. Hibbard, T.N.: A generalization of context-free determinism. Inf. Control **11**(1/2), 196–238 (1967)
3. Hopcroft, J., Ullman, J.: Introduction to Automata Theory, Languages, and Computation. Addison-Wesley, Reading (1979)
4. Jirásková, G., Okhotin, A.: On the state complexity of operations on two-way finite automata. Inf. Comput. **253**, 36–63 (2017). https://doi.org/10.1016/j.ic.2016.12.007
5. Kunc, M., Okhotin, A.: State complexity of union and intersection for two-way non-deterministic finite automata. Fundamenta Informaticae **110**(1–4), 231–239 (2011). https://doi.org/10.3233/FI-2011-540
6. Leiss, E.L.: Succinct representation of regular languages by Boolean automata. Theor. Comput. Sci. **13**, 323–330 (1981). https://doi.org/10.1016/S0304-3975(81)80005-9
7. Maslov, A.N.: Estimates of the number of states of finite automata. In: Doklady Akademii Nauk, vol. 194, pp. 1266–1268. Russian Academy of Sciences (1970)
8. Pighizzini, G.: Limited automata: properties, complexity and variants. In: Hospodár, M., Jirásková, G., Konstantinidis, S. (eds.) DCFS 2019. LNCS, vol. 11612, pp. 57–73. Springer, Cham (2019). https://doi.org/10.1007/978-3-030-23247-4_4
9. Pighizzini, G., Pisoni, A.: Limited automata and regular languages. Int. J. Found. Comput. Sci. **25**(7), 897–916 (2014)
10. Pighizzini, G., Pisoni, A.: Limited automata and context-free languages. Fundam. Inform. **136**(1–2), 157–176 (2015)
11. Pighizzini, G., Prigioniero, L.: Limited automata and unary languages. Inf. Comput. **266**, 60–74 (2019)
12. Pighizzini, G., Prigioniero, L., Sádovský, Š.: 1-limited automata: witness languages and techniques. J. Autom. Lang. Comb. (2022, to appear)
13. Wagner, K.W., Wechsung, G.: Computational Complexity. D. Reidel Publishing Company, Dordrecht (1986)
14. Yamakami, T.: Behavioral strengths and weaknesses of various models of limited automata. In: Catania, B., Královič, R., Nawrocki, J., Pighizzini, G. (eds.) SOFSEM 2019. LNCS, vol. 11376, pp. 519–530. Springer, Cham (2019). https://doi.org/10.1007/978-3-030-10801-4_40

15. Yu, S., Zhuang, Q.: On the state complexity of intersection of regular languages. SIGACT News **22**(3), 52–54 (1991). https://doi.org/10.1145/126537.126543
16. Yu, S., Zhuang, Q., Salomaa, K.: The state complexities of some basic operations on regular languages. Theor. Comput. Sci. **125**(2), 315–328 (1994). https://doi.org/10.1016/0304-3975(92)00011-F

Binomial Complexities and Parikh-Collinear Morphisms

Michel Rigo⑩, Manon Stipulanti$^{(\boxtimes)}$⑩, and Markus A. Whiteland⑩

Department of Mathematics, University of Liège, Liège, Belgium
{m.rigo,m.stipulanti,mwhiteland}@uliege.be

Abstract. Inspired by questions raised by Lejeune, we study the relationships between the k and $(k+1)$-binomial complexities of an infinite word; as well as the link with the usual factor complexity. We show that pure morphic words obtained by iterating a Parikh-collinear morphism, i.e., a morphism mapping all words to words with bounded abelian complexity, have bounded k-binomial complexity. We further study binomial properties of the images of aperiodic binary words in general, and Sturmian words in particular, by a power of the Thue–Morse morphism.

Keywords: Binomial complexity · Powers of Thue–Morse morphism · Morphic words

1 Introduction

When interested in the combinatorial structure of an infinite word $\mathbf{x}$ over a finite alphabet A, it is often useful to study its language $\mathcal{L}(\mathbf{x})$, i.e., the set of its factors, and in particular to inspect factors of a given length n. We let $\mathcal{L}_n(\mathbf{x})$ denote $\mathcal{L}(\mathbf{x}) \cap A^n$. The usual *factor complexity* function $\mathsf{p}_{\mathbf{x}} \colon \mathbb{N} \to \mathbb{N}$ counts the number $\#\mathcal{L}_n(\mathbf{x})$ of words of length n occurring in $\mathbf{x}$. This is a highly useful notion: for instance, ultimately periodic words are characterized by a bounded factor complexity and Sturmian words are exactly those words satisfying $\mathsf{p}_{\mathbf{x}}(n) = n+1$ for all n [1, §10]. However, to highlight particular combinatorial properties of the infinite word of interest, other complexity functions such as abelian [16], k-abelian [8], cyclic [2], privileged [15], and k-binomial [17] complexities have been introduced. In most cases, one considers the quotient of the language $\mathcal{L}(\mathbf{x})$ by a convenient equivalence relation $\sim$ and the corresponding complexity function therefore maps $n \in \mathbb{N}$ to $\#(\mathcal{L}_n(\mathbf{x})/\sim)$. For instance, a binary (non-periodic) word is balanced if and only if its abelian complexity is equal to the constant function 2. This paper focuses on the binomial complexity introduced in [17] and which is the central theme of Lejeune's thesis [9]. The notion is based on the binomial equivalence relations, which have both theoretical and practical importance in the sciences (see, e.g., [5] and [4] and references therein).

M. Stipulanti and M. Whiteland—Supported by the FNRS Research grants 1.B.397.20F and 1.B.466.21F respectively. M. Whiteland dedicates this paper to the memory of his father Alan Whiteland (1940–2021).

© Springer Nature Switzerland AG 2022
V. Diekert and M. Volkov (Eds.): DLT 2022, LNCS 13257, pp. 251–262, 2022.
https://doi.org/10.1007/978-3-031-05578-2_20

1.1 Binomial Coefficients and Complexity Functions

General references about word combinatorics can be found in [1,12]. For any integer $k \geq 1$, we let A^k (resp., $A^{\leq k}$; resp., $A^{<k}$) denote the set of words of length exactly (resp., at most; resp., less than) k over A. We use A^* (resp., A^+) for the semi-group of finite words (resp., non-empty finite words) over A equipped with concatenation. We let ε denote the empty word. The length of a word w is denoted by $|w|$, and the number of occurrences of a letter a in w is denoted by $|w|_a$. Writing $A = \{a_1, \ldots, a_k\}$ and fixing the order $a_1 < a_2 < \cdots < a_k$ on the letters, the *Parikh vector* of a word $w \in A^*$ is defined as the column vector $\Psi(w) = (|w|_{a_1}, |w|_{a_2}, \ldots, |w|_{a_k})^\mathsf{T}$. Let $u, w \in A^*$. The *binomial coefficient* $\binom{u}{w}$ of u and w is the number of times w occurs as a subsequence of u, i.e., writing $u = u_1 \cdots u_n$ with $u_i \in A$,

$$\binom{u}{w} = \# \left\{ i_1 < i_2 < \cdots < i_{|w|} : u_{i_1} u_{i_2} \cdots u_{i_{|w|}} = w \right\}.$$

By convention, $\binom{u}{\varepsilon} = 1$. See, e.g., [12, Sect. 6] for more on binomial coefficients. We make the important distinction between a *factor* and a *subword* of a word. The former is a contiguous subsequence of a word, while the latter is just a subsequence. Let $k \geq 1$ be an integer. Two words $u, v \in A^*$ are *k-binomially equivalent*, and we write $u \sim_k v$, if $\binom{u}{x} = \binom{v}{x}$ for all $x \in A^{\leq k}$. Observe that the word u is obtained as a permutation of the letters in v if and only if $u \sim_1 v$. In this case, we say that u and v are *abelian equivalent*.

Definition 1. *Let $k \geq 1$ be an integer. The k-binomial complexity function of an infinite word $\mathbf{x}$ is defined as $b_{\mathbf{x}}^{(k)} : \mathbb{N} \to \mathbb{N}$, $n \mapsto \#(\mathcal{L}_n(\mathbf{x})/\sim_k)$.*

For all $k \geq 1$, $u \sim_{k+1} v$ implies $u \sim_k v$. Thus, for all n, we have

$$b_{\mathbf{x}}^{(1)}(n) \leq b_{\mathbf{x}}^{(2)}(n) \leq \cdots \leq b_{\mathbf{x}}^{(k)}(n) \leq b_{\mathbf{x}}^{(k+1)}(n) \leq \cdots \leq p_{\mathbf{x}}(n). \tag{1}$$

The k-binomial complexity function has been studied for particular infinite words: for $k \geq 2$, the k-binomial complexity of Sturmian words coincides with their factor complexity [17] and the same property holds true for the Tribonacci word [11]. Recently, the 2-binomial complexity of generalized Thue–Morse words was also computed [13]. The k-binomial complexity of the Thue–Morse word $\mathbf{t}$, the fixed point of the morphism $0 \mapsto 01$, $1 \mapsto 10$, is bounded by a constant (depending on k) [10], and more generally bounded k-binomial complexity holds for any fixed point of a prolongable Parikh-constant morphism f [17], i.e., $\Psi(f(a)) = \Psi(f(b))$ for all letters a, b.

1.2 Questions Addressed in This Paper

In this work, we generalize the above property of the fixed points of Parikh-constant morphisms to what we call *Parikh-collinear* morphisms f: for all letters a, b, there is a rational number $r_{a,b}$ such that $\Psi(f(a)) = r_{a,b}\Psi(f(b))$.

Such morphisms were characterized in [3]; see Theorem 9. In Sect. 3, we provide a new characterization of these morphisms in terms of the binomial complexity: they map all words with bounded k-binomial complexity to words with bounded $(k+1)$-binomial complexity. Finally, Corollary 11 shows that fixed points of Parikh-collinear morphisms have bounded k-binomial complexity.

For all $j \geq 1$, the exact value of $b_t^{(j)}(n)$ computed in [10] is given by

$$b_t^{(j)}(n) = \begin{cases} p_t(n) & \text{if } n \leq 2^j - 1; \\ 3 \cdot 2^j - 3, & \text{if } n \equiv 0 \pmod{2^j} \text{ and } n \geq 2^j; \\ 3 \cdot 2^j - 4, & \text{otherwise.} \end{cases} \tag{2}$$

We show in Theorem 14 that such a behavior is not specific to t, but appears for a large class of words. More precisely, let φ be the Thue–Morse morphism. For any aperiodic binary word $\mathbf{y}$, the word $\mathbf{x} = \varphi^k(\mathbf{y})$ is such that $b_{\mathbf{x}}^{(j)}(n) = b_t^{(j)}(n)$ for all $j \leq k$ and $n \geq 2^j$.

In general, not much is known about the general behavior or the properties that can be expected for the k-binomial complexity. In particular, computing the k-binomial complexity of a particular infinite word remains challenging. It would also be desirable to compare in some ways k and $(k+1)$-binomial complexities of a word. For two functions $f, g \colon \mathbb{N} \to \mathbb{N}$, we write $f \prec g$ when $f(n) \leq g(n)$ for all n and $f(n) < g(n)$ holds for infinitely many $n \in \mathbb{N}$. Our reflection is driven by the following questions inspired by Lejeune's questions [9, pp. 115–117] that are natural to consider in view of (1).[1]

Question A. *Does there exist an infinite word* $\mathbf{w}$ *such that, for all* $k \geq 1$, $b_{\mathbf{w}}^{(k)}$ *is unbounded and* $b_{\mathbf{w}}^{(k)} \prec b_{\mathbf{w}}^{(k+1)}$? *If yes, is there a (pure) morphic such word* $\mathbf{w}$?

From (1), notice that $b_{\mathbf{w}}^{(k)}$ is unbounded for all $k \geq 1$ if and only if the abelian complexity $b_{\mathbf{w}}^{(1)}$ is unbounded. Even though the Thue–Morse word t has $b_t^{(k)} \prec b_t^{(k+1)}$ for all $k \geq 1$, $b_t^{(k)}$ remains bounded (recall (2)). So t is not a satisfying answer to Question A. It is however easy to check that the binary Champernowne word c, that is, the concatenation of the binary representations of the non-negative integers, has the required property of Question A. In Sect. 2, we provide several, more structured, words having this property. (The content of Sect. 3 was discussed above.)

Section 4 is about binomial properties of powers of φ. Going further than (2), we also study the $(k+1)$- and $(k+2)$-binomial complexity of words of the form $\mathbf{x} = \varphi^k(\mathbf{y})$ with $\mathbf{y}$ aperiodic. In Sect. 4.1 we prove Theorem 14 discussed above. In Sect. 4.2, we characterize $(k+1)$-binomial equivalence in $\mathbf{x}$ with Proposition 22. As a consequence we obtain $b_{\mathbf{x}}^{(k)} \prec b_{\mathbf{x}}^{(k+1)}$. These considerations are motivated by the question of whether it is possible that the factor complexity coincides with $b^{(k)}$ for some k (dismissing the trivial cases of periodic words).

---

[1] We define $\prec$ deliberately with "infinitely many" rather than "all large enough": for the period-doubling word $\mathbf{pd}$ (fixed point of $0 \mapsto 01$, $1 \mapsto 00$) there exist infinitely many n and m such that $b_{\mathbf{pd}}^{(2)}(n) = p_{\mathbf{pd}}(n)$ and $b_{\mathbf{pd}}^{(2)}(m) < p_{\mathbf{pd}}(m)$ [9, Prop. 4.5.1].

Question B. *For each $\ell \geq 1$, does there exist a word $\mathbf{w}_\ell$ such that $b_{\mathbf{w}_\ell}^{(1)} \prec b_{\mathbf{w}_\ell}^{(2)} \prec \cdots \prec b_{\mathbf{w}_\ell}^{(\ell-1)} \prec b_{\mathbf{w}_\ell}^{(\ell)} = p_{\mathbf{w}_\ell}$? If yes, is there a (pure) morphic such word $\mathbf{w}_\ell$?*

Putting together results from Sects. 4 and 5 we answer Question B: Theorem 14 and Proposition 22 provide a word $\mathbf{x}$ for which $b_{\mathbf{x}}^{(1)} \prec b_{\mathbf{x}}^{(2)} \prec \cdots \prec b_{\mathbf{x}}^{(k-1)} \prec b_{\mathbf{x}}^{(k)} \prec b_{\mathbf{x}}^{(k+1)}$, while assuming that $\mathbf{y}$ above is Sturmian, we show that $b_{\mathbf{x}}^{(k+2)} = p_{\mathbf{x}}$. We mention that powers of φ applied to Sturmian words are studied (among other words) in [6]. Our construction leads to words with bounded abelian complexity. Question B is then strengthened in Sect. 5 where we ask for words with unbounded abelian complexity. We give a pure morphic answer when $\ell = 3$.

This paper is an extended abstract of the preprint [18] which is 26-page long. Due to the page limitation, here most of the proofs, auxiliary results, and remarks have been omitted. However, the short and long versions have similar structures; we hope this helps the reader to navigate between the two versions.

2   Several Answers to Question A

We collect a minimal amount of results on k-binomial equivalence. First note that $\sim_k$ is a congruence, i.e., for $u, v, x, y \in A^*$, $u \sim_k v$ and $x \sim_k y$ implies $ux \sim_k vy$. Second, we have a *cancellation property* [10, Lem. 10]: for $u, v, w \in A^*$, we have $vu \sim_k wu \Leftrightarrow v \sim_k w \Leftrightarrow uv \sim_k uw$. Finally, let $k \geq 2$ and $x, y \in A^*$ such that $|x| = |y|$. Then $xy \sim_k yx$ if and only if $x \sim_{k-1} y$ [19, Theorem 3.5]. For the next results, see [14], [10, Lemmas 30 and 31].

Theorem 2. *Let $\varphi \colon 0 \mapsto 01, 1 \mapsto 10$ be the Thue–Morse morphism. For all $k \geq 1$, we have $\varphi^k(0) \sim_k \varphi^k(1)$ and $\varphi^k(0) \not\sim_{k+1} \varphi^k(1)$.*

Lemma 3 (Transfer lemma). *Let $k \geq 1$. Let u, v, v' be three non-empty words such that $|v| = |v'|$. We have $\varphi^{k-1}(u)\varphi^k(v) \sim_k \varphi^k(v')\varphi^{k-1}(u)$.*

Observe that the Champernowne word, already mentioned in the introduction, is not morphic, nor is it uniformly recurrent. In the rest of the section we provide more "structured" words answering Question A. Let φ be the Thue–Morse morphism and define the morphism g by $a \mapsto a0\alpha$, $0 \mapsto \varphi(0)$, $1 \mapsto \varphi(1)$, $\alpha \mapsto \alpha^2$. Let $\mathbf{g} = g^\omega(a) := \lim_{n\to\infty} g^n(a) = a \prod_{j=0}^\infty \varphi^j(0)\alpha^{2^j}$.

Proposition 4. *We have that $b_{\mathbf{g}}^{(1)}$ is unbounded and $b_{\mathbf{g}}^{(k)} \prec b_{\mathbf{g}}^{(k+1)}$ for all $k \geq 1$.*

Proof. The first claim follows from the fact that $\{|u|_\alpha : u \in \mathcal{L}_n(\mathbf{g})\} = \{0, \ldots, n\}$. For the second claim, let k be fixed and take $u_n = \varphi^k(0)\alpha^n$ and $v_n = \varphi^k(1)\alpha^n$ for each $n \in \mathbb{N}$; we have $u_n \sim_k v_n$ but $u_n \not\sim_{k+1} v_n$ by Theorem 2. Consequently $b_{\mathbf{g}}^{(k)} \prec b_{\mathbf{g}}^{(k+1)}$ for all $k \geq 1$. $\qquad\square$

We modify the above construction to obtain a morphic binary word.

Proposition 5. *With the coding $\tau : a \mapsto \varepsilon, 0 \mapsto 0, 1 \mapsto 1, \alpha \mapsto 1$, we have that $b_{\tau(\mathbf{g})}^{(1)}$ is unbounded and $b_{\tau(\mathbf{g})}^{(k)} \prec b_{\tau(\mathbf{g})}^{(k+1)}$ for all $k \geq 1$.*

We note that none of the above words are *uniformly recurrent* (a word **x** is uniformly recurrent if for each $x \in \mathcal{L}(\mathbf{x})$ there exists $N \in \mathbb{N}$ such that x appears in all factors in $\mathcal{L}_N(\mathbf{x})$). We recall a particular construction from Grillenberger [7] for uniformly recurrent words having arbitrary entropy. The word of interest is constructed as follows. Define $D_0 = \{0, 1\}$. Assuming D_k is constructed, let u_k be the product of words of D_k in lexicographic order, assuming $0 < 1$. Define then $D_{k+1} := u_k D_k^2$. Now the sequence $(u_k)_{k \in \mathbb{N}}$ converges to a uniformly recurrent word $\mathbf{u} = 0100010101100111 \cdots$.

Lemma 6. *Let $k \geq 1$. If, for some $j \geq 0$, D_j contains two words u, v, such that $u \sim_k v$ and $u \not\sim_{k+1} v$, then D_{j+1} contains words x, y, z and w such that $x \sim_k y$ but $x \not\sim_{k+1} y$; $z \sim_{k+1} w$ but $z \not\sim_{k+2} w$.*

Proposition 7. *The map $\mathsf{b}_{\mathbf{u}}^{(1)}$ is unbounded and $\mathsf{b}_{\mathbf{u}}^{(k)} \prec \mathsf{b}_{\mathbf{u}}^{(k+1)}$ for all $k \geq 1$.*

Proof. First we show that $\mathsf{b}_{\mathbf{u}}^{(1)}$ is unbounded. Assume, for some $j \geq 0$, that D_j contains words u, v with $|u|_0 - |v|_0 = 2^j$ (this holds for $j = 0$). Then by definition D_{j+1} contains the words $x = u_j uu$ and $y = u_j vv$, for which $|x|_0 - |y|_0 = 2(|u|_0 - |v|_0) = 2^{j+1}$. This observation suffices for the claim.

We then prove the second part of the statement. Observe that D_1 contains the words 0101 and 0110, which are abelian equivalent, but not 2-binomially equivalent (as $\binom{0101}{01} = 3$ and $\binom{0110}{01} = 2$). The above lemma then implies that for all $k \geq 1$ and for all $j \geq k$, the set D_j contains words that are k-binomially equivalent, but not $(k + 1)$-binomially equivalent. The claim follows. □

3   An Interlude: Parikh-Collinear Morphisms

In this section, we show that, given an infinite fixed point of a prolongable Parikh-collinear morphism, its k-binomial complexity is bounded for each k.

Definition 8 (Parikh-collinear morphisms). *A morphism $f \colon A^* \to B^*$ is Parikh-collinear if, for all $a, b \in A$, $\Psi(f(b)) = r_{a,b}\Psi(f(a))$ for some $r_{a,b} \in \mathbb{Q}$.*

Theorem 9 ([3, Theorem 11]). *A morphism $f \colon A^* \to B^*$ maps all infinite words to words with bounded abelian complexity if and only if it is Parikh-collinear.*

We extend the above theorem to the following one, where 0-*binomial complexity* has to be understood as the trivial complexity function corresponding to the "equal length" equivalence relation.

Theorem 10. *A morphism $f \colon A^* \to B^*$ maps, for all $k \geq 0$, all words with bounded k-binomial complexity to words with bounded $(k + 1)$-binomial complexity if and only if it is Parikh-collinear.*

Before proving this result in Sect. 3.2, let us mention a straightforward consequence, which generalizes [17, Theorem 13] from Parikh-constant to Parikh-collinear morphisms.

Corollary 11. *Let* $\mathbf{z}$ *be a fixed point of a Parikh-collinear morphism. For any* $k \geq 1$ *there exists a constant* $C_{\mathbf{z},k} \in \mathbb{N}$ *such that* $b_{\mathbf{z}}^{(k)}(n) \leq C_{\mathbf{z},k}$ *for all* $n \in \mathbb{N}$.

Proof. Let $f \colon A^* \to A^*$ be a Parikh-collinear morphism whose fixed point is $\mathbf{z}$. Since $f(\mathbf{z}) = \mathbf{z}$, Theorem 9 implies that $\mathbf{z}$ has bounded abelian complexity. For any $k \geq 1$, we have that $\mathbf{z} = f(f^{k-1}(\mathbf{z}))$ implying that $\mathbf{z}$ has bounded k-binomial complexity by induction and the previous theorem. $\qquad\square$

3.1 A Characterization of Parikh-Collinear Morphisms

In proving Theorem 10, we characterize Parikh-collinear morphisms as follows.

Proposition 12. *Let* $f \colon A^* \to B^*$ *be a morphism. The following are equivalent.*

(i) For all $k \geq 2$ *and* $u, v \in A^*$, $u \sim_{k-1} v$ *implies* $f(u) \sim_k f(v)$.
(ii) There exists an integer $k \geq 2$ *such that for all* $u, v \in A^*$, $u \sim_{k-1} v$ *implies* $f(u) \sim_k f(v)$.
(iii) For all $u, v \in A^*$, $u \sim_1 v$ *implies* $f(u) \sim_2 f(v)$.
(iv) f *is Parikh-collinear.*

3.2 Proof of Theorem 10

The next technical result can be extracted from the proof of [3, Theorem 12].

Lemma 13. *Let* $\mathbf{x}$ *be a an infinite word over A with bounded abelian complexity. Let* $f \colon A^* \to B^*$ *be a morphism and assume* $\mathbf{y} = f(\mathbf{x})$ *is an infinite word. Then for all* $c \in \mathbb{N}$ *there exists* $D_{\mathbf{x},c} \in \mathbb{N}$ *such that if* $\big||f(u)| - |f(v)|\big| \leq c$, *for some* $u, v \in \mathcal{L}(\mathbf{x})$, *then* $\big||u| - |v|\big| \leq D_{\mathbf{x},c}$.

Proof (of Theorem 10). If $f \colon A \to B^*$ maps all words with bounded 0-binomial complexity (i.e., all words) to words with bounded 1-binomial complexity, then f is Parikh-collinear by Theorem 9.

Assume thus that f is Parikh-collinear. Theorem 9 implies that f maps all words (i.e., all words with bounded 0-binomial complexity) to words with bounded 1-binomial complexity. Let then $k \geq 1$ and let $\mathbf{x}$ be a word with bounded k-binomial complexity. Let $n \in \mathbb{N}$. Any length-n factor of $f(\mathbf{x})$ can be written as $pf(u)s$, where the word u is a factor of $\mathbf{x}$, p is a suffix of $f(a)$ and s is a prefix of $f(b)$ for some letters $a, b \in A$. Here $n - 2m < |f(u)| \leq n$, where $m := \max_{a \in A} |f(a)|$. The $(k+1)$-binomial equivalence class of $pf(u)s$ is completely determined by the words p, s, and the k-binomial equivalence class of $f(u)$, which itself is determined by the k-binomial equivalence class of u by Proposition 12.

The former two words p and s are drawn from a finite set, as their lengths are bounded by the constant m (depending on f). The length of u can be chosen from an interval whose length is uniformly bounded in n. Indeed, assume we have equal length factors $w = pf(u)s$ and $w' = p'f(v)s'$. As observed above, $n \geq |f(u)|$ and $|f(v)| > n - 2m$, so that $\big||f(u)| - |f(v)|\big| < 2m$.

Applying Lemma 13 (by assumption, $\mathbf{x}$ has bounded k-binomial complexity and thus, $\mathbf{x}$ has bounded abelian complexity by (1)) there exists a bound D such that $\big||u| - |v|\big| \leq D$ uniformly in n. Since the number of k-binomial equivalence classes in $\mathbf{x}$ of each length is uniformly bounded by assumption, and the number of admissible lengths for u above is bounded, we conclude that the number of choices for the k-binomial equivalence class of u is bounded. We have shown that the number of $(k+1)$-binomial equivalence classes among factors of length n in $f(\mathbf{x})$ is determined from a bounded amount of information (not depending on n), as was to be shown. $\square$

4   Binomial Properties of the Thue–Morse Morphism

In this section, we consider binomial complexities of powers of the Thue-Morse morphism φ on aperiodic binary words. Repeated application of Theorem 10 shows that, for any $k \geq 1$ and any binary word $\mathbf{y}$, the k-binomial complexity function of the word $\varphi^k(\mathbf{y})$ is bounded. We will prove the sharper result:

Theorem 14. *Let j, k be integers with $1 \leq j \leq k$ and let $\mathbf{y}$ be an aperiodic binary word. Let $\mathbf{x} = \varphi^k(\mathbf{y})$. For all $n \geq 2^j$, we have $\mathsf{b}_{\mathbf{x}}^{(j)}(n) = \mathsf{b}_{\mathbf{t}}^{(j)}(n)$ which is given by (2) and, for $n < 2^j$, $\mathsf{b}_{\mathbf{x}}^{(j)}(n) = \mathsf{p}_{\mathbf{x}}(n)$.*

This is a generalization of [10, Theorem 6], which says that, for all $j \geq 1$, the j-binomial complexity of the Thue–Morse word $\mathbf{t}$ is given by (2). Notice it implies that $\mathsf{b}_{\mathbf{x}}^{(1)} \prec \mathsf{b}_{\mathbf{x}}^{(2)} \prec \cdots \prec \mathsf{b}_{\mathbf{x}}^{(k)}$. The aim of Sect. 4.2 is to go one step further and get $\mathsf{b}_{\mathbf{x}}^{(k)} \prec \mathsf{b}_{\mathbf{x}}^{(k+1)}$. To do so, we characterize k-binomial and $(k+1)$-binomial equivalence among factors of $\mathbf{x}$ (Theorem 19 and Proposition 22).

4.1   The First k Binomial Complexities

Before proving Theorem 14, we require the following general lemma about aperiodic binary words. There, we let $\overline{}$ denote the complementation morphism defined by $\overline{a} = 1 - a$, for $a \in \{0, 1\}$.

Lemma 15. *Let $\mathbf{z}$ be an aperiodic binary word. Then for all $n \geq 2$ we have $\mathcal{L}_n(\mathbf{z}) \cap L \neq \emptyset$ for each $L \in \{0A^*1, 1A^*0, 0A^*0 \cup 1A^*1\}$. Furthermore, for all $n \geq 2$ and $a \in \{0, 1\}$, we have $(\mathcal{L}_n(\mathbf{z}) \cap aA^*a) \cup (\mathcal{L}_{n+1}(\mathbf{z}) \cap \overline{a}A^*\overline{a}) \neq \emptyset$.*

Definition 16. *Let $j \geq 0$. For any factor u of $\varphi^j(\mathbf{y})$ of length at least $2^j - 1$ there exist $a, b \in \{0, 1\}$ and $z \in \{0, 1\}^*$ with $azb \in \mathcal{L}(\mathbf{y})$ such that $u = p\varphi^j(z)s$ for some proper suffix p of $\varphi^j(a)$ and some proper prefix s of $\varphi^j(b)$. (Note that z could be empty.) The triple $(p, \varphi^j(z), s)$ is called a φ^j-factorization[2] of u. The word azb (resp., zb; az; z) is said to be the corresponding φ^j-ancestor of u when p, s are non-empty (resp., $p = \varepsilon$ and $s \neq \varepsilon$; $p \neq \varepsilon$ and $s = \varepsilon$; $p = s = \varepsilon$).*

---

[2] We warn the reader that the term φ-factorization has a different meaning in [10]. Our φ^j-factorization corresponds to their "factorization of order j".

Since the words $\varphi^j(0)$ and $\varphi^j(1)$ begin with different letters, we notice that if $s \neq \varepsilon$ in a φ^j-factorization of a word, then the letter b is uniquely determined. Similarly the jth images of the letters end with distinct letters, whence the letter a is uniquely determined once $p \neq \varepsilon$.

Proof (of Theorem 14). Let $j \in \{1, \ldots, k\}$. Notice all factors of length at most $2^j - 1$ of $\mathbf{x} = \varphi^k(\mathbf{y})$ occur already in the Thue–Morse word $\mathbf{t}$: such factors appear in factors of the form $\varphi^j(ab)$, $ab \in \mathcal{L}(\mathbf{y})$. Since $\varphi^j(ab)$ appears in the Thue–Morse word for all $a, b \in \{0, 1\}$, it follows from (2) that all such words are pairwise j-binomially non-equivalent. Hence we have shown that $\mathsf{b}_{\mathbf{x}}^{(j)}(n) = \mathsf{p}_{\mathbf{x}}(n)$ for $n \leq 2^j - 1$.

In the remaining of the proof we let $n \geq 2^j$. We show that $\mathcal{L}_n(\mathbf{t})/\sim_j = \mathcal{L}_n(\mathbf{x})/\sim_j$ by double inclusion, which suffices for the claim since Theorem 14 holds true for $\mathbf{x} = \mathbf{t}$.

Let $u \in \mathcal{L}(\mathbf{x})$; we show that there exists $v \in \mathcal{L}(\mathbf{t})$ such that $u \sim_j v$. To this end, let $\mathbf{z} = \varphi^{k-j}(\mathbf{y})$ so that $\mathbf{x} = \varphi^j(\mathbf{z})$. Let u have φ^j-factorization $p\varphi^j(u')s$ with φ^j-ancestor $au'b \in \mathcal{L}(\mathbf{z})$. The Thue–Morse word contains a factor $av'b$, where $|v'| = |u'|$ (see, e.g., [10, Proposition 33]). It follows that $\mathbf{t}$ contains the factor $v := p\varphi^j(v')s$. Now $u \sim_j v$ because $\varphi^j(u') \sim_j \varphi^j(v')$ by Theorem 2.

Let then $u \in \mathcal{L}(\mathbf{t})$ have φ^j-factorization $p\varphi^j(u')s$ with φ^j-ancestor $au'b \in \mathcal{L}(\mathbf{t})$. As before we show that there exists $v \in \mathcal{L}(\mathbf{x})$ such that $u \sim_j v$. By the previous lemma, $\mathbf{z}$ contains, at each length, factors from both the languages $0A^*1$ and $1A^*0$. Hence, if a and b above are distinct, we may argue as in the previous paragraph to obtain the desired conclusion. Assume thus that $a = b$. Again the previous lemma says that $\mathbf{z}$ contains a factor of length $|u'| + 2$ in the language $1A^*1 \cup 0A^*0$. Assume without loss of generality that it contains a factor from $0A^*0$. Then, if $a = b = 0$, we may again argue as in the previous paragraph. So assume now that $a = b = 1$ and $\mathcal{L}_{|u|'+2}\,\mathbf{z} \cap 1A^*1 = \emptyset$. Notice that by the previous lemma, $\mathcal{L}_{|u|'+2}\,\mathbf{z} \cap 0A^*0 \neq \emptyset$ and, further, $\mathcal{L}_{|u|'+2\pm1}\,\mathbf{z} \cap 0A^*0 \neq \emptyset$. To conclude with the proof, we have four cases to consider depending on the length of p and s which can be less or equal, or greater than 2^{j-1}.

Case 1: Assume that p is a suffix of $\varphi^{j-1}(0)$ and s is a prefix of $\varphi^{j-1}(1)$. For all v' such that $|v'| = |u'| - 1$, $\varphi^j(u') \sim_j \varphi^j(v'1)$ by Theorem 2. By the Transfer Lemma (Lemma 3), $\varphi^j(v'1) \sim_j \varphi^{j-1}(1)\varphi^j(v')\varphi^{j-1}(0)$. Consequently

$$u \sim_j p\varphi^{j-1}(1)\varphi^j(v')\varphi^{j-1}(0)s =: v$$

where $p\varphi^{j-1}(1)$ is a suffix of $\varphi^j(0)$ and $\varphi^{j-1}(0)s$ is a prefix of $\varphi^j(0)$. Hence v is a factor of $\varphi^j(0v'0)$. Recall that a factor of the form $0v'0$ appears in $\mathbf{z}$ by assumption, and thus $\varphi^j(0v'0)$ appears in $\mathbf{x}$. To recap, we have shown a factor v of $\mathbf{x}$ j-binomially equivalent to u.

The three other cases where $2^{j-1} \leq |p| < 2^j$ or $2^{j-1} \leq |s| < 2^j$ are similar. □

4.2 The $(k + 1)$-Binomial Complexity

The previous subsection dealt with the j-binomial equivalence in $\mathbf{x} = \varphi^k(\mathbf{y})$, where $\mathbf{y}$ is an aperiodic binary word and $j \leq k$. Here, we are concerned with the $(k + 1)$-binomial equivalence in such words. To this end, we take a closer look at the k-binomial equivalence in $\mathbf{x}$. First, we have a closer look at the φ^j-factorizations of a word and in particular at the associated prefixes and suffixes.

Definition 17 ([10, Definition 43]). *Let $j \geq 1$. Let us define the equivalence relation $\equiv_j$ on $A^{<2^j} \times A^{<2^j}$ by $(p_1, s_1) \equiv_j (p_2, s_2)$ whenever there exists $a \in A$ such that one of the following situations occurs:*

1. $|p_1| + |s_1| = |p_2| + |s_2|$ *and*
   (a) $(p_1, s_1) = (p_2, s_2)$;
   (b) $(p_1, \varphi^{j-1}(a)s_1) = (p_2\varphi^{j-1}(a), s_2)$;
   (c) $(p_2, \varphi^{j-1}(a)s_2) = (p_1\varphi^{j-1}(a), s_1)$;
   (d) $(p_1, s_1) = (s_2, p_2)$
   $\qquad = (\varphi^{j-1}(a), \varphi^{j-1}(\bar{a}))$;

2. $\big||p_1| + |s_1| - (|p_2| + |s_2|)\big| = 2^j$ *and*
   (a) $(p_1, s_1) = (p_2\varphi^{j-1}(a), \varphi^{j-1}(\bar{a})s_2)$;
   (b) $(p_2, s_2) = (p_1\varphi^{j-1}(a), \varphi^{j-1}(\bar{a})s_1)$.

The next lemma is essentially [10, Lemmas 40 and 41] (except that with an arbitrary word $\mathbf{y}$ instead of the Thue–Morse word $\mathbf{t}$, we cannot use the fact that $\mathbf{t}$ is overlap-free, so factors such as 10101 may appear in $\mathbf{y}$). To each φ^j-factorization there is a natural corresponding φ^{j-1}-factorization, though two φ^j-factorizations may correspond to the same φ^{j-1}-factorization. The next lemma says that in such a case the φ^j-factorizations are related.

Lemma 18. *Let $j \geq 1$. Let u be a factor of $\varphi^j(\mathbf{y})$ such that $|u| \geq 2^j - 1$ with a φ^j-factorization of the form $(p, \varphi^j(z), s)$ and $z_0 z z_{n+1}$ being the corresponding φ^j-ancestor (where according to Definition 16 z_0, z_{n+1} or z could be empty). The factor u has a unique φ^j-factorization if and only if the word $z_0 z z_{n+1}$ contains both letters 0 and 1. Otherwise stated, the φ^j-factorization is not unique if and only if u is a factor of $\varphi^{j-1}(m)$ with $m \in (01)^* \cup (10)^* \cup 1(01)^* \cup 0(10)^*$. Moreover, when the φ^j-factorization is not unique, i.e., if there is another φ^j-factorization $(p', \varphi^j(z'), s')$, then $(p, s) \equiv_j (p', s')$.*

We have the following theorem, the proof of which is essentially the proof of [10, Theorem 48]. Indeed, the lemmas leading to its proof do not require that the factors u and v are from the Thue–Morse word, only that they have φ^j-factorizations.[3]

Theorem 19. *Let $\mathbf{y}$ be an aperiodic binary word. Let $k \geq j \geq 1$. Let u and v be equal-length factors of $\mathbf{x} = \varphi^k(\mathbf{y})$ with φ^j-factorizations $u = p_1\varphi^j(z)s_1$ and $v = p_2\varphi^j(z')s_2$. Then $u \sim_j v$ if and only if $(p_1, s_1) \equiv_j (p_2, s_2)$.*

We then turn to the $(k + 1)$-binomial equivalence in $\mathbf{x}$.

Lemma 20. *Let u, v be two binary words of equal length. For $k \geq 1$, $u \not\sim_1 v$ implies $\varphi^k(u) \not\sim_{k+1} \varphi^k(v)$. Moreover, if $u \sim_1 v$, for $k \geq 1$, $u \not\sim_2 v$ implies $\varphi^k(u) \not\sim_{k+2} \varphi^k(v)$.*

---

[3] We note that [10, Theorem 48] is stated for $j \geq 3$. The case $j = 1$ is trivial. The case $j = 2$ is obtained by looking closely at the proof of [10, Theorem 34].

We define some notation regarding factors of $\mathbf{y}$. For $n \geq 1$ we let $\mathcal{S}(n) = \mathcal{L}_n(\mathbf{y})$. Further, for all $a, b \in \{\varepsilon, 0, 1\}$ such that $ab \neq \varepsilon$, we define $\mathcal{S}_{a,b}(n) = \mathcal{L}_{n+|ab|}(\mathbf{y}) \cap aA^*b$. We call these sets *factorization classes of order n*.

Consider now a factor u of $\varphi(\mathbf{y})$. We associate with u some factorization classes as follows. Let $a\varphi(u')b$ be the φ-factorization of u with φ-ancestor $au'b \in \mathcal{L}(\mathbf{y})$. If $ab = \varepsilon$, we associate the factorization class $\mathcal{S}(|u'|)$. For $ab \neq \varepsilon$, we have that u is a factor of $\varphi(\bar{a}u'b)$. In this case we associate the factorization class $\mathcal{S}_{\bar{a},b}(|u'|)$. If u is associated with a factorization class $\mathcal{T}$, we write $u \models \mathcal{T}$, otherwise we write $u \not\models \mathcal{T}$.

Observe that $u \models \mathcal{S}(n)$ implies that $|u| = 2n$. Also, for $ab \neq \varepsilon$, $u \models \mathcal{S}_{a,b}(n)$ implies that $|u| = 2n + |ab|$. Notice also that a factor u of $\varphi(\mathbf{y})$ can be associated with several factorization classes: take, e.g., $(10)^\ell 1 = 1(01)^\ell$ which is associated with both $\mathcal{S}_{\varepsilon,1}(\ell)$ and $\mathcal{S}_{0,\varepsilon}(\ell)$, or $(01)^{\ell+1} = 0(10)^\ell 1$ which is associated with both $\mathcal{S}(\ell+1)$ and $\mathcal{S}_{1,1}(\ell)$.

Lemma 21. *Let $u, v \in \mathcal{L}(\varphi(\mathbf{y}))$ be such that $u \sim_2 v$. If $u \models \mathcal{T}$ for some factorization class $\mathcal{T}$, then $v \models \mathcal{T}$. Furthermore, a factor u of $\varphi(\mathbf{y})$ is associated with distinct factorization classes if and only if $u \in L = (01)^* \cup (10)^* \cup 1(01)^* \cup 0(10)^*$.*

The next result characterizes $(k+1)$-binomial equivalence in $\mathbf{x} = \varphi^k(\mathbf{y})$ when $\mathbf{y}$ is an arbitrary binary word.

Proposition 22. *Let u and v be factors of length at least $2^k - 1$ of $\mathbf{x}$ with the φ^k-factorizations $u = p_1\varphi^k(z)s_1$ and $v = p_2\varphi^k(z')s_2$. Then $u \sim_{k+1} v$ and $u \neq v$ if and only if $z \sim_1 z'$, $z' \neq z$, and $(p_1, s_1) = (p_2, s_2)$.*

Notice that the proposition claims that those factors of $\mathbf{x}$ having more than one φ^k-factorization are $(k+1)$-binomially equivalent only to themselves (in $\mathcal{L}(\mathbf{x})$).

Proof. The "if"-part of the statement follows by a repeated application of Proposition 12 on the Thue–Morse morphism together with the fact that the morphism is injective.

Assume that $u \sim_{k+1} v$ for some distinct factors. It follows that $u \sim_k v$, which implies that $(p_1, s_1) \equiv_k (p_2, s_2)$ by Theorem 19. Next we show that $(p_1, s_1) = (p_2, s_2)$ and $z \sim_1 z'$. We have the following case distinction from Definition 17:

(1.a): We have that $(p_1, s_1) = (p_2, s_2)$. By deleting the common prefix p_1 and suffix s_1, we are left with the equivalent statement $\varphi^k(z) \sim_{k+1} \varphi^k(z')$. If $z \not\sim_1 z'$, then we have a contradiction with Lemma 20. The desired result follows in this case.

In the remaining cases, we assume towards a contradiction that $(p_1, s_1) \neq (p_2, s_2)$.

(1.b): Suppose that $(p_1, s_2) = (p_2\varphi^{k-1}(a), \varphi^{k-1}(a)s_1)$. Deleting the common prefixes p_2 and suffixes s_1, we are left with $\varphi^{k-1}(a\varphi(z)) \sim_{k+1} \varphi^{k-1}(\varphi(z')a)$. Now $a\varphi(z) \sim_1 \varphi(z')a$, but $a\varphi(z) \not\sim_2 \varphi(z')a$ by Lemma 21 (otherwise $a\varphi(z) = \varphi(z')a$ and thus $u = v$ contrary to the assumption). Lemma 20 then implies that $\varphi^{k-1}(a\varphi(z)) \not\sim_{k+1} \varphi^{k-1}(\varphi(z')a)$, which is a contradiction.

(2.a): Suppose that $(p_1, s_1) = (p_2\varphi^{k-1}(a), \varphi^{k-1}(\bar{a})s_2)$. After removing common prefixes and suffixes, we are left with $\varphi^{k-1}(a\varphi(z)\bar{a}) \sim_{k+1} \varphi^{k-1}(\varphi(z'))$.

We have that $a\varphi(z)\overline{a} \sim_1 \varphi(z')$, but by Lemma 21 $a\varphi(z)\overline{a} \not\sim_2 \varphi(z')$ (otherwise $z = \overline{a}^\ell$ and $z' = a^{\ell+1}$, implying that $u = v$, a contradiction). This is again a contradiction by Lemma 20.

The rest of the cases go exactly the same way. $\square$

Corollary 23. *Let* $\mathbf{x} = \varphi^k(\mathbf{y})$, *where* $\mathbf{y}$ *is an arbitrary aperiodic binary word. We have* $b_{\mathbf{x}}^{(1)} \prec b_{\mathbf{x}}^{(2)} \prec \ldots \prec b_{\mathbf{x}}^{(k)} \prec b_{\mathbf{x}}^{(k+1)}$.

Proof. Recall that $\mathbf{y}$ contains arbitrarily long factors of the form $\overline{a}za$, $a \in \{0,1\}$. Therefore $\mathbf{x}$ contains the k-binomially equivalent (by Lemma 3) factors $\varphi^{k-1}(a)\varphi^k(z)$ and $\varphi^k(z)\varphi^{k-1}(a)$. However, by Proposition 22 these factors are either not $(k+1)$-binomially equivalent, or $\varphi^{k-1}(a)\varphi^k(z) = \varphi^k(z)\varphi^{k-1}(a)$. The latter happens when $\varphi^k(z) = \varphi^{k-1}(a)^\ell$ for some $\ell \geq 0$ (indeed, it is not hard to show that $\varphi(w)$ is primitive whenever w is), and thus only when $\ell = 0$ and $z = \varepsilon$. This observation suffices for showing $b_{\mathbf{x}}^{(k)} \prec b_{\mathbf{x}}^{(k+1)}$. The rest of the claim follows by Theorem 14. $\square$

5   Answer to Question B and Beyond

Theorem 24. *Let* φ *be the Thue–Morse morphism and* $\mathbf{s}$ *a Sturmian word. For each* $k \geq 0$, *the word* $\mathbf{s}_k = \varphi^k(\mathbf{s})$ *has* $b_{\mathbf{s}_k}^{(1)} \prec b_{\mathbf{s}_k}^{(2)} \prec \cdots \prec b_{\mathbf{s}_k}^{(k+1)} \prec b_{\mathbf{s}_k}^{(k+2)} = p_{\mathbf{s}_k}$.

In particular, putting the Fibonacci word for $\mathbf{s}$, the family $\mathbf{s}_k$ answers Question B positively with morphic words.

Proof. Observe that $\mathbf{s}_k$ has bounded $(k+1)$-binomial complexity as a straightforward application of Theorem 10 (because $\mathbf{s}$ has bounded abelian complexity), and thus $b_{\mathbf{s}_k}^{(k+1)} \prec p_{\mathbf{s}_k}$. By Corollary 23, we need only to show that $b_{\mathbf{s}_k}^{(k+2)} = p_{\mathbf{s}_k}$.

Let u and v be distinct factors of $\mathbf{s}_k$. Assume they are $(k+2)$-binomially equivalent. By Proposition 22, we have that $u = p\varphi^k(z)s$, $v = p\varphi^k(z')s$ with $z \sim_1 z'$. If $z \neq z'$, then $z \not\sim_2 z'$ by [17, Theorem 7]. But then Lemma 20 implies that $\varphi^k(z) \not\sim_{k+2} \varphi^k(z')$, contradicting the assumption. Hence we deduce that $z = z'$, but then $u = v$ contrary to the assumption. $\square$

Notice that the binomial complexities $b_{\mathbf{s}_k}^{(1)}, \ldots, b_{\mathbf{s}_k}^{(k+1)}$ are bounded. To go beyond Question B, one might require that $b_{\mathbf{w}}^{(1)}$ is unbounded. To this end we provide the following partial answer.

Theorem 25. *The word* $\mathbf{h} = 0112122122212222122222\cdots$ *fixed point of the morphism* $0 \mapsto 01$, $1 \mapsto 12$, *and* $2 \mapsto 2$ *is such that its abelian complexity* $b_{\mathbf{h}}^{(1)}$ *is unbounded and* $b_{\mathbf{h}}^{(1)} \prec b_{\mathbf{h}}^{(2)} \prec b_{\mathbf{h}}^{(3)} = p_{\mathbf{h}}$.

References

1. Allouche, J.P., Shallit, J.: Automatic Sequences Theory, Applications, Generalizations. Cambridge University Press, Cambridge (2003). https://doi.org/10.1017/CBO9780511546563

2. Cassaigne, J., Fici, G., Sciortino, M., Zamboni, L.Q.: Cyclic complexity of words. J. Comb. Theory Ser. A **145**, 36–56 (2017). https://doi.org/10.1016/j.jcta.2016.07.002

3. Cassaigne, J., Richomme, G., Saari, K., Zamboni, L.Q.: Avoiding Abelian powers in binary words with bounded Abelian complexity. Int. J. Found. Comput. S. **22**(4), 905–920 (2011). https://doi.org/10.1142/S0129054111008489

4. Cheraghchi, M., Gabrys, R., Milenkovic, O., Ribeiro, J.: Coded trace reconstruction. IEEE Trans. Inf. Theory **66**(10), 6084–6103 (2020). https://doi.org/10.1109/TIT.2020.2996377

5. Dudík, M., Schulman, L.J.: Reconstruction from subsequences. J. Comb. Theory, Ser. A **103**(2), 337–348 (2003). https://doi.org/10.1016/S0097-3165(03)00103-1

6. Frid, A.: Applying a uniform marked morphism to a word. Discrete Math Theor. Comput. Sci. **3**(3), 125–139 (1999). https://doi.org/10.46298/dmtcs.255

7. Grillenberger, C.: Constructions of strictly ergodic systems. I. Given entropy. Z. Wahrscheinlichkeit. **25**, 323–334 (1973). https://doi.org/10.1007/BF00537161

8. Karhumäki, J., Saarela, A., Zamboni, L.Q.: On a generalization of Abelian equivalence and complexity of infinite words. J. Comb. Theory Ser. A **120**(8), 2189–2206 (2013). https://doi.org/10.1016/j.jcta.2013.08.008

9. Lejeune, M.: On the k-binomial equivalence of finite words and k-binomial complexity of infinite words. Ph.D. thesis, University of Liège (2021). http://hdl.handle.net/2268/259266

10. Lejeune, M., Leroy, J., Rigo, M.: Computing the k-binomial complexity of the Thue-Morse word. J. Comb. Theory Ser. A **176**, 44 (2020). https://doi.org/10.1016/j.jcta.2020.105284

11. Lejeune, M., Rigo, M., Rosenfeld, M.: Templates for the k-binomial complexity of the Tribonacci word. Adv. Appl. Math. **112**, 26 (2020). https://doi.org/10.1016/j.aam.2019.101947

12. Lothaire, M.: Combinatorics on Words. Cambridge Mathematical Library. Cambridge University Press, Cambridge (1997). https://doi.org/10.1017/CBO9780511566097

13. Lü, X.T., Chen, J., Wen, Z.X., Wu, W.: On the 2-binomial complexity of the generalized Thue-Morse words (2021, preprint). https://arxiv.org/abs/2112.05347

14. Ochsenschläger, P.: Binomialkoeffizienten und shuffle-zahlen. T.H. Darmstadt, Technischer bericht, Fachbereicht Informatik (1981)

15. Peltomäki, J.: Introducing privileged words: Privileged complexity of Sturmian words. Theor. Comput. Sci. **500**, 57–67 (2013). https://doi.org/10.1016/j.tcs.2013.05.028

16. Rigo, M.: Relations on words. Indag Math. New Ser. **28**(1), 183–204 (2017). https://doi.org/10.1016/j.indag.2016.11.018

17. Rigo, M., Salimov, P.: Another generalization of abelian equivalence: binomial complexity of infinite words. Theor. Comput. Sci. **601**, 47–57 (2015). https://doi.org/10.1016/j.tcs.2015.07.025

18. Rigo, M., Stipulanti, M., Whiteland, M.A.: Binomial complexities and Parikh-collinear morphisms (2022, preprint). https://arxiv.org/abs/2201.04603

19. Whiteland, M.A.: Equations over the k-binomial monoids. In: Lecroq, T., Puzynina, S. (eds.) WORDS 2021. LNCS, vol. 12847, pp. 185–197. Springer, Cham (2021). https://doi.org/10.1007/978-3-030-85088-3_16

Rational Index of Languages
with Bounded Dimension of Parse Trees

Ekaterina Shemetova[1,3,4], Alexander Okhotin[1(✉)],
and Semyon Grigorev[2,4]

[1] Department of Mathematics and Computer Science, St. Petersburg State
University, 14th Line V. O., 29, Saint Petersburg 199178, Russia
alexander.okhotin@spbu.ru
[2] Department of Mathematics and Mechanics, St. Petersburg State University,
7/9 Universitetskaya nab., Saint Petersburg 199034, Russia
[3] St. Petersburg Academic University, ul. Khlopina, 8,
Saint Petersburg 194021, Russia
[4] JetBrains Research, Primorskiy prospekt 68-70, Building 1,
St. Petersburg 197374, Russia
{ekaterina.shemetova,semyon.grigorev}@jetbrains.com

Abstract. The rational index ρ_L of a language L is an integer function,
where $\rho_L(n)$ is the maximum length of the shortest string in $L \cap R$,
over all regular languages R recognized by n-state nondeterministic finite
automata (NFA). This paper investigates the rational index of languages
defined by (context-free) grammars with bounded tree dimension, and
shows that it is of polynomial in n. More precisely, it is proved that
for a grammar with tree dimension bounded by d, its rational index is
$O(n^{2d})$, and that this estimation is asymptotically tight, as there exists
a grammar with rational index $\Theta(n^{2d})$.

Keywords: Dimension of a parse tree · Strahler number · Rational
index · Context-free languages · CFL-reachability

1 Introduction

The notion of a rational index was introduced by Boasson, Courcelle and Nivat [3]
as a complexity measure for context-free languages. The rational index ρ_L of a
language L is an integer function, where $\rho_L(n)$ is the maximum length of the
shortest string in a language of the form $L \cap R$, where R is a regular language
recognized by n-state nondeterministic finite automata (NFA), and the maximum is taken over all such languages R with $L \cap R \neq \varnothing$. The rational index
plays an important role in determining the parallel complexity of practical problems, such as the CFL-reachability problem and the more general Datalog query
evaluation.

The *CFL-reachability problem* is stated as follows: for a context-free grammar
G given an NFA A over the same alphabet, determine whether $L(G) \cap L(A)$

Research supported by the Russian Science Foundation, project 18-11-00100.

V. Diekert and M. Volkov (Eds.): DLT 2022, LNCS 13257, pp. 263–273, 2022.
https://doi.org/10.1007/978-3-031-05578-2_21

is non-empty. With A is regarded as a labelled graph, this is a kind of graph reachability problem with path constraints given by context-free languages. This is an important problem used in static code analysis [16] and graph database query evaluation [19].

The CFL-reachability problem is P-complete already for a fixed context-free grammar [9]. The question on the parallel complexity of this problem was investigated by Ullman and Van Gelder [17] in a much more general case, with a rich logic for database queries instead of grammars, and it was proved that under an assumption called the *polynomial fringe property* the problem is decidable in NC [17]. In the special case of grammars, the result of Ullman and Van Gelder [17] gives an NC^2 algorithm for the CFL-reachability problem, under the assumption that the grammar's rational index is polynomial.

Theoretical properties of the rational index have received some attention in the literature. Pierre and Farinone [15] proved that for every algebraic number $\gamma \geqslant 1$, a language with the rational index in $\Theta(n^\gamma)$ exists. An upper bound on the rational index, shown by Pierre [14], is $2^{\Theta(n^2/\ln n)}$, and this bound is reached on the Dyck language on two pairs of parentheses. For several important subfamilies of grammars, such as the linear and the one-counter languages, there are polynomial upper bounds on the rational index, which imply that the CFL-reachability problem is in NC^2; they can be proved to lie in NL by direct methods not involving the rational index [11,12].

In this paper we investigate the rational index of a generalization of linear languages: the *languages of bounded tree dimension*, that is, those defined by grammars with a certain limit on branching in the parse trees. The notion of tree dimension is well-known in the literature under different names: Chytil and Monien [6] use the term *k-caterpillar trees*, Esparza et al. [7] call this the *Strahler number* of a tree and mention numerous applications and alternative names for this notion, while Luttenberger and Schlund [13] use the term *tree dimension*, which is adopted in this paper.

Linear languages are languages of tree dimension 1, and their rational index is known to be $O(n^2)$ [3]. It can be derived from the work of Chytil and Monien [6] that languages of tree dimension bounded by d have rational index $O(n^{2d})$: this is explained in Sect. 3 of this paper. The new result of this paper, presented in Sect. 4, is that, for every d, there is a language of tree dimension bounded by d with rational index $\Theta(n^{2d})$. Some implications of this result are presented in Sect. 5: the rational index is asymptotically determined for *super-linear languages* [4], and some bounds are obtained for *languages of bounded oscillation* [8,18].

2   Definitions

A *(context-free) grammar* is a quadruple $G = (\Sigma, N, R, S)$, where Σ is an alphabet; N is a set of nonterminal symbols; R is a set of rules, each of the form $A \to \alpha$, with $A \in N$ and $\alpha \in (\Sigma \cup N)^*$; and $S \in N$ is the start symbol. A parse tree is a tree, in which every leaf is labelled with a symbol from Σ, while every internal node is labelled with a nonterminal symbol $A \in N$ and has an associated rule $A \rightarrow X_1 \ldots X_\ell \in R$, so that the node has ℓ ordered children labelled with $X_1, \ldots, X_\ell$. The language defined by each nonterminal symbol $A \in N$, denoted by $L_G(A)$, is the set of all strings $w \in \Sigma^*$, for which there exists a parse tree, with A as a root and with the leaves forming the string w, The language defined by the grammar is $L(G) = L_G(S)$.

A grammar G is said to be is in the *Chomsky normal form*, if all rules of R are of the form $A \rightarrow BC$, with $B, C \in N$, or of the form $A \rightarrow a$, with $a \in \Sigma$.

A *nondeterministic finite automaton* (NFA) is a quintuple $\mathcal{A} = (\Sigma, Q, Q_0, \delta, F)$, where Q is a finite set of states, Σ is a finite set of input symbols, $Q_0 \subseteq Q$ is the set of initial states, $\delta \colon Q \times \Sigma \rightarrow 2^Q$ is the transition function, $F \subseteq Q$ is the set of accepting states. It accepts a string $w = a_1 \ldots a_n$ if there is a sequence of states $q_0, \ldots, q_n \in Q$ with $q_0 \in Q_0$, $q_i \in \delta(q_{i-1}, a_i)$ for all i, and $q_n \in F$. The language of all strings accepted by $\mathcal{A}$ is denoted by $L(\mathcal{A})$.

For a language L over an alphabet Σ, its rational index ρ_L is a function defined as follows:

$$\rho_L(n) = \max_{\substack{\mathcal{A}:\text{NFA with } n \text{ states} \\ L \cap L(\mathcal{A}) \neq \varnothing}} \min_{w \in L \cap L(\mathcal{A})} |w|$$

Tree Dimension. For each node v in a parse tree t, its *dimension* $\dim v$ is an integer representing the amount of branching in its subtree. It is defined inductively: a leaf v has dimension 0. For an internal node v, if one of its children $v_1, v_2, \ldots, v_k$, with $k \geqslant 1$, has a greater dimension than all the others, then v has the same dimension, and if there are multiple children of maximum dimension, then the dimension of v is greater by one.

$$\dim v = \begin{cases} \max_{i \in \{1,\ldots,k\}} \dim v_i & \text{if there is a unique maximum} \\ \max_{i \in \{1,\ldots,k\}} \dim v_i + 1 & \text{otherwise} \end{cases}$$

The dimension of a parse tree t, denoted by $\dim t$, is the dimension of its root.

Definition 1 (Grammars of bounded tree dimension). *A grammar G is of d-bounded tree dimension if every parse tree t of G has $\dim t \leqslant d$, where d is some constant. This constant is called the dimension of G, denoted by $\dim G = d$.*

Classical transformation to the Chomsky normal form preserves the class of grammars of d-bounded tree dimensions. Languages defined by such grammars are called *languages of d-bounded tree dimension*.

3   Upper Bound on the Rational Index

The first result of this paper is that, if the dimension of trees in a grammar is bounded by a constant d, then the rational index of its language is bounded by $O(n^{2d})$, where the constant factor depends upon the grammar.

Theorem 1. *Let G be a grammar of d-bounded tree dimension, and let $\mathcal{A}$ be an NFA with n states, with non-empty intersection $L(G) \cap L(\mathcal{A})$. Then the length of the shortest string in $L(G) \cap L(\mathcal{A})$ is at most $|G|^d n^{2d}$, where $|G| = \sum_{A \to \alpha}(|\alpha| + 2)$ is number of symbols used for the description of the grammar.*

The main component of the proof is the following lemma by Chytil and Monien [6], which they used in their study of unambiguous grammars of finite index.

Lemma 1 (Chytil and Monien [6, Lemma 7]). *Let $G = (\Sigma, N, R, S)$ be a grammar, let m be the maximal length of the right-hand side of its rules, and assume that there exists a parse tree of dimension $d \geqslant 1$ in this grammar. Then the grammar defines some string of length at most $(|N|(m-1)+1)^d$.*

The proof proceeds by simplifying the tree of dimension d by removing paths beginning and ending with the same nonterminal symbol. This contraction results in a parse tree of a bounded size, which has the same dimension d [6].

Proof (of Theorem 1). A given grammar G is first transformed to the Chomsky normal form, resulting in a grammar $G' = (\Sigma, N', R', S')$ with the same bound on the dimension of parse trees and with at most $|G|$ nonterminal symbols.
   Next, a grammar G'' for the language $L(G) \cap L(\mathcal{A})$ is obtained from G' and $\mathcal{A}$ by the classical construction by Bar-Hillel et al. [2], which produces $|N'| \cdot n^2 + 1$ nonterminal symbols: these are triples of the form (A, p, q), where $A \in N'$ and p, q are two states of the automaton, as well as a new start symbol. The grammar G'' is still in the Chomsky normal form, that is, the maximum length of a right-hand side of any rule is $m = 2$. Since G'' defines at least one string, there exists a parse tree of dimension at most d. Then, by Lemma 1, the length of the shortest string defined by this grammar is at most $(|N'| \cdot n^2 + 1 + 1)^d \leqslant (|G| \cdot n^2)^d$. $\square$

4   Lower Bound on the Rational Index

The upper bound $O(n^{2d})$ on the rational index of a language defined by a grammar with tree dimension bounded by d has a matching lower bound $\Omega(n^{2d})$. It is first established for a convenient infinite set of values of n, to be extended to arbitrary n in the following.

Lemma 2. *For every $d \geqslant 1$, there is a grammar G of bounded tree dimension d, such that for every $n \geqslant 2^{d+1}$ divisible by 2^d there is an n-state NFA $\mathcal{B}$, such that the shortest string w in $L(G) \cap L(\mathcal{B})$ is of length at least $\frac{1}{2^{d^2+3d-3}} n^{2d}$.*

Proof. The grammar and the automaton are constructed inductively on d, for every d and only for n divisible by 2^d. Each constructed NFA shall have a unique initial state, which is also the unique accepting state.

Basis. $dim(G) = 1$. The family of languages having dimension $d = 1$ coincides with the family of linear languages. Let G be a linear grammar with the rules $S \to aSb \mid ab$, which defines the language $L(G) = \{a^i b^i \mid i \geqslant 1\}$.

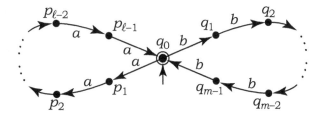

Fig. 1. NFA $\mathcal{B}$ defined in Lemma 2 for $d = 1$.

For every $n \geqslant 4$ divisible by $2^d = 2$, let $\ell = \frac{n}{2}$, $m = \frac{n}{2} + 1$. Then ℓ and m are coprime integers. Define an NFA $\mathcal{B}$ over the alphabet $\{a, b\}$, which consists of two cycles sharing one node, q_0, which is both the initial and the unique accepting state. The cycle of length ℓ has all transitions by a, and the other by b, as shown in Fig. 1. The automaton has $\ell + m - 1 = n$ states.

Every string in $L(G) \cap L(\mathcal{B})$ is of the form $a^i b^i$, with $i \geqslant 1$. For the automaton to accept it, i must be divisible both by ℓ and by m. Since the cycle lengths are relatively prime, the shortest string w with this property has $i = \ell m$, and is accordingly of length $2\ell m$. Its growth with n is estimated as follows.

$$|w| = 2\ell m = 2\frac{n}{2} \cdot \left(\frac{n}{2} + 1\right) = \frac{1}{2}n^2 + n$$

This example is well-known to the community [10, 19].

Inductive Step. $dim(G) = d$.
By the induction hypothesis, there is a grammar $\widehat{G} = (\widehat{\Sigma}, \widehat{N}, \widehat{R}, \widehat{S})$ of bounded dimension $dim(\widehat{G}) = d - 1$, which satisfies the statement of the lemma. The new grammar $G = (\Sigma, N, R, S)$ of dimension d is defined over the alphabet $\Sigma = \widehat{\Sigma} \cup \{a, b, c\}$, where $a, b, c \notin \widehat{\Sigma}$ are new symbols. It uses nonterminal symbols $N = \widehat{N} \cup \{S, A\}$, adding two new nonterminals $A, S \notin \widehat{N}$ to those in $\widehat{G}$, where S is the new initial symbol. Its set of rules includes all rules from $\widehat{G}$ and the following new rules.

$$S \rightarrow ASc \mid Ac$$
$$A \rightarrow aAb \mid a\widehat{S}b$$

To see that the dimension of the new grammar is greater by 1 than the dimension of $\widehat{G}$, first consider the dimension of any parse tree t with the root labeled by the nonterminal A, shown in Fig. 2(right). The dimension of the $\widehat{S}$-subtree at the bottom is at most $d - 1$ by the properties of $\widehat{G}$. This dimension is inherited by all A-nodes in the tree, because their remaining children are leaves.

Now consider the dimension of a complete parse tree t with the start symbol S in the root, as in Fig. 2(left). All A-subtrees in this tree have dimension at most $d - 1$. Then the bottom S-subtree, which uses the rule $S \rightarrow Ac$, also has

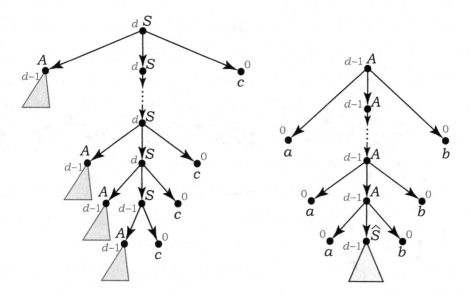

Fig. 2. Parse trees for S and for A, annotated with dimensions of their vertices

dimension at most $d - 1$. Every S-subtree higher up in the tree uses a rule $S \to ASc$, and its dimension is at most d, because getting a higher dimension would require two subtrees of dimension d, which is never the case.

Now, for every $n \geqslant 2^{d+1}$ divisible by 2^d, the goal is to construct an n-state NFA over the alphabet Σ, so that the shortest string w in $L(G) \cap L(\mathcal{B})$ is of length at least $\frac{1}{2^{d^2+3d-3}}n^{2d}$. Since the number $\frac{n}{2}$ is at least 2^d and is divisible by 2^{d-1}, the induction hypothesis for the grammar $\widehat{G}$ asserts that there is an NFA $\widehat{\mathcal{B}} = (\widehat{Q}, \widehat{\Sigma}, \widehat{\delta}, \widehat{q_0}, \{\widehat{q_0}\})$, with $\frac{n}{2}$ states, with the shortest string $\widehat{w}$ in $L(\widehat{G}) \cap L(\widehat{\mathcal{B}})$ of length $\frac{1}{2^{(d-1)^2+3(d-1)-3}}(\frac{n}{2})^{2(d-1)}$.

The desired n-state NFA $\mathcal{B} = (\Sigma, Q, q_0, \delta, \{q_0\})$ is constructed as follows. Let $\ell = \frac{n}{4}$ and $m = \frac{n}{4} + 1$, these are two coprime integers. The set of states of $\mathcal{B}$ contains all $\frac{n}{2}$ states from $\widehat{Q}$, in which $\mathcal{B}$ it operates as $\widehat{\mathcal{B}}$, and $m + \ell - 1 = \frac{n}{2}$ new states forming a cycle of length ℓ and a chain of length m, which share a state.

$$Q = \widehat{Q} \cup \{p_1, \ldots, p_{\ell-1}, q_0, \ldots, q_{m-1}\}$$

The new initial state q_0 has a transition by a leading to the initial state of $\widehat{\mathcal{B}}$, from where one can return to q_1 by b.

$$\delta(q_0, a) = \{\widehat{q_0}\}$$
$$\delta(\widehat{q_0}, b) = \{q_1\}$$

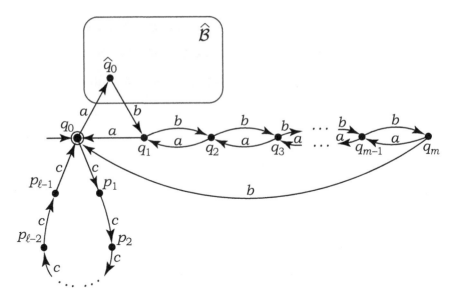

Fig. 3. NFA $\mathcal{B}$ defined in Lemma 2 for d, which incorporates NFA $\widehat{\mathcal{B}}$ for $d - 1$.

There is a chain of transitions by a from q_{m-1} to q_0, and another chain b in the opposite direction, from q_1 to q_{m-1} and back to q_0.

$$\delta(q_i, a) = \{q_{i-1}\}, \qquad \text{with } 1 \leqslant i \leqslant m - 1$$
$$\delta(q_i, b) = \{q_{i+1}\}, \qquad \text{with } 1 \leqslant i \leqslant m - 2$$
$$\delta(q_{m-1}, b) = \{q_0\}$$

There is a cycle by c in the states $q_0, p_1, \ldots, p_{\ell-1}$; for uniformity, denote $p_0 = q_0$.

$$\delta(p_i, c) = \{p_{i+1 \bmod \ell}\}, \qquad \text{with } 0 \leqslant i \leqslant \ell - 1$$

The general form of $\mathcal{B}$ is shown in Fig. 3.

Let w be the shortest string in $L(G) \cap L(\mathcal{B})$. Consider how w is formed. Start state is q_0. According to the grammar rule $S \to ASc \mid Ac$, the string w should start with a substring u in $L_G(A)$. There is the only one outgoing edge labeled with a, so the next state is $\widehat{q}_0$. The next part of w should be a symbol a or a string v in $L(\widehat{G})$. As there is no outgoing edge labeled with a, the string v is the shortest string in $L(\widehat{G}) \cap L(\widehat{\mathcal{B}})$, and, hence, $v = \widehat{w}$. Now the first part of w is $a\widehat{w}$. To complete a substring derived by the nonterminal A, there is only one possible transition, which is an edge from $\widehat{q}_0$ to q_1 labeled with b. The next substring should be symbol c (the rule $S \to Ac$) or a string derived by A. The only suitable transition here is from q_1 to q_0 by a, so a substring in $L(A)$ is started. Again, to complete the string generated by A, one goes to the state q_2, and w now starts with $a\widehat{w}baa\widehat{w}bb$. By the construction of NFA $\mathcal{B}$, this process continues until one comes to the state q_0 without starting a substring derived by the nonterminal A (notice that such substrings are the shortest possible). Clearly, it happens after m iterations. Then it is left to read m symbols c by going from q_0 to q_0. But m and ℓ are coprime, so to balance the number of substrings derived by the nonterminal A and the number of symbols c, one needs to repeat the first cycle ℓ times and the second cycle m times.

Accordingly, the shortest string w has the following structure. Let w_i be the shortest string such that there exists computation $q_{i-1} \overset{w_i}{\leadsto} q_i$ ($q_{m-1} \overset{w_m}{\leadsto} q_0$ for w_m) for $1 \leqslant i \leqslant m$ in $\mathcal{B}$ and $w_i \in L(A)$. Notice that $w_i = a w_{i-1} b$ and $w_0 = \widehat{w}$, and there exists computation $q_0 \overset{w_1}{\leadsto} q_1 \overset{w_2}{\leadsto} q_2 \overset{w_3}{\leadsto} \ldots \overset{w_{m-1}}{\leadsto} q_m \overset{w_m}{\leadsto} q_0$ in $\mathcal{B}$.

Considering the above and the rules $S \to ASc \mid Ac$ of the grammar G, the string w is of the following form:

$$w = \left(\prod_{i=1}^{m} w_i \right)^{\ell} c^{m\ell}$$

Then the length of w can be bounded as follows.

$$|w| = \left(\sum_{i=1}^{m} |w_i| \right) \ell + \ell m = \left(\sum_{i=1}^{m} (|\widehat{w}| + 2i) \right) \ell + \ell m \geqslant \ell m |\widehat{w}|$$

Using the lower bound on the length of $\widehat{w}$, the desired lower bound on the length of w is obtained.

$$\ell m |\widehat{w}| \geqslant \frac{n}{4} \cdot \frac{n}{4} \cdot \frac{1}{2^{(d-1)^2 + 3(d-1) - 3}} \left(\frac{n}{2} \right)^{2(d-1)} =$$

$$= \frac{n^2}{16} \cdot \frac{1}{2^{d^2 + d - 5}} \cdot \frac{n^{2d-2}}{2^{2d-2}} = \frac{1}{2^{d^2 + 3d - 3}} n^{2d}$$

$\square$

Theorem 2. *For every $d \geqslant 1$, there is a grammar G of bounded tree dimension d, such that for every $n \geqslant 2^{d+1}$ there is an n-state NFA $\mathcal{B}$, such that the shortest string w in $L(G) \cap L(\mathcal{B})$ is of length at least $\frac{1}{2^{d^2+d-3} 3^{2d}} n^{2d}$.*

Proof. Let G be the grammar given for d by Lemma 2. Let $2^d r \leqslant n < 2^d (r+1)$, for some integer r. Then $r \geqslant 2$ (for otherwise $n < 2^{d+1}$), and $2^d r \geqslant 2^{d+1}$.

Since $2^d r$ is divisible by 2^d, by Lemma 2, there is an NFA $\mathcal{B}$ with $2^d r \leqslant n$ states, such that the length of the shortest string w in $L(G) \cap L(\mathcal{B})$ is at least $\frac{1}{2^{d^2+3d-3}} (2^d r)^{2d}$. This is the desired n-state NFA.

The inequality $n < 2^d (r+1)$ implies that $n < 2^d \frac{3r}{2}$, because $r+1$ is at most $\frac{3r}{2}$ for $r \geqslant 2$. Then $2^d r > \frac{2}{3} n$, and the lower bound on the length of w is expressed as a function of n as follows.

$$|w| \geqslant \frac{1}{2^{d^2+3d-3}} (2^d r)^{2d} \geqslant \frac{1}{2^{d^2+3d-3}} \left(\frac{2}{3} n \right)^{2d} = \frac{1}{2^{d^2+d-3} 3^{2d}} n^{2d}$$

$\square$

For finite automata with fewer than 2^{d+1} states, no lower bounds are given, as the construction in the proof relies on having sufficiently long cycles in the automata.

Overall, the rational index of grammars with tree dimension bounded by d is $\Theta(n^{2d})$ in the worst case.

5   Rational Indices for Some Language Families

Superlinear Languages. A grammar $G = (\Sigma, N, R, S)$ is *superlinear* (Brzozowski [4]) if its nonterminal symbols split into two classes, $N = N_{lin} \cup N_{nonlin}$, where rules for each nonterminal $A \in N_{lin}$ are of the form $A \to uBv$ or $A \to w$, with $B \in N_{lin}$, $u, v, w \in \Sigma^*$, while rules for a nontermial $A \in N_{nonlin}$ are of the form $A \to \alpha B\beta$, with $B \in N$ and $\alpha, \beta \in (\Sigma \cup N_{lin})^*$. A language is *superlinear* if it is generated by some superlinear grammar.

Corollary 1. *For every superlinear grammar G, the rational index $\rho_{L(G)}$ is at most $|G|^2 \cdot n^4$.*

Proof. Parse trees in a superlinear grammar G have dimension at most 2. Then, by Theorem 1, the rational index $\rho_{L(G)}$ is bounded by $|G|^2 \cdot n^4$.

Turning to a lower bound, note that the grammar constructed in Theorem 2 for $d = 2$ is actually superlinear.

Corollary 2. *There exists a superlinear grammar G with rational index $\rho_{L(G)}(n) \geq \frac{1}{648} n^4$.*

Bounded-Oscillation Languages. The notion of oscillation of runs in pushdown automata, applicable to Turing machines with auxiliary pushdown tape, was introduced by Wechsung [18]. *Languages with oscillation bounded by k* are then a generalization of the linear languages (as one-turn pushdown automata are those with oscillation bounded by $k = 1$).

This family was later studied by Ganty and Valput [8], who introduced the corresponding notion of oscillation in parse trees of grammars.

Among other results, they prove that oscillation of a parse tree is closely related to its dimension.

Lemma 3 (Ganty and Valput [8]). *Let $G = (\Sigma, N, R, S)$ be a grammar in the Chomsky normal form, and let t be a parse tree in G. Then, $\operatorname{osc} t - 1 \leq \dim t \leq 2 \operatorname{osc} t$.*

Thus, k-bounded-oscillation grammars have dimension of parse trees bounded by $2k$, and Theorem 1 gives the following upper bound on the rational index of these languages.

Corollary 3. *Let L be a k-bounded-oscillation language. Then $\rho_L(n) = O(n^{4k})$.*

6 Conclusion and Open Problems

Languages of bounded tree dimension were proved to have polynomial rational index. This implies, in particular, that the CFL-reachability problem and Datalog query evaluation for these languages is in NC, and the degree of the polynomial becomes a constant factor for the circuit depth.

There is another family of languages which has polynomial rational index, *the one-counter languages*. Their rational index is known to be $O(n^2)$ [5]. Could this class be generalized in the same manner as linear languages, preserving the polynomial order of the rational index? One can consider the Polynomial Stack Lemma by Afrati et al. [1], where some restriction on the PDA stack contents is given, or investigate the properties of the substitution closure of the one-counter languages, which is known to have polynomial rational index [3].

Acknowledgment. The authors are grateful to the anonymous reviewers for numerous helpful remarks and suggestions, and particularly for alerting the authors of the work by Chytil and Monien [6].

References

1. Afrati, F., Papadimitriou, C.: The parallel complexity of simple chain queries. In: Proceedings of the Sixth ACM SIGACT-SIGMOD-SIGART Symposium on Principles of Database Systems, PODS 1987, pp. 210–213. ACM, New York (1987). https://doi.org/10.1145/28659.28682
2. Bar-Hillel, Y., Perles, M., Shamir, E.: On formal properties of simple phreise structure grammars. STUF Lang. Typol. Univ. **14**(1–4), 143–172 (1961). https://doi.org/10.1524/stuf.1961.14.14.143
3. Boasson, L., Courcelle, B., Nivat, M.: The rational index: a complexity measure for languages. SIAM J. Comput. **10**(2), 284–296 (1981)
4. Brzozowski, J.A.: Regular-like expressions for some irregular languages. In: 9th Annual Symposium on Switching and Automata Theory (swat 1968), pp. 278–286 (1968). https://doi.org/10.1109/SWAT.1968.24
5. Chistikov, D., Czerwinski, W., Hofman, P., Pilipczuk, M., Wehar, M.: Shortest paths in one-counter systems. Log. Methods Comput. Sci. **15**(1) (2019). https://doi.org/10.23638/LMCS-15(1:19)2019
6. Chytil, M.P., Monien, B.: Caterpillars and context-free languages. In: Choffrut, C., Lengauer, T. (eds.) STACS 1990. LNCS, vol. 415, pp. 70–81. Springer, Heidelberg (1990). https://doi.org/10.1007/3-540-52282-4_33
7. Esparza, J., Luttenberger, M., Schlund, M.: A brief history of strahler numbers. In: Dediu, A.-H., Martín-Vide, C., Sierra-Rodríguez, J.-L., Truthe, B. (eds.) LATA 2014. LNCS, vol. 8370, pp. 1–13. Springer, Cham (2014). https://doi.org/10.1007/978-3-319-04921-2_1
8. Ganty, P., Valput, D.: Bounded-oscillation pushdown automata. Electron. Proc. Theor. Comput. Sci. **226**, 178–197 (2016). https://doi.org/10.4204/eptcs.226.13
9. Greenlaw, R., Hoover, H.J., Ruzzo, W.L.: Limits to Parallel Computation: P-completeness Theory. Oxford University Press Inc., New York (1995)
10. Hellings, J.: Path results for context-free grammar queries on graphs. CoRR abs/1502.02242 (2015)

11. Holzer, M., Kutrib, M., Leiter, U.: Nodes connected by path languages. In: Mauri, G., Leporati, A. (eds.) DLT 2011. LNCS, vol. 6795, pp. 276–287. Springer, Heidelberg (2011). https://doi.org/10.1007/978-3-642-22321-1_24

12. Komarath, B., Sarma, J., Sunil, K.S.: On the complexity of L-reachability. In: Jürgensen, H., Karhumäki, J., Okhotin, A. (eds.) DCFS 2014. LNCS, vol. 8614, pp. 258–269. Springer, Cham (2014). https://doi.org/10.1007/978-3-319-09704-6_23

13. Luttenberger, M., Schlund, M.: Convergence of newton's method over commutative semirings. Inf. Comput. **246**, 43–61 (2016). https://doi.org/10.1016/j.ic.2015.11.008

14. Pierre, L.: Rational indexes of generators of the cone of context-free languages. Theor. Comput. Sci. **95**(2), 279–305 (1992). https://doi.org/10.1016/0304-3975(92)90269-L

15. Pierre, L., Farinone, J.M.: Context-free languages with rational index in $\theta(n^\gamma)$ for algebraic numbers γ. RAIRO - Theor. Inf. Appl. - Informatique Théorique et Appl. **24**(3), 275–322 (1990)

16. Reps, T.W.: Program analysis via graph reachability. Inf. Softw. Technol. **40**, 701–726 (1997)

17. Ullman, J.D., Van Gelder, A.: Parallel complexity of logical query programs. In: 27th Annual Symposium on Foundations of Computer Science (sfcs 1986), pp. 438–454 (Oct 1986). https://doi.org/10.1109/SFCS.1986.40

18. Wechsung, G.: The oscillation complexity and a hierarchy of context-free languages. In: Fundamentals of Computation Theory, FCT 1979, Proceedings of the Conference on Algebraic, Arthmetic, and Categorial Methods in Computation Theory, Berlin/Wendisch-Rietz, Germany, 17–21 September 1979, pp. 508–515 (1979)

19. Yannakakis, M.: Graph-theoretic methods in database theory. In: Rosenkrantz, D.J., Sagiv, Y. (eds.) Proceedings of the Ninth ACM SIGACT-SIGMOD-SIGART Symposium on Principles of Database Systems, Nashville, Tennessee, USA, April 2–4, 1990, pp. 230–242. ACM Press (1990). https://doi.org/10.1145/298514.298576

Measuring Power of Locally Testable Languages

Ryoma Sin'ya[✉]

Akita University, Akita, Japan
ryoma@math.akita-u.ac.jp

Abstract. A language L is said to be $\mathcal{C}$-measurable, where $\mathcal{C}$ is a class of languages, if there is an infinite sequence of languages in $\mathcal{C}$ that converges to L. In this paper we investigate the measuring power of LT the class of all locally testable languages. Although each locally testable language only can check some local property (prefix, suffix, and infix of some bounded length), it is shown that many non-locally-testable languages are LT-measurable. In particular, we show that the measuring power of locally testable languages coincides with the measuring power of unambiguous polynomials. We also examine the measuring power of some fragments of unambiguous polynomials.

1 Introduction

A language L is called *star-free* if it can be represented as a finite combination of Boolean operations and concatenation of finite languages, and L is called *locally testable* if it is a finite Boolean combination of languages of the form uA^*, A^*v and A^*wA^*. After the celebrated Schützenberger's theorem giving an algebraic characterisation [18] and McNaughton–Papert theorem giving a logical characterisation [10] of star-free languages, both algebraic and logical counterparts of many fragments of star-free languages are deeply well-investigated: see a survey [6] or [11] for example. In particular, McNaughton [9], Zalcstein [24], and Brzozowski–Simon [4] showed that it is decidable whether a given regular language is locally testable by giving an algebraic counterpart. Although the definition of locally testable languages is quite simple, this result is non-trivial and a proof relies on a deep algebraic decomposition theory.

In this paper, we shed new light on the fragments of star-free languages by using *measurability* which is a measure theoretic notion on formal languages. $\mathcal{C}$-measurability for a class $\mathcal{C}$ of languages is introduced by [21] and it was used for classifying non-regular languages by using regular languages. A language L is said to be $\mathcal{C}$-measurable if there is an infinite sequence of languages in $\mathcal{C}$ that converges to L. Roughly speaking, L is $\mathcal{C}$-measurable means that it can be approximated by a language in $\mathcal{C}$ with *arbitrary high precision*: the notion of "precision" is formally defined by the density of formal languages. Hence that a language L is not REG-measurable, where REG is the class of all regular languages, means that L has a complex shape so that it can not be approximated

© Springer Nature Switzerland AG 2022
V. Diekert and M. Volkov (Eds.): DLT 2022, LNCS 13257, pp. 274–285, 2022.
https://doi.org/10.1007/978-3-031-05578-2_22

by regular languages. While the membership problem for a given language L and a class $\mathcal{C}$ asks the existence of single language $K \in \mathcal{C}$ such that $L = K$, the $\mathcal{C}$-measurability asks the existence of an infinite sequence of languages in $\mathcal{C}$ that converges to L. In this sense, measurability is much more difficult than the membership problem and its analysis is a challenging task. For example, the author [22] showed that, for the class SF of all star-free languages, the class of all SF-measurable regular languages strictly contains SF but does not contain some regular languages. However, the decidability of SF-measurability is still unknown.

Instead of the class of all regular languages or star-free languages, in this paper we consider LT-measurability where LT is the class of all locally testable languages and also consider measuring power of three other fragments of star-free languages: the class UPol of all *unambiguous polynomials*, the class PT of all *piecewise testable* languages and the class AT of all *alphabet testable* languages. The main results of this paper are briefly summarised as follows.

(1) LT-measurability and UPol-measurability are equivalent (Theorem 6 and Theorem 7).
(2) AT- and PT-measurability are strictly weaker than LT-measurability and decidable for regular languages (Theorem 8, Theorem 9–11).

The result (1) is the first example of two *incomparable* subclasses of regular languages with the *same measuring power*. The result (2) (PT-measurability, in particular) is the first non-trivial examples of subclasses of regular languages with *decidable measurability*. Historically, locally testable languages [10] and unambiguous polynomials [17] are originally introduced with two different motivations: **"locality"** versus **"unambiguity"**. But interestingly, they have the same measuring power.

The structure of this paper is as follows. Section 2 provides preliminaries including density, measurability and definitions of fragments of star-free languages. The measuring power of LT, UPol and AT, PT are investigated in Sect. 3 and Sect. 4, respectively. A summary of all results and future work are described in Sect. 5.

2   Preliminaries

This section provides the precise definitions of density, measurability and local varieties of regular languages. REG_A denotes the family of all regular languages over an alphabet A. We assume that the reader has a standard knowledge of automata theory including the concept of syntactic monoids (*cf.* [8]).

2.1   Density of Formal Languages

For a set X, we denote by $\#(X)$ the cardinality of X. We denote by $\mathbb{N}$ and $\mathbb{Z}$ the set of natural numbers including 0 and the set of integers, respectively. For an alphabet A, we denote the set of all words (all non-empty words, respectively) over A by A^* (A^+, respectively). We write $|w|$ for the length of w and $A^{\leq n}$ for the set of all words of length less than or equal to n. For a word $w \in A^*$ and a letter $a \in A$, $|w|_a$ denotes the number of occurrences of a in w. We denote by $\mathrm{alph}(w) = \{a \mid |w|_a > 0\}$ the set of all letters contained in w. A word v is said to be a subword of a word w if $w = xvy$ for some $x, y \in A^*$. For a language $L \subseteq A^*$, we denote by $\overline{L} = A^* \setminus L$ the complement of L. A language L is said to be *dense* if $L \cap A^* w A^* \neq \emptyset$ holds for any $w \in A^*$. L is not dense means $L \cap A^* w A^* = \emptyset$ for some word w by definition, and such word w is called a forbidden word of L.

Definition 1 (*cf.* [2]). *The* density $\delta_A(L)$ *of* $L \subseteq A^*$ *is defined as*

$$\delta_A(L) = \lim_{n \to \infty} \frac{1}{n} \sum_{k=0}^{n-1} \frac{\#(L \cap A^k)}{\#(A^k)}$$

if its exists, otherwise we write $\delta_A(L) = \bot$. L *is called* null *if* $\delta_A(L) = 0$, *and conversely* L *is called* co-null *if* $\delta_A(L) = 1$.

Example 1. It is known that every regular language has a rational density (*cf.* [16]) and it is computable. Here we explain two examples of (co-)null languages.

(1) For each word w, the language $A^* w A^*$, the set of all words that contain w as a subword, is of density 1 (co-null). This fact is sometimes called *infinite monkey theorem*. A language L having a forbidden word w is always null: this means $A^* w A^* \subseteq \overline{L}$ and $\delta_A(A^* w A^*) \leq \delta_A(\overline{L})$ which implies $\delta_A(\overline{L}) = 1$ by infinite monkey theorem.

(2) The semi-Dyck language $D = \{\varepsilon, ab, aabb, abab, aaabbb, \ldots\}$ over $A = \{a, b\}$ is dense but null. This follows from the fact that $\#(D \cap A^{2n})$ equals the n-th Catalan number whose asymptotic formula is $\Theta(4^n / n^{3/2})$.

As explained above, "dense" does not imply "not null". But these two notions are equivalent for regular languages as the following theorem says. We denote by ZO_A the family of all null or co-null regular languages over A (ZO stands for "zero-one").

Theorem 1 (*cf.* [16]). *A regular language* L *is not null if and only if* L *is dense.*

2.2 Measurability of Formal Languages

The notion of "measurability" on formal languages is defined by a standard measure theoretic approach as follows.

Definition 2 [21]. *Let* $\mathcal{C}_A$ *be a family of languages over* A. *For a language* $L \subseteq A^*$, *we define its* $\mathcal{C}_A$ *-inner-density* $\underline{\mu}_{\mathcal{C}_A}(L)$ *and* $\mathcal{C}_A$-*outer-density* $\overline{\mu}_{\mathcal{C}_A}(L)$ *over* A *as*

$$\underline{\mu}_{\mathcal{C}_A}(L) = \sup\{\delta_A(K) \mid K \subseteq L, K \in \mathcal{C}_A, \delta_A(K) \neq \bot\} \text{ and}$$
$$\overline{\mu}_{\mathcal{C}_A}(L) = \inf\{\delta_A(K) \mid L \subseteq K, K \in \mathcal{C}_A, \delta_A(K) \neq \bot\}, \text{respectively.}$$

A language L is said to be $\mathcal{C}_A$ *-measurable if* $\underline{\mu}_{\mathcal{C}_A}(L) = \overline{\mu}_{\mathcal{C}_A}(L)$ *holds. We say that an infinite sequence* $(L_n)_n$ *of languages over* A *converges to* L *from inner (from outer, respectively) if* $L_n \subseteq L$ *($L_n \supseteq L$, respectively) for each* n *and* $\lim_{n \to \infty} \delta_A(L_n) = \delta_A(L)$.

Example 2 [21]. *The semi-Dyck language* $D = \{\varepsilon, ab, aabb, abab, aaabbb, \ldots\}$ *over* $A = \{a, b\}$ *is REG-measurable. We notice that there is no regular language* L *such that* $\delta_A(L) = 0$ *and* $D \subseteq L$, *since any null regular language has a forbidden word but* D *has no forbidden word. Hence we should construct an infinite sequence* $(L_k)_k$ *of different regular languages that converges to* D *from outer. This can be done by letting* $L_k = \{w \in A^* \mid |w|_a = |w|_b \mod k\}$. *Clearly,* $D \subseteq L_k$ *holds and it can be shown that* $\delta_A(L_k) = 1/k$ *holds. Hence* $\delta_A(L_k)$ *tends to zero if* k *tends to infinity. We will see this type of languages* L_k *again in the next section.*

For a family $\mathcal{C}_A$ *of languages over* A, *we define its Carathéodory extension and regular extension as* $\mathrm{Ext}_A(\mathcal{C}_A) = \{L \subseteq A^* \mid L \text{ is } \mathcal{C}_A\text{-measurable}\}$ *and* $\mathrm{RExt}_A(\mathcal{C}_A) = \mathrm{Ext}_A(\mathcal{C}_A) \cap \mathrm{REG}_A$, *respectively. We say that "*$\mathcal{C}_A$ *has a stronger measuring power than* $\mathcal{D}_A$*" if* $\mathrm{Ext}_A(\mathcal{C}_A) \supseteq \mathrm{Ext}_A(\mathcal{D}_A)$ *holds.*

Theorem 2 [22]. *Let* $\mathcal{C}_A \subseteq \mathrm{REG}_A$ *be a family of regular languages over* A. *Then* $L \in \mathrm{REG}_A$ *is* $\mathcal{C}_A$*-measurable if and only if* L *satisfies the following Carathéodory's condition:*

$$\forall X \subseteq A^* \qquad \overline{\mu}_{\mathcal{C}_A}(X) = \overline{\mu}_{\mathcal{C}_A}(X \cap L) + \overline{\mu}_{\mathcal{C}_A}(X \cap \overline{L}).$$

Moreover, this is equivalent to $\overline{\mu}_{\mathcal{C}_A}(L) + \overline{\mu}_{\mathcal{C}_A}(\overline{L}) = 1$ *(the case* $X = A^*$ *in the above condition).*

2.3   Fragments of Star-Free Languages

In this paper we examine measuring power of several subclasses of star-free languages equipping rich closure properties. For a family $\mathcal{C}_A$ of languages over A, we denote by $\mathscr{B}\mathcal{C}_A$ the Boolean closure of $\mathcal{C}_A$. Then the class LT_A of all locally testable languages can be defined as $\mathrm{LT}_A = \mathscr{B}\{wA^*, A^*w, A^*wA^* \mid w \in A^*\}$. A family of regular languages over A is called *local variety* [1] over A if it is closed under Boolean operations and left-and-right quotients. The reason why we focus on this type of families is that, the notion of measurability is well-behaved on Boolean operations and quotients as the following theorem says.

Theorem 3 [22]. Ext_A *is a closure operator, i.e., it satisfies the following three properties for each* $\mathcal{C} \subseteq \mathcal{D} \subseteq 2^{A^*}$: *(extensive)* $\mathcal{C} \subseteq \mathrm{Ext}_A(\mathcal{C})$, *(monotone)* $\mathrm{Ext}_A(\mathcal{C}) \subseteq \mathrm{Ext}_A(\mathcal{D})$, *and (idempotent)* $\mathrm{Ext}_A(\mathrm{Ext}_A(\mathcal{C})) = \mathrm{Ext}_A(\mathcal{C})$. *Moreover,* RExt_A *is a closure operator over the class of all local varieties of regular languages over* A, *i.e.,* $\mathcal{C}_A$*-measurability is preserved under Boolean operations and quotients for any local variety* $\mathcal{C}_A$.

Example 3. By Theorem 1, for any regular language L in ZO_A, L or its complement has a forbidden word, which implies $\emptyset \subsetneq L \subseteq \overline{A^*wA^*}$ or $A^*wA^* \subseteq L \subseteq A^*$. This fact and infinite monkey theorem implies that $\mathrm{ZO}_A \subseteq \mathrm{RExt}_A(\mathscr{B}\{A^*wA^* \mid w \in A^*\})$ holds. On the other hand, $\mathscr{B}\{A^*wA^* \mid w \in A^*\} \subseteq \mathrm{ZO}_A$ holds because ZO_A forms a local variety (*cf.* [20]). Moreover, it was shown that $\mathrm{RExt}_A(\mathrm{ZO}_A) = \mathrm{ZO}_A$ in [22]. By combining these facts with Theorem 3 we have the following chain of inclusion: $\mathrm{ZO}_A \subseteq \mathrm{RExt}_A(\mathscr{B}\{A^*wA^* \mid w \in A^*\}) \subseteq \mathrm{RExt}_A(\mathrm{ZO}_A) = \mathrm{ZO}_A$ where the second inclusion $\subseteq$ follows from the monotonicity of RExt_A.

The corresponding notion of a family of finite monoids is called *local pseudovariety* [1], and there is a natural one-to-one correspondence between the class of all local varieties and the class of all local pseudovarieties [7]. The class SF_A of all star-free languages over A forms a local variety and its corresponding local pseudovariety is the class of all aperiodic monoids [18]. Thanks to Theorem 3, the regular extension $\mathrm{RExt}_A(\mathrm{SF}_A)$ of star-free languages is also a local variety. The following theorem says that RExt_A extends SF_A non-trivially, while it does not for ZO_A.

Theorem 4 [22]. $\mathrm{SF}_A \subsetneq \mathrm{RExt}_A(\mathrm{SF}_A) \subsetneq \mathrm{REG}_A$ *if* $\#(A) \geq 2$.

The class LT_A of all locally testable languages over A is also a local variety.

The characterisation given in [4,9,24] says that L is locally testable if and only if *its syntactic semigroup is locally idempotent and commutative* (see the full version [23] for more details).

We end this section by giving precise definitions of three additional subclasses of star-free languages. We denote by AT_A the Boolean combination of languages of the form B^* where $B \subseteq A$ (AT stands for "alphabet testable", *cf.* [15]). This class also can be represented as $\mathrm{AT}_A = \mathscr{B}\{A^*aA^* \mid a \in A\}$ and hence $\mathrm{AT}_A \subsetneq \mathrm{LT}_A$. AT_A forms a (finite) local variety, and its corresponding local pseudovariety is idempotent and commutative monoids (*cf.* [6]). Clearly, the density of every language in AT_A is either zero or one, thus we have $\mathrm{AT}_A \subseteq \mathrm{ZO}_A$. A language L is called *monomial* if it is of the form $A_0^*a_1A_1^*a_2A_2^* \cdots A_{n-1}^*a_nA_n^*$ where each $a_i \in A$, $A_i \subseteq A$ and $n \geq 0$. A monomial defined above is said to be *simple* if $A_i = A$ for each i. For $w = a_1a_2 \cdots a_n$ we denote by L_w the simple monomial $A^*a_1A^*a_2A^* \cdots A^*a_nA^*$. A language is called *piecewise testable* if it can be represented as a finite Boolean combination of simple monomials. The class PT_A of all piecewise testable languages over A forms a local variety. The corresponding local pseudovariety of PT_A is the class of all $\mathcal{J}$-trivial monoids [19]. A monomial $L = A_0^*a_1A_1^* \cdots a_nA_n^*$ is *unambiguous* if for all $w \in L$ there exists exactly one factorisation $w = w_0a_1w_1 \cdots a_nw_n$ where each i satisfies $w_i \in A_i^*$. A language is an *unambiguous polynomial* if it is a finite disjoint union of unambiguous monomials. The family UPol_A of all unambiguous polynomials over A forms a local variety [17]. In particular, the complement of an unambiguous polynomial is also an unambiguous polynomial. This fact plays a key role in the next section. By definition we have the following chain of inclusion $\mathrm{AT}_A \subsetneq \mathrm{PT}_A \subsetneq \mathrm{UPol}_A \subsetneq \mathrm{SF}_A$ and every inclusion is strict. We also notice that PT_A (UPol_A, respectively) and LT_A are incomparable. For example,

$A^*abaA^* \in \mathrm{LT}_A \setminus \mathrm{PT}_A$ (because the syntactic monoid of A^*abaA^* is not $\mathcal{J}$-trivial) and $L_{aba} = A^*aA^*bA^*aA^* \in \mathrm{PT}_A \setminus \mathrm{LT}_A$ (because the syntactic monoid of L_{aba} is not locally idempotent). Every $\mathcal{J}$-trivial finite monoid has a zero element, and a language whose syntactic monoid has a zero is of density zero or one (*cf.* [20]), thus we have $\mathrm{PT}_A \subsetneq \mathrm{ZO}_A$.

3 Measuring Power of Locally Testable Languages

In this section we examine the measuring power of locally testable languages: what kind of languages are LT_A-measurable and what are not? First we show there are "many" LT_A-measurable languages.

Proposition 1. *For any language $L \subseteq A^*$, A^*LA^* is LT_A-measurable.*

Proof. If $L = \emptyset$ then $A^*LA^* = \emptyset$ is in LT_A. If $L \neq \emptyset$, we can choose $w \in L$ and the ideal language $A^*wA^* \subseteq A^*LA^*$ is co-null by infinite monkey theorem. Hence $\underline{\mu}_{\mathrm{LT}_A}(A^*LA^*) = 1$ *i.e.*, $A^*LA^* \in \mathrm{Ext}_A(\mathrm{LT}_A)$. □

If A contains two distinct letters a and b, then the subword relation $x \sqsubseteq y$ ($\Leftrightarrow$ "x is a subword of y") has an infinite antichain in A^*, *e.g.*, $\{ab^na \mid n \geq 0\}$. Two different subsets L_1 and L_2 of this infinite antichain produce two different languages $A^*L_1A^*$ and $A^*L_2A^*$. Hence the above theorem implies there are uncountably many LT_A-measurable languages. In fact, in [22], a stronger statement was shown as follows[1].

Theorem 5 [22]. *For any real number $\alpha \in [0,1]$ there is a LT_A-measurable language with density α if $\#(A) \geq 2$.*

The following proposition says that languages with modulo counting, which were used for the convergent sequence to the semi-Dyck language in Example 2, are LT_A-*im*measurable (see the full version [23] for details).

Proposition 2. *The language $L_k = \{w \in A^* \mid |w|_a = |w|_b \mod k\}$ over $A = \{a, b\}$ is LT_A-immeasurable for any $k \geq 2$.*

The next theorem says that LT_A has a stronger measuring power than UPol_A.

Theorem 6. $\mathrm{Ext}_A(\mathrm{LT}_A) \supseteq \mathrm{Ext}_A(\mathrm{UPol}_A)$ *for any A.*

We use the following simple lemma for proving this theorem.

Lemma 1. *The concatenation LK of two null regular languages L and K is also null.*

---

[1] [22] considered REG_A-measurability instead of LT_A-measurability, but the convergent sequence constructed in the proof of Theorem 5 is actually a sequence of locally testable languages.

Proof. By Theorem 1, L and K have some forbidden words $u, v \in A^*$, *i.e.*, $L \subseteq \overline{A^* u A^*}$ and $K \subseteq \overline{A^* v A^*}$. Then uv is a forbidden word of LK as follows. For any word $w \in A^* uv A^*$ and any factorisation $w = xy$, either x contains u or y contains v as a subword. This means $x \notin L$ or $y \notin K$, thus w is not in LK. $\square$

Proof (of Theorem 6). By the monotonicity and idempotency of Ext_A (Theorem 3), it is enough to show $\text{UPol}_A \subseteq \text{Ext}_A(\text{LT}_A)$: this implies $\text{Ext}_A(\text{UPol}_A) \subseteq \text{Ext}_A(\text{Ext}_A(\text{LT}_A)) = \text{Ext}_A(\text{LT}_A)$. Let $L = \biguplus_{i=1}^{k} M_i$ be an unambiguous polynomial where each M_i is an unambiguous monomial and $\uplus$ represents the disjoint union.

We show that, for each monomial M_i, $\mu_{\underline{\text{LT}}_A}(M_i) = \delta_A(M_i)$ holds, *i.e.*, we can construct a convergent sequence $(L_{i,j})_j$ of locally testable languages to M_i *from inner*: $L_{i,j} \subseteq M_i$ for each j and $\lim_{j \to \infty} \delta_A(L_{i,j}) = \delta_A(M_i)$. If M_i is null, then clearly we can take $L_{i,j} = \emptyset$ for each j. Hence we assume M_i is not null. In this case, M_i should be of the form $M_i = A_0^* a_1 A_1^* \cdots A_{n-1}^* a_n A_n^*$ and $(\star)$ there is a *unique* ℓ satisfying $A_\ell = A$. We show $(\star)$. Notice that at least one ℓ satisfies $A_\ell = A$, because if not every A_ℓ^* and every a_ℓ is clearly null and hence these concatenation M_i is also null by Lemma 1. Suppose there are two $\ell < \ell'$ with $A_\ell = A_{\ell'} = A$. In this case the word $(a_1 \cdots a_n)^2 \in M_i$ has two different factorisations:

$$(\varepsilon, a_1, \ldots, a_\ell, \underbrace{a_{\ell+1} \cdots a_n a_1 \cdots a_\ell}_{A_\ell^*}, a_{\ell+1}, \ldots, a_{\ell'}, \underbrace{\varepsilon}_{A_{\ell'}^*}, a_{\ell'+1}, \ldots, a_n, \varepsilon)$$

$$(\varepsilon, a_1, \ldots, a_\ell, \underbrace{\varepsilon}_{A_\ell^*}, a_{\ell+1}, \ldots, a_{\ell'}, \underbrace{a_{\ell'+1} \cdots a_n a_1 \cdots a_{\ell'}}_{A_{\ell'}^*}, a_{\ell'+1}, \ldots, a_n, \varepsilon)$$

This contradicts with the unambiguity of M_i. Hence $(\star)$ is true and we can write $M_i = PA^*S$ where $P = A_0^* a_1 A_1^* \cdots A_{\ell-1}^* a_\ell$ and $S = a_{\ell+1} A_{\ell+1}^* \cdots A_{n-1}^* a_n A_n^*$. Because M_i is unambiguous, for each word $w \in M_i$, there is a unique factorisation $w = xyz$ where $x \in P$, $y \in A^*$ and $z \in S$, respectively. Hence, for any $n \geq 0$, we have

$$\frac{\#(M_i \cap A^n)}{\#(A^n)} = \frac{\#(\{(x,y,z) \in P \times A^* \times S \mid |xyz| = n\})}{\#(A^n)} = \frac{\#\left(\biguplus_{(x,z) \in U_n} xA^* z \cap A^n\right)}{\#(A^n)}$$

$$= \frac{\sum_{(x,z) \in U_n} \#(xA^* z \cap A^n)}{\#(A^n)} = \sum_{(x,z) \in U_n} \#(A)^{-|xz|} \tag{1}$$

holds where $U_n = \{(x,z) \in P \times S \mid |x| + |z| \leq n\}$. Because the sequence $(\#(M_i \cap A^n)/\#(A^n))_n$ is bounded above by 1 and non-decreasing, the limit of (1) exists, say $\lim_{n \to \infty}(1) = \alpha$. In general, if a sequence converges to some value, then its average also converges to the same value. Hence we have $\delta_A(M_i) = \alpha$. For each $j \in \mathbb{N}$, the language $L_{i,j} = \bigcup_{(x,z) \in U_j} xA^* z$ is locally testable, because (i) for each $x, z \in A^*$, $xA^* z = (xA^* \cap A^* z) \setminus \{w \in A^* \mid |w| < |x| + |z|\}$ is locally testable, and (ii) U_j is finite. Moreover, $L_{i,j} \subseteq M_i$ for each j and $\delta_A(L_{i,j}) = \sum_{(x,z) \in U_j} \#(A)^{-|xz|}$. Hence $\lim_{j \to \infty} \delta_A(L_{i,j}) = \alpha = \delta_A(M_i)$, *i.e.*,

$\underline{\mu}_{\mathrm{LT}_A}(M_i) = \delta_A(M_i)$. This fact implies that $\underline{\mu}_{\mathrm{LT}_A}(L) = \delta_A(L)$ because we have the following equality:

$$\underline{\mu}_{\mathrm{LT}_A}(L) = \underline{\mu}_{\mathrm{LT}_A}\left(\biguplus_{i=1}^{k} M_i\right) = \sum_{i=1}^{k} \underline{\mu}_{\mathrm{LT}_A}(M_i) = \sum_{i=1}^{k} \delta_A(M_i) = \delta_A(L).$$

Next we show $\underline{\mu}_{\mathrm{LT}_A}(\overline{L}) = \delta_A(\overline{L})$. Notice that the complement of L is also an unambiguous polynomial since UPol_A is a local variety. Thus $\overline{L} = \biguplus_{i=1}^{k'} M_i'$ holds for some unambiguous monomials M_i'. Hence we can conclude that $\underline{\mu}_{\mathrm{LT}_A}(\overline{L}) = \delta_A(\overline{L}) = 1 - \delta_A(L)$ which implies $\underline{\mu}_{\mathrm{LT}_A}(L) + \underline{\mu}_{\mathrm{LT}_A}(\overline{L}) = 1$. Because LT_A is closed under complementation, we have $\underline{\mu}_{\mathrm{LT}_A}(K) = 1 - \overline{\mu}_{\mathrm{LT}_A}(\overline{K})$ for any K. Thus $\overline{\mu}_{\mathrm{LT}_A}(L) + \overline{\mu}_{\mathrm{LT}_A}(\overline{L}) = 1$, i.e., L is LT_A-measurable by Theorem 2. □

Next we show the reverse inclusion of Theorem 6. This direction is more easy.

Theorem 7. $\mathrm{Ext}_A(\mathrm{UPol}_A) \supseteq \mathrm{Ext}_A(\mathrm{LT}_A)$ *for any* A.

Proof. By the monotonicity and idempotency of Ext_A (Theorem 3), this is equivalent to $\mathrm{LT}_A \subseteq \mathrm{Ext}_A(\mathrm{UPol}_A)$. Moreover, UPol_A-measurability is preserved under Boolean operations by Theorem 3, we only have to show that wA^*, A^*w and A^*wA^* are all UPol_A-measurable for each $w \in A^*$. Let $w = a_1 \cdots a_n$ where each $a_i \in A$.

First we show $wA^* \in \mathrm{Ext}_A(\mathrm{UPol}_A)$. This is easy because the language $wA^* = \emptyset^* a_1 \emptyset^* a_2 \emptyset^* \cdots \emptyset^* a_n A^*$ itself is actually an unambiguous polynomial. Similarly, we also have $A^*w \in \mathrm{UPol}_A$.

Next we show $A^*wA^* \in \mathrm{Ext}_A(\mathrm{UPol}_A)$. This language is not in UPol_A in general. For example, A^*abA^* is not an unambiguous polynomial if $A = \{a, b, c\}$ (*cf.* [6]). Since the case $w = \varepsilon$ is trivial, we assume $w = a_1 \cdots a_n$ where $a_i \in A$ and $n \geq 1$. Define $W_k = (A^k \backslash K_k)wA^*$ where $K_k = \{u \in A^k \mid ua_1 \cdots a_{n-1} \in A^*wA^*\}$ for each $k \geq 0$. Intuitively, W_k is the set of all words in which w *firstly* appears at the position $k + 1$ as a subword. W_k is in UPol_A for each k, because it can be written as $W_k = \biguplus_{v \in (A^k \backslash K_k)} vwA^*$, where each vwA^* is an unambiguous polynomial as shown above, which means that this disjoint finite union W_k is also an unambiguous polynomial. Clearly, $W_i \cap W_j = \emptyset$ and $\delta_A(W_i) > 0$ for each $i \neq j$, thus we have $\biguplus_{k \geq 0} W_k = A^*wA^*$ and hence $\lim_{n \to \infty} \delta_A\left(\biguplus_{k \geq 0}^{n} W_k\right) = 1$ i.e., $\underline{\mu}_{\mathrm{UPol}_A}(A^*wA^*) = 1$. Thus $A^*wA^* \in \mathrm{Ext}_A(\mathrm{UPol}_A)$. □

Combining Theorem 6 and Theorem 7, we have the following equivalence.

Corollary 1. $\mathrm{Ext}_A(\mathrm{LT}_A) = \mathrm{Ext}_A(\mathrm{UPol}_A)$ *for each* A.

We showed that LT_A has a certain measuring power, but yet we do not know whether LT_A-measurability on REG_A is decidable or not. We only know that $\mathrm{RExt}_A(\mathrm{LT}_A)$ forms a local variety thanks to Theorem 3.

4 Measuring Power of Alphabet and Piecewise Testable Languages

For any alphabet A, AT_A is a finite family of regular languages, hence we can decide, for a given regular language $L \subseteq A^*$, whether L is AT_A-measurable or not: enumerate every pair (L_1, L_2) of languages in AT_A and check $L_1 \subseteq L \subseteq L_2$ and $\delta_A(L_1) = \delta_A(L_2) = \delta_A(L)$ holds. But the next theorem gives us a more simpler way to check AT_A-measurability than this naïve approach.

Theorem 8. *A co-null language $L \subseteq A^*$ is AT_A-measurable if and only if L contains $\bigcap_{a \in A} A^* a A^*$.*

Proof. Clearly, $\bigcap_{a \in A} A^* a A^* \in \mathrm{AT}_A$ and $\delta_A(\bigcap_{a \in A} A^* a A^*) = 1$ holds. Thus any language $L \supseteq \bigcap_{a \in A} A^* a A^*$ is AT_A-measurable. If $L \not\supseteq \bigcap_{a \in A} A^* a A^*$, then any subset of L in AT_A is null, because every language in AT_A not containing $\bigcap_{a \in A} A^* a A^*$ is a subset of $\bigcup_{B \subsetneq A} B^*$ and hence it is clearly null. $\square$

We notice that the above theorem also gives a characterisation of null AT_A-measurable languages: because AT_A is closed under complementation, L is AT_A-measurable if and only if $\overline{L}$ is AT_A-measurable by Theorem 2. Hence a null language $L \subseteq A^*$ is AT_A-measurable if and only if $\overline{L}$ contains $\bigcap_{a \in A} A^* a A^*$. The latter condition is equivalent to the following: $\mathrm{alph}(w) \neq A$ for any $w \in L$.

Next we give a simple different characterisation of PT-measurability. The following lemma can be considered as a specialised version of Theorem 1 (a regular language is co-null if and only if it contains an ideal language $A^* w A^*$) to piecewise testable languages. Notice that $A^* w A^* \subseteq L_w$ always holds hence L_w is more "larger" than $A^* w A^*$.

Lemma 2. *A piecewise testable language $L \in \mathrm{PT}_A$ is co-null if and only if it contains a simple monomial.*

Proof. ($\Leftarrow$:) this is trivial: every simple monomial L_w is co-null by infinite monkey theorem.
($\Rightarrow$:) Let $L \in \mathrm{PT}_A$ be a co-null piecewise testable language. By definition of PT_A, L can be written as a finite Boolean combination of simple monomials, hence it can be written as a disjunctive normal form $L = I_1 \cup \cdots \cup I_n$ where $n \geq 1$ and each I_i is the intersection of some simple monomials or complements of simple monomials. $\delta_A(L) = 1$ implies that, at least one I_i is the intersection of some simple monomials (otherwise $\delta_A(L) = 0$), say $I_i = L_{w_1} \cap \cdots \cap L_{w_k}$. Hence we can conclude that L contains a simple monomial $L_{w_1 \cdots w_k} \subseteq I_i \subseteq L$. $\square$

Theorem 9. *A co-null language $L \subseteq A^*$ is PT_A-measurable if and only if $L_w \subseteq L$ holds for some $w \in A^*$.*

Proof. ($\Leftarrow$): trivial.
($\Rightarrow$): L is PT_A-measurable means there is a convergent sequence $(L_k)_k$ of piecewise testable languages to L from inner. This means that, for some $i \geq 0$, $\delta_A(L_j) = 1$ holds for any $j \geq i$ because the density of each L_k is either zero or one. By Lemma 2, L_j contains a simple monomial L_{w_j} for each $j \geq i$. Hence $L_{w_i} \subseteq L_i \subseteq L$, in particular. $\square$

Table 1. Correspondence of language-algebra-logic and summary of our results.

Language	Algebra	Logic	Measurability
SF	Aperiodic	FO	$SF \subsetneq \mathrm{RExt}_A(SF) \subsetneq REG$ [22]
LT	Locally idempotent and commutative		$\mathrm{Ext}_A(LT) = \mathrm{Ext}_A(UPol)$
UPol	**DA**	FO^2	
PT	$\mathcal{J}$-trivial	$\mathbb{B}\Sigma_1$	$PT \subsetneq \mathrm{RExt}_A(PT) \subsetneq ZO$ L is PT-measurable iff L or $\overline{L}$ contains a simple monomial
AT	Idempotent and commutative	FO^1	$AT \subsetneq \mathrm{RExt}_A(AT) \subsetneq \mathrm{RExt}_A(PT)$ L is AT-measurable iff L or $\overline{L}$ contains $\bigcap_{a \in A} A^* a A^*$

We notice that the above theorem also gives a characterisation of null PT_A-measurable languages: because PT_A is closed under complementation, L is PT_A-measurable if and only if $\overline{L}$ is PT_A-measurable by Theorem 2. By using Lemma 2, we can also show that the measuring power of PT_A is strictly weaker than ZO_A as follows.

Theorem 10. $PT_A \subsetneq \mathrm{RExt}_A(PT_A) \subsetneq ZO_A$ *if* $\#(A) \geq 2$.

Proof (sketch). By using some combinatorial reasoning, we can show that $A^* w A^* \notin \mathrm{Ext}_A(PT_A)$ holds for any word $w \in A^*$ satisfying $|w| \geq 3$. See the full version [23] for the proof. $\square$

Finally, we give an algebraic characterisation of PT_A-measurability based on Theorem 9. We notice that the syntactic monoid of every co-null regular language has the zero element 0 (*cf.* [20]). We use Green's $\mathcal{J}$-relation $=_{\mathcal{J}}$ and $<_{\mathcal{J}}$ on a monoid M defined by $x =_{\mathcal{J}} y \Leftrightarrow MxM = MyM$ and $x <_{\mathcal{J}} y \Leftrightarrow MxM \subsetneq MyM$, respectively (*cf.* [8]).

Theorem 11. *A co-null regular language* $L \subseteq A^*$ *is* PT_A-*measurable if and only if* $(\diamond)$ *for every* $x \in M \setminus \{0\}$ *there is a letter* $a \in A$ *such that* $x'\eta(a) <_{\mathcal{J}} x'$ *for every* $x' =_{\mathcal{J}} x$, *where* $\eta : A^* \to M$ *and* M *is the syntactic morphism and monoid of* L, *respectively.*

Due to the space limitation, we omit the proof of the above theorem (see the full version [23] for details). Because the syntactic monoid of every regular language is finite, the condition $(\diamond)$ is decidable.

Corollary 2. PT_A-*measurability is decidable for* REG_A.

5 Summary and Future Work

For simplicity, in this section we only consider alphabets with two more letters, and omit the subscript A for denoting local varieties. Table 1 shows algebraic and logical counterparts of local varieties we considered (left) and a summary of our results (right). Here FO^n stands for first-order logic with n-variables and $\mathbb{B}\Sigma_1$ is the Boolean closure of existential first-order logic. The hierarchy of languages is strictly decreasing top down excluding that LT and UPol (PT, respectively) are incomparable. All algebraic and logical counterparts in Table 1 are nicely described in a survey [6], with the sole exception LT [4, 9, 24].

Our future work are two kinds.

(1) Prove or disprove $\mathrm{Ext}_A(\mathrm{LT}) \subsetneq \mathrm{Ext}_A(\mathrm{SF})$.
(2) Prove or disprove the decidability of LT-measurability.

To show the decidability, perhaps we can use some known techniques related to locally testable languages, for example, the so-called *separation problem* for a language class $\mathcal{C}$: for a given pair of regular languages (L_1, L_2), is there a language L in $\mathcal{C}$ such that $L_1 \subseteq L$ and $L \cap L_2 = \emptyset$ (L "separates" L_1 and L_2)? It is known that the separation problem for PT, LT, and SF are all decidable [12–14]. Theorem 8 and Theorem 9 says that, AT-measurability and PT-measurability does not rely on the existence of an infinite convergent sequence, but relies on the existence of a *single* language $\cap_{a \in A} A^* a A^*$ and L_w as a subset, respectively. But from Theorem 5, we can observe that, the situation of LT-measurability is essentially different: LT-measurability heavily relies on the existence of an *infinite sequence* of different locally testable languages. Because the density of every regular language is rational (*cf.* [16]), for each LT_A-measurable language L with an irrational density, there is no single pair of regular languages (L_1, L_2) such that $L_1 \subseteq L \subseteq L_2$ and $\delta_A(L_1) = \delta_A(L_2) = \delta_A(L)$.

Between SF and LT, there is a fine-grained infinite hierarchy called the *dot-depth hierarchy* originally introduced by Cohen and Brzozowski [5] in 1970. For a family $\mathcal{C}$ of languages, we denote by $\mathcal{MC} = \{L_1 \cdots L_k \mid k \geq 1, L_1, \ldots, L_k \in \mathcal{C}\} \cup \{\{\varepsilon\}\}$ the monoid closure of $\mathcal{C}$. The dot-depth hierarchy starts with the family $\mathcal{B}_0$ of all finite or co-finite languages, and continues as $\mathcal{B}_{i+1} = \mathcal{BMB}_i$ for each $i \geq 0$. Brzozowski and Knast [3] showed that this infinite hierarchy is strict: $\mathcal{B}_i \subsetneq \mathcal{B}_{i+1}$ for each $i \geq 0$. By definition, we have $\mathrm{SF} = \bigcup_{i \geq 0} \mathcal{B}_i$, and actually, we also have $\mathcal{B}_0 \subsetneq \mathrm{LT} \subsetneq \mathcal{B}_1$ because each of wA^*, A^*w and A^*wA^* is obtained by concatenating a finite language $\{w\}$ and a co-finite language A^*. Although the dot-depth hierarchy was introduced in a half-century before and much ink has been spent on it, the decidability of the membership problem for $\mathcal{B}_i$ is open for $i \geq 3$ and the research on this topic is still active: see a survey [11] or a recent progress given by Place and Zeitoun [15] that shows the decidability of the separation problem for $\mathcal{B}_2$, which implies the decidability of membership of $\mathcal{B}_2$. The equation $\mathrm{Ext}_A(\mathrm{LT}) = \mathrm{Ext}_A(\mathrm{SF})$ means that the dot-depth hierarchy *collapses* via Ext_A. But if not, it might be interesting to consider the new hierarchy $\mathcal{B}_0 = \mathrm{Ext}_A(\mathcal{B}_0) \subsetneq \mathrm{Ext}_A(\mathcal{B}_1) \subseteq \mathrm{Ext}_A(\mathcal{B}_2) \subseteq \cdots \subseteq \mathrm{Ext}_A(\mathrm{SF})$.

Acknowledgements. I am grateful to my colleague Y. Yamaguchi (Osaka University) and Y. Nakamura (Tokyo Tech) for useful discussions. I also thank to anonymous reviewers for many valuable comments. This work was supported by JSPS KAKENHI Grant Number JP19K14582 and JST ACT-X Grant Number JPMJAX210B, Japan.

References

1. Adámek, J., Milius, S., Myers, R.S.R., Urbat, H.: Generalized eilenberg theorem I: local varieties of languages. In: Foundations of Software Science and Computation Structures. pp. 366–380 (2014)
2. Berstel, J., Perrin, D.: Theory of codes, Pure and Applied Mathematics, vol. 117. Academic Press Inc. (1985)
3. Brzozowski, J., Knast, R.: The dot-depth hierarchy of star-free languages is infinite. J. Comput. Syst. Sci. **16**, 37–55 (1978)
4. Brzozowski, J., Simon, I.: Characterizations of locally testable events. Disc. Math. **4**, 243–271 (1973)
5. Cohen, R.S., Brzozowski, J.: Dot-depth of star-free events. J. Comput. Syst. Sci. **5**, 1–16 (1971)
6. Diekert, V., Gastin, P., Kufleitner, M.: A survey on small fragments of first-order logic over finite words. Int. J. Found. Comput. Sci. **19**(3), 513–548 (2008)
7. Gehrke, M., Grigorieff, S., Pin, J.: Duality and equational theory of regular languages. In: Automata, Languages and Programming, pp. 246–257 (2008)
8. Lawson, M.V.: Finite Automata. Chapman and Hall/CRC (2004)
9. McNaughton, R.: Algebraic decision procedures for local testability. Math. Syst. Theory **8**, 60–76 (1974)
10. McNaughton, R., Papert, S.: Coutner-Free Automata. The MIT Press (1971)
11. Pin, J.: The dot-depth hierarchy, 45 years later. In: The Role of Theory in Computer Science - Essays Dedicated to Janusz Brzozowski, pp. 177–202 (2017)
12. Place, T., van Rooijen, L., Zeitoun, M.: Separating regular languages by locally testable and locally threshold testable languages. In: IARCS Annual Conference on Foundations of Software Technology and Theoretical Computer Science, pp. 363–375 (2013)
13. Place, T., van Rooijen, L., Zeitoun, M.: Separating regular languages by piecewise testable and unambiguous languages. In: Mathematical Foundations of Computer Science 2013, pp. 729–740 (2013)
14. Place, T., Zeitoun, M.: Separating regular languages with first-order logic. Log. Methods Comput. Sci. **12**(1) (2016)
15. Place, T., Zeitoun, M.: Separation for dot-depth two. Log. Methods Comput. Sci. **17** (2021)
16. Salomaa, A., Soittola, M.: Automata Theoretic Aspects of Formal Power Series. Springer, New York (1978)
17. Schützenberger, M.: Sur le produit de concaténation non ambigu. Semigroup Forum **13**, 47–75 (1976)
18. Schützenberger, M.P.: On finite monoids having only trivial subgroups. Inf. Control **8**(2), 190–194 (1965)
19. Simon, I.: Piecewise testable events. In: Automata Theory and Formal Languages, pp. 214–222 (1975)
20. Sin'ya, R.: An automata theoretic approach to the zero-one law for regular languages. In: Games, Automata, Logics and Formal Verification, pp. 172–185 (2015)
21. Sin'ya, R.: Asymptotic approximation by regular languages. In: Current Trends in Theory and Practice of Computer Science, pp. 74–88 (2021)
22. Sin'ya, R.: Carathéodory extensions of subclasses of regular languages. In: Developments in Language Theory, pp. 355–367 (2021)
23. Sin'ya, R.: Measuring power of locally testable languages (full version) (2022). http://www.math.akita-u.ac.jp/~ryoma/misc/LTmeasure_full.pdf
24. Zalcstein, Y.: Locally testable languages. J. Comput. Syst. Sci. **6**(2), 151–167 (1972)

The Power Word Problem in Graph Products

Florian Stober and Armin Weiß[(✉)]

FMI, Universität Stuttgart, Stuttgart, Germany
`armin.weiss@fmi.uni-stuttgart.de`

Abstract. The power word problem of a group G asks whether an expression $p_1^{x_1} \ldots p_n^{x_n}$, where the p_i are words and the x_i binary encoded integers, is equal to the identity of G. We show that the power word problem in a fixed graph product is AC^0-Turing-reducible to the word problem of the free group F_2 and the power word problem of the base groups. Furthermore, we look into the uniform power word problem in a graph product, where the dependence graph and the base groups are part of the input. Given a class of finitely generated groups $\mathcal{C}$, the uniform power word problem in a graph product can be solved in $\mathsf{AC}^0(\mathsf{C}_=\mathsf{L}^{\mathrm{PowWP}\mathcal{C}})$. As a consequence of our results, the uniform knapsack problem in graph groups is NP-complete.

1 Introduction

The word problem is among the most fundamental algorithmic questions in group theory: given a word in the generators of a finitely generated (f.g.) group, decide whether that word is equal to the identity in the group. More than a century ago, Max Dehn [3] recognized the importance of the word problem with its consequences for topology. Moreover, with the discovery of finitely presented groups with undecidable word problem [16], it also became a topic of active study within computer science. Indeed, there are many natural classes of groups where the word problem is efficiently decidable. One of the most prominent examples is the class of linear groups (groups embeddable into matrix groups over some field): here the word problem can be decided in LOGSPACE [10].

In recent years variants of the word problem have been studied where the input is given in compressed form. Most notably is the so-called *compressed word problem* (or circuit value problem) where the input is given as a straight-line program (a context free grammar producing a single word). The compressed word problem is not only interesting on its own, but it naturally appears when solving the word problem in certain automorphism groups or semidirect products [11, Sect. 4.2].

The power word problem is somehow in between the word problem and the compressed word problem. It has been proposed by Lohrey and the second author [13]. The input is given as two lists $(p_1, \ldots, p_n)$ and $(x_1, \ldots, x_n)$, where p_i is a word in the generators of the group and x_i is a binary encoded

© Springer Nature Switzerland AG 2022
V. Diekert and M. Volkov (Eds.): DLT 2022, LNCS 13257, pp. 286–298, 2022.
https://doi.org/10.1007/978-3-031-05578-2_23

integer for each i. The problem is to decide whether $p_1^{x_1} \ldots p_n^{x_n}$ is equal to the identity in the group. In [13] it is shown that the power word problem for a finitely generated free group is AC^0-Turing reducible to the word problem in the free group with two generators F_2; thus, essentially, it is as difficult as the word problem.

In this work we study the power word problem in right-angled Artin groups and graph products: A right-angled Artin group (RAAG, also known as graph group or free partially commutative group) is a free group subject to relations that certain generators commute. Graph products, introduced by Green in 1990 [7], are a generalization of RAAGs: Let $(\mathcal{L}, I)$ be an undirected simple graph and let G_α for $\alpha \in \mathcal{L}$ be groups. The graph product $\mathrm{GP}(\mathcal{L}, I; (G_\alpha)_{\alpha \in \mathcal{L}})$ is the free product of the G_α subject to the relations that $g \in G_\alpha$ and $h \in G_\beta$ commute whenever $(\alpha, \beta) \in I$. In particular, RAAGs are graph products where all G_α are infinite cyclic groups.

Since RAAGs are linear, their word problem can be solved in LOGSPACE (see [4]). Moreover, in [9], Kausch gave precise a characterization of the complexity of the word problem in RAAGs and graph products – both in the case that the group is fixed and in the case that the (in-)dependence graph and the groups G_α are part of the input (the so-called *uniform* case). In the easiest case, for a fixed RAAG, he showed that the word problem is uAC^0-Turing reducible to the word problem in the free group with two generators F_2 (meaning that it can be decided by a DLOGTIME-uniform family of bounded-depth, polynomial-size Boolean circuits with oracle gates for the word problem in F_2). Moreover, in the uniform case for graph products, the word problem can be solved in the counting logspace class $\mathsf{C_=L}$ with an oracle for the (uniform) word problem in the base groups G_α.

On the other hand, the compressed word problem of a fixed graph product is polynomial time Turing reducible to the compressed word problem in the base groups [8]. Moreover, for a fixed non-abelian RAAG it is P-complete [11]. Whether the uniform compressed word problem in RAAGs is solvable in polynomial time is posed as an open problem in [12].

In this work, we fill the gaps in this picture for the power word problem in RAAGs and graph products. Our approach follows the outline in [13] for free groups. However, we have to overcome several additional difficulties: indeed, the main part of this paper considers the construction and computation of a certain "nice" normal form of the input. Moreover, we have to adapt the algorithms and proofs from [9] for the word problem to some restricted variant of the power word problem. Altogether this leads us to the following results:

Theorem A. *Let* $G = \mathrm{GP}(\mathcal{L}, I; (G_\alpha)_{\alpha \in \mathcal{L}})$ *be a graph product of f.g. groups. The power word problem in* G *can be decided in* uAC^0 *with oracle gates for the word problem in* F_2 *and for the power word problems in the base groups* G_α.

Theorem B. *Let* $\mathcal{C}$ *be a non-trivial class of f.g. groups. The uniform power word problem for graph products, where* $(\mathcal{L}, I)$ *and* $G_\alpha \in \mathcal{C}$ *for* $\alpha \in \mathcal{L}$ *are part of the input, can be decided in* $\mathsf{uAC}^0(\mathsf{C_=L}^{\mathrm{PowWP}\mathcal{C}})$ *where* $\mathrm{PowWP}\mathcal{C}$ *denotes the uniform power word problem for the class* $\mathcal{C}$.

Theorem C. *Let G be a RAAG. The power word problem in G is* uAC^0-*Turing reducible to the word problem in the free group F_2 and, thus, in* LOGSPACE.

The knapsack problem is a classical optimization problem. Myasnikov et al. have formulated the decision variant of the knapsack problem for a group G: Given $g_1, \ldots, g_n, g \in G$, decide whether there are $x_1, \ldots, x_n \in \mathbb{Z}$, with $x_i \geq 0$, such that $g_1^{x_1} \ldots g_n^{x_n} =_G g$ [15]. Lohrey and Zetzsche have studied the knapsack problem for fixed RAAGs showing, in particular, that, if G is a RAAG with independence relation I and I contains an induced subgraph C_4 or P_4, then the knapsack problem for G is NP-complete [14]. However, membership of the uniform version of the knapsack problem for RAAGs in NP remained open. Using our results on the power word problem, solves this missing piece:

Corollary D. *The uniform knapsack problem for RAAGs is* NP-*complete: On the input of a RAAG G (given as alphabet Σ and independence relation $I \subseteq \Sigma \times \Sigma$) and $g_1, \ldots, g_n, g \in G$, it can be decided in* NP *whether there are $x_1, \ldots, x_n \in \mathbb{Z}$ with $x_i \geq 0$ such that $g_1^{x_1} \ldots g_n^{x_n} =_G g$.*

This work is based on the first author's master thesis [17]. Due to lack of space most proofs are omitted and can be found in the full version on arXiv [18].

2 Preliminaries

The free monoid over Σ is the set of words $M(\Sigma) = \Sigma^*$ together with the concatenation operation. Its identity is the empty word 1. An element of Σ is called a *letter*. For $A \subseteq \Sigma$ we write $|w|_A$ for the number of letters from A in w and we set $|w| = |w|_\Sigma$. A word $w = w_1 \cdots w_n$ has *period* k if $w_i = w_{i+k}$ for all i.

Let M be a monoid. We write $x =_M y$ to indicate equality in M (as opposed to equality as words). Let $x =_M uwv$ for some $x, u, w, v \in M$; we say u is a *prefix*, w is a *factor* and v is a *suffix* of x. We call u a *proper prefix* if $u \neq x$. Similarly, w is a *proper factor* if $w \neq x$ and v is a *proper suffix* if $v \neq x$. An element $u \in M$ is *primitive* if $u \neq_M v^k$ for any $v \in M$ and $k > 1$. Two elements $u, v \in M$ are transposed if there are $x, y \in M$ such that $u =_M xy$ and $v =_M yx$. We call u and v conjugate if they are related by a series of transpositions. For a transposition in a free monoid we use the term *cyclic permutation*. Two elements $u, v \in \Sigma^*$ are related by a cyclic permutation if and only if they are conjugate.

By F_2 we denote the free group with two generators. If G is a group, then $u, v \in G$ are conjugate if and only if there is a $g \in G$ such that $u = g^{-1}vg$.

The (Power) Word Problem. Let G be a group with a presentation over the alphabet Σ. The *word problem* WP(G) is to decide for $w \in \Sigma^*$ whether $w =_G 1$. A *power word* is a word $w = w_1^{x_1} \ldots w_n^{x_n}$, where $w_1, \ldots, w_n \in \Sigma^*$ is a list of words and $x_1, \ldots, x_n \in \mathbb{Z}$ is a list of binary encoded integers. The *power word problem* PowWP(G) is to decide whether $w =_G 1$, where w is a power word. Let $\mathcal{C}$ be a class of groups. We write PowWP$\mathcal{C}$ for the uniform power word problem for groups in $\mathcal{C}$: on input of $G \in \mathcal{C}$ and a power word w, decide whether $w =_G 1$.

Circuit Complexity. A language $L \subseteq \{0,1\}^*$ is AC^0-Turing-reducible to $K \subseteq \{0,1\}^*$ if there is a family of constant-depth, polynomial-size Boolean circuits with oracle gates for K deciding L. More precisely, $L \subseteq \{0,1\}^*$ belongs to $\mathsf{AC}^0(K)$ if there exists a family $(C_n)_{n \geq 0}$ of circuits which, apart from the input gates $x_1, \ldots, x_n$ are built up from *not, and, or,* and *oracle gates* for K (which output 1 if and only if their input is in K). All gates may have unbounded fan-in, but there is a polynomial bound on the number of gates and wires and a constant bound on the depth (length of a longest path). In the following, we only consider $\mathsf{DLOGTIME}$-uniform $\mathsf{AC}^0(K)$ for which we write $\mathsf{uAC}^0(K)$. For more details on these definitions we refer to [19]. We may use oracle gates from a finite class of languages $\mathcal{C}$. We write $A \in \mathsf{uAC}^0(\mathcal{C})$ to indicate that A can be decided in uAC^0 with oracle gates for the problems in $\mathcal{C}$.

Counting Complexity Classes. Counting complexity classes are built on the idea of counting the number of accepting and rejecting paths of a Turing machine. For a non-deterministic Turing machine M, let accept_M denote the number of accepting paths and let reject_M denote the number of rejecting paths. We define $\mathrm{gap}_M = \mathrm{accept}_M - \mathrm{reject}_M$. The class of functions GapL and the class of languages $\mathsf{C_=L}$ are defined as follows:

$$\mathsf{GapL} = \left\{ \mathrm{gap}_M \ \middle| \ \begin{array}{l} M \text{ is a non-deterministic, logarithmic space-bounded} \\ \text{Turing machine} \end{array} \right\}$$

$$\mathsf{C_=L} = \{L \mid \text{There is } f \in \mathsf{GapL} \text{ with } \forall_{w \in \Sigma^*} : w \in L \Longleftrightarrow f(w) = 0\}$$

We write GapL^K and $\mathsf{C_=L}^K$ to denote the corresponding classes where the Turing machine M is equipped with an oracle for the language K. We have the following relationships of $\mathsf{C_=L}$ with other complexity classes – see e.g., [1].

$$\mathsf{uTC}^0 = \mathsf{uAC}^0(\mathrm{WP}(\mathbb{Z})) \subseteq \mathsf{uAC}^0(\mathrm{WP}(F_2)) \subseteq \mathsf{LOGSPACE} \subseteq \mathsf{NL} \subseteq \mathsf{C_=L} \subseteq \mathsf{uAC}^0(\mathsf{C_=L})$$

Rewriting Systems. A relation $S \subseteq \Sigma^* \times \Sigma^*$, where Σ is an alphabet, is called a rewriting system. We use the notation $l \to r$ to denote that $(l, r) \in S$. Based on the set S, the rewriting relation $\underset{S}{\Longrightarrow}$ is defined by $u \underset{S}{\Longrightarrow} v$ whenever there are $(l, r) \in S$ and $p, q \in \Sigma^*$ with $u = plq$ and $v = prq$.

Let $\underset{S}{\overset{+}{\Longrightarrow}}$ define the transitive closure and $\underset{S}{\overset{*}{\Longrightarrow}}$ the reflexive, transitive closure. We write $u \underset{S}{\overset{\leq k}{\Longrightarrow}} v$ to denote that u can be rewritten to v using at most k steps. We say that a word $w \in \Sigma^*$ is *irreducible* w.r.t. S if there is no $v \in \Sigma^*$ with $w \underset{S}{\Longrightarrow} v$. The set of irreducible words is denoted as $\mathrm{IRR}(S) = \{w \in \Sigma^* \mid w \text{ is irreducible}\}$.

Partially Commutative Monoids. Let $M(\Sigma) = \Sigma^*$ be the free monoid over the set of generators Σ. Let $I \subseteq \Sigma \times \Sigma$ be symmetric and irreflexive. The partially commutative monoid defined by (Σ, I) is $M(\Sigma, I) = M(\Sigma)/\{ab = ba \mid (a, b) \in I\}$. Thus, the relation I describes which generators commute; it is called the *commutation relation* or *independence relation*. The relation $D = (\Sigma \times \Sigma) \setminus I$ is called *dependence relation* and (Σ, D) is called *dependence graph*.

Elements of a partially commutative monoid can represented by a directed acyclic graph: Let $w = u_1 \ldots u_n$ with $u_i \in \Sigma$. Each u_i is a node in the graph; there is an edge $u_i \to u_j$ if and only if $i < j$ and $(u_i, u_j) \in D$. Some $v \in M(\Sigma, I)$ is said to be *connected* if this directed acyclic graph is weakly connected – or, equivalently, if the induced subgraph of (Σ, D) consisting only of the letters occurring in v is connected.

Graph Products. Let $(G_\alpha)_{\alpha \in \mathcal{L}}$ be a family of groups and $I \subseteq \mathcal{L} \times \mathcal{L}$ be an irreflexive, symmetric relation (the *independence relation*). The graph product $\mathrm{GP}(\mathcal{L}, I; (G_\alpha)_{\alpha \in \mathcal{L}})$ is defined as the free product of the G_α modulo the relations that G_α and G_β commute whenever $(\alpha, \beta) \in I$.

For representing a graph product, we use $\Gamma = \bigcup_{\alpha \in \mathcal{L}}(G_\alpha \setminus \{1\})$ as an alphabet. For $u, v \in G_\alpha$ we write $[uv]$ for the element obtained by multiplying uv in G_α (whereas uv denotes the two-letter word in Γ^*). For $w \in \Gamma$ we define $\mathrm{alph}(w) = \alpha$ where $w \in G_\alpha$. For $\mathbf{u} = u_1 \ldots u_k \in \Gamma^*$ we define $\mathrm{alph}(\mathbf{u}) = \{\mathrm{alph}(u_1), \ldots, \mathrm{alph}(u_k)\}$. We extend the independence relation over $\Gamma \times \Gamma$ by $I = \{(u, v) \mid (\mathrm{alph}(u), \mathrm{alph}(v)) \in I\}$ and even further to $I \subseteq \Gamma^* \times \Gamma^*$ by $(u, v) \in I$ whenever $(\alpha, \beta) \in I$ for all $\alpha \in \mathrm{alph}(u)$ and $\beta \in \mathrm{alph}(v)$. Hence,

$$\mathrm{GP}(\mathcal{L}, I; (G_\alpha)_{\alpha \in \mathcal{L}}) = \langle \Gamma \mid uv = [uv] \text{ for } u, v \in G_\alpha; \ uv = vu \text{ for } (u, v) \in I \rangle.$$

We represent elements of a graph product by elements of the corresponding partially commutative monoid $M(\Gamma, I)$. The canonical homomorphism $M(\Gamma, I) \to \mathrm{GP}(\mathcal{L}, I; (G_\alpha)_{\alpha \in \mathcal{L}})$ is surjective. A *reduced* representative of a group element is a representative of minimum length. Equivalently, $w \in M(\Gamma, I)$ is reduced if there is no two-letter factor uv of w such that $\mathrm{alph}(u) = \mathrm{alph}(v)$. A word $w \in \Gamma^*$ is called *reduced* if its image in $M(\Gamma, I)$ is reduced. A word $w \in \Gamma^*$ is called *cyclically reduced* if for all $u, v \in \Gamma^*$ with $w =_{M(\Gamma,I)} uv$ the transposed word vu is reduced. Let S be the rewriting system for G defined by the following relations, where $a, b \in \Gamma$ and $\mathbf{u} \in \Gamma^*$.

$$(*) \qquad a\mathbf{u}b \to [ab]\mathbf{u} \qquad\qquad \text{if } \mathrm{alph}(a) = \mathrm{alph}(b) \text{ and } (a, \mathbf{u}) \in I$$

Then $w \in \Gamma^*$ is reduced if and only if $w \in \mathrm{IRR}(S)$. Moreover, observe that $w \overset{*}{\underset{S}{\Rightarrow}} 1$ if and only if $w =_G 1$.

Remark 1. Let $G = \mathrm{GP}(\mathcal{L}, I; (G_\alpha)_{\alpha \in \mathcal{L}})$ and $M = M(\Gamma, I)$ and $u, v \in \Gamma^*$. If $u =_M v$, then also $u =_G v$. Moreover, if u and v are reduced, then $u =_M v$ if and only if $u =_G v$.

Since Γ might be an infinite alphabet, for inputs of algorithms, we need to encode elements of Γ over a finite alphabet. For $\alpha \in \mathcal{L}$ let Σ_α be an alphabet for G_α (i.e., there is a surjective homomorphism $\Sigma_\alpha^* \to G_\alpha$). Then, clearly every element of Γ can be represented as a word over $\Sigma = \bigcup_{\alpha \in \mathcal{L}} \Sigma_\alpha$. However, in general, representatives are not unique. To decide whether two words $w, v \in \Sigma_\alpha^*$ represent the same element of Γ is the word problem of G_α.

3   Cyclic Normal Forms and Conjugacy

In this section we develop various tools concerning combinatorics on traces, which later we will use to solve the power word problem in graph products. In particular, we aim for some special kind of cyclic normal forms ensuring uniqueness within a conjugacy class (see Definition 9 below).

Cyclic Normal Forms. By $\leq_{\mathcal{L}}$ we denote a linear order on the set $\mathcal{L}$. The *length-lexicographic normal form* of $g \in G$ is the reduced representative $\mathrm{nf}_G(g) = w \in \Gamma^*$ for g that is lexicographically smallest. Note that this normal form is on the level of Γ. Each letter of Γ still might have different representations over the alphabet Σ as outlined above.

Definition 2. *Let $w \in \Gamma^*$ be cyclically reduced. We say w is a* cyclic normal form *if w is a length-lexicographic normal form and all its cyclic permutations are length-lexicographic normal forms.*

Cyclic normal forms have been introduced in [2] for RAAGs. Moreover, by [2], given an element which has a cyclic normal form, it can be computed in linear time. In Theorem 5 below, we show that they also exist in the case of graph products and that they also can be computed efficiently. It is easy to see that every cyclic normal form is connected. To emphasize the importance of being connected we always require connectedness explicitly in the following. In particular, not every element has a cyclic normal form. Moreover, there can be more than one cyclic normal per conjugacy class; however, by Definition 2 they are all cyclic permutations of each other.

We make use of the following two crucial properties of cyclic normal forms. The proof of Lemma 3 follows the outline given in [2] for RAAGs.

Lemma 3. *Let $u, v \in \Gamma^*$ be cyclic normal forms with $|\mathrm{alph}(u)|, |\mathrm{alph}(v)| \geq 2$. Then u and v are conjugate (in G) if and only if u is a cyclic permutation of v (as words over Γ).*

Lemma 4. *Let $u \in \Gamma^*$ be a connected cyclic normal form with $|\mathrm{alph}(u)| \geq 2$. If $u =_G v^k$, then $\mathrm{nf}_G(v)$ is a cyclic normal form and $u = \mathrm{nf}_G(v)^k$ (as words).*

Theorem 5. *In the following, w is cyclically reduced and connected.*

- *Let $G = \mathrm{GP}(\mathcal{L}, I; (G_\alpha)_{\alpha \in \mathcal{L}})$ be a graph product of f.g. groups. Then computing a cyclic normal form conjugate to $w \in \Gamma^*$ is in $\mathsf{uAC}^0 \subseteq \mathsf{uAC}^0(\mathrm{WP}(F_2))$.*
- *Given a non-trivial class of f.g. groups $\mathcal{C}$, computing a cyclic normal form conjugate to $w \in G = \mathrm{GP}(\mathcal{L}, I; (G_\alpha)_{\alpha \in \mathcal{L}})$, where $(\mathcal{L}, I)$ and $G_\alpha \in \mathcal{C}$ for $\alpha \in \mathcal{L}$ are part of the input, is in $\mathsf{uAC}^0(\mathrm{NL})$.*

Proof (Sketch). We use the fact, that if v is cyclically reduced and conjugate to $w^{|\mathcal{L}|}$, then v is already transposed to $w^{|\mathcal{L}|}$ (which is not true for w itself). Therefore, the cyclic normal form can be computed with the following algorithm:

1. Compute the length-lexicographic normal form $\tilde{w} = \mathrm{nf}_G(w^{|\mathcal{L}|})$.
2. Let $\tilde{w} = ydz$, where $d \in G_\alpha$ such that α is maximal w.r.t. $\leq_{\mathcal{L}}$, $y \in (\Gamma \setminus G_\alpha)^*$ and $z \in \Gamma^*$. Compute the cyclic permutation dzy. That is, rotate the first occurrence of d to the front.
3. Compute the length-lexicographic normal form of dzy. We have $\mathrm{nf}_G(dzy) = u^{|\mathcal{L}|}$, where u is a cyclic normal form conjugate to w.

The complexity of the algorithm is dominated by the computation of the length-lexicographic normal form, see [9, Theorem 6.3.7, Theorem 6.3.13]. □

Factors, Suffixes and Prefixes. Next we characterize the shape of a prefix, suffix or factor of a power in a graph product (denoted as above). We write $\sigma = |\mathcal{L}|$.

Lemma 6. *Let $p \in M(\Gamma, I)$ be connected, $k \in \mathbb{N}$. Then we have:*

   (i) Every prefix w of p^k can be written as $w = p^x w_1 \cdots w_s$ where $x \in \mathbb{N}$, $s < \sigma$ and w_i is a proper prefix of p for each i.

   (ii) Every suffix u of p^k can be written as $u = u_1 \cdots u_t p^x$ where $x \in \mathbb{N}$, $t < \sigma$ and u_i is a proper suffix of p for each i.

   (iii) Given a factor v of p^k at least one of the following is true.

       $- v = u_1 \cdots u_a v_1 \ldots v_b w_1 \cdots w_c$ where $a, b, c \in \mathbb{N}$, $a + b + c \leq 2\sigma - 2$, u_i is a proper suffix of p for $1 \leq i \leq a$, v_i is a proper factor of p for $1 \leq i \leq b$ and w_i is a proper prefix of p for $1 \leq i \leq c$.

       $- v = u_1 \cdots u_a p^b w_1 \cdots w_c$ where $a, b, c \in \mathbb{N}$, $a, c < \sigma$, u_i is a proper suffix of p for $1 \leq i \leq a$ and w_i is a proper prefix of p for $1 \leq i \leq c$.

Projections to Free Monoids. We define a projection to a direct product of free monoids similar to [20]. Let $\mathcal{A} = \{\Gamma_\alpha \cup \Gamma_\beta \mid (\alpha, \beta) \in D\}$. For $A_i \in \mathcal{A}$ let $\pi_i : \Sigma^* \to A_i^*$ be the projection to the free monoid A_i^*, defined by $\pi_i(a) = a$ for $a \in A_i$ and $\pi_i(a) = 1$ otherwise. We define $\Pi : \Sigma^* \to A_1^* \times \cdots \times A_k^*, w \mapsto (\pi_1(w), \ldots, \pi_k(w))$. By Lemma 1 the following lemmata presented in [5] for trace monoids hold also for reduced representatives in graph products of groups.

Lemma 7. *Proposition 1.2 in [5] For reduced $u, v \in M = M(\Sigma, I)$ we have $u =_M v$ if and only if $\Pi(u) = \Pi(v)$.*

Lemma 8. *Proposition 1.7 in [5] Let $w \in M = M(\Sigma, I)$ and $t > 1$. Then, there is $u \in \Sigma^*$ with $w =_M u^t$ if and only if there is $a \in \prod A_i^*$ with $\Pi(w) = a^t$.*

   We are going to apply Lemma 8 to the situation that $G = \mathrm{GP}(\mathcal{L}, I; (G_\alpha)_{\alpha \in \mathcal{L}})$ is a graph product of groups and $w \in \Gamma^*$ is cyclically reduced, connected, and $|\mathrm{alph}(w)| \geq 2$. Then, for any $t > 1$ there exists $u \in \Gamma^*$ with $w =_G u^t$ if and only if there is $a \in \prod A_i^*$ with $\Pi(w) = a^t$.

Conjugacy. For the rest of this section let $G = \mathrm{GP}(\mathcal{L}, I; (G_\alpha)_{\alpha \in \mathcal{L}})$ be a graph product and $M(\Gamma, I)$ as in Sect. 2.

Definition 9. *Let Ω be the set of all $w \in \Gamma^*$ satisfying the following properties:*

- $|\mathrm{alph}(w)| \geq 2$ *(in particular, $w \neq_G 1$),*
- w *is cyclically reduced,*
- w *represents a primitive and connected element of $M(\Gamma, I)$,*
- w *is a cyclic normal form,*
- w *is minimal w.r.t. $\leq_{\mathcal{L}}$ among its cyclic permutations and the cyclic permutations of a cyclic normal form of its inverse.*

The crucial property of Ω is that each $w \in \Omega$ is a unique representative for its conjugacy class and the conjugacy class of its inverse. That leads us to the following theorem, which is central to solving the power word problem in graph products (see Lemma 15 below). The intuition behind it is that, if there are two powers p^x and q^y, where $p, q \in \Omega$ and $q \neq p$, then in $p^x q^y$ only a small number of letters can cancel out. Conversely, if a sufficiently large suffix of p^x cancels with a prefix of q^y, then $p = q$. In the end this will allow us to decrease all the exponents of p simultaneously as described in Definition 18.

Theorem 10. *Let $p, q \in \Omega$, and $x, y \in \mathbb{Z}$. Moreover, let u be a factor of p^x and v a factor of q^y (read as elements of $M(\Gamma, I)$) such that u and v are reduced and $uv =_G 1$. If $|u| = |v| > 2\sigma(|p| + |q|)$, then $p = q$.*

We derive Theorem 10 from the following lemma. Note that, if p is connected, p^x is cyclically reduced if and only if p is cyclically reduced and $|\mathrm{alph}(p)| \geq 2$.

Lemma 11. *Let $p, q, v \in M(\Gamma, I)$, $x, y \in \mathbb{N}$ such that p and q are primitive and connected, p^x and q^y are cyclically reduced and have a common factor v. If p^2 and q^2 are factors of v, then for all i the projections $\pi_i(p)$ and $\pi_i(q)$ are conjugate as words.*

Proof. We define $J_v = \{i \mid \mathrm{alph}(A_i) \subseteq \mathrm{alph}(v)\} = \{i \mid |\mathrm{alph}(\pi_i(v))| = 2\}$. For each $i \in J_v$ we write $\pi_i(p) = \tilde{p}_i^{s_i}$ and $\pi_i(q) = \tilde{q}_i^{r_i}$ where $\tilde{p}_i$ and $\tilde{q}_i$ are primitive. As v is a common factor of p^x and q^y, its projection $\pi_i(v)$ is a common factor of $\pi_i(p^x) = \tilde{p}_i^{s_i x}$ and $\pi_i(q^y) = \tilde{q}_i^{r_i y}$. Thus $\pi_i(v)$ has periods $|\tilde{p}_i|$ and $|\tilde{q}_i|$. Since p^2 is a factor of v, $\pi_i(p^2)$ is a factor of $\pi_i(v)$. This yields the lower bound $2|\tilde{p}_i|$, and by symmetry $2|\tilde{q}_i|$, on the length of $\pi_i(v)$. Combining those we obtain $|\pi_i(v)| \geq \max\{2|\tilde{p}_i|, 2|\tilde{q}_i|\} \geq |\tilde{p}_i| + |\tilde{q}_i| \geq |\tilde{p}_i| + |\tilde{q}_i| - 1$. By the theorem of Fine and Wilf [6], we have that $|\tilde{p}_i| = |\tilde{q}_i|$. As p is a factor of v, we have that $\tilde{p}_i$ is a factor of $\pi_i(v)$ and transitively of $\pi_i(q^y) = \tilde{q}_i^{r_i \cdot y}$. Hence, $\tilde{p}_i$ and $\tilde{q}_i$ are conjugate.

Assume that for some i we have $s_i \neq r_i$. Then, there are λ and μ such that $\lambda s_i = \mu r_i$. W.l.o.g. let $\mu \neq 1$ and $\gcd\{\lambda, \mu\} = 1$. Now μ divides s_i. Let J be the set of indices $j \in J_v$ such that $\lambda s_j = \mu r_j$. Clearly $i \in J$. Let ℓ be an index such that $A_\ell \cap A_j \neq \emptyset$ for some j in J. Let $\{\alpha\} = \mathrm{alph}(A_\ell \cap A_j)$. We write $|w|_\alpha$ for $|w|_{\Gamma_\alpha}$. We have $s_\ell |\tilde{p}_\ell|_\alpha = |p|_\alpha = s_j |\tilde{p}_j|_\alpha$. Similarly, we have $r_\ell |\tilde{q}_\ell|_\alpha = |q|_\alpha = r_j |\tilde{q}_j|_\alpha$, which is equivalent to $r_\ell |\tilde{p}_\ell|_\alpha = r_j |\tilde{p}_j|_\alpha$ (as $\tilde{p}_i$ and $\tilde{q}_i$ are conjugate for all i). Thus, we obtain $\lambda s_\ell |\tilde{p}_\ell|_\alpha = \lambda s_j |\tilde{p}_j|_\alpha = \mu r_j |\tilde{p}_j|_\alpha = \mu r_\ell |\tilde{p}_\ell|_\alpha$. Since $|\tilde{p}_\ell|_\alpha \neq 0$, we conclude $\lambda s_\ell = \mu r_\ell$. As p is connected, we obtain $J = J_v$ by induction. Thus, every s_i is divisible by μ, and by Lemma 8 we can write $p = u^\mu$ contradicting p being primitive. □

The proof idea for Theorem 10 is as follows: We use the length bound from Lemma 6 in order to show that the requirements of Lemma 11 are satisfied. After applying that lemma, we show that p and q are conjugate using Lemma 7. From the definition of Ω it follows that $p = q$.

4   The Power Word Problem in Graph Products

In this section we show our main results: In order to solve the power word problem, we follow the outline of [13]. In particular, our proof also consists of three major steps:

– In a preprocessing step we replace all powers with powers of elements of Ω.
– We define a symbolic rewriting system which we use to prove correctness.
– We define the shortened word, replacing each exponent with a smaller one, bounded by a polynomial in the input.

Here we rely on Theorem 10 instead of [13, Lemma 11], an easy fact about words.

The Simple Power Word Problem. Let $G = \mathrm{GP}(\mathcal{L}, I; (G_\alpha)_{\alpha \in \mathcal{L}})$ be a graph product. A *simple power word* is a word $w = w_1^{x_1} \ldots w_n^{x_n}$, where $w_1, \ldots, w_n \in \Gamma$ (each w_i again encoded over the finite alphabet Σ as outlined above) and $x_1, \ldots, x_n \in \mathbb{Z}$ is a list of binary encoded integers. Note that this is more restrictive than a power word: We only allow powers of elements in a single base group. The *simple power word problem* SPowWP(G) is to decide whether $w =_G 1$, where w is a simple power word.

The following result on the complexity of the simple power word problem is obtained by using the corresponding algorithm for the word problem [9, Theorem 5.6.5, Theorem 5.6.14] and replacing the oracles for the word problem of the base groups with oracles for the power word problem in the base groups.

Proposition 12. *For the simple power word problem the following holds.*

– *Let* $G = \mathrm{GP}(\mathcal{L}, I; (G_\alpha)_{\alpha \in \mathcal{L}})$ *be a graph product of f.g. groups. Then* SPowWP$(G) \in \mathsf{uAC}^0(\{\mathrm{WP}(F_2)\} \cup \{\mathrm{PowWP}(G_\alpha) \mid \alpha \in \mathcal{L}\})$.
– *Let* $\mathcal{C}$ *be a non-trivial class of f.g. groups,* $G = \mathrm{GP}(\mathcal{L}, I; (G_\alpha)_{\alpha \in \mathcal{L}})$. *Then* SPowWP$(\mathrm{GP}\mathcal{C}) \in \mathsf{C_=L}^{\mathrm{PowWP}\mathcal{C}}$.

Preprocessing. Let $G = \mathrm{GP}(\mathcal{L}, I; (G_\alpha)_{\alpha \in \mathcal{L}})$ be a graph product of f.g. groups. We define the alphabet $\tilde{\Gamma} = \Gamma \times \mathbb{Z}$, where (v, z) represents the letter v^z. Note that $\tilde{\Gamma}$ is the alphabet of the simple power word problem in G. During preprocessing, the input is transformed into the form $w = u_0 p_1^{x_1} u_1 \ldots p_n^{x_n} u_n$, where $p_i \in \Omega$ and $u_i \in \tilde{\Gamma}^*$ for all i. We denote the uniform word problem for graph products with base groups in $\mathcal{C}$ by WPGP$\mathcal{C}$.

Lemma 13. *The preprocessing can be reduced to the word problem.*

- *Let $G = \mathrm{GP}(\mathcal{L}, I; (G_\alpha)_{\alpha \in \mathcal{L}})$ be a graph product of f.g. groups. Then computing the preprocessing is in $\mathsf{uAC}^0(\mathrm{WP}(G), \mathrm{WP}(F_2))$.*
- *Let $\mathcal{C}$ be a non-trivial class of f.g. groups. Given $w \in G = \mathrm{GP}(\mathcal{L}, I; (G_\alpha)_{\alpha \in \mathcal{L}})$, where $(\mathcal{L}, I)$ and $G_\alpha \in \mathcal{C}$ for $\alpha \in \mathcal{L}$ are part of the input, the preprocessing can be done in $\mathsf{uAC}^0(\mathrm{WPGP}\mathcal{C}, \mathsf{NL})$.*

Proof (Sketch). The preprocessing consists of several steps: First, we reduce all powers p_i cyclically using [9, Lemma 7.3.4]. Next, we replace the powers by their connected components. In the uniform case this requires to solve the (undirected) path connectivity problem, which is in NL. Then, we replace each element of Γ by a unique (but arbitrary) representative over the finite alphabet Σ. After that we can compute cyclic normal forms using Theorem 5 and replace a power p by the lexicographically minimal cyclic permutation of this cyclic normal form. By Lemma 3 this gives us a unique representative for each conjugacy class. Finally, in order to make the powers primitive, we use Lemma 4. $\qquad\square$

Symbolic Rewriting System. Let $G = \mathrm{GP}(\mathcal{L}, I; (G_\alpha)_{\alpha \in \mathcal{L}})$ be a graph product of f.g. groups. Let S denote the rewriting system from $(*)$. As before, let $\sigma = |\mathcal{L}|$. Recall that we defined the input alphabet of the simple power word problem $\tilde{\Gamma} = \Gamma \times \mathbb{Z}$. A letter $(v, z) \in \tilde{\Gamma}$ is interpreted as v^z. In $\tilde{\Gamma}^*$ we can have powers of individual letters (which are words in the base groups), but not powers of words containing letters from multiple base groups. For some $x \in \mathbb{Z} \setminus \{0\}$ we denote by $\mathrm{sgn}\, x \in \{\pm 1\}$ the sign of x. We define the alphabet Δ by $\Delta' = \bigcup_{p \in \Omega} \Delta_p$, $\Delta'' = \tilde{\Gamma}$ and $\Delta = \Delta' \cup \Delta''$, where for $p \in \Omega$

$$\Delta_p = \left\{ \beta_i p^x \alpha \;\middle|\; \begin{array}{l} x \in \mathbb{Z}, \\ \alpha \in \mathrm{IRR}(S) \text{ is a prefix of } p^{\sigma \mathrm{sgn}\, x} \text{ and } p \text{ is no prefix of } \alpha, \\ \beta \in \mathrm{IRR}(S) \text{ is a suffix of } p^{\sigma \mathrm{sgn}\, x} \text{ and } p \text{ is no suffix of } \beta. \end{array} \right\}$$

The rewriting system T over Δ^* is defined by the following rules, where $\beta p^x \alpha$, $\delta p^y \gamma$, $\delta q^y \gamma \in \Delta'$; $a, b \in \Delta''$; $r \in \Delta''^*$; $d, e \in \mathbb{Z}$; $0 \le k \le \sigma$; $a_i \in \Delta''$ and $\mathbf{u} \in \Delta^*$.

(1) $\quad \beta p^x \alpha \mathbf{u} \delta p^y \gamma \to \beta p^{x+y+d} \gamma \mathbf{u}$ $\qquad\qquad\qquad$ if $\alpha \delta \xRightarrow{*}_S p^d$ and $(p, \mathbf{u}) \in I$

(2) $\quad \beta p^x \alpha \mathbf{u} \delta p^y \gamma \to \beta p^{x-d} \alpha' a_1 \cdots a_k \mathbf{u} \delta' p^{y-e} \gamma$

$\qquad\qquad$ if $((\nexists c \in \mathbb{Z}: \alpha \delta \xRightarrow{*}_S p^c)$ or $(p, \mathbf{u}) \notin I)$, $\beta p^x \alpha \mathbf{u} \in \mathrm{IRR}(S)$,

$\qquad\qquad \mathbf{u} \delta p^y \gamma \in \mathrm{IRR}(S)$, and $p^x \alpha \mathbf{u} \delta p^y \xRightarrow{+}_S p^{x-d} \alpha' a_1 \cdots a_k \mathbf{u} \delta' p^{y-e} \in \mathrm{IRR}(S)$

(3) $\quad \beta p^x \alpha \mathbf{u} \delta q^y \gamma \to \beta p^{x-d} \alpha' a_1 \cdots a_k \mathbf{u} \delta' q^{y-e} \gamma$

$\qquad\qquad$ if $p \ne q$ and $p^x \alpha \mathbf{u} \delta q^y \xRightarrow{+}_S p^{x-d} \alpha' a_1 \cdots a_k \mathbf{u} \delta' q^{y-e} \in \mathrm{IRR}(S)$

(4) $\quad \beta p^x \alpha \to r$ $\qquad\qquad\qquad\qquad\qquad$ if $\beta \alpha \xRightarrow{*}_S r \in \mathrm{IRR}(S)$ and $x = 0$

(5) $\quad a \mathbf{u} \beta p^x \alpha \to a' \mathbf{u} \beta' p^{x-d} \alpha$

$\qquad\qquad\qquad$ if $(a, \mathbf{u}) \in I$ and $a \mathbf{u} \beta p^x \xRightarrow{+}_S a' \mathbf{u} \beta' p^{x-d} \in \mathrm{IRR}(S)$

(6) $\beta p^x \alpha \mathbf{u} a \to \beta p^{x-d} \alpha' \mathbf{u} a'$

$$\text{if } (a, \mathbf{u}) \in I \text{ and } p^x \alpha \mathbf{u} a \xRightarrow[s]{+} p^{x-d} \alpha' \mathbf{u} a' \in \text{IRR}(S)$$

(7) $a \mathbf{u} b \to r \mathbf{u}$ if $(a, \mathbf{u}) \in I, \text{alph}(a) = \text{alph}(b), \text{ and } r = [ab]$

We define the projection $\pi : \Delta^* \to \Gamma^*$ by $\pi(a, k) = a^k$ for $(a, k) \in \Delta''$ and $\pi(\beta p^x \alpha) = \beta p^x \alpha$ for $\beta p^x \alpha \in \Delta'$ and $\pi(w) = \pi(w_1) \dots \pi(w_n)$. We write $|w|_\Gamma$ to emphasize that w is a word in Γ^*. The following fact about T is crucial.

Lemma 14. *For $u \in \Delta^*$ we have $\pi(u) =_G 1$ if and only if $u \xRightarrow[T]{*} 1$.*

Lemma 15. *The following length bounds hold:*

- *Rule (2): $|d| \le 2\sigma$ and $|e| \le 2\sigma$*
- *Rule (3): $|d| \le 4\sigma |q|_\Gamma$ and $|e| \le 4\sigma |p|_\Gamma$*
- *Rule (4): $|r|_\Gamma < 2(\sigma - 1)|p|_\Gamma$*

Proof (Sketch). When applying rule (3) we have a suffix α'' of $p^x \alpha$ and a prefix δ'' of δq^y such that $\alpha'' \delta'' \xRightarrow[s]{*} a_1 \dots a_k$. The Γ-length of the suffix of α'' that cancels with a prefix of δ'' is at most $2\sigma(|p|_\Gamma + |q|_\Gamma)$. Otherwise, we would have $p = q$ by Theorem 10 contradicting $p \ne q$. Considering the additional letters $a_1, \dots, a_k$ and the bound on $|\alpha'|_\Gamma$ from Lemma 6, we obtain $|d| < 4\sigma |q|_\Gamma$. The bound on $|e|$ follows by symmetry. The other cases follow similarly.   □

Let $w = w_1 \dots w_n \in \Delta^*$. We define $\mu(w) = \max\{|p|_\Gamma \mid w_i = \beta p^x \alpha \in \Delta'\}$ and $\lambda(w) = |w|_{\Delta''} + \sum_{w_i = \beta p^x \alpha \in \Delta'} |p|_\Gamma$.

Lemma 16. *If $w \xRightarrow[T]{*} v$, then $w \xRightarrow[T]{\le k} v$ with $k = 10\sigma^3 \lambda(w) + 6\sigma^2 \lambda(w)^2$.*

Proof (Sketch). Rules (4) and (1) clearly can be applied at most $|w|_{\Delta'}$ times in total. We conclude that rules (2) and (3) can be applied at most $2\sigma |w|_{\Delta'}$ times. Hence, at most $(2\sigma^2 + 2(\sigma - 1)\lambda(w))\lambda(w)$ new letters from Δ'' are created. Rule (7) reduces the length of the word, rules (5) and (6) either reduce the length or preserve it. One can compute an upper bound of $10\sigma^3 \lambda(w) + 6\sigma^2 \lambda(w)^2$ on the total number of rules which can be applied.   □

The Shortened Word. In this section we describe the shortening process. It is an almost verbatim repetition of [13] and we keep it as short as possible. Let $u \in \Delta^*$ and $p \in \Omega$. We write $u = u_0 \beta_1 p^{y_1} \alpha_1 u_1 \dots \beta_m p^{y_m} \alpha_m u_m$ with $u_i \in (\Delta \setminus \Delta_p)^*$ and $\beta_i p^{y_i} \alpha_i \in \Delta_p$. We define $\eta_p = \sum_{j=1}^m y_j$ and $\eta_p^i = \sum_{j=1}^i y_j$. The following lemma follows from the bounds given in Lemma 15.

Lemma 17. *Let $u, v \in \Delta^*$ and $u \xRightarrow[T]{} v$. For every prefix v' of v there is a prefix u' of u such that $|\eta_p(u') - \eta_p(v')| \le 4\sigma\mu(u)$ for all $p \in \Omega$.*

If the applied rule is neither (1) nor (4), then for all $p \in \Omega$ and $0 \le i \le m$ we have $|\eta_p^i(u) - \eta_p^i(v)| \le 4\sigma\mu(u)$.

From Lemma 16 we know that, if $\pi(u) =_G 1$, then $u \overset{\leq k}{\underset{T}{\Longrightarrow}} 1$ with $k = 10\sigma^3\lambda(w) + 6\sigma^2\lambda(w)^2$. By Lemma 17, each application of a rule changes $\eta_p(\cdot)$ by at most $4\sigma\mu(u) \leq 4\sigma\lambda(u)$. Thus, the partial sums of the exponents change by at most $K = 40\sigma^3\lambda(u)^2(\sigma + \lambda(u))$. We define a set of intervals $\mathcal{C}_{u,p}^K$ that will be carved out of the exponents during the shortening process. Let $\{c_1, \ldots, c_\ell\} = \{\eta_p^i(u) \mid 0 \leq i \leq m\}$ be the ordered set of the $\eta_p^i(u)$, i. e., $c_1 < \cdots < c_\ell$ and define

$$(1) \qquad \mathcal{C}_{u,p}^K = \{[c_i + K, c_{i+1} - K] \mid 1 \leq i < \ell, c_{i+1} - c_i \geq 2K\}.$$

Definition 18. *Write* $\mathcal{C}_{u,p}^K = \{[l_j, r_j] \mid 1 \leq j \leq \ell\}$ *(with* l_j *in increasing order). The shortened version of* u *is* $\mathcal{S}_{\mathcal{C}_{u,p}^K}(u) = u_0\beta_1 p^{z_1}\alpha_1 u_1 \ldots \beta_m p^{z_m}\alpha_m u_m$.

The new exponents are given by $z_i = y_i - \operatorname{sgn}(y_i) \cdot \sum_{j \in C_i} d_j$, *where* $d_j = r_j - l_j + 1$ *and* C_i *is the set of intervals to be removed from* y_i, *defined by*

$$C_i = \begin{cases} \{j \mid 1 \leq j \leq k, \eta_p^{i-1}(u) < l_j \leq r_j < \eta_p^i(u)\} & \text{if } y_i > 0, \\ \{j \mid 1 \leq j \leq k, \eta_p^i(u) < l_j \leq r_j < \eta_p^{i-1}(u)\} & \text{if } y_i < 0. \end{cases}$$

Lemma 19. $\pi(u) =_G 1$ *if and only if* $\pi(\mathcal{S}_{\mathcal{C}_{u,p}^K}(u)) =_G 1$.

Lemma 20. *Let* $\mathcal{S}_{\mathcal{C}_{u,p}^K}(u) = u_0\beta_1 p^{z_1}\alpha_1 u_1 \ldots \beta_m p^{z_m}\alpha_m u_m$ *for some* $u \in \Delta^*$. *Then* $|z_i| \leq 80m\sigma^3\lambda(u)^2(\sigma + \lambda(u))$ *for* $1 \leq i \leq m$.

Solving the Power Word Problem. Now we are ready for the proofs of our main results from the introduction:

Proof of Theorem A. By Lemma 13 the preprocessing can be done in uAC^0 with oracles for the word problems in G and F_2 (thus, by [9, Theorem 5.6.5, Theorem 5.6.14] in $\mathsf{uAC}^0(\mathrm{WP}(F_2), (\mathrm{WP}(G_\alpha))_{\alpha \in \mathcal{L}}) \subseteq \mathsf{uAC}^0(\mathrm{WP}(F_2), (\mathrm{PowWP}(G_\alpha))_{\alpha \in \mathcal{L}}))$. The shortening procedure can be computed in parallel for each $p \in \{p_i \mid 1 \leq i \leq n\}$. It requires iterated additions, which is in $\mathsf{uTC}^0 \subseteq \mathsf{uAC}^0(\mathrm{WP}(F_2))$. By Lemma 20 the exponents of the shortened word are bounded by a polynomial. We write the shortened word as a simple power word of polynomial length and solve the simple power word problem, which by Lemma 12, is in $\mathsf{uAC}^0(\mathrm{WP}(F_2), (\mathrm{PowWP}(G_\alpha))_{\alpha \in \mathcal{L}})$. $\qquad\square$

The proof of Theorem B is analogous to the proof of Theorem A using the respective statements of the lemmas for the uniform case.

Proof of Corollary D. By [14, Theorem 3.11], there is a polynomial $p(N)$, where N is the input length, such that if there is a solution, then there is a solution $x_1, \ldots, x_n$ with $x_i \leq 2^{p(N)}$. Therefore, we can guess a potential solution within the bound in NP. From Theorem B it follows that the uniform power word problem in RAAGs can be decided in P. Hence, the uniform knapsack problem can be decided in NP. Finally, NP-completeness follows immediately from the NP-completeness of the knapsack problem for a certain fixed RAAGs, which has been shown in [14]. $\qquad\square$

Note that this proof even shows NP-completeness of the slightly more general problem of uniformly solving exponent equations for RAAGs as defined in [14].

References

1. Allender, E.: Arithmetic circuits and counting complexity classes. Compl. Comput. Proofs Quaderni di Matematica **13**, 33–72 (2004)
2. Crisp, J., Godelle, E., Wiest, B.: The conjugacy problem in subgroups of right-angled artin groups. J. Topol. **2**(3), 442–460 (2009)
3. Dehn, M.: Über unendliche diskontinuierliche gruppen. Math. Ann. **71**(1), 116–144 (1911)
4. Diekert, V., Kausch, J., Lohrey, M.: Logspace computations in Coxeter groups and graph groups. Contemp. Math. (Amer. Math. Soc.) **582**, 77–94 (2012)
5. Duboc, C.: On some equations in free partially commutative monoids. Theoret. Comput. Sci. **46**, 159–174 (1986)
6. Fine, N.J., Wilf, H.S.: Uniqueness theorems for periodic functions. Proc. Am. Math. Soc. **16**(1), 109–114 (1965)
7. Green, E.R.: Graph products of groups. Ph.D. thesis, University of Leeds (1990)
8. Haubold, N., Lohrey, M., Mathissen, C.: Compressed decision problems for graph products and applications to (outer) automorphism groups. IJAC **22**(08), 218–230 (2012)
9. Kausch, J.: The parallel complexity of certain algorithmic problems in group theory. Ph.D. thesis (2017). http://dx.doi.org/10.18419/opus-9152
10. Lipton, R.J., Zalcstein, Y.: Word problems solvable in logspace. J. ACM **24**, 522–526 (1977)
11. Lohrey, M.: The Compressed Word Problem for Groups. SM, Springer, New York (2014). https://doi.org/10.1007/978-1-4939-0748-9
12. Lohrey, M., Schleimer, S.: Efficient computation in groups via compression. In: Diekert, V., Volkov, M.V., Voronkov, A. (eds.) CSR 2007. LNCS, vol. 4649, pp. 249–258. Springer, Heidelberg (2007). https://doi.org/10.1007/978-3-540-74510-5_26
13. Lohrey, M., Weiß, A.: The power word problem. In: MFCS 2019, Proceedings. LIPIcs, vol. 138, pp. 43:1–43:15 (2019). https://doi.org/10.4230/LIPIcs.MFCS.2019.43
14. Lohrey, M., Zetzsche, G.: Knapsack in graph groups. Theor. Comput. Syst. **62**(1), 192–246 (2018)
15. Myasnikov, A., Nikolaev, A., Ushakov, A.: Knapsack problems in groups. Math. Comput. **84**(292), 987–1016 (2015)
16. Novikov, P.S.: On the algorithmic unsolvability of the word problem in group theory (1955)
17. Stober, F.: The power word problem in graph groups. Master's thesis (2021). https://doi.org/10.18419/opus-11768
18. Stober, F., Weiß, A.: The power word problem in graph products .https://arxiv.org/abs/2201.06543 (2022)
19. Vollmer, H.: Introduction to Circuit Complexity. Springer, Berlin (1999)
20. Wrathall, C.: The word problem for free partially commutative groups. J. Symb. Comput. **6**(1), 99–104 (1988)

On One-Counter Positive Cones of Free Groups

Zoran Šunić[(✉)] ⓘ

Department of Mathematics, Hofstra University, Hempstead, NY 11549, USA
zoran.sunic@hofstra.edu

Abstract. We present an uncountable family of (left) orders on non-cyclic free groups of finite rank. The orders are encoded by bi-infinite words of certain form. Each order in the family is the limit of explicitly constructed orders whose positive cones are represented by one-counter, thus context-free, languages. As an application, we provide three explicit constructions of Cantor spaces of orders on free groups. The first two consist of orders extending the lexicographic and the short-lex order, respectively, on the free monoid, while the third one consists of discrete orders. Each has a dense subset of orders with one-counter positive cones.

Keywords: Order · Free group · Positive cone · Infinite word · Short-lex · Lexicographic · Discrete order · One-counter language · Cantor space

1 Introduction

General Setting. All group and semigroup orders in the text are left orders.

Let $\mathcal{A}$ be a finite alphabet with $|\mathcal{A}| = k \geq 2$, and F_k the free group over $\mathcal{A}$. Let $\mathcal{A}^{-1} = \{a^{-1} \mid a \in \mathcal{A}\}$ be a disjoint copy of $\mathcal{A}$ consisting of formal inverses, and e a symbol outside of $\mathcal{A} \sqcup \mathcal{A}^{-1} = \mathcal{A}^{\pm}$.

Language Complexity of Orders. Our goal is to provide a countable set of left orders on the free group F_k, each of which is explicitly defined by a one-counter language over $\mathcal{A}^{\pm}$. First, an uncountable set of orders is defined, encoded by certain bi-infinite words, and then it is shown that each order encoded by an eventually orientably periodic word (definition given later) can be described by a one-counter language. Both the space of words and the space of orders admit natural topologies and our construction respects them – the function associating orders to bi-infinite words is continuous. As an application, we provide an explicit construction of an infinite set of orders on F_k, forming a Cantor space, extending the lexicographic order on the free monoid over $\mathcal{A}$, another Cantor space of orders extending the short-lex order, and a third one consisting of discrete orders, all three having dense subsets of orders with one-counter positive cones.

An *order* $\leq$ on a group G is a total order on the set G that is compatible with the (left) multiplication, that is, for all $f, g, h \in G$, if $f \leq g$, then $hf \leq hg$.

© Springer Nature Switzerland AG 2022
V. Diekert and M. Volkov (Eds.): DLT 2022, LNCS 13257, pp. 299–311, 2022.
https://doi.org/10.1007/978-3-031-05578-2_24

If $\leq$ is an order on G, then the set of positive elements $P = \{g \in G \mid e < g\}$ is the *positive cone* of G with respect to $\leq$. The positive cone P is a semigroup such that $G = P^{-1} \sqcup \{e\} \sqcup P$, and $f \leq g$ if and only if $f^{-1}g \in P$. Conversely, if P is a subsemgroup of G such that $G = P^{-1} \sqcup \{e\} \sqcup P$, an order can be defined on G by $f \leq g$ if and only if $f^{-1}g \in P$. Under this order, P is the set of positive elements of G. Thus, any order is determined by its positive cone.

It is known since the 1940s [7] that the free group F_k admits orders, and in fact, two-sided orders, but unlike the case of free monoids, where many comprehensible orders are in everyday use (lexicographic and short-lex come to mind), explicitly defined orders on free groups are rare and far between. We may follow the work of Shibmireva [7], which relies on the earlier work of Magnus [4], and define orders on F_k as follows. The lower central series quotients $\gamma_1(F_k)/\gamma_2(F_k)$, $\gamma_2(F_k)/\gamma_3(F_k)$, ... are free abelian groups of finite rank. Place an order on each, independently, as desired (say, lexicographic by coordinates). A nontrivial element g is declared positive in F_k if it is positive in the quotient $\gamma_{n-1}(F_k)/\gamma_n(F_k)$, where n is the smallest positive integer such that $g \notin \gamma_n(F_k)$. This gives a continuum of orders on F_k, but they are not particularly easy to work with on the concrete level, as deciding if a concrete element g is positive requires collecting commutators until one reaches the correct n.

Example 1. We provide, informally, two examples of orders on F_2, with base $\mathcal{A} = \{a, b\}$, whose positive cones can be recognized by the two machines in Fig. 1. The machines read the input from right to left. The initial states of the machines are not shown. We assume that each machine has a register, initially set at 0, that can store an integer and it can add or subtract from it. After reading the initial letter (A and B stand for a^{-1} and b^{-1}, respectively), the machine enters one of the four states a, b, A, or B through the dotted arrows, depending on the initial letter. It continues by reading the next letter of the input, it follows the arrow to the next state, depending on the letter, and so on. Each time it follows an arrow, it adds the amount indicated on the arrow to the register. The machine works like this until it reads the entire input, and it accepts the word if the number in the register at that moment is positive.

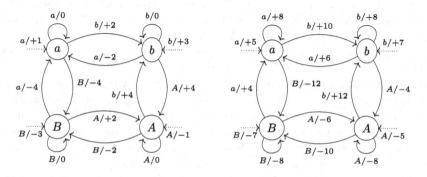

Fig. 1. Two machines recognizing positive cones in F_2

The only words that can be read completely by the two machines are the freely reduced words over $\mathcal{A}^{\pm}$. We claim that the sets of reduced words that are accepted by these machines represent positive cones in F_2 (proved later). The point of the examples is to show simple, explicit representations of positive cones of F_2 in the flavor of our general results. Deciding if $f < g$ amounts to calculating the reduced form of $f^{-1}g$, running it through the corresponding machine and checking if the sign of the number in the register at the end is positive.

We compare ba and aba. The sequence of configurations of the form (remaining input) $\times$ state $\times$ register, on input $ABaba$, for the machine on the left, is

$$(ABaba, \text{start}, 0) \vdash (ABab, a, 1) \vdash (ABa, b, 3) \vdash (AB, a, 1) \vdash (A, B, -3) \vdash (\varepsilon, A, -1),$$

implying that $ABaba$ is rejected, it is a negative element of F_2, and $ba > aba$. On the other hand, $ABaba$ is accepted by the machine on the right and, under that order, $ba < aba$. As we will see later, the orders represented by these machines extend the lexicographic and the short-lex order, respectively, from the free monoid $\{a, b\}^*$ to the free group F_2. This claim is consistent with our calculation showing that $ba > aba$ in the first case, and $ba < aba$ in the second.

Definition 1. *Let L be a language over the finite alphabet Σ. An order $\leq$ on a finitely generated group G is represented by L if there is a homomorphism $\Sigma^* \to G$ such that the image of L is the positive cone P of G with respect to $\leq$.*

We may classify orders by the complexity of the languages that represent them and call an order, or its corresponding positive cone, regular if it admits representation by a regular language, context-free if it admits a representation by a context-free language, and so on.

Calegary [2] showed that no regular language representing a positive cone of a hyperbolic 3-manifold group can consist of geodesics. Hermiller and the author [3] showed that no order on a free product of orderable groups is regular. In particular, no order on the free group F_k is regular. On the positive side, Rourke and Wiest [6] showed that certain mapping class groups admit regular orders. Su [8] showed that $\mathcal{R}$, the class of groups admitting regular orders is closed under subgroups of finite index. Antolín, Rivas, and Su [1] showed that $\mathcal{R}$ is closed under extensions and wreath products. In particular, all orderable poly-cyclic groups admit regular orders.

We contribute to the positive side by constructing explicitly the first countable family of one-counter (hence, context-free) orders on the free group F_k. Given that no order on a free product is regular [3], a representation by one-counter languages seems to be the best one can do. Only finitely many examples of one-counter orders on F_k have appeared previously in [9,10], where it was stated, without a proof, that those finitely many examples yield one-counter positive cones (the emphasis in those works was elsewhere, the orders in [10] were constructed and analyzed by using quasi-morphisms).

Our our one-counter machines are deterministic, read the input from right to left, at most one symbol at a time. The bottom of the stack symbol is immutable and unmovable, it is present from the beginning, and the machine can check its presence at the top of the stack, that is, the machine can test if the stack is empty. If an attempt is made to pop from an empty stack the machine stops.

2 One-Counter Languages and Bi-infinite Words

A partial function ϕ on $\mathbb{Z}$ is *eventually right-periodic* with period p if there exists $n_0 \in \mathbb{Z}$ such that for each $n \geq n_0$, either $\phi(n) = \phi(n+p)$ or none of n and $n+p$ is in the domain of ϕ. Left-periodicity is defined in analogous way. A function is eventually periodic if it is both left and right-periodic (the periods may differ).

Let Φ be a finite set of partial functions on $\mathbb{Z}$ and $\phi \in \Phi$. The *jump* function $\delta_\phi : \mathbb{Z} \to \mathbb{Z}$ of ϕ is defined, wherever ϕ is, by $\delta_\phi(n) = \phi(n) - n$. Let n be in the domain of $\phi\gamma$, where $\phi \in \Phi$ and γ is a composition of members of Φ. Then

$$\phi\gamma(n) = \gamma(n) + \delta_\phi(\gamma(n)). \tag{1}$$

Let n_0 be in the domain of $\gamma = \phi_m \cdots \phi_1$, and set $n_i = \phi_i \cdots \phi_1(n_0)$, and $\delta_i = \delta_{\phi_i}$, for $i = 0, \ldots, m$. Equation (1) implies

$$\phi_m \cdots \phi_1(n_0) = n_{m-1} + \delta_m(n_{m-1}) = n_0 + \delta_1(n_0) + \delta_2(n_1) + \cdots + \delta_m(n_{m-1}). \tag{2}$$

Proposition 1. *Let Φ be a finite set of partial functions on $\mathbb{Z}$ such that their jump functions are eventually periodic. Let R be any regular language over Φ such that, for every word $\gamma = \phi_m \cdots \phi_1 \in R$, the number 0 is in the domain of γ. Then, the language $L = \{\gamma \in R \mid \gamma(0) > 0\}$ is a one-counter language.*

Our construction of orders on F_k relies on intervals that are spaced apart on $\mathbb{R}$. To simplify the presentation, we use bi-infinite words indexed by the set $\mathbb{Z}_{\text{odd}}$ of odd integers (we are "reserving" the even numbers for the requisite spacing).

The set $\underline{S}$ of *spaced words* over $\mathcal{A}^\pm$ consists of all odd-indexed bi-infinite words $\xi = \ldots \xi_{-3}\, \xi_{-1}\, \xi_1\, \xi_3\, \xi_5\, \xi_7 \ldots$, where $\xi_i \in \mathcal{A}^\pm$, for $i \in \mathbb{Z}_{\text{odd}}$, such that:

- ξ has infinitely many appearances of each letter in $\mathcal{A}^\pm$, in both directions,
- exactly one appearance of each letter from $\mathcal{A}^\pm$ in ξ is marked.

A *spaced sequence* is a strictly increasing bi-infinite sequence $(x_n)_{n \in \mathbb{Z}}$ of odd integers. Such sequences do not have lower or upper bounds. A *spaced structure* over $\mathcal{A}^\pm$ is a function $I : \mathcal{A}^\pm \times \mathbb{Z} \to \mathbb{Z}_{\text{odd}}$ such that:

- I is bijective,
- for every $a \in \mathcal{A}^\pm$, the sequence $(I_{a,n})_{n \in \mathbb{Z}}$ is spaced.

For $a \in \mathcal{A}^\pm$, let $I_a = \{I_{a,n} \mid n \in \mathbb{Z}\}$ and I_e be the set of even integers. The $2k+1$ sets I_a, for $a \in \mathcal{A}^\pm \sqcup \{e\}$, partition $\mathbb{Z}$, since I is bijective. Setting $\xi_i = a$ if and only if $i \in I_a$ and marking, for each $a \in \mathcal{A}^\pm$, the appearance of a in *position* (the index in ξ) $I_{a,0}$ provides a well defined spaced word ξ over $\mathcal{A}^\pm$. Conversely, given a spaced word $\xi \in \underline{S}$, we construct a spaced structure I over $\mathcal{A}^\pm$ as follows. For $a \in \mathcal{A}^\pm$, set $I_{a,0}$ to be the position of the marked occurrence of a in ξ.

For $n > 0$, set $I_{a,n}$ to be the position of the nth occurrence of a in ξ to the right of the marked position $I_{a,0}$ and, for $n < 0$, the position of the $|n|$th occurrence of a to the left of the marked position $I_{a,0}$. Thus, whenever either a spaced word or a spaced structure is given, we consider that its counterpart is given too.

Example 2. Let $\mathcal{A} = \{a, b\}$. An example of a corresponding pair of a spaced word ξ and a spaced structure I is shown below. The top two rows in the table provide a spaced word ξ, with marked positions indicated by underlined letters. Row 1, 3, and 4 provide the function I. Each pair in a singe column in rows 3 and 4 is a pair in $\mathcal{A}^{\pm} \times \mathbb{Z}$, and the corresponding entry in the same column in row i is its value under I. E.g., consider the b-columns, indicated by boldface.

i	...	-9	-7	-5	-3	-1	**1**	**3**	**5**	7	9	11	13	**15**	17	19	...
ξ	...	a^{-1}	a	$\underline{b}^{-1}$	b	a	$\underline{b}$	a	b	a	$\underline{a}^{-1}$	b^{-1}	$\underline{a}$	a	b	a	a^{-1}	...
$\mathcal{A}^{\pm}$	...	a^{-1}	a	b^{-1}	b	a	b	a	b	a	a^{-1}	b^{-1}	a	a	b	a	a^{-1}	...
$\mathbb{Z}$	...	-1	-3	0	-1	-2	0	-1	1	0	1	0	1	2	2	1	...

We have $I_{b,-1} = -3$, $I_{b,0} = 1$, $I_{b,1} = 5$, and $I_{b,2} = 15$, because the marked b appears in position 1 in ξ, the first and second b to the right of it are in positions 5 and 15, and the first b to the left of it is in position -3.

For $a \in \mathcal{A}^{\pm}$ and $n \in \mathbb{Z}$, denote $I'_{a,n} = \{i \mid I_{a,n-1} < i < I_{a,n}\}$ and $I'_a = \mathbb{Z} \setminus I_a$. For $a \in \mathcal{A}^{\pm}$, let $\phi_a : \mathbb{Z} \to \mathbb{Z}$ be the partial function defined, only on $I'_{a^{-1}}$, by

$$\phi_a(i) = \begin{cases} I_{a,n}, & a \in \mathcal{A}, \\ I_{a,n-1}, & a \in \mathcal{A}^{-1}, \end{cases} \qquad \text{for all } i \text{ with } I_{a^{-1},n-1} < i < I_{a^{-1},n}.$$

In other words, ϕ_a is characterized by the property

$$\phi_a(I'_{a^{-1},n}) = \begin{cases} I_{a,n}, & a \in \mathcal{A}, \\ I_{a,n-1} & a \in \mathcal{A}^{-1}. \end{cases} \tag{3}$$

Example 3. We continue our previous example. The black vertices in the row labeled $\phi_{b^{-1}}$ in Fig. 2 are elements of I_b. All other integers, including the even

$\phi_{b^{-1}}$	...												...
i	...	-5	-3	-1	1	3	5	7	9	11	13	15	...
ξ	...	$\underline{b}^{-1}$	b	a	$\underline{b}$	a	b	$\underline{a}^{-1}$	b^{-1}	$\underline{a}$	a	b	...
n	...	0	-1	-2	0	-1	1	0	1	0	1	2	...

Fig. 2. The partial function $\phi_{b^{-1}} : I'_b \to \mathbb{Z}$

ones, are in I'_b, the domain of ϕ_{b-1}. We have $I_{b,1} = 5$, $I_{b,2} = 15$, and $I_{b-1,1} = 9$. Thus, for all $i \in I'_{b,2} = \{6, 7, 8, 9, 10, 11, 12, 13, 14\}$, we have $\phi_{b-1}(i) = 9$ (the even numbers are not shown in the figure). Similarly, $I'_{b,1} = \{2, 3, 4\}$ and $\phi_{b-1}(2) = \phi_{b-1}(3) = \phi_{b-1}(4) = I_{b-1,0} = -5$.

An *orientable* word over $\mathcal{A}^{\pm}$ is a word of length $2k$ in which very letter from $\mathcal{A}^{\pm}$ appears exactly once. There are exactly $(2k)!$ such words. An eventually orientably periodic word in $\underline{S}$ is a word ξ that is eventually periodic and its periods on both ends are products of orientable words. Let $\underline{S}^o$ be the set of eventually orientably periodic words in $\underline{S}$.

Proposition 2. *If ξ is an eventually orientably periodic, spaced word in $\underline{S}$, then the jump functions δ_a corresponding to ϕ_a, for $a \in \mathcal{A}^{\pm}$, are eventually periodic.*

Example 4. It is not sufficient to require that ξ is eventually periodic in order to obtain eventually periodic jump functions. For instance, let $\mathcal{A} = \{a, b\}$ and $\xi = \cdots a a^{-1} a^{-1} b b^{-1} . \underline{a} \underline{a}^{-1} a^{-1} \underline{b} \underline{b}^{-1} a a^{-1} a^{-1} b b^{-1} \cdots$, where the dot indicates the position between -1 and 1. Then $\phi_a(10n+1) = 20n+1$, which implies $\delta_a(10n + 1) = 10n$, and the jump function is not even bounded.

3   Free Groups and Bi-infinite Words

Line Homeomorphisms and Spaced Sequences of Intervals. By a half-open interval we mean a non-empty interval of the form $[x, x')$, for some $x, x' \in \mathbb{R}$ with $x < x'$. A bi-infinite sequence $(J_n)_{n \in \mathbb{Z}}$, where $J_n = [x_n, x'_n)$, of half-open intervals in $\mathbb{R}$ is *spaced* if $\ldots < x_{-1} < x'_{-1} < x_0 < x'_0 < x_1 < x'_1 < x_2 < x'_2 < \ldots$ and the sequences of endpoints are not bounded at either end. Let $J'_n = [x'_{n-1}, x_n)$ be the half-open interval of points between J_{n-1} and J_n. The sequence $(J'_n)_{n \in \mathbb{Z}}$ is also spaced and the intervals in the two sequences together partition $\mathbb{R}$. Let $(K_n)_{n \in \mathbb{Z}}$, where $K_n = [y_n, y'_n)$, be another spaced sequence of half-intervals, with complementary sequence $(K'_n)_{n \in \mathbb{Z}}$, where $K'_n = [y'_{n-1}, y_n)$.

For $n \in \mathbb{Z}$, let $\alpha_n : J_n \to K'_{n+1}$ and $\alpha'_n : J'_n \to K_n$ be any homeomorphisms (e.g., linear). Since $\alpha_n(x_n) = y'_n$ and $\alpha'_n(x'_n) = y_{n+1}$, the maps α_n, α'_n combine in an order preserving homeomorphism $\alpha : \mathbb{R} \to \mathbb{R}$ (Fig. 3), given by

$$\alpha(x) = \begin{cases} \alpha_n(x), \text{if } x \in J_n, \text{ for some } n \in \mathbb{Z}, \\ \alpha'_n(x), \text{if } x \in J'_n, \text{ for some } n \in \mathbb{Z}. \end{cases}$$

We say that α is defined by the sequence $(J_n)_{n \in \mathbb{Z}}$ in the domain and the sequence $(K_n)_{n \in \mathbb{Z}}$ in the codomain. For $n \in \mathbb{Z}$, we have

$$\alpha(J_n) = K'_{n+1} \quad \text{and} \quad \alpha(J'_n) = K_n. \tag{4}$$

A *spaced interval structure* over $\mathcal{A}^{\pm}$ is a function J from $\mathcal{A}^{\pm} \times \mathbb{Z}$ to the set of half-open intervals on $\mathbb{R}$ such that, for each $a \in \mathcal{A}^{\pm}$, $(J_{a,n})_{n \in \mathbb{Z}}$ is a spaced sequence of half-open intervals, the sequences are strongly disjoint (meaning that, whenever $(a, n) \neq (a', n')$, the closures of the intervals $J_{a,n}$ and $J_{a',n'}$ are disjoint), and the complement $J_e = \mathbb{R} \setminus \bigcup_{a \in \mathcal{A}^{\pm}} \bigcup_{n \in \mathbb{Z}} J_{a,n}$ contains 0. Denote $J_a = \bigcup_{n \in \mathbb{Z}} J_{a,n}$. For $a \in \mathcal{A}^{\pm}$, let $(J'_{a,n})_{n \in \mathbb{Z}}$ be the spaced sequence complementary to $(J_{a,n})_{n \in \mathbb{Z}}$ and denote $J'_a = \bigcup_{n \in \mathbb{Z}} J'_{a,n}$.

Free Groups of Line Homeomorphisms and Spaced Interval Structures. For $a \in \mathcal{A}$, define the order preserving automorphism $\alpha_a : \mathbb{R} \to \mathbb{R}$ based on the sequence $(J_{a^{-1},n})_{n \in \mathbb{Z}}$ in the domain and $(J_{a,n})_{n \in \mathbb{Z}}$ in the codomain. Let $\alpha : F_k \to \mathrm{Homeo}_+(\mathbb{R})$ be the extension of the map $a \mapsto \alpha_a$, for $a \in \mathcal{A}$, to a homomorphism, that is, for $a \in \mathcal{A}^{\pm}$, we have $\alpha_{a^{-1}} = \alpha_a^{-1}$, and if $g = a_m \ldots a_1$ is a word over $\mathcal{A}^{\pm}$, then $\alpha_g = \alpha_{a_m} \cdots \alpha_{a_1} \in \mathrm{Homeo}_+(\mathbb{R})$.

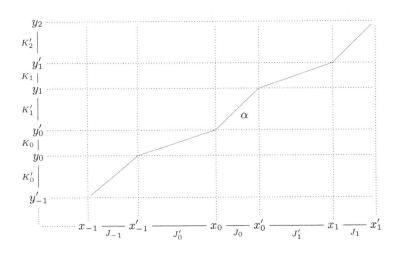

Fig. 3. The order preserving homeomorphism $\alpha : \mathbb{R} \to \mathbb{R}$

By (4), for every $a \in \mathcal{A}$, we have $\alpha_a(J_{a^{-1}}) = J'_a$ and $\alpha_a(J'_{a^{-1}}) = J_a$, implying $\alpha_a^{-1}(J'_a) = J_{a^{-1}}$ and $\alpha_a(J'_{a^{-1}}) = J_a$. Since $\alpha_a^{-1} = \alpha_{a^{-1}}$, we may summarize the last two equalities into a single one by saying that, for all $a \in \mathcal{A}^{\pm}$,

$$\alpha_a(J'_{a^{-1}}) = J_a. \tag{5}$$

Since, for any two distinct letters a and b in $\mathcal{A}^{\pm}$, we have $J_a \cap J_b = \varnothing$, the Ping-Pong Lemma and (5) imply that the group generated by α_a, for $a \in \mathcal{A}$, is free with basis $\{\alpha_a \mid a \in \mathcal{A}\}$. In other words, $\alpha : F_k \to \mathrm{Homeo}_+(\mathbb{R})$ is injective, that is, α is a faithful representation of the free group F_k.

If $a_m \cdots a_1$ is a nonempty reduced word over $\mathcal{A}^{\pm}$, the equality (5) shows that

$$\alpha_{a_m \cdots a_1}(J'_{a_1^{-1}}) \subseteq J_{a_m}. \tag{6}$$

Indeed, $\alpha_{a_m \cdots a_2 a_1}(J'_{a_1^{-1}}) = \alpha_{a_m \cdots a_2}(J_{a_1}) \subseteq \alpha_{a_m \cdots a_2}(J'_{a_2^{-1}}) = \alpha_{a_m \cdots a_3}(J_{a_2}) \subseteq \ldots \subseteq J_{a_m}$. By (6), we have

$$\alpha_{a_m \cdots a_1}(J_e) \subseteq \alpha_{a_m \cdots a_1}(J'_{a_1^{-1}}) \subseteq J_{a_m} \subseteq \mathbb{R} \setminus J_e. \tag{7}$$

Proposition 3. *The homomorhpism* $\alpha : F_k \to \mathrm{Homeo}_+(\mathbb{R})$, *extending the map* $a \to \alpha_a$, *for* $a \in \mathcal{A}$, *is a faithful representation of* F_k *in* $\mathrm{Homeo}_+(\mathbb{R})$. *The free group* F_k *acts freely on the orbit of 0 (in fact, any point in* J_e*).*

Proof. By (7), no element in J_e is fixed by any nontrivial element of F_k.

Orders on Free Groups and Bi-infinite Words. Any faithful representation α of F_k in $\mathrm{Homeo}_+(\mathbb{R})$ with a free orbit at 0, defines a left order on F_k by

$$g > e \iff \alpha_g(0) > 0. \tag{8}$$

(This is true for any group with a faithful left action on $\mathbb{R}$ by order-preserving homeomorphisms, and a free orbit at 0.) This order can also be expressed by saying $g < h$ if and only of $\alpha_g(0) < \alpha_h(0)$. The equalities (4), for $a \in \mathcal{A}$ and $n \in \mathbb{Z}$, give $\alpha_a(J_{a^{-1},n}) = J'_{a,n+1}$ and $\alpha_a(J'_{a^{-1},n}) = J_{a,n}$, which implies

$$\alpha_a(J'_{a^{-1},n}) = \begin{cases} J_{a,n}, & \text{if } a \in \mathcal{A}, \\ J_{a,n-1}, & \text{if } a \in \mathcal{A}^{-1}. \end{cases} \tag{9}$$

At this point, a comparison of (3) and (9) is inevitable and we make the connection explicit. For distinct $a, b \in \mathcal{A}^\pm$ and $n \in \mathbb{Z}$, there are only finitely many intervals $J_{a,i}$ that lie between $J_{b,n-1}$ and $J_{b,n}$. Therefore, J_e is a countable union of half-open intervals that are strongly disjoint, and one of these J_e-intervals contains 0. Thus, all intervals $J_{a,n}$, for $a \in \mathcal{A}^\pm$ and $n \in \mathbb{Z}$, together with the intervals partitioning J_e can be arranged in a single bi-infinite sequence

$$\cdots < T_{-3} < T_{-2} < T_{-1} < T_0 < T_1 < T_2 < T_3 < \cdots \tag{10}$$

that agrees with the order of these intervals on $\mathbb{R}$ (intervals further to the right get higher index), the union of the even numbered intervals is J_e, the interval J_0 contains 0, and $\cup_{n\in\mathbb{Z}} T_n = \mathbb{R}$.

We can think of J as a bijection from $\mathcal{A}^\pm \times \mathbb{Z}$ to the set of odd-indexed intervals in (10), inducing a bijection $I : \mathcal{A}^\pm \times \mathbb{Z} \to \mathbb{Z}_{\mathsf{odd}}$, given by

$$I_{a,n} = p \iff J_{a,n} = T_p. \tag{11}$$

The function I is a spaced structure over $\mathcal{A}^\pm$ and, therefore J induces a spaced word ξ in $\underline{\mathcal{S}}$. Conversely, let ξ be a spaced word in $\underline{\mathcal{S}}$ and I the corresponding spaced structure. We can construct a corresponding spaced interval structure as follows. Set $T_i = [i, i+1)$, for $i \in \mathbb{Z}$ and define J on $\mathcal{A}^\pm \times \mathbb{Z}$ by using (11). The spaced sequences of intervals $(J_{a,n})_{n\in\mathbb{Z}}$, for $a \in \mathcal{A}^\pm$, satisfy the conditions for a spaced interval structure. Therefore, we obtain a faithful representation $\alpha : F_k \to \mathrm{Homeo}_+(\mathbb{R})$, along with the corresponding order given by (8).

Lemma 1. *Let $I \leftrightarrow J$ be a corresponding pair of a spaced structure and a spaced interval structure. For $a \in \mathcal{A}^\pm$ and $i, j \in \mathbb{Z}$,*

$$i \in I'_{a^{-1}} \iff T_i \subseteq J'_{a^{-1}} \qquad \text{and} \qquad \phi_a(i) = j \iff \alpha_a(T_i) \subseteq T_j.$$

As a consequence, for every nonempty, reduced word $a_m \ldots a_1$ over $\mathcal{A}^\pm$,

$$\phi_{a_m} \cdots \phi_{a_1}(0) = p \iff \alpha_{a_m} \cdots \alpha_{a_1}(T_0) \subseteq T_p.$$

Proof. By (11), for $a \in \mathcal{A}^{\pm}$ and $i, n \in \mathbb{Z}$, we have $I_{a^{-1},n-1} < i < I_{a^{-1},n}$ if and only if $J_{a^{-1},n-1} < T_i < J_{a^{-1},n}$, that is, $i \in I'_{a^{-1},n}$ if and only if $T_i \subseteq J'_{a^{-1},n}$. Therefore,

$$\phi_a(i) = \begin{cases} I_{a,n}, & a \in \mathcal{A}, \\ I_{a,n-1}, & a \in \mathcal{A}^{-1} \end{cases} \iff \alpha_a(T_i) \subseteq \begin{cases} J_{a,n}, & a \in \mathcal{A}, \\ J_{a,n-1}, & a \in \mathcal{A}^{-1}. \end{cases}$$

Corollary 1. *Let $I \leftrightarrow J$ be a corresponding pair of a spaced structure and a spaced interval structure. For nonempty, reduced words $g = a_m \ldots a_1$ over $\mathcal{A}^{\pm}$,*

$$g > e \iff \alpha_g(0) > 0 \iff \phi_g(0) > 0.$$

Proof. We have $\alpha_g(0) > 0$ if and only if $\alpha_g(0) \in T_p$ for some positive p. On the other hand, $\alpha_g(0) \in T_p$ if and only if $\alpha_g(T_0) \subseteq T_p$ if and only if $\phi_g(0) = p$. $\quad\square$

Theorem 1. *If ξ is an eventually orientably periodic, spaced word in $\underline{S}$, then the positive cone of the order on F_k defined by ξ is a one-counter language.*

Proof. By Proposition 2 and Proposition 1, the jump functions are eventually periodic and $L = \{a_m \cdots a_1 \in R \mid \phi_{a_m} \cdots \phi_{a_1}(0) > 0\}$ is a one-counter language, where R is the regular language of reduced words over $\mathcal{A}^{\pm}$. By Corollary 1, the language L represents the positive cone of the order defined by ξ. $\quad\square$

Example 5. The examples in the introduction correspond, respectively, to the periodic words $\cdots a\,b\,B\,A\,.\,\underline{a}\,\underline{b}\,\underline{B}\,\underline{A}\,a\,b\,B\,A \cdots$ and $\cdots B\,A\,a\,b\,.\,\underline{B}\,\underline{A}\,\underline{a}\,\underline{b}\,B\,A\,a\,b \cdots$. From our perspective here, the finitely many examples in [10], exactly $(2k)!$ of them, correspond to the periodic words $\cdots W.WW \cdots$, where W is an orientable word, with marked letters in the W to the right of the decimal point.

4 The Space of Orders Defined by Spaced Words

Let $\mathcal{LF}$ be the metric space of all left orders on F_k. Two orders are close if their positive cones agree on large balls around e. For concreteness, we can define the distance between two distinct orders $\leq$ and $\leq'$ on F_k to be $1/2^{\ell}$, where ℓ is the length of the shortest reduced word which is in one of the positive cones $P_{\leq}$ and $P_{\leq'}$, but not in the other. It is known that this set has the structure of a Cantor space [5] (recall that a *Cantor space* is a compact, metrizable, totally disconnected, perfect space; it is unique, up to a homeomorphism). Let $\underline{\mathcal{L}}$ and $\underline{\mathcal{L}}^o$ be the subspaces of $\mathcal{LF}$ defined by the words in $\underline{S}$ and $\underline{S}^o$, respectively.

Theorem 2. *The map ord from the space of words $\underline{S}$ to the space of orders $\mathcal{LF}$ associating orders to spaced words is continuous. The range of this map, $\underline{\mathcal{L}}$, is a totally disconnected, metric space with no isolated points. The space $\underline{\mathcal{L}}^o$ is also a totally disconnected, metric space with no isolated points, it is dense in $\underline{\mathcal{L}}$ and its topological closure $\mathrm{Cl}(\underline{\mathcal{L}}^o)$ in $\mathcal{LF}$ is a Cantor space.*

The following lemma provides a criterion to decide which way the jump $\delta_a(i)$ goes when i is in the domain of ϕ_a. Note that part (iii) of the lemma shows, among other things, that the set $\{\phi_g(0) \mid g \in F_k\}$ is unbounded in both directions.

Lemma 2. *Let ξ be a word in $\underline{S}$ with associated interval structure I, and $a \in \mathcal{A}$.*
(i) Let $i \in I'_{a^{-1},n'}$ and n be the smallest integer such that $i \leq I_{a,n}$. Then

$$
\begin{aligned}
\phi_a(i) &< i & &\text{if } n' < n, \\
\phi_a(i) &> i & &\text{if } n' > n, \\
\phi_a(i) &> i & &\text{if } n' = n \text{ and } \xi_i \neq a, \\
\phi_a(i) &= i & &\text{if } n' = n \text{ and } \xi_i = a.
\end{aligned}
$$

(ii) Let $i \in I'_{a,n}$ and n' be the smallest integer such that $i < I_{a^{-1},n'}$. Then

$$
\begin{aligned}
\phi_{a^{-1}}(i) &< i & &\text{if } n' > n, \\
\phi_{a^{-1}}(i) &> i & &\text{if } n' < n, \\
\phi_{a^{-1}}(i) &< i & &\text{if } n' = n \text{ and } \xi_i \neq a^{-1}, \\
\phi_{a^{-1}}(i) &= i & &\text{if } n' = n \text{ and } \xi_i = a^{-1}.
\end{aligned}
$$

(iii) Let $i \in I'_{a^{-1},n'} \cap I'_{a,n}$. Then

$$
\begin{aligned}
\phi_a(i) &< i < \phi_{a^{-1}}(i), & &\text{if } n' < n, \\
\phi_{a^{-1}}(i) &< i < \phi_a(i), & &\text{if } n' \geq n.
\end{aligned}
$$

Proof (Proof of Theorem 2). Let ξ be a word in $\underline{S}$ and $\leq$ the corresponding order on F_k. The set B_r of elements of length at most r in F_k is finite, making the set $B_r(0) = \{g(0) \mid g \in B_r\}$ finite. Let n be a positive integer such that $B_r(0) \subseteq \bigcup_{i=-n}^{n} T_i$. Increase n, if necessary, to make sure that $\{-n, \ldots, n\}$ includes all positions of the marked letters. Any bi-infinite word ξ' that agrees with ξ in the positions up to n, in both directions, defines an order that agrees with ξ on the entire ball B_r. Thus, ord is continuous.

The spaces $\underline{\mathcal{L}}$ and $\underline{\mathcal{L}}^o$ are totally disconnected, metric spaces, since they are subspaces of the totally disconnected, metric space $\underline{\mathcal{LF}}$.

Let us show that $\underline{\mathcal{L}}$ has no isolated points. Let ξ be a word in $\underline{S}$ and r a positive number. We need to show that there exists a word ξ' in $\underline{S}$ that agrees with ξ on the positions up to r, in both directions, and that defines an order that is different from the order $\leq$ defined by ξ. Without loss of generality, increase r to make sure that the positions of all marked letters are included (this makes the open set around ξ smaller). We will show that there exists two words ξ' and ξ'' in $\underline{S}^o$ that agree with ξ on the positions up to $\pm r$ and that define two different orders (thus, at least one of them must be different from the order defined by ξ). This will also show that $\underline{\mathcal{L}}^o$ has no isolated points.

Fix two distinct elements a and b in $\mathcal{A}$ and a reduced group word h over $\{a, b\}$ such that $\phi_g(0) = j > r$ and all the jumps along the way are positive (possible by Lemma 2). Without loss of generality, assume that $\xi_j \in \{a, a^{-1}\}$ (jump once more to the right, if necessary). Let $g = b'h$, where $b' \in \{b, b^{-1}\}$ is chosen so that $\phi_g(0) = i > j$. Let n and n' be the unique integers such that

$i \in I'_{a-1,n'} \cap I'_{a,n}$. Define ξ' to be equal to ξ on positions up to $\pm i$, and extend ξ' beyond $\pm i$ in an eventually orientably periodic way. Define ξ'' to be equal to ξ up to $\pm j$, insert $d = |n - n'| + 1$ symbols a^{-1} between positions j and i in ξ, if $n' < n$, or d symbols a between positions j and i in ξ, if $n' \geq n$, and extend ξ'' in an eventually orientably periodic way beyond positions $\pm(i + d)$. The count of b' letters in ξ'' between positions j and $i + d$ is the same as the count of b' letters between j and i in ξ, since only $a^{\pm}$ letters are inserted. Thus, we have

$$\xi'_i = b', \quad \phi_g^{\xi'}(0) = \phi_g(0) = i \qquad \text{and} \qquad \xi''_{i+d} = b', \quad \phi_g^{\xi''}(0) = \phi_g(0) + d = i + d.$$

Since ξ and ξ' agree on positions up to $\pm i$, the $a^{\pm}$ counts at position i are the same for ξ and ξ', which means that the directions of the jumps $\delta_a^\xi(i)$ and $\delta_a^{\xi'}(i)$ agree. On the other hand, the insertion of the $a^{\pm}$ letters, whichever case it may be, in ξ'' in positions between j and i causes the relative count between $a^{\pm}$ to flip. Therefore, the directions of the jumps $\delta_a^\xi(i)$ and $\delta_a^{\xi''}(i+d)$ disagree, implying that the directions of the jumps $\delta_g^{\xi'}(i)$ and $\delta_g^{\xi''}(i+d)$ disagree. This implies that $\phi_{ag}^{\xi'}(0) < \phi_g^{\xi'}(0)$ if and only if $\phi_{ag}^{\xi}(0) > \phi_g^{\xi''}(0)$, that is, $ag <' g$ if and only if $ag >'' g$. Thus, the orders defined by ξ' and ξ'' are different.

Finally, since $\underline{\mathcal{L}}^o$ does not have isolated points, its closure $\mathrm{Cl}(\underline{\mathcal{L}}^o)$ in $\mathcal{LF}$ does not have isolated points either. Nonempty, closed subspaces of the Cantor space without isolated points are themselves Cantor spaces, completing the proof.

We distinguish a special subspace of $\underline{\mathcal{S}}$. Namely, let $\mathcal{S}$ be the set of words in $\underline{\mathcal{S}}$ with marking on the first occurrence of each letter in $\mathcal{A}^{\pm}$ at a positive index.

For the remainder of the text we set $\mathcal{A} = \{a_1, \ldots, a_k\}$.

Let $\mathcal{W}_{\mathsf{lex}}$ be the set of $k!$ oriented words of the form $a_1 \ldots a_k w$, $\mathcal{S}_{\mathsf{lex}}$ the Cantor space of all bi-infinite words over $\mathcal{W}_{\mathsf{lex}}$ viewed as a subspace of $\mathcal{S}$, and $\mathcal{L}_{\mathsf{lex}}$ the space of orders induced by the words in $\mathcal{S}_{\mathsf{lex}}$. Let $\mathcal{S}_{\mathsf{lex}}^o$ be the space of eventually periodic words in $\mathcal{S}_{\mathsf{lex}}$ and $\mathcal{L}_{\mathsf{lex}}^o$ the corresponding space of orders.

Proposition 4. *The Cantor space of bi-infinite words $\mathcal{S}_{\mathsf{lex}}$ and the space of orders $\mathcal{L}_{\mathsf{lex}}$ are homeomorphic under* ord. *All orders in $\mathcal{L}_{\mathsf{lex}}$ are extensions of the lexicographic order on the free monoid $\mathcal{A}^*$ based on $a_1 < \ldots < a_k$. Each order in the dense subset $\mathcal{L}_{\mathsf{lex}}^o$ of $\mathcal{L}_{\mathsf{lex}}$ has a one-counter positive cone.*

Proof. Let $\xi = \cdots W_{-1}.W_0 W_1 W_2 \cdots$, where $W_i = a_1 \cdots a_k w_i \in \mathcal{W}_{\mathsf{lex}}$. For $a, b \in \mathcal{A}$ and $n \in \mathbb{Z}$, we have $I_{a^{-1},n-1} < I_{b,n} < I_{a^{-1},n}$, implying $\phi_a(I_{b,n}) = I_{a,n}$. Since $I_{a^{-1},-1} < 0 < I_{a^{-1},0}$, we have $\phi_a(0) = I_{a,0}$. Thus, for $a \in \mathcal{A}$ and $u \in \mathcal{A}^*$, we have $\phi_{au}(0) = I_{a,0} > 0$, showing that all nonempty words in $\mathcal{A}^*$ are greater than the empty word. If $a, b \in \mathcal{A}$ are two distinct letters with $a < b$, then for any words $u, v \in \mathcal{A}^*$, we have $\phi_{au}(0) = I_{a,0} < I_{b,0} = \phi_{bv}(0)$, showing that $au < bv$. Therefore the order $\leq$ induced by ξ on $\mathcal{A}^*$ is the lexicographic order.

The restriction of ord on $\mathcal{S}_{\mathsf{lex}}$ and $\mathcal{L}_{\mathsf{lex}}$ is surjective by definition, and continuous by Theorem 2. We claim that it is injective. Suppose ξ is as above, and $\xi' = \cdots W_{-1}.W_0 \cdots W_{n-1} W_n' W_{n+1}' \cdots$ is a word in $\mathcal{S}_{\mathsf{lex}}$ different from ξ. Assume, further, that ξ and ξ' differ for the first time at some $n \geq 0$ (the argument when $n < 0$ is similar). For $0 \leq i \leq n - 1$, let $w_i = b_{i,1}^{-1} \cdots b_{i,k}^{-1}$. By Lemma 2,

$$\phi_{a_k b_{i,1} b_{i,k}^{-1} a_1^{-1}}(I_{a_k,i}) = \phi_{a_k b_{i,1} b_{i,k}^{-1}}(I_{a_1^{-1},i}) = \phi_{a_k b_{i,1}}(I_{b_{i,k}^{-1},i}) = \phi_{a_k}(I_{b_{i,1},i+1}) = I_{a_k,i+1}.$$

Therefore, for the reduced word $h = (a_k b_{n-1,1} b_{n-1,k}^{-1} a_1^{-1}) \cdots (a_k b_{0,1} b_{0,k}^{-1} a_1^{-1}) a_k$ of length $4n + 1$, we have $\phi_h(0) = \phi_h'(0) = I_{a_k,n}$, since ξ and ξ' agree on $W_0, \ldots, W_{n-1}$. Since $W_n = a_1 \cdots a_k w_n$ and $W_n' = a_1 \cdots a_k w_n'$ are different, there exist two different letters $a, b \in \mathcal{A}$ such that a^{-1} is before b^{-1} in w_n and b^{-1} is before a^{-1} in w_n'. Consider the reduced words $g = a^{-1} a_1^{-1} h$ and $bg = ba^{-1} a_1^{-1} h$. The positions $\phi_g(0)$ and $\phi_g'(0)$ are the positions of the occurrence of a^{-1} in w_n and w_n', respectively, while $\phi_{bg}(0) = I_{b,n}$ and $\phi_{bg}'(0) = I_{b,n+1}$. Therefore, $bg < g$ and $bg >' g$, showing that the orders defined by ξ and ξ' are different. Since bijective continuous maps from a Cantor space to a Hausdorff space are homeomorphisms, $\mathcal{S}_{\text{lex}}$ and $\mathcal{L}_{\text{lex}}$ are homeomorphic.

The argument above showing injectivity shows also that $\mathcal{L}_{\text{lex}}^o$ is dense in $\mathcal{L}_{\text{lex}}$, since we can alter ξ at any position n and then extend it periodically beyond n.

The proofs of the remaining two propositions are similar to the one above, but less technical (the construction of the elements proving injectivity is simpler).

Let $\mathcal{W}_{\text{s.lex}}$ be the set of $k!$ oriented words of the form $wa_1 \ldots a_k$, $\mathcal{S}_{\text{s.lex}}$ the Cantor space of all bi-infinite words over $\mathcal{W}_{\text{s.lex}}$ viewed as a subspace of $\mathcal{S}$, and $\mathcal{L}_{\text{s.lex}}$ the space of orders induced by the words in $\mathcal{S}_{\text{s.lex}}$. Let $\mathcal{S}_{\text{s.lex}}^o$ be the space of eventually periodic words in $\mathcal{S}_{\text{s.lex}}$ and $\mathcal{L}_{\text{s.lex}}^o$ the corresponding space of orders.

Proposition 5. *The Cantor space of bi-infinite words $\mathcal{S}_{\text{s.lex}}$ and the space of orders $\mathcal{L}_{\text{s.lex}}$ are homeomorphic under ord. All orders in $\mathcal{L}_{\text{s.lex}}$ are extensions of the short-lex order on the free monoid $\mathcal{A}^*$ based on $a_1 < \ldots < a_k$. Each order in the dense subset $\mathcal{L}_{\text{s.lex}}^o$ of $\mathcal{L}_{\text{s.lex}}$ has a one-counter positive cone.*

Example 6. The two orders in the introduction are extensions of the lexicographic and the short-lex order, respectively, on the free monoid $\{a, b\}^*$, based on $a < b$.

Let $\mathcal{W}_{\text{dis}}$ be the set of $(2 * (k-1))!$ oriented words of the form $a_1 w a_1^{-1}$, $\mathcal{S}_{\text{dis}}$ the Cantor space of all bi-infinite words over $\mathcal{W}_{\text{dis}}$ viewed as a subspace of $\mathcal{S}$, and $\mathcal{L}_{\text{dis}}$ the space of orders induced by the words in $\mathcal{S}_{\text{dis}}$. Let $\mathcal{S}_{\text{dis}}^o$ be the space of eventually periodic words in $\mathcal{S}_{\text{dis}}$ and $\mathcal{L}_{\text{dis}}^o$ the corresponding space of orders. An order is *discrete* if its positive cone has a smallest element.

Proposition 6. *The Cantor space of bi-infinite words $\mathcal{S}_{\text{dis}}$ and the space of orders $\mathcal{L}_{\text{dis}}$ are homeomorphic under ord. All orders in $\mathcal{L}_{\text{dis}}$ are discrete with smallest positive element a_1. Each order in the dense subset $\mathcal{L}_{\text{dis}}^o$ of $\mathcal{L}_{\text{dis}}$ has a one-counter positive cone.*

References

1. Antolín, Y., Su, H.L., Rivas, C.: Regular left-orders on groups (2021). (to appear in Journal of Combinatorial Algebra)
2. Calegari, D.: Problems in foliations and laminations of 3-manifolds. In: Topology and geometry of manifolds (Athens, GA, 2001), Proceedings of Symposia in Pure Mathematics, vol. 71, pp. 297–335. American Mathematical Society, Providence, RI (2003)

3. Hermiller, S., Šunić, Z.: No positive cone in a free product is regular. Internat. J. Algebra Comput. **27**(8), 1113–1120 (2017)
4. Magnus, W.: Beziehungen zwischen Gruppen und Idealen in einem speziellen Ring. Math. Ann. **111**(1), 259–280 (1935)
5. Rivas, C.: Left-orderings on free products of groups. J. Algebra **350**, 318–329 (2012)
6. Rourke, C., Wiest, B.: Order automatic mapping class groups. Pacific J. Math. **194**(1), 209–227 (2000)
7. Shimbireva, H.: On the theory of partially ordered groups. Rec. Math. [Mat. Sbornik] N.S. **20**(62), 145–178 (1947)
8. Su, H.L.: Formal language convexity in left-orderable groups. Internat. J. Algebra Comput. **30**(7), 1437–1456 (2020)
9. Šunić, Z.: Explicit left orders on free groups extending the lexicographic order on free monoids. C. R. Math. Acad. Sci. Paris **351**(13–14), 507–511 (2013)
10. Šunić, Z.: Orders on free groups induced by oriented words (2013). http://arxiv.org/abs/1309.6070

Kolmogorov Complexity Descriptions of the Exquisite Behaviors of Advised Deterministic Pushdown Automata

Tomoyuki Yamakami[✉]

Faculty of Engineering, University of Fukui, 3-9-1 Bunkyo, Fukui 910-8507, Japan
TomoyukiYamakami@gmail.com

Abstract. Kolmogorov complexity has proven itself to be a useful, practical tool in obtaining numerous impossibility results in various scientific fields. Among them are the non-regularity and non-deterministic-context-freeness of formal languages by Li and Vitányi [SIAM J. Comput., 24 (1995) 398–410] and Glier [SIAM J. Comput., 32 (2003) 1389–1394], who proposed the so-called KC-DCF(L) lemma. This is viewed as a Kolmogorov complexity analogue of a pumping lemma (or an iteration theorem) for deterministic context-free (dcf) languages but it is not applicable to "advised" dcf languages, composed of accepted strings by one-way deterministic pushdown automata helped by external sources of additional information, called advice. To amend the lack of applicability, we propose a new practical form of the lemma for those advised dcf languages. This new lemma provides another criterion, which is incomparable to the KC-DCF(L) lemma of Li, Vitányi, and Glier.

1 Background and Our Challenges

1.1 Kolmogorov Complexity Approaches to Formal Languages

The notion of *Kolmogorov complexity* dates back to the mid 1960s. This complexity expresses the minimal size of a program p that, along with an auxiliary input w, produces a target string x on a fixed universal Turing machine. The notation $C(x \mid w)$ is commonly used to denote this minimal size and it is further abbreviated as $C(x)$ if w is the empty string. Since a different choice of a universal Turing machine only adds up an additive constant, this complexity measure turns out to be a robust notion and it thus works out as a basis of *algorithmic information theory*. Kolmogorov complexity has found numerous applications to a wide range of scientific fields. Refer to, e.g., the textbook [5] for more information on such applications.

In 1995, Li and Vitányi [4] sought a direct application of Kolmogorov complexity to formal languages and automata theory, in particular, targeting two fundamental families REG and DCFL of *regular languages* and *deterministic context-free (dcf) languages*. These languages are precisely recognized by one-way deterministic finite automata (or 1dfa's, for short) and *one-way deterministic pushdown*

V. Diekert and M. Volkov (Eds.): DLT 2022, LNCS 13257, pp. 312–324, 2022.
https://doi.org/10.1007/978-3-031-05578-2_25

automata (or 1dpda's). Li and Vitányi firstly gave a Kolmogorov-complexity characterization of regular languages in a similar way as the Myhill-Nerode characterization. They further proposed a new Kolmogorov-complexity-theoretical property (called the *KC-DCF(L) lemma*) for dcf languages (unfortunately, this property was faulty and amended later by Glier [2]). This lemma roughly asserts the following. Let x, y, z, u, v_n, w_n denote strings for all $n \in \mathbb{N}$ and fix a 1dpda M, where $\mathbb{N}$ is the set of natural numbers. The succinct notation $\gamma(w)$ indicates the stack content of M obtained just after reading input w. Given a constant $c > 0$, there exists another constant $c' > 0$ such that, for any sufficiently large number $n \in \mathbb{N}$, Conditions (i)–(iii) below imply $C(w_n) < c'$. (i) For any factorization $v_n = v'v''$, $C(v'' \mid \alpha) \leq c$ follows, where α is any stack content, which provides "essentially" the same computation as the stack content obtained by processing $xu^n v'$. (ii) $C(v_n) \geq 2 \log \log n$. (iii) $C(w_n \mid \gamma(xu^n yv_n z)) \leq c$. Moreover, the size of $\gamma(xu^n yv_n z)$ is upper-bounded by a certain constant [8]. This lemma is useful to verify that, e.g., the language $Pal = \{w \mid w \in \{a, b\}^*, w = w^R\}$ of palindromes is not in DCFL, where w^R expresses the *reverse* of w. Other applications of Kolmogorov complexity to pushdown automata were also found in the past literature, e.g., [14].

Because of the usefulness of Kolmogorov complexity, it has been expected to expand the scope of the application of Kolmogorov complexity within the field of formal languages and automata theory. In this paper, we look for another interesting application to "advised" pushdown automata, which will be reviewed in the subsequent subsection in details.

1.2   Advice-Aided Languages and Pumping Lemmas

It is often practical to provide an external source of supplemental information, dubbed generally as *advice*, to an underlying machine in order to significantly enhance the power of its computation. Among various forms of such advice, we are particularly interested in the advice discussed first in 1982 by Karp and Lipton [3] for underlying Turing machines. The advice notion of Karp and Lipton was later adapted to formal languages and automata theory. The past literature, e.g., [1, 9, 10, 13] has proposed several feasible advised models for one-way read-once[1] finite automata as well as pushdown automata, depending on how to access given advice. In the advice model of [9, 11], however, we feed an advice string to an underlying one-way read-once finite automaton in such a way that this machine simultaneously reads the advice string given in parallel to a standard input string and it processes them promptly. This simultaneous processing is achieved by splitting the machine's single input tape into two *tracks*, one of which (i.e., upper track) holds the standard input and the other (i.e., lower track) holds the advice string, and a single tape head scans both tracks simultaneously from left to right along the input tape. Even with such a restrictive access to advice, advised finite automata can exhibit a significant enhancement of computational

---

[1] A one-way tape is *read-once* if its tape head never moves back to the left and, whenever it reads a non-ε symbol, it must move to the right.

power over non-advised machines. As a quick demonstration of the power of advice in this model, it is possible to show that, with the help of resourceful advice, appropriately-designed finite automata can recognize even non-context-free languages, such as $L_{3ab} = \{a^n b^n c^n, a^n b^{2n} c^n \mid n \geq 0\}$ and $3Pal = \{wc^n w^R \mid w \in \{a,b\}^n, n \geq 0\}$. The rest of this paper will focus on this particular advice model for its simplicity and an easy handling of its computations in hope that a number of related questions to be solved. We therefore leave the study of other advice models to the avid reader.

An advised version of the family DCFL of dcf languages, called *advised dcf languages*, refers to languages recognized by 1dpda's helped by externally given advice (see Sect. 3 for their precise definition). The notation DCFL/n denotes the family of all advised dcf languages. We wonder whether DCFL/n is closed under union or intersection since so is DCFL.

In the past literature [9–13], two special advised language families, REG/n and CFL/n, which are obtained respectively from REG and the family CFL of all context-free languages by supplying advice in parallel to input strings using two tracks, were discussed extensively, where "n" symbolizes the use of advice strings whose sizes match the size n of each input.

Whereas numerous properties of context-free languages are naturally transferred to the advise setting (e.g., union and concatenation), certain important properties do not hold for advised context-free languages. For instance, a structural property, known as a pumping lemma (or an iteration theorem), is useful to demonstrate the non-membership of a given language to CFL; however, it does not work in the advice setting because, if input size changes, so does advice, causing an underlying pushdown automaton to behave quite differently. We thus need to seek out a similar useful lemma for advised languages. For this purpose, there are a few lemmas, such as the interchange lemma [7] and the swapping lemma [10,13], which may serve in the nondeterministic case. Unfortunately, the deterministic case lacks similar lemmas.

1.3   Overview of Main Contributions

The main goal of this paper is to prove a new structural property of advised dcf languages, which can be described in terms of Kolmogorov complexity.

When we attempt to disprove the membership question of a given language to CFL, a traditional approach takes an appropriate use of practical but technical tools, known as pumping lemmas, interchange lemma [7], and swapping lemma [10, Lemma 3.1] and [13, Corollary 4.2].

As for an advised variant of CFL, CFL/n, the *swapping lemma* has been proven to be useful in disproving the membership of several languages to CFL/n [10,13]. However, there is no such lemma known for the advised dcf languages. We therefore wish to seek for a new practical means to handle those languages.

In this paper, in the spirit of [2,4], we look for a practical structural property for advised 1dpda's in terms of Kolmogorov complexity. Recall that an underlying idea of the KC-DCF(L) lemma of Li and Vitányi [4] and Glier [2] comes from pumping lemmas; however, those pumping lemmas do not serve well for advised machines because pumped strings change their sizes. We thus hope to develop a different but practical lemma that properly works for DCFL/n. In Sect. 4.1, we will propose such a lemma, dubbed as the *KC-DCFL/n lemma*, in order to disprove that a given language belongs to DCFL/n. Three applications of this new lemma will be given in Sect. 4.2.

All omitted proofs will be included in a complete version of this paper.

2   Preparation: Notions and Notation

2.1   Numbers, Alphabets, Languages, etc.

The notation $\mathbb{N}$ denotes the set of all *natural numbers* (i.e., nonnegative integers) and $\mathbb{N}^+$ denotes $\mathbb{N} - \{0\}$. Given two integers m, n with $m \leq n$, $[m, n]_{\mathbb{Z}}$ denotes the *integer interval*, which is the set $\{m, m+1, m+2, \ldots, n\}$. In particular, for any $n \in \mathbb{N}^+$, $[1, n]_{\mathbb{Z}}$ is abbreviated as $[n]$. For a set A, $\mathcal{P}(A)$ denotes the *power set* of A.

A finite nonempty set of "symbols" or "letters" is called an *alphabet*. We denote by ε the *empty string* of length 0. For a string w and a number $i \in [\|w\|]$, the ith symbol of w is expressed as $(w)_{(i)}$. For convenience, we additionally set $(w)_{(0)} = \varepsilon$. A subset of Σ^* is a *language* over alphabet Σ. Given a number $n \in \mathbb{N}$ and a language L, $L^{=n}$ (resp., $L^{\leq n}$) consists of all strings of length exactly n (resp., at most n) in L. Note that L coincides with $\bigcup_{n \in \mathbb{N}} L^{=n}$.

Given three strings x, y, and z, if $z = xy$, then x is a *prefix* of z and y is a *suffix* of z. In this case, we write $x \sqsubseteq z$. The notation $Pref(L)$ denotes the language $\{z \in \Sigma^* \mid \exists w \in L[z \sqsubseteq w]\}$. Obviously, $L \subseteq Pref(L)$ follows. Given a language L over Σ and a string $x \in \Sigma^*$, the *left quotient* $x\backslash L$ is the language $\{y \in \Sigma^* \mid xy \in L\}$ and the *right quotient* L/x is $\{z \in \Sigma^* \mid zx \in L\}$. Given a string w, a tuple $(x_1, x_2, \ldots, x_n)$ of strings is said to be a *factorization* of w if $w = x_1 x_2 \cdots x_n$. In this case, $(x_1, x_2, \ldots, x_n)$ is also said to *build* the string w.

2.2   One-Way Deterministic Pushdown Automata

The main topic of this paper is *one-way deterministic pushdown automata* (or 1dpda's, for short). A 1dpda M is formally defined as a nonuple $(Q, \Sigma, \{\dashv\}, \Gamma, \delta, q_0, \bot, Q_{acc}, Q_{rej})$, where Q is a finite set of inner states, Σ is an input alphabet, Γ is a stack alphabet, q_0 is the initial (inner) state, Q_{acc} and Q_{rej} are sets of accepting states and of rejecting states, respectively. Let $\check{\Sigma} = \Sigma \cup \{\dashv\}$, $\check{\Sigma}_\varepsilon = \check{\Sigma} \cup \{\varepsilon\}$, and $\Gamma^{(-)} = \Gamma - \{\bot\}$. Let $Q_{halt} = Q_{acc} \cup Q_{rej}$ and call any element in Q_{halt} a *halting (inner) state*. Moreover, δ maps $(Q - Q_{halt}) \times \check{\Sigma}_\varepsilon \times \Gamma$ to $\mathcal{P}(Q \times \Gamma^{\leq e})$, where the minimum number e is called the *push size* of M, and δ must satisfy the following *deterministic requirement*: (i) $|\delta(q, \sigma, a)| \leq 1$ for any $(q, \sigma, a) \in (Q - Q_{halt}) \times \check{\Sigma}_\varepsilon \times \Gamma$ and (ii) if $\delta(q, \varepsilon, a) \neq \varnothing$, then $\delta(q, \sigma, a) = \varnothing$ for all $\sigma \in \check{\Sigma}$. For readability, we express a transition of the form $(p, z) \in \delta(q, \sigma, a)$ as $\delta(q, \sigma, a) = (p, z)$ in the rest of this paper. The entire content of a stack (briefly called the *stack content*) is viewed as a string $a_0 a_1 a_2 \cdots a_n$ with $a_n = \bot$, a topmost stack symbol a_0, and $a_i \in \Gamma^{(-)}$ for any $i \in [n]$. We assume that $\perp$ is not pushed or removed at any moment. The *stack height* of a stack content γ is the length $|\gamma|$. The notation $\gamma^{(top)}$ denotes the topmost stack symbol of γ and $\gamma^{(-)}$ denotes γ except for $\gamma^{(top)}$. Initially, an input string with the endmarker $\dashv$ is given to an input tape, and a tape head is stationed at the leftmost tape cell with $\perp$ in the stack.

A *configuration* of M on input x is a triplet (q, w, z) such that q is an inner state in Q, w is a suffix of $\tilde{x}$, and z is a stack content, where $\tilde{x}$ denotes $x \dashv$. The *initial configuration* of M on x is of the form $(q_0, \tilde{x}, \perp)$. We write $(q, w, z) \vdash_M (q', w', z')$ if M starts with a configuration (q, w, z) and changes it to another one (q', w', z') by a single application of δ. If M takes a number of consecutive steps (including 0 step) from (q, w, z) to (q', w', z'), we use the notation $(q, w, z) \vdash_M^* (q', w', z')$. When $w = w'$, we call this move (or a transition) an ε-*move* (or an ε-transition). There may be a consecutive series of possible ε-moves (called an ε-*chain*) after each non-ε-move. It is sometimes useful to group an entire ε-chain together. For this purpose, we write $(q, w, z) \vdash_{\varepsilon, M} (q', w', z')$ if (q', w', z') is reached from (q, w, z) by a single non-ε-move followed by a chain of all possible ε-moves.

In this paper, M *accepts* (resp., *rejects*) x if there exist an inner state $q' \in Q_{acc}$ (resp., $q' \in Q_{rej}$) and a stack content β for which $(q_0, \tilde{x}, \perp) \vdash_M^* (q', \varepsilon, \beta)$. Given an input string x, we write $\gamma(x)$ to denote the stack content obtained by M on x just after reading off x and performing the subsequent ε-chain. A language L is said to be *recognized* by M if (i) for any $x \in L$, M accepts x and (ii) for any $x \in \Sigma^* - L$, M rejects x. We write $L(M)$ to express the language that is recognized by M.

A *stack history* refers to a sequence of stack contents of M produced in a given computation. Such a stack history induces a sequence of stack heights. In a computation, it suffices to focus our attention on a certain portion of a stack content. We introduce the notion of surface behavior of a 1dpda. A *surface behavior* of M during a computation $(q, w, z) \vdash_{\varepsilon, M}^* (q', w', z')$ is represented by a series of transitions $\tau_1, \tau_2, \ldots, \tau_m$ of the form $\delta(p, \sigma, a) = (r, s)$ applied sequentially to produce this computation. If the surface behavior of M on y starting in inner state q with stack content γ is the same as that of M on y starting in inner state q and stack content α, then we say that these two computations are *essentially the same computation*.

In practice, we need to pay our attention only to 1dpda's of a restricted form called "ideal shape". A one-way pushdown automaton is *in an ideal shape* [16] if each transition must be one of the following forms: (1) scanning $\sigma \in \Sigma^*$, preserve the topmost stack symbol, (2) scanning $\sigma \in \Sigma^*$, push a new symbol without changing any other symbol in the stack, (3) scanning $\sigma \in \Sigma^*$, pop the topmost stack symbol, (4) without scanning an input symbol (i.e., ε-move), pop the topmost symbol, and additionally (5) the stack operation (4) comes only after either (3) or (4). Notice that a machine in an ideal shape has push size of 2.

It is known that every 1dpda can be transformed into an equivalent 1dpda whose transitions are restricted to the ideal shape (see, e.g., [16] for more information and references). To simplify the analysis of 1dpda's in the proof of Lemma 5, because of the use of the right endmarker ⊣, it is possible to demand that the stack of a 1dpda becomes empty (except for ⊥) when it halts.

Lemma 1. *Let N denote any 1dpda that always halts. There exists a 1dpda M that satisfies the following three conditions: (i) $L(M) = L(N)$, (ii) M is in an ideal shape, and (iii) M always enters a halting state with the empty stack.*

2.3 Kolmogorov Complexity Primer

We fix a *universal deterministic Turing machine* equipped with a read-only input tape (or two input tapes for auxiliary inputs), multiple rewritable work tapes, and a write-once[2] output tape. Such a universal machine U takes, as a bundle of inputs, an appropriately-defined encoding of a "program" (or a Gödel number) p and a string w, simulates the program p on w, produces an output string x on its output tape, and finally halts. In this case, we briefly write $U(p, w) = x$.

The *conditional Kolmogorov complexity* of a string x conditional to auxiliary information w, denoted by $C(x \mid w)$, is defined to be $\min_{p \in \Sigma^*}\{|p| : U(p, w) = x\}$. In particular, when $w = \varepsilon$, we succinctly write $C(x)$ and call it the *(unconditional) Kolmogorov complexity* of x.

For more details, the reader should refer to a textbook, e.g., [5].

3 Advised Computation

An *advice function* is a map h from $\mathbb{N}$ to Θ^* for a certain fixed alphabet Θ (called an *advice alphabet*). Each value of such a function is called an *advice string*. Notice that we do not require the "computability" of such advice functions. Initially, an advice string is provided onto the lower track of the input tape whereas the upper track keeps a standard input string. To express a parallel combination of these two strings, we use the *track notation* of [9]. Given two alphabets Σ and Θ, we define a new alphabet $\Sigma_\Theta = \{[\begin{smallmatrix}\sigma\\\tau\end{smallmatrix}] \mid \sigma \in \Sigma, \tau \in \Theta\}$. For two strings $x = x_1 x_2 \cdots x_n \in \Sigma^*$ and $y = y_1 y_2 \cdots y_n \in \Theta^*$, we express $[\begin{smallmatrix}x_1\\y_1\end{smallmatrix}][\begin{smallmatrix}x_2\\y_2\end{smallmatrix}] \cdots [\begin{smallmatrix}x_n\\y_n\end{smallmatrix}]$ as $[\begin{smallmatrix}x\\y\end{smallmatrix}]$. An *advised 1dpda* takes strings over Σ_Θ as "actual" inputs given to its input tape.

A function $h : \mathbb{N} \to \Theta^*$ is *length-preserving* if $|h(n)| = n$ for all $n \in \mathbb{N}$. When a length-preserving advice function $h : \mathbb{N} \to \Theta^*$ is given, an advised 1dpda M reads a string of the form $[\begin{smallmatrix}x\\h(|x|)\end{smallmatrix}]$ written on its input tape, either accepts or rejects the standard input x, and halts. A language recognized by such an advised 1dpda together with an advice function is called an *advised dcf language*. Since advice is not in general computable, some advised languages are not even recursive. Let DCFL/n denote the family of all advised dcf languages, where

---

[2] A tape is called *write-once* if its tape head never moves back to the left and, exactly when it writes a non-blank symbol, it moves to the right.

the symbol "n" refers to the fact that the length of advice strings matches the corresponding input length n.

Lemma 2. *Given a language L, $L \in \mathrm{DCFL}/n$ iff there exist a set $A \in \mathrm{DCFL}$ and a length-preserving advice function h such that $L = \{x \mid [\substack{x \\ h(|x|)}] \in A\}$.*

Lemma 3. *DCFL/n is closed under union and intersection with languages in REG/n.*

To promote the reader's understanding on the expressive power of advised dcf languages, we provide a quick example below.

Example 4. Consider the non-context-free language $3Pal = \{wc^n w^R \mid n \geq 0, w \in \{a, b\}^n\}$. We define $h(n) = 0^k 10^{k-1} 10^{k-1}$ if $n = 3k \geq 3$, and $h(n) = 1^n$ otherwise. The desired advised 1dpda takes $[\substack{x \\ h(|x|)}]$ and checks the first symbol, say, a of $h(|x|)$. If $a = 1$, then M rejects immediately. Otherwise, M stores w with the help of 0^k, M skips c^n using 10^{k-1}, and M compares the stack content with the rest of the input symbol by symbol. This concludes that $3Pal$ belongs to DCFL/n.

4   Kolmogorov Complexity Approach

4.1   Key Lemma – KC-DCFL/n Lemma

The KC-DCF(L)-lemma of Li and Vitányi [4] and Glier [2] plays as an alternative tool to a pumping lemma for DCFL. Unfortunately, there is no structural property known for DCFL/n. Our key lemma (Lemma 5) is at this moment the only "generic" way to demonstrate the non-membership of a large number of languages to DCFL/n, some of which will be exemplified in Sect. 4.2.

Recall from Lemma 2 that every language L in DCFL/n is recognized by an appropriate choice of a language A in DCFL and a length-preserving advice function h. Given two languages L_1 and L_2, we say that a program p *decides L_1 over L_2* if (i) p always halts on any input string x in L_2 and (ii) p accepts input strings x if x is in $L_1 \cap L_2$, and p rejects x if x is in $L_2 - L_1$ [2]. We do not demand that, for any input not in L_2, p halts and makes a correct decision.

In what follows, we assume a natural enumeration of all strings in $\{0, 1\}^*$ according to the "lexicographic" order, by which we first sort strings by length and then lexicographically.

Lemma 5 (KC-DCFL/n Lemma). *Let L be any language in DCFL/n over an input alphabet Σ and let h denote a length-preserving advice function from $\mathbb{N}$ to Θ^* for an advice alphabet Θ witnessing "$L \in \mathrm{DCFL}/n$". Given any two positive constants c and d, there exists another constant $c' > 0$ such that, for any sufficiently large number $n \in \mathbb{N}$ and for any strings $x, y, u, v \in \Sigma^*$ and $e_1, e_2, e_3, e_4 \in \Theta^*$ satisfying $|xuvy| = n$, $h(n) = e_1 e_2 e_3 e_4$, $|x| = |e_1|$, $|u| = |e_2|$, $|v| = |e_3|$, and $|y| = |e_4|$, if the following six conditions are all met, then we conclude that $C(y \mid x, e_1, e_2, e_3, e_4) \leq c'$.*

1. *For any factorizations* $u = u'u''$ *and* $e_2 = e_2'e_2''$ *with* $|e_2'| = |u'|$, *let* $\tau = (x, u', v, y, e_1, e_2', e_3, e_4)$ *and let* p_τ *denote any program that enumerates all strings* $z \in \Sigma^{|e_2''|}$ *in* $xu' \backslash L^{=n}/vy$ *in the lexicographic order with the help of the auxiliary input* e_2''. *It then follows that* $C(u'' \mid p_\tau, e_2'') \leq c$.

2. *For any factorizations* $x = x'x''$ *and* $e_2 = e_2'e_2''$, $C(u \mid x'', e_1'', e_2', e_2'') + d \geq |u|$ *holds, where* $e_1 = e_1'e_1''$ *with* $|e_1'| = |x'|$.

3. *For any factorizations* $v = v'v''$ *and* $e_3 = e_3'e_3''$ *with* $|e_3'| = |v'|$, *let* $\tau = (x, u, v', y, e_1, e_2, e_3', e_4)$ *and let* p_τ *be any program that decides* $xuv' \backslash L^{=n}/y$ *over* $\text{Pref}(\{v''\})$ *with the help of the auxiliary inputs* e_3''. *It then follows that* $C(v'' \mid p_\tau, e_3'') \leq c$.

4. $C(v \mid e_2, e_3) + d \geq C(u \mid e_2)$.

5. *For any factorizations* $x = x'x''$ *and* $v = v'v''$, $C(v'' \mid x', e_1', e_3'') + d \geq \log\log|v'|$ *follows, where* $e_1 = e_1'e_2''$ *and* $e_3 = e_3'e_3''$ *with* $|e_1'| = |x'|$ *and* $|e_3'| = |v'|$.

6. *Let* $\tau = (x, u, v, e_1, e_2, e_3)$ *and let* p_τ *be any program that enumerates all strings* $z \in \Sigma^{|e_4|}$ *in* $xuv \backslash L^{=n}$ *in the lexicographic order with the help of the auxiliary input* e_4. *It then follows that* $C(y \mid p_\tau, e_4) \leq c$.

Note that the partitioned portions, e_i, e_i', e_i'', of the advice string are used as auxiliary inputs in $C(u \mid e_2)$, $C(v \mid e_2, e_3)$, etc. They also provide necessary information on the lengths, $|e_i|, |e_i'|, |e_i''|$, for the algorithmic constructions of u, v, and x.

The KC-DCFL/n Lemma is a direct consequence of the stronger statement given as Lemma 6. This statement concerns with the behavior of an underlying 1dpda with the help of advice.

In the following description, we always assume that a 1dpda M has the form $(Q, \Sigma, \{\dashv\}, \Gamma, \delta, q_0, \bot, Q_{acc}, Q_{rej})$ and satisfies all the properties of Lemma 1.

Lemma 6 (Stronger Form). *Let M denote any 1dpda over an input alphabet Σ using a set Q of inner states and let h denote a length-preserving advice function from $\mathbb{N}$ to Θ^* for an advice alphabet Θ. Assume that M satisfies all the properties of Lemma 1. Given any two positive constants c and d, there exists another constant $c' > 0$ such that, for any sufficiently large number $n \in \mathbb{N}$ and for any strings $x, y, u, v \in \Sigma^*$ and $e_1, e_2, e_3, e_4 \in \Theta^*$ satisfying $|xuvy| = n$, $h(n) = e_1e_2e_3e_4$, $|x| = |e_1|$, $|u| = |e_2|$, $|v| = |e_3|$, and $|y| = |e_4|$, if the following six conditions are all met, then $C(y \mid x, e_1, e_2, e_3, e_4) \leq c'$ follows.*

1. *For any factorizations* $u = u'u''$ *and* $e_2 = e_2'e_2''$ *with* $|u'| = |e_2'|$, *any two stack contents* α *and* β, *and any two inner states* q *and* q', *if* $(q, [\begin{smallmatrix} u'' \\ e_2'' \end{smallmatrix}], \gamma([\begin{smallmatrix} xu' \\ e_1e_2' \end{smallmatrix}])) \vdash_{\varepsilon,M}^*$ $(q', \varepsilon, \gamma([\begin{smallmatrix} xu \\ e_1e_2 \end{smallmatrix}]))$ *and* $(q, [\begin{smallmatrix} u'' \\ e_2'' \end{smallmatrix}], \alpha) \vdash_{\varepsilon,M}^* (q', \varepsilon, \beta)$ *are essentially the same computation, then* $C(u'' \mid q, \alpha, \beta, e_2'') \leq c$ *follows.*

2. *For any factorizations* $x = x'x''$ *and* $e_2 = e_2'e_2''$, $C(u \mid x'', e_1'', e_2', e_2'') + d \geq |u|$ *holds, where* $e_1 = e_1'e_1''$ *with* $|e_1'| = |x'|$.

3. *For any factorizations* $v = v'v''$ *and* $e_3 = e_3'e_3''$ *with* $|v'| = |e_3'|$, *any stack contents* α *and* β, *and any two inner states* q *and* q', *if* $(q, [\begin{smallmatrix} v'' \\ e_3'' \end{smallmatrix}], \gamma([\begin{smallmatrix} uv' \\ e_2e_3' \end{smallmatrix}])) \vdash_{\varepsilon,M}^*$ $(q', \varepsilon, \gamma([\begin{smallmatrix} uv \\ e_2e_3 \end{smallmatrix}]))$ *and* $(q, [\begin{smallmatrix} v'' \\ e_3'' \end{smallmatrix}], \alpha) \vdash_{\varepsilon,M}^* (q', \varepsilon, \beta)$ *are essentially the same computation, then* $C(v'' \mid q, \alpha, e_3'') \leq c$ *follows. (Note that β is not included.)*

4. $C(v \mid e_3) + d \geq C(u \mid e_2)$.
5. For any factorizations $x = x'x''$ and $v = v'v''$, $C(v'' \mid x', e_1', e_3'') + d \geq \log\log|v''|$ follows, where $e_1 = e_1'e_2''$ and $e_3 = e_3'e_3''$ with $|e_1'| = |x'|$ and $|e_3'| = |v'|$.
6. $C(y \mid \gamma([\begin{smallmatrix} xuv \\ e_1e_2e_3 \end{smallmatrix}]), e_4) \leq c$.

Here is a brief, informal, and intuitive description of why the lemma is correct. While processing $[\begin{smallmatrix} u \\ e_2 \end{smallmatrix}]$, the stack of M grows because x is irrelevant to u due to Condition 2. By Condition 1, we can algorithmically construct u from (q', α, β, e_2) on a certain Turing machine. From $C(u \mid q', \alpha, \beta, e_2) = O(1)$, $C(u \mid x, e_2) + d \geq |u|$, and $C(q') = O(1)$, it follows that $C(\alpha, \beta) \geq |u|$ and $C(u \mid \alpha, \beta, e_2) = O(1)$. Conditions 3–4 then imply that, while processing v, β is erased from the stack. After the processing of xuv, the stack content has only the information associated with x. Thus, we can construct $\gamma([\begin{smallmatrix} xuv \\ e_1e_2e_3 \end{smallmatrix}])$ from (x, e_1, e_2). Since Condition 5 yields $C(y \mid \gamma([\begin{smallmatrix} xuv \\ e_1e_2e_3 \end{smallmatrix}]), e_4) = O(1)$, we conclude that $C(y \mid x, e_1, e_2, e_3, e_4) = O(1)$.

Let us prove the KC-DCFL/n lemma using Lemma 6.

Proof of Lemma 5. Given L and h in the premise of Lemma 5, we take a language $A \in \mathrm{DCFL}$ for which $L = \{w \mid [\begin{smallmatrix} w \\ h(|w|) \end{smallmatrix}] \in A\}$ by Lemma 2. We then take a 1dpda M that recognizes A with the properties of Lemma 1. Let $xuvy$ be any string in L of length n. Let $h(n) = e_1e_2e_3e_4$ with $|e_1| = |x|$, $|e_2| = |u|$, $|e_3| = |v|$, and $|e_4| = |y|$. Assume that Conditions 1–6 of the lemma hold. Our goal is to draw a conclusion of $C(y \mid x, e_1, e_2, e_3, e_4) \leq c'$. For this purpose, it suffices to demonstrate that Conditions 1, 3, and 6 of Lemma 6 are all met since the desired conclusion then follows.

To show Condition 1 of Lemma 6, we assume that (*) $(q, [\begin{smallmatrix} u'' \\ e_2'' \end{smallmatrix}], \gamma([\begin{smallmatrix} xu \\ e_1e_2' \end{smallmatrix}])) \vdash_{\varepsilon,M}^*$ $(q', \varepsilon, \gamma([\begin{smallmatrix} xu \\ e_1e_2' \end{smallmatrix}]))$ and $(q, [\begin{smallmatrix} u'' \\ e_2'' \end{smallmatrix}], \alpha) \vdash_{\varepsilon,M}^* (q', \varepsilon, \beta)$ are essentially the same computation. Based on the given tuple $\tau = (x, u', v, y, e_1, e_2', e_3, e_4)$, we construct a program p_τ as follows. By taking all strings $z \in \Sigma^{|e_2''|}$ one by one in the lexicographic order, using τ, we first run M on $[\begin{smallmatrix} z \\ e_2'' \end{smallmatrix}]$, starting with the inner state q and α, and then check if the resulting inner state and track content are respectively q' and β. If so, then we output z; otherwise, we continue the process. Note by (*) that, if a string $z \in \Sigma^{|e_2''|}$ satisfies $(q, [\begin{smallmatrix} z \\ e_2'' \end{smallmatrix}], \alpha) \vdash_{\varepsilon,M}^* (q', \varepsilon, \beta)$, then z must belong to $xu'\backslash L^{=n}/vy$ because of $xuvy \in L^{=n}$. Thus, p_τ can enumerate all strings $z \in \Sigma^{|e_2''|}$ in $xu'\backslash L^{=n}/vy$. By Condition 1 of Lemma 5, this implies that $C(u'' \mid p_\tau, e_2'') \leq c$. Note that p_τ is algorithmically constructed from $(q, q', \alpha, \beta, e_2'')$. Since q and q' are expressed by a constant number of bits, we thus conclude that $C(p_\tau \mid q, q', \alpha, \beta, e_2'') = O(1)$. Combining this with $C(u'' \mid p_\tau, e_2'') \leq c$, we obtain $C(u'' \mid q, q', \alpha, \beta, e_2'') = O(1)$, as requested.

Next, we target Condition 3. Let $v = v'v''$ and $e_3 = e_3'e_3''$ be arbitrary factorizations with $|e_3'| = |v'|$. Assume that $(q, [\begin{smallmatrix} v'' \\ e_3'' \end{smallmatrix}], \gamma([\begin{smallmatrix} uv' \\ e_2e_3' \end{smallmatrix}])) \vdash_{\varepsilon,M}^* (q', \varepsilon, \gamma([\begin{smallmatrix} uv \\ e_2e_3 \end{smallmatrix}]))$ and $(q, [\begin{smallmatrix} v'' \\ e_3'' \end{smallmatrix}], \alpha) \vdash_{\varepsilon,M}^* (q', \varepsilon, \beta)$ are essentially the same computation. Letting $\tau = (x, u, v', y, e_1, e_2, e_3', e_4)$, we define a program p_τ as follows. Given any prefix $z \sqsubseteq v''$, we start with the inner state q and the stack content α, run M on

$[\begin{smallmatrix} z \\ e_3'' \end{smallmatrix}]$, and check if $|z| = |e_3''|$. If M finally enters q', then we accept the input; otherwise, we reject it. As a result, p_τ can be constructed from (q, q', α, e_3''), implying $C(p_\tau \mid q, q', \alpha, e_3'') = O(1)$. Condition 3 of Lemma 5 then ensures that $C(v'' \mid p_\tau, e_3'') \leq c$. It thus follows that $C(v'' \mid q, q', \alpha, e_3'') = O(1)$.

Finally, we show Condition 6. Given $\gamma = \gamma([\begin{smallmatrix} xuv \\ e_1 e_2 e_3 \end{smallmatrix}])$ and e_4, we need to reconstruct y. Letting $\tau = (x, u, v, e_1, e_2, e_3)$, we design a program p_τ in the following fashion. Let q denote the inner state obtained after running M on $[\begin{smallmatrix} xuv \\ e_1 e_2 e_3 \end{smallmatrix}]$. Starting with (q, γ) and the auxiliary input e_4, we run M on $[\begin{smallmatrix} z \\ e_4 \end{smallmatrix}]$ for all strings $z \in \Sigma^{|e_4|}$ to check if $z \in xuv \backslash L^{=n}$ and we then enumerate all such z's in the lexicographic order. By Condition 5 of Lemma 5, we then conclude that $C(y \mid p_\tau, e_4) \leq c$. Since p_τ is built from (q, γ) and q is expressed by a constant number of bits, we obtain $C(y \mid \gamma, e_4) = O(1)$. □

4.2   Applications of the KC-DCFL/n Lemma

Hereafter, we demonstrate how to apply the KC-DCFL/n lemma (Lemma 5) to two example languages. These examples suggest the usefulness of the KC-DCFL/n lemma for advised dcf languages.

Example 7. Let $Double = \{x\#yu\#v \mid n \in \mathbb{N}, x, y, u, v \in \{0, 1\}^n, (x \neq u^R \vee y \neq v^R)\}$. Note that $Double \in$ CFL/n. We intend to show that $Double \notin$ DCFL/n. Let $FORM = \{x\#yu\#v \mid n \in \mathbb{N}, x, y, u, v \in \{0, 1\}^n\}$ and $D = \overline{Double} \cap FORM$. Since $Double \in$ DCFL/n implies $\overline{Double} \in$ DCFL/n and thus $D \in$ DCFL/n by Lemma 3, it suffices to prove that $D \notin$ DCFL/n.

Toward a contradiction, we assume that $D \in$ DCFL/n via a length-preserving advice function h. Take a sufficiently large number n satisfying $C(n) \geq n$. Consider any string of the form $z\#wz^R\#w^R$ with $|z| = |w| = n$. Let $\hat{n} = 4n + 2$ and let $h(\hat{n}) = e_1 e_2 e_3 e_4$ with $e_1 = \varepsilon$, $|e_2| = 2n + 1$, $|e_3| = n$, and $|e_4| = n + 1$. Without loss of generality, each e_i has an "indicator" (i.e., a special advice symbol) that marks the end of e_i. This indicator signals an underlying 1dpda where to partition $h(\hat{n})$.

Take z and w of length n to satisfy that $C(z \mid h(\hat{n})) \geq |z|$, $C(w \mid h(\hat{n})) \geq |w|$, $C(z \mid w, h(\hat{n})) \geq |z|/2$, and $C(w \mid z, h(\hat{n})) \geq |w|/2$. We set $x = \varepsilon$, $u = z\#w$, $v = z^R$, and $y = \#w^R$. Note that $|xuvy| = \hat{n}$. Our goal is to show that Conditions 1–6 of Lemma 5 are all satisfied. By the definition, Conditions 2, 4, and 5 hold.

To see Condition 1, let $\tau = (x, u', v, y, e_1, e_2', e_3, e_4)$ and let p_τ denote any program that enumerates all $z \in \Sigma^{|e_2''|}$ in $xu'\backslash L^{=\hat{n}}/vy$ in the lexicographic order. Since u'' is uniquely determined from (u', vy), we can construct u'' uniquely by running p_τ. Thus, $C(u'' \mid p_\tau, e_2'') = O(1)$ follows.

Next, we look into Condition 3. For $\tau = (x, u, v', y, e_1, e_2, e_4)$, p_τ denotes any program that decides $xuv'\backslash L^{=\hat{n}}/y$ over $Pref(\{v''\})$. Let $z = z_1 z_2$ with $|z_2| = |v'|$. Note that $xuv' = z\#wz_2^R$ and $v'' = z_1^R$. Since v'' is uniquely determined from xuv', we can construct v'' from p_τ and e_3''.

For Condition 6, let $\tau = (x, u, v, e_1, e_2, e_3)$ and let p_τ be any program that enumerates all $z \in \Sigma^{|e_4|} \cap xuv\backslash L^{=\hat{n}}$ in the lexicographic order. Notice that y is uniquely determined from u. Hence, we obtain $C(y \mid p_\tau, e_4) = O(1)$.

Lemma 5 then concludes that $C(y \mid x, e_1, e_2, e_3, e_4) \leq c'$. However, since $x = \varepsilon$, it follows that $C(y \mid x, e_1, e_2, e_3, e_4) \geq C(w \mid e_1, e_2, e_3, e_4) - O(1) \geq |w| - O(1) = n - O(1)$, a contradiction. We thus obtain the desired consequence.

Example 8. Our next example is the *pattern matching* of the form $PM = \{x \# y x^R z \mid x, y, z \in \{0,1\}^*\}$. We obtain $PM \in \mathrm{CFL}$ by simply guessing the location of x^R. We want to prove that $PM \notin \mathrm{DCFL}/n$. Toward a contradiction, we assume that $PM \in \mathrm{DCFL}/n$, witnessed by an appropriate length-preserving advice function h.

As in the premise of Lemma 5, we take any sufficiently large number n satisfying $C(n) \geq \log n$, and we set $\hat{n} = 6n + 1$. Consider four strings x, u, v, y such that $|x| = n$, $u = r\#$, $v = r^R$, $y = x^R r^R s_1$, and $\hat{y} = s_2 r^R x^R$ for appropriate strings r, s_1, and s_2 of length n. Let $w = xuvy$ and $\hat{w} = xuv\hat{y}$. Moreover, we set $h(\hat{n}) = e_1 e_2 e_3 e_4$ with $|e_1| = |x|$, $|e_2| = |u|$, $|e_3| = |v|$, and $|e_4| = |y|$. Note that $|w| = |\hat{w}| = \hat{n}$ and $xuv = xr\#r^R$. Clearly, $y, \hat{y} \in xuv \backslash L^{=\hat{n}}$ holds for any s_1 and s_2 of length n.

We choose x and r for which $C(x \mid e_1, e_4) \geq n$ and $C(r \mid x, e_1, e_2, e_3, e_4) \geq n$. It then follows that $C(y \mid x, e_1, e_2, e_3, e_4) + d \geq C(r \mid x, e_1, e_2, e_3, e_4) \geq n$ for a certain constant $d > 0$.. To utilize Lemma 5, we need to check that Conditions 1–5 of the lemma are all satisfied. Conditions 2, 4, and 5 are trivial by the choice of u and v. In what follows, we thus need to focus our attention only on Conditions 1, 3, and 6.

Firstly, we show Condition 1. Let $u = u' u''$ and $e_2 = e_2' e_2''$ with $|u'| = |e_2'|$. Let $\tau = (x, u', v, y, e_1, e_2', e_3, e_4)$ and let p_τ denote any program that enumerates all $z \in \Sigma^{|e_2''|}$ in $xu' \backslash L^{=\hat{n}}/vy$. Since u'' has $\#$, we can easily identify u'' by checking $\#$ and running p_τ. Thus, we obtain $C(u'' \mid p_\tau, e_2'') = O(1)$.

Condition 3 is shown as follows. Consider two arbitrary factorizations $v = v' v''$ and $e_3 = e_3' e_3''$ with $|e_3'| = |v'|$. Since $v = r^R$, let $v' = r_2^R$ and $v'' = r_1^R$ with $r = r_1 r_2$. Let $\tau = (x, u, v', y, e_1, e_2, e_4)$ and let p_τ denote a program that decides $xuv' \backslash L^{=\hat{n}}/y$ over $Pref(\{v''\})$. Note that $xuv' = xr\#r_2^R$. Since v'' is uniquely determined from v' and u, we can algorithmically construct v'' from p_τ and e_3''. Thus, we obtain $C(v'' \mid p_\tau, e_3'') = O(1)$.

Finally, we show Condition 6. By setting $s_1 = s_2 = 0^n$, if we are given a program p_τ that enumerates all $z \in \Sigma^{|e_4|}$ in $xuv \backslash L^{=\hat{n}}$ with an auxiliary input e_4, then we can produce y and $\hat{y}$ uniquely. Therefore, $C(y \mid p_\tau, e_4)$ is $O(1)$.

Lemma 5 then yields the inequality $C(y \mid x, e_1, e_2, e_3, e_4) \leq c'$. This obviously contradicts the choice of y. Therefore, PM is outside of DCFL/n.

To close this subsection, we show a non-closure property of DCFL/n by a direct application of the key lemma. This property naturally reflects the specific trait of DCFL regarding non-closure properties; however, we also remark that a non-closure property of DCFL/n does not in general follows directly from that of DCFL.

Proposition 9. DCFL/n *is neither closed under intersection nor closed under union.*

Proof. Consider two languages $K_1 = \{x\#yu\#v \mid n \in \mathbb{N}, x, y, u, v \in \{0,1\}^n, x \neq u\}$ and $K_2 = \{x\#yu\#v \mid n \in \mathbb{N}, x, y, u, v \in \{0,1\}^n, y \neq v^R\}$. Clearly, both K_1 and K_2 are in DCFL/n. Since the union $K_1 \cup K_2$ equals *Double*, this union is not in DCFL/n, as shown in Example 7. The case of intersection follows from the fact that DCFL/n is closed under complementation. $\square$

There are still numerous questions that have been left open regarding the power and limitation of advice in automata theory. For instance, consider the collection DCFL(k) of all intersections of k dcf languages [15]. It is known that DCFL(k) $\neq$ DCFL($k + 1$) for all $k \in \mathbb{N}^+$ [6] (re-proven in [15]). By expanding DCFL(k), we can define DCFL(k)/n. For every $k \in \mathbb{N}^+$, is DCFL($k + 1$)/n different from DCFL(k)/n? A similar question was raised in [13] for CFL(k)/n.

References

1. Damm, C., Holzer, M.: Automata that take advice. In: Wiedermann, J., Hájek, P. (eds.) MFCS 1995. LNCS, vol. 969, pp. 149–158. Springer, Heidelberg (1995). https://doi.org/10.1007/3-540-60246-1_121
2. Glier, O.: Kolmogorov complexity and deterministic context-free languages. SIAM J. Comput. **32**, 1389–1394 (2003)
3. Karp, R.M., Lipton, R.: Turing machines that take advice. Enseign. Math. **28**, 191–209 (1982)
4. Li, M., Vitányi, P.: A new approach to formal language theory by Kolmogorov complexity. SIAM J. Comput. **24**, 398–410 (1995)
5. Li, M., Vitányi, P.: An Introduction to Kolmogorov Complexity and Its Applications, 3rd edn. Springer (2008)
6. Liu, L.W., Weiner, P.: An infinite hierarchy of intersections of context-free languages. Math. Syst. Theory **7**, 185–192 (1973)
7. Ogden, W., Ross, R.J., Winklmann, K.: An "interchange lemma" for context-free languages. SIAM J. Comput. **14**, 410–415 (1985)
8. Rubtsov, A.A.: A structural lemma for deterministic context-free languages. In: Hoshi, M., Seki, S. (eds.) DLT 2018. LNCS, vol. 11088, pp. 553–565. Springer, Cham (2018). https://doi.org/10.1007/978-3-319-98654-8_45
9. Tadaki, K., Yamakami, T., Lin, J.C.H.: Theory of one-tape linear-time Turing machines. Theor. Comput. Sci. **411**, 22–43 (2010)
10. Yamakami, T.: Swapping lemmas for regular and context-free languages (2008). arXiv:0808.4122
11. Yamakami, T.: The roles of advice to one-tape linear-time Turing machines and finite automata. Int. J. Found. Comput. Sci. **21**, 941–962 (2010)
12. Yamakami, T.: Immunity and pseudorandomness of context-free languages. Theor. Comput. Sci. **412**, 6432–6450 (2011)
13. Yamakami, T.: Pseudorandom generators against advised context-free languages. Theor. Comput. Sci. **613**, 1–27 (2016)
14. Yamakami, T.: One-way bounded-error probabilistic pushdown automata and Kolmogorov complexity. In: Charlier, É., Leroy, J., Rigo, M. (eds.) DLT 2017. LNCS, vol. 10396, pp. 353–364. Springer, Cham (2017). https://doi.org/10.1007/978-3-319-62809-7_27

15. Yamakami, T.: Intersection and union hierarchies of deterministic context-free languages and pumping lemmas. In: Leporati, A., Martín-Vide, C., Shapira, D., Zandron, C. (eds.) LATA 2020. LNCS, vol. 12038, pp. 341–353. Springer, Cham (2020). https://doi.org/10.1007/978-3-030-40608-0_24. Extended at arXiv:2112.09383
16. Yamakami, T.: The no endmarker theorem for one-way probabilistic pushdown automata (2021). arXiv:2111.02688

Author Index